AF739963

Mechanical and Electromagnetic Vibrations and Waves

Mechanical and Electromagnetic Vibrations and Waves

Tamer Bécherrawy

iSTE

WILEY

First published 2012 in Great Britain and the United States by ISTE Ltd and John Wiley & Sons, Inc.

Apart from any fair dealing for the purposes of research or private study, or criticism or review, as permitted under the Copyright, Designs and Patents Act 1988, this publication may only be reproduced, stored or transmitted, in any form or by any means, with the prior permission in writing of the publishers, or in the case of reprographic reproduction in accordance with the terms and licenses issued by the CLA. Enquiries concerning reproduction outside these terms should be sent to the publishers at the undermentioned address:

ISTE Ltd
27-37 St George's Road
London SW19 4EU
UK

www.iste.co.uk

John Wiley & Sons, Inc.
111 River Street
Hoboken, NJ 07030
USA

www.wiley.com

© ISTE Ltd 2012

The rights of Tamer Bécherrawy to be identified as the author of this work have been asserted by him in accordance with the Copyright, Designs and Patents Act 1988.

Library of Congress Cataloging-in-Publication Data

Bécherrawy, Tamer.
Mechanical and electromagnetic vibrations and waves / Tamer Becherrawy.
p. cm.
Includes bibliographical references and index.
ISBN 978-1-84821-283-1
1. Electromagnetic fields--Mathematical models. 2. Electromagnetic waves--Mathematical models. 3. Electrodynamics--Mathematical models. 4. Oscillations--Mathematical models. 5. Engineering mathematics. I. Title.
QC665.E4B38 2012
531'.32--dc23

2011043541

British Library Cataloguing-in-Publication Data
A CIP record for this book is available from the British Library
ISBN: 978-1-84821-283-1

Printed and bound in Great Britain by CPI Group (UK) Ltd., Croydon, Surrey CR0 4YY

MIX
Paper from responsible sources
FSC www.fsc.org FSC® C013604

Table of Contents

Preface

Oscillatory and wave phenomena are encountered in almost all branches of physics: mechanics, geophysics, electromagnetism, optics, quantum physics, etc. Some of them were first observed in antiquity, but their scientific study only started in the 17th Century. The phenomena include mechanical vibrations and waves, electromagnetic vibrations and waves, matter waves, etc. Electromagnetic vibrations and waves were discovered in the 19th Century, while matter waves were discovered in the 20th Century. Each branch of physics has its own concepts, and even its own proper mathematical language. Nevertheless, all types of vibrations and waves share several common properties: modes, similar forms of energy, superposition, interference, diffraction, etc.

The purpose of this book is to study oscillatory and wave phenomena at the undergraduate level. It was not conceived with the intended application as a textbook for a specific physics course. Some sections, indicated by an asterisk (*), may prove difficult and may be omitted without loss of continuity.

Chapter 1 introduces the basic concepts and studies some examples of vibrations of mechanical and electromagnetic systems with one or several degrees of freedom. Chapter 2 studies the superposition of vibrations and introduces Fourier analysis. Chapter 3 analyzes forced vibrations and resonances. Chapter 4 introduces the basic notions of waves in infinite media: wave equations and their solutions, energy density and energy transfer, etc. Chapter 5 is devoted to the study of mechanical waves (elastic waves, sound and surface waves). In Chapter 6, we summarize the basic laws of electromagnetism and analyze the electromagnetic waves in insulators, conductors and plasmas. Reflection and refraction are studied in Chapter 7, interference and diffraction are studied in Chapter 8 and finally standing waves and waveguides, in Chapter 9. This book shall not study the emission of waves or optical setups.

The required mathematical techniques are introduced as the need arises. Appendix A aids understanding by summarizing the principal mathematical formulas, integrals and vector analysis. We tried to use clear notations by assigning similar symbols for the various physical quantities: a boldfaced symbol for a vector quantity, an italic symbol for a scalar quantity or a component of a vector quantity, an underlined symbol for a complex quantity, and script symbol for a curve, a surface, a volume and some special quantities. Physical quantities of the same type are referred to by symbols with different indexes: for instance, $f_{(Fr)}$, $f_{(ez)}$, $\mathbf{F}_{(E)}$, etc., for the different types of force. The energy is designated by U to avoid confusion with the components of the electric field **E**. The frequency is represented by $\tilde{\nu}$, instead of the usual Greek symbol ν, to avoid its confusion with the velocity v.

A unit vector is often represented by **e**, while the unit vectors of the axes are represented by $\mathbf{e}_x$, $\mathbf{e}_y$ and $\mathbf{e}_x$. In order to write summations in a condensed form, we sometimes designate the Cartesian coordinates x, y and z by x_1, x_2 and x_3 respectively, and the components of a vector **V** by $V_1 \equiv V_x$, $V_2 \equiv V_y$ and $V_3 \equiv V_z$. The partial derivative of $u(x, y, z, t)$ with respect to time is represented by $\dot{u}$ or $\partial_t u$ and its partial derivatives by $\partial_x u$ for $\partial u/\partial x$, $\partial^2{}_{xt}u$ for $\partial^2 u/\partial x\ \partial t$, etc. We also use the notation $\partial_i u$ for the partial derivatives $\partial u/\partial x_i$ and $\partial_i V_j$ for $\partial V_j/\partial x_i$ (i and $j = 1, 2, 3$).

Each chapter ends with a *Summary* section for the principal results of the chapter, and a section entitled *Problem solving suggestions*, which contains remarks or possible errors to be avoided, approximation methods and further clarifications. For training students, each chapter contains some *examples* that are worked out in detail and two kinds of exercises: *conceptional questions*, a selection of discussion questions designed to develop the understanding of the physical concepts, often without a need for calculations; and *problems*, which are ordered according to the sections of the chapter and arranged in approximate levels of difficulty (an asterisk (*) indicates a problem of some difficulty, two asterisks (**) indicates a problem with some connectional or computational difficulties. The answers to most of the problems are given in a special addendum entitled *Answers to the Problems*, which enables students to check their results.

I hope that this text makes the subject more accessible for students, and that it is utilized as a good teaching aid for professors.

T. BÉCHERRAWY
November 2011

Chapter 1

Free Oscillations

In this chapter we introduce the basic notions of free oscillations. Starting with a study of the differential equation governing the undamped vibrations, its general solution and its trigonometric, complex and phasor representations, we then progress to the equation of damped oscillations and its solutions. We analyze some simple oscillating systems with one degree of freedom by emphasizing the notion of energy which, in modern physics, is considered to be a more fundamental quantity than forces. We generalize these results to systems undergoing small displacements or variations of the state "back-and-forth" near an equilibrium position. Afterwards, we analyze systems with two or several degrees of freedom.

1.1. Oscillations and waves, period and frequency

Vibrations or *oscillations* are motions or changes in the state of physical systems back-and-forth on both sides of an equilibrium position that are repeated more or less regularly in time. *Waves* are vibrations that propagate from one region to another. We encounter vibratory and wave phenomena in almost all branches of physics: mechanics, geophysics, electromagnetism, optics, quantum physics, etc. We consider in this book two kinds of vibrations: *mechanical vibrations* (of a pendulum, a string, etc.) and *electromagnetic vibrations* (of electric circuits, radio waves, etc.).

Vibrations are *free* if, after an initial excitation, the system oscillates subject to its own internal forces but no-external forces. On the other hand, the vibrations are said to be *forced* if the external force continues to sustain the oscillation of the system. The external force is called the *driving force*.

Vibrations of a system are *periodic* if the system returns exactly to the same state after each time interval T, called the *period* of vibration. Any physical quantity u of the system takes the same value after time intervals T, $2T$, $3T$, etc. Figure 1.1a shows the variation of a periodic function u in time. This periodicity may be expressed mathematically by the relation

$$u(t) = u(t+T) = u(t+2T) = ... = u(t+nT) \ldots \qquad [1.1]$$

The *frequency* $\tilde{\nu}$ is the number of complete vibrations in unit time, thus

$$\tilde{\nu} = 1/T. \qquad [1.2]$$

In the *International System of Units* (*SI*), the period is expressed in *seconds* (s) and the frequency in s^{-1}, called *hertz* (Hz). For high frequencies, we use *kilohertz* (kHz = 10^3 Hz), *megahertz* (MHz = 10^6 Hz) and *gigahertz* (GHz = 10^9 Hz).

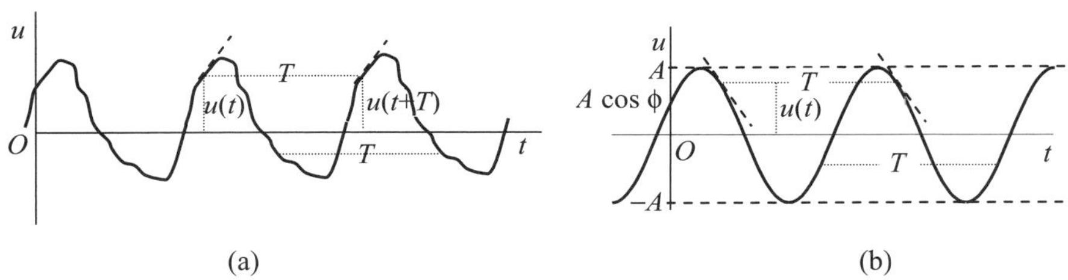

Figure 1.1. *a) Periodic vibration; b) simple harmonic vibration* $u = A\cos(\omega t + \phi)$

1.2. Simple harmonic vibrations: differential equation and linearity

Periodic vibrations are referred to as *harmonic vibrations* by analogy to musical sounds. The simplest periodic vibration is represented by a sine or a cosine function known as a *simple harmonic function*

$$u = A\cos(\omega t + \phi). \qquad [1.3]$$

A is the *amplitude* and ω is the *angular frequency* (often called *frequency* for short). $(\omega t + \phi)$ is the *phase* at time t and ϕ is the *initial phase* (called *phase*, for short). The phase has the dimension of angles, and is therefore expressed in *radians* (rad); while the angular frequency ω is in *radians per second* (rad/s). A and u have the same dimensions, so they are expressed in the same units. The simple harmonic function [1.3] is illustrated in Figure 1.1b.

The values of the sinusoidal function are repeated if the phase varies by 2π (or $2n\pi$ with n as an integer). Thus, u has a period T such that $\omega(t + T) + \phi = \omega t + \phi + 2\pi$; hence, the relationships between ω, T and $\tilde{\nu}$:

$$\omega = 2\pi/T = 2\pi\tilde{\nu} \tag{1.4}$$

By differentiating expression [1.3] twice with respect to time, we obtain:

$$\dot{u} \equiv du/dt = -A\omega \sin(\omega t + \phi) = A\omega \cos(\omega t + \phi + \pi/2) \tag{1.5}$$

$$\ddot{u} \equiv d^2u/dt^2 = -A\omega^2 \cos(\omega t + \phi) = A\omega^2 \cos(\omega t + \phi + \pi). \tag{1.6}$$

Therefore, u is a solution of the differential equation of simple harmonic oscillations

$$\ddot{u} + \omega^2 u = 0. \tag{1.7}$$

This is a second-order homogeneous differential equation. ω is the *natural angular frequency* (also called the *normal angular frequency*). It depends on the physical characteristics of the oscillating system (masses, internal forces, etc.). Expression [1.3] is the *general solution* or the *normal mode* of oscillation. Any oscillation of the free system may be written in this form and any expression of this form is a possible state of oscillation. The *constants of integration* A and ϕ depend on the initial excitation of the system, that is, the *initial conditions* which are the values of u and $\dot{u}$ at a given time t_o. For instance, if the system is set in oscillation at $t_o = 0$, we have the conditions $u(0) = A \cos\phi$ and $\dot{u}(0) = -A\omega \sin\phi$, from which we deduce that

$$A = \sqrt{u(0)^2 + \dot{u}(0)^2/\omega^2}\,, \tag{1.8}$$

$$\cos\phi = u(0)/A, \qquad \sin\phi = -\dot{u}(0)/\omega A\,. \tag{1.9}$$

The relationships in [1.9] determine the phase ϕ. We may also write

$$\tan\phi = -\dot{u}(0)/\omega u(0)\,. \tag{1.10}$$

However, this relationship determines ϕ only up to π. Instead of [1.3], we may use the expression $u = A \sin(\omega t + \phi')$. By adding or subtracting $\pi/2$ or π from the phase, it is possible to write any simple harmonic vibration in the form of [1.3] with a positive amplitude A.

The equation of oscillations [1.7] is *linear* and *homogeneous*. Therefore, it has the important property of satisfying the *superposition principle*: If $u(t)$ is a solution and C is an arbitrary constant, $Cu(t)$ is also a solution. Likewise, if $u_1(t)$ and $u_2(t)$ are two solutions, any linear superposition with arbitrary coefficients C_1 and C_2

$$u(t) = C_1\, u_1(t) + C_2\, u_2(t) \tag{1.11}$$

is also a solution of the equation. The initial conditions of $u(t)$ are linear combinations of the initial conditions of $u_1(t)$ and $u_2(t)$ with the same coefficients C_1 and C_2. The superposition principle plays a crucial role in many branches of physics.

Equation [1.7] may be written in the form $D.u = 0$, where $D = d^2/dt^2 + \omega^2$ or a more complicated form. D is called an *operator*. It transforms a function u to another function $D.u$. If D verifies the condition $D.(C_1u_1 + C_2u_2) = C_1\, D.u_1 + C_2\, D.u_2$, it is said to be *linear*. If u_1 and u_2 are two solutions of the linear and homogeneous differential equation, $D.u = 0$, any linear superposition $u = C_1u_1 + C_2u_2$ is also a solution, for any values of the constants C_1 and C_2. If u_1 and u_2 are two independent solutions of the second-order differential equation $Du = 0$, u is the general solution of this equation, because it depends on two arbitrary and independent constants C_1 and C_2. A differential equation $D.u = f$ is not homogeneous because it contains a term f that is independent of u.

The general solution of differential equation [1.7] may be written in one of the following equivalent forms

$$u(t) = A\cos(\omega t + \phi) \tag{1.12}$$
$$= A'\sin(\omega t + \phi') \tag{1.13}$$
$$= A_1\cos(\omega t) + A_2\sin(\omega t). \tag{1.14}$$

Each of these expressions depends on two independent parameters that are determined by the initial conditions. The relationships between these parameters are:

$$A' = A \text{ and } \phi' = \phi + \pi/2 \tag{1.15}$$
$$A_1 = A\cos\phi = A\sin\phi', \qquad A_2 = -A\sin\phi = A\cos\phi'. \tag{1.16}$$

Note that, in form [1.14], A_1 and A_2 may be positive, negative or zero. If the amplitudes A and A' are chosen to be positive, we have the following relationships:

$$A = \sqrt{A_1^2 + A_2^2}, \qquad \tan\phi = -A_2/A_1, \qquad \tan\phi' = A_1/A_2. \tag{1.17}$$

EXAMPLE 1.1. Write the expression $u = 3\cos(\omega t) + 2\cos(\omega t - \pi/3)$ in the form $u = A\cos(\omega t + \phi)$.

SOLUTION – Using the addition formula for $\cos(\omega t - \pi/3)$, we find

$$u = 3\cos(\omega t) + 2\cos\frac{\pi}{3}\cos(\omega t) + 2\sin\frac{\pi}{3}\sin(\omega t) = 4\cos(\omega t) + 1.732\sin(\omega t).$$

We may write u in the form $A\cos(\omega t + \phi) = A\cos\phi\cos(\omega t) - A\sin\phi\sin(\omega t)$ if $A\cos\phi = 4$ and $A\sin\phi = -1.732$. By squaring both sides of these equations and adding them, we find $A^2 = 4^2 + 1.732^2$. Hence $A = 4.359$, $\cos\phi = 0.9177$ and $\sin\phi = -0.3974$. Thus we deduce that $\phi = -0.4086$ rad and $x = 4.359\cos(\omega t - 0.4086)$.

1.3. Complex representation and phasor representation

A *complex variable* (designated by an underlined symbol) may be written in the algebraic form

$$\underline{z} = x + iy, \qquad \text{with} \qquad x = \mathcal{R}e\,\underline{z} \qquad \text{and} \qquad y = \mathcal{I}m\,\underline{z}, \qquad [1.18]$$

where $i^2 \equiv -1$. The *real part* of $\underline{z}$ is x and its *imaginary part* is y. A complex number $\underline{z}$ is usually represented by a point of coordinates x and y in the Oxy plane called an *Argand diagram* (Figure 1.2a). We may also use the exponential form in terms of the polar coordinates

$$\underline{z} = \rho\cos\phi + i\rho\sin\phi = \rho e^{i\phi}, \qquad \text{where } \rho \equiv |\underline{z}| = \mathit{modulus}\,\underline{z} \qquad \text{and } \phi = \mathit{phase}\,\underline{z}. \quad [1.19]$$

$\rho \equiv |\underline{z}|$ is the *modulus* of $\underline{z}$ and ϕ is the *phase* of $\underline{z}$, where we have used Euler equation (see section A.5 of Appendix A)

$$e^{i\phi} = \cos\phi + i\sin\phi. \qquad [1.20]$$

The two representation are related by the equations

$$x = \rho\cos\phi, \quad y = \rho\sin\phi; \quad \rho = \sqrt{x^2 + y^2}, \quad \cos\phi = x/\rho, \quad \sin\phi = y/\rho. \qquad [1.21]$$

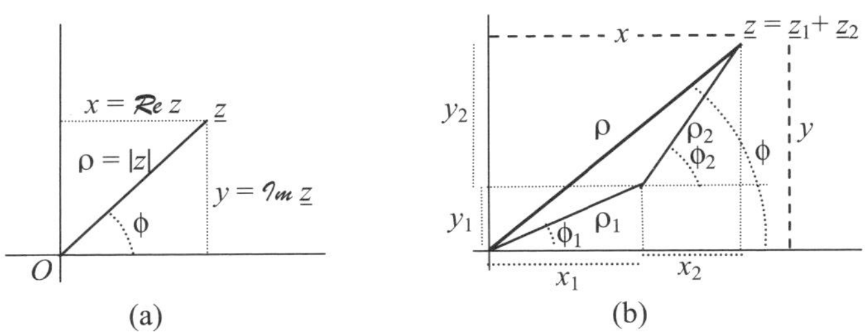

Figure 1.2. *a) Argand diagram; b) sum of two complex numbers $\underline{z}_1$ and $\underline{z}_2$*

The sum of two complex numbers is easily evaluated using the algebraic form (Figure 1.2b)

$$\underline{z}_1 + \underline{z}_2 = (x_1 + \mathrm{i}y_1) + (x_2 + \mathrm{i}y_2) = (x_1 + x_2) + \mathrm{i}(y_1 + y_2), \qquad [1.22]$$

while their product and ratio are easily evaluated using the exponential form:

$$\underline{z}_1\,\underline{z}_2 = (\rho_1\, e^{\mathrm{i}\phi_1})\,(\rho_2\, e^{\mathrm{i}\phi_2}) = \rho_1\,\rho_1\, e^{\mathrm{i}(\phi_1+\phi_2)} \qquad [1.23]$$

$$\underline{z}_1/\underline{z}_2 = (\rho_1\, e^{\mathrm{i}\phi_1})/(\rho_2\, e^{\mathrm{i}\phi_2}) = (\rho_1/\rho_1)\; e^{\mathrm{i}(\phi_1-\phi_2)}. \qquad [1.24]$$

It is easy to verify that differential equation [1.7] has the *general complex solution*

$$\underline{u}(t) = \underline{C}\, e^{\mathrm{i}\omega t} \qquad \text{with} \qquad \underline{C} \equiv C\, e^{\mathrm{i}\alpha}, \qquad [1.25]$$

where $\underline{C}$ is the *complex amplitude* and $C = |\underline{C}|$ is its modulus. This expression depends on two real parameters C and α, as it should for any general solution of a second order differential equation. By taking the real part, we find

$$u(t) = \mathcal{R}e\; \underline{u}(t) = \mathcal{R}e\,[C\, e^{\mathrm{i}(\omega t\, +\, \alpha)}] = C\cos(\omega t + \alpha). \qquad [1.26]$$

Comparing this with the expression $u = A\cos(\omega t + \phi)$, we deduce that

$$A = |\,\underline{C}\,| = \textit{modulus}\ \underline{C}, \qquad \phi = \alpha = \textit{phase}\ \underline{C}. \qquad [1.27]$$

Thus, the amplitude and the phase of the real solution u are respectively the modulus and the phase of the complex amplitude $\underline{C}$ of the complex solution.

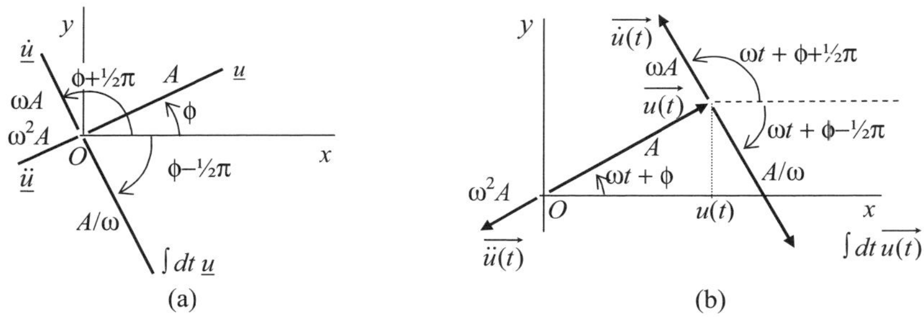

Figure 1.3. *Two similar representations of a simple harmonic function, its derivatives and its primitive: a) complex exponential representation and b) phasor representation*

The derivative of the function $\underline{u}$ and its primitive are:

$$\dot{\underline{u}} = \mathrm{i}\omega\, \underline{u}(t) = \omega A\, e^{\mathrm{i}(\omega t + \phi + \pi/2)} = \omega\, \underline{A}\, e^{\mathrm{i}\pi/2}\, e^{\mathrm{i}\omega t} \qquad [1.28]$$

$$\int dt\, \underline{u}(t) = \underline{u}/\mathrm{i}\omega = (A/\omega)\, e^{\mathrm{i}(\omega t + \phi - \pi/2)} = (\underline{A}/\omega)\, e^{-\mathrm{i}\pi/2}\, e^{\mathrm{i}\omega t}\,. \qquad [1.29]$$

Thus, to differentiate the simple harmonic function $\underline{u}$, we simply multiply it by iω, which is equivalent to multiplying the amplitude of $\underline{u}$ by ω and adding $\pi/2$ to the phase of $\underline{u}$ (Figure 1.3a). Similarly, the primitive of $\underline{u}$ with respect to time is obtained by dividing it by iω, which is equivalent to dividing the amplitude of $\underline{u}$ by ω and subtracting $\pi/2$ from its phase. As these functions depend on time only through the factor $e^{\mathrm{i}\omega t}$, the complex exponential functions are much easier to use than trigonometric functions.

In the case of linear oscillation equations like [1.7], it is often more convenient (and sometime inevitable) to determine the complex solution $\underline{u}$ and take, at the end, its real part $\mathcal{Re}\ \underline{u}$ as the real solution, which effectively represents the physical oscillation. More generally, we may evaluate any physical quantity, that is linear in u, using the complex function $\underline{u}$ and then taking its real part to represent the physical quantity. Note that this procedure is not valid for physical quantities that are nonlinear in u (for instance, the energy which is quadratic in u and $\dot{u}$). Such quantities do not obey the superposition principle. So, they must be evaluated using real quantities.

There is another geometrical representation of the simple harmonic oscillations, which is useful in some simple situations. Consider the vector $\overrightarrow{u(t)}$ of length A, which makes an angle (ωt + ϕ) with the x-axis (Figure 1.3b). This vector rotates counterclockwise around O at a rate ω, starting from the angle ϕ at $t = 0$. The *phasor diagram* (or rotating *vector diagram*) considers the function $u = A\cos(\omega t + \phi)$ as the projection of $\overrightarrow{u(t)}$ onto the x-axis.

The time derivative of $u(t)$, given by equation [1.5], is represented by the vector of length ωA, which has a phase lead of $\pi/2$ over $u(t)$, and its second derivative is represented by the vector of length $\omega^2 A$, which has a phase lead of π over $u(t)$. The primitive of $u(t)$ is represented by the vector $\int dt\,\overrightarrow{u(t)}$ of length A/ω which has a phase lag of $\pi/2$ over $u(t)$. If all the physical quantities have the same frequency ω, we represent the corresponding phasors at $t = 0$. Their representation at time t is simply obtained by a rotation of the whole figure through an angle ωt.

Two simple harmonic functions are said to be *isochrones* if they have the same frequency. The corresponding phasors $\overrightarrow{u_1(t)}$ and $\overrightarrow{u_2(t)}$ rotate around O with the same angular velocity ω. Therefore, they form a constant angle $\phi_1 - \phi_2$.

– u_1 and u_2 are said to be *in phase* if they have equal phases ($\phi_1 = \phi_2$). In this case, their phasors are collinear and in the same direction.

– u_1 and u_2 are said to be *in opposite phase* if $\phi_2 - \phi_1 = \pm\pi$. In this case, their phasors are collinear but in opposite directions.

– u_1 and u_2 are said to be *in quadrature* if $\phi_2 - \phi_1 = \pm\pi/2$. In this case, their phasors are perpendicular.

EXAMPLE 1.2. **a)** Using the complex exponential representation, evaluate the sum of the simple harmonic functions $u_1 = A\cos(\omega t)$ and $u_2 = B\sin(\omega t)$. **b)** Derive the same result using the phasor diagram.

SOLUTION –

a) The functions u_1 and u_2 are respectively the real parts of $\underline{u}_1 = A\,e^{i\omega t}$ and $\underline{u}_2 = B\,e^{i(\omega t - \pi/2)} = -iB\,e^{i\omega t}$. Their sum is $\underline{u} = \underline{u}_1 + \underline{u}_2 = \underline{C}\,e^{i\omega t}$, where the complex amplitude is $\underline{C} = A - iB$ which we may cast in the form $C\,e^{i\phi}$ with

$$C = \sqrt{A^2 + B^2}, \qquad \cos\phi = A/C \quad \text{and} \quad \sin\phi = -B/C.$$

The complex function $\underline{u}$ can then be written as $\underline{u} = C\,e^{i(\omega t + \phi)}$ and, by taking its real part, we get $u = C\cos(\omega t + \phi)$.

b) The functions u_1 and u_2 are respectively represented by the phasors $\overrightarrow{u_1}$ and $\overrightarrow{u_2}$ of lengths A and B, and making the angles ωt and $(\omega t - \pi/2)$ with the x-axis (Figure 1.4a). Thus $\overrightarrow{u_1}$ is orthogonal to $\overrightarrow{u_2}$. Their vector sum $\vec{u}$ is the hypotenuse of a right triangle. Pythagoras' theorem gives its magnitude $C = \sqrt{A^2 + B^2}$. The phasor $\vec{u}$ along with $\overrightarrow{u_1}$ makes an angle ϕ such that $\tan\phi = -B/A$. This angle is equal to the phase lead of u over u_1. For instance, if A and B are positive, ϕ lies between $-\pi/2$ and 0; then u effectively has a phase lag over u_1. As u_1 and u_2 have the same frequency, we may draw the phasor diagram at $t = 0$ as can be seen in Figure 1.4b.

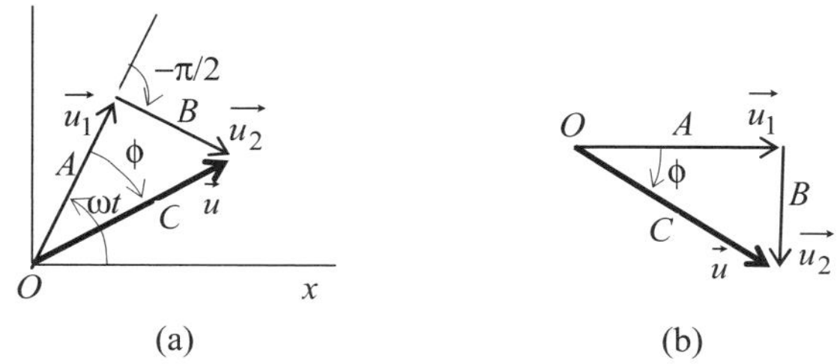

Figure 1.4. *Example 1.2*

1.4. Point mass subject to a force $-Kx$

Consider a point-like body of mass m confined to movement along the x-axis and subject to a force F, which depends on the position x according to the linear relation

$$F = -Kx, \tag{1.30}$$

where K is a positive *force constant*. As the force vanishes at the origin ($x = 0$), this point is an equilibrium position of the body. On the other hand, as F and x have opposite signs, F always tends to bring the system back to its equilibrium configuration. For this reason, it is called a *restoring force*. Thus, the origin $x = 0$ is a *stable equilibrium position*. An elastic spring, fixed at one end, ideally exerts this type of force on a point-like m attached to the other end, if it is displaced by a distance x from its equilibrium position (Figure 1.5). In this case, K is the spring *force constant*.

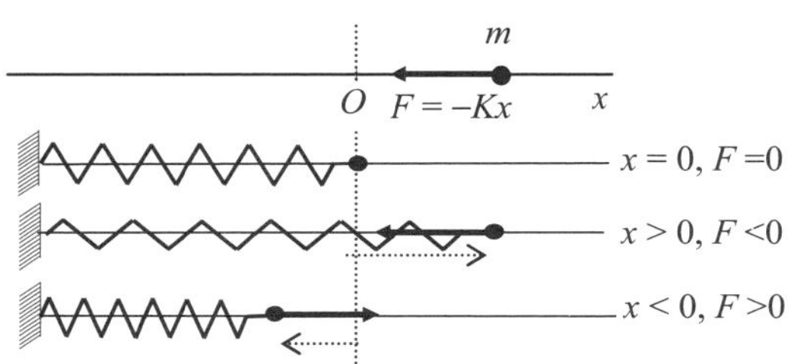

Figure 1.5. *Point-like mass subject to a force –Kx exerted by a spring of force constant K*

The equation of motion of the body is $F = m\ddot{x}$. Hence, by using [1.30],

$$\ddot{x} + (K/m)\, x = 0. \tag{1.31}$$

This is the equation of simple harmonic oscillations [1.7] with a natural frequency

$$\omega = \sqrt{K/m}\,. \qquad [1.32]$$

The general solution of equation [1.31] is [1.3], for the position and its corresponding velocity is [1.5]. In order to impose the initial conditions, it should be noted that the position of the body and its velocity cannot have any discontinuities because this would correspond to an infinite force or an infinite rate of energy transfer.

The energy of this mechanical oscillator may have two forms: *kinetic energy* $U_{(C)}$ and *potential energy* $U_{(P)}$. The kinetic energy is

$$U_{(C)} = \tfrac{1}{2}\, m\, \dot{x}^2 = \tfrac{1}{2}\, KA^2 \sin^2(\omega t + \phi). \qquad [1.33]$$

The potential energy $U_{(P)}$ at position x is the energy of the body if it is without velocity at this position. To write down the expression of $U_{(P)}$, let us assume that the body is initially at rest at x_o and that it is displaced to x slowly enough so that the velocity and consequently the body's kinetic energy remain negligible in the course of the motion. When the body has a displacement ξ from the equilibrium position O, we must exert upon it a force $F^{(ex)} = K\xi$, which is the exact opposite of restoring force [1.30]. Thus, a displacement $d\xi$ requires a work $dU^{(ex)} = F^{(ex)}\, d\xi = K\xi\, d\xi$ and the total work to achieve the displacement from x_o to x is

$$\int_{x_o}^{x} dx\, dU^{(ex)} = \int_{x_o}^{x} d\xi\; K\xi = \tfrac{1}{2}\, Kx^2 - \tfrac{1}{2}Kx_o^{\,2}. \qquad [1.34]$$

This result is still valid if the initial velocity and the final velocity are equal, regardless of the intermediate positions and velocities of the body. If the body is maintained at position x, this work remains stored in the oscillator as potential energy. Note that this energy does not depend on the body. It depends only on the displacement of the end point of the spring. Therefore, it is a state function of the spring. We also note that expression [1.34] only allows us to define the variation of the potential energy as the position of the end point of the spring varies from x_o to x. Thus, the potential energy is defined up to the addition of an arbitrary constant. We remove this arbitrariness by taking $U_{(P)} = 0$ at the equilibrium position $x = 0$. Thus, if the end point of the spring is displaced a distance x, the potential energy of the spring is

$$U_{(P)} = \tfrac{1}{2}\, Kx^2 = \tfrac{1}{2}\, KA^2 \cos^2(\omega t + \phi). \qquad [1.35]$$

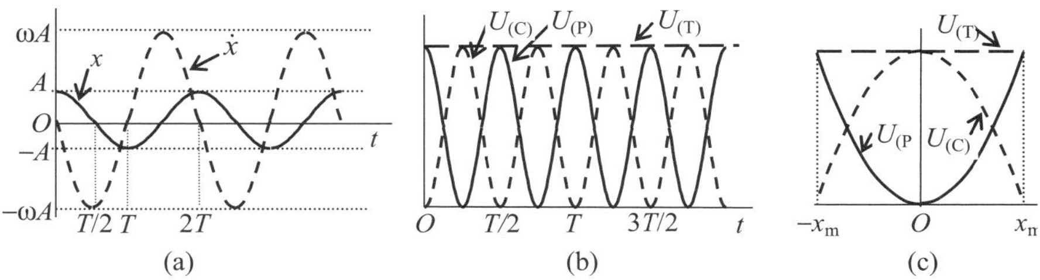

Figure 1.6. *a) The displacement x and the velocity $\dot{x}$ of an oscillator as functions of time; b) the energies $U_{(C)}$, $U_{(P)}$ and $U_{(T)}$ as functions of time; c) $U_{(C)}$ and $U_{(P)}$ as functions of x*

The *total energy* of the oscillator is

$$U_{(T)} = U_{(C)} + U_{(P)} = \frac{1}{2} KA^2. \qquad [1.36]$$

It is conserved in time as it should be for any isolated system. Indeed, we have assumed that there are no forces other than the spring restoring force $-Kx$, which is an internal force of the spring-mass system, and we have ignored all dissipative forces like air resistance, internal frictions in the spring, etc. Figures 1.6a and 1.6b illustrate the variations of the position x, velocity $\dot{x}$, kinetic energy $U_{(C)}$, potential energy $U_{(P)}$ and total energy $U_{(T)}$ as functions of time, in the case of a body which is initially displaced to x_o and released without initial velocity (thus, $A = x_o$ and $\phi = 0$). At $t = 0$, the energy is totally potential in form and equal to $\frac{1}{2}KA^2$. After the release, the displacement and the potential energy decrease while the velocity and the kinetic energy increase. At $t = T/4$, the body is back in the equilibrium position O, which corresponds to the potential energy equal to zero and the kinetic energy reaching its maximum value $\frac{1}{2}KA^2$ as the body is moving with maximum velocity ωA. After that, the body continues its motion toward the negative values of x, the potential energy increases and the kinetic energy decreases. As it reaches the point $x = -A$, all the energy is again in the potential form, the kinetic energy is zero and so is the velocity. Then, the body returns to the origin under the influence of the restoring force. Thus, the energy of the oscillator is transformed periodically from one form to the other without any change in the total energy.

Figure 1.6c illustrates $U_{(P)} = \frac{1}{2}Kx^2$ and $U_{(C)} = U_{(T)} - U_{(P)}$ as functions of x. The stable equilibrium position ($x = 0$) is a minimum of the potential energy and this is a general property of physical systems. Indeed, as the force is related to the potential energy by the equation $F = -dU_{(P)}/dx$, the equilibrium requires that $dU_{(P)}/dx = 0$ for $x = 0$. This means that $x = 0$ is a minimum or a maximum of $U_{(P)}$. The equilibrium is *stable* if, for $x > 0$, for instance, the force is negative. This is the case if $x = 0$ corresponds to a minimum of $U_{(P)}$. On the other hand, for a given value of the total energy $U_{(T)}$, the line drawn at height $U_{(T)}$ meets the $U_{(P)}$ curve at two symmetric turning points corresponding to the displacements $\pm x_m$. These are the boundaries of the accessible domain of displacement: they correspond to $U_{(C)} = 0$ and the body which is instantaneously at rest and, because the force attracts it toward O, it accelerates back. The condition $U_{(P)} = \frac{1}{2}Kx_m^2 = U_{(T)}$ gives the turning points $\pm x_m = \pm\sqrt{2U_{(T)}/K}$.

EXAMPLE 1.3. Using the law of conservation of energy, establish the equation of motion of a body of mass m attracted to the origin O with a force $F = -Kx$. What is the frequency of oscillations if m = 100 g and $K = 4.0 \times 10^3$ N/m? What is the maximum displacement if the mass is brought to the position x = 5 cm and thrown with a velocity 10 m/s in the positive x direction?

SOLUTION – The total energy of this oscillator is $U_{(T)} = \frac{1}{2}m\dot{x}^2 + \frac{1}{2}Kx^2$. As $U_{(T)}$ is constant, the time-derivative of the equation $U_{(T)}$ = constant gives

$$\tfrac{1}{2}m(2\dot{x}\ddot{x}) + \tfrac{1}{2}K(2x\dot{x}) = 0 \quad \Rightarrow \quad \dot{x}(\ddot{x} + \omega^2 x) = 0, \quad \text{where } \omega^2 = K/m.$$

This equation is identically satisfied if $\dot{x} = 0$ (which corresponds to the body at rest) or if $\ddot{x} + \omega^2 x = 0$ (which is the equation of simple harmonic oscillations). Thus, the oscillation frequency is $\tilde{\nu} = \sqrt{K/m}\,/2\pi = 32$ Hz .

If the body is thrown from point $x(0)$ = 5 cm with a velocity of $\dot{x}(0)$ = 10 m/s, its total energy is $U_{(T)} = \frac{1}{2}m\dot{x}^2 + \frac{1}{2}Kx^2 = 10$ J. As it reaches its maximum displacement x_m, its velocity vanishes and its energy is totally in the potential form. Thus, at this position, $\frac{1}{2}Kx_m^2 = U_{(T)}$ and $x_m = \sqrt{2U_{(T)}/K} = 7.1$ cm. This result may also be obtained by writing the explicit solution, which corresponds to the given initial conditions (see section 1.2). Equation [1.8] gives the amplitude of the oscillations $A = \sqrt{x(0)^2 + \dot{x}(0)^2/\omega^2} = 7.1$ cm which is equal to x_m.

1.5. Angular oscillations

Consider a rigid body confined to rotate around the z-axis (Figure 1.7a). Let P be a plane containing the z-axis which is fixed with respect to the rotating body. We specify the angular position of the body around the z-axis using the angle θ that the plane P makes with its position P_o when the body is in equilibrium. The equation of rotation of the body around Oz is $J\ddot{\theta} = \Gamma_z$, where J is the moment of inertia of the body about the z-axis and Γ_z is the z-component of the moment of the external forces, which act on the body. The angle θ is algebraic according to the right-hand rule (i.e. θ is positive if the right-hand fingers curl in the direction of rotation as the thumb points in the z-direction). We consider the case of a body subject to a restoring moment of force $\Gamma_z = -\hat{K}\,\theta$, where $\hat{K}$ is a positive constant. Γ_z tends to bring the body back to the stable equilibrium position $\theta = 0$ if the body is rotated. Thus, the equation of motion $J\ddot{\theta} = \Gamma_z$ leads to the oscillation equation

$$\ddot{\theta} + \omega^2\theta = 0, \qquad \text{where} \quad \omega = \sqrt{\hat{K}/J}\,. \qquad [1.37]$$

The general solution for $\theta(t)$ and the angular speed $\dot{\theta}$ are, respectively, [1.3] and [1.5]. The kinetic energy of the rotating body is $U_{(C)} = \tfrac{1}{2} J\,\dot{\theta}^2$. The potential energy $U_{(P)}$ is the work required to rotate the body from the equilibrium position $\theta = 0$ to the position θ. An analysis similar to that of the mass subject to the restoring force $-Kx$ shows that $U_{(P)} = \tfrac{1}{2}\hat{K}\,\theta^2$. Thus, the total energy of the oscillator is

$$U_{(T)} = U_{(C)} + U_{(P)} = \tfrac{1}{2} J\omega^2 A^2 \sin^2(\omega t + \phi) + \tfrac{1}{2}\hat{K} A^2 \cos^2(\omega t + \phi) = \tfrac{1}{2}\hat{K} A^2. \qquad [1.38]$$

It is conserved in the course of the motion as is to be expected for any isolated system.

A simple example of an angular oscillator is the torsional pendulum. This is a rigid body attached to the lower end of a cylindrical wire whose other end is fixed (Figure 1.7a). If the body is rotated through an angle θ, the wire is twisted and it exerts on the body a restoring moment of force $\Gamma_z = -\hat{K}\,\theta$. The restoring force constant (or *torsion constant*) is $\hat{K} = \mu R^4/2L$, where R is the radius of the wire and L is its length. μ is a characteristic constant of the wire material called the *shear modulus* (see section 5.2).

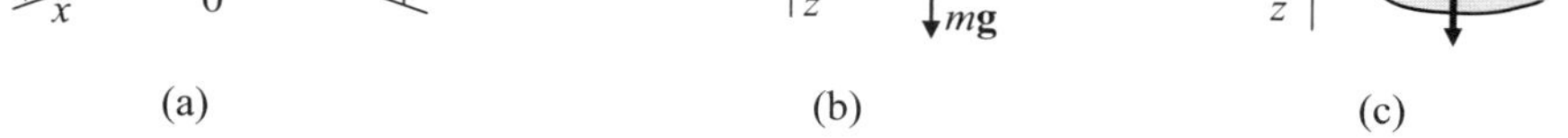

Figure 1.7. *Angular oscillations: a) torsional pendulum, b) simple pendulum and c) physical pendulum. The angle* θ *is positive in these positions*

The gravity pendulum is another example of an angular oscillator. Consider a *simple pendulum*, that is, a point-like mass m suspended from the end M of a thin and inextensible thread of length L. The mass m is confined to oscillate within a vertical plane Oxz (Figure 1.7b). The position of the mass is determined by the angle θ of the thread with the downward vertical axis Oz. θ is algebraic according to the right-hand rule about the horizontal axis of rotation Oy. Let $\mathbf{e}_x$, $\mathbf{e}_y$ and $\mathbf{e}_z$ be the unit vectors of the coordinate axes. In this case, the restoring force is the weight $\mathbf{f} = mg\,\mathbf{e}_z$. Its moment with respect to the fixed end O of the thread is $\mathbf{\Gamma} = \mathbf{OM} \times \mathbf{f} = -mgL \sin\theta\, \mathbf{e}_y$. As $\Gamma_y = -mgL \sin\theta$ and θ have opposite signs, Γ_y is a restoring moment of force. The moment of inertia of the mass about Oy is $J = mL^2$. Thus the equation of motion $J\ddot{\theta} = \Gamma_z$ leads to the equation $mL^2\ddot{\theta} = -mgL \sin\theta$. Hence,

$$\ddot{\theta} + (g/L)\sin\theta = 0. \quad [1.39]$$

A physical pendulum is a rigid body, which is confined to oscillate Ozx about a horizontal axis Oy under the influence of its own weight (Figure 1.7c). Let G be the center of mass of the body, m its total mass and J its moment of inertia about Oy. We set $OG = L$. The body weight is equivalent to a single force $mg\,\mathbf{e}_z$ applied at G. Its moment with respect to O is $\mathbf{\Gamma} = \mathbf{OG} \times mg\,\mathbf{e}_z = -mgL \sin\theta\,\mathbf{e}_y$ and the equation of rotation is $J\ddot{\theta} = -\,mgL\sin\theta$. Hence,

$$\ddot{\theta} + (mgL/J)\sin\theta = 0. \quad [1.40]$$

The equations of [1.39] for the simple pendulum and [1.40] for the physical pendulum are not linear. However, if θ is small, we may make the approximation $\sin\theta \cong \theta$, the equation of motion [1.40] then takes the approximate linear form

$$\ddot{\theta} + \omega^2\theta = 0, \qquad \text{where} \qquad \omega = \sqrt{mgL/J}\,. \quad [1.41]$$

This is the equation of a simple harmonic oscillator of frequency $\omega = \sqrt{mgL/J}$ and period $T = 2\pi\sqrt{J/mgL}$. In the case of a simple pendulum, $J = mL^2$. Hence,

$$\omega = \sqrt{g/L}\,, \qquad T = 2\pi\sqrt{L/g}\,. \quad [1.42]$$

The period of small oscillations does not depend on the amplitude or the mass m of the simple pendulum or the mass density of a physical pendulum if it is homogeneous. This is a consequence of the property of the weight being proportional to the mass.

1.6. Damped oscillations

A real oscillator is always subject to forces, which dissipate a part of the oscillator energy as heat, or in other forms. Friction forces acting on solids and viscosity forces in fluids are examples of such *dissipative forces*. As the energy is proportional to the square of the amplitude, the amplitude decreases gradually until the oscillator returns to equilibrium; the oscillator is then said to be *damped*.

For instance, consider a body which oscillates on the x-axis. The friction force is a function of the velocity $\dot{x}$, which vanishes in the limit $\dot{x} \rightarrow 0$. The simplest form of this force is $F_{(\mathrm{fr})} = -b\,\dot{x}$, where b is the *friction coefficient*, which is always positive. The (–) sign indicates that $F_{(\mathrm{fr})}$ points in the opposite direction to the velocity. This type of friction is called a *viscous friction*. In this case, the equation of motion $m\,\ddot{x} = \Sigma\,F$ takes the form

$$\ddot{x} + 2\beta\,\dot{x} + \omega_o{}^2 x = 0, \qquad [1.43]$$

where $\beta \equiv b/2m$ is the *damping constant* and $\omega_o = \sqrt{K/m}$ is the *natural frequency* (i.e. the frequency of oscillations in the absence of friction). To solve the differential equation [1.43], we try a solution of the form $x = \underline{C}\,e^{-\eta t}$, where η is a constant to be determined. Substituting this expression into [1.43], we find that η must be a root of the *characteristic equation*

$$\eta^2 - 2\beta\eta + \omega_o{}^2 = 0. \qquad [1.44]$$

The form of the solution of [1.43] depends on the nature of the roots of the characteristic equation.

A) Overdamping

This case corresponds to a strong friction force ($\beta > \omega_o$). So, characteristic equation [1.44] has two real and positive roots $\eta_1 = \beta - \sigma$ and $\eta_2 = \beta + \sigma$, where

$$\sigma = \sqrt{\beta^2 - \omega_o{}^2}\,. \qquad [1.45]$$

Thus, equation of motion [1.43] has two particular solutions $x_i = C_i\ e^{-\eta_i t}$, which decrease exponentially with time. As these solutions are independent, differential equation [1.43] is linear and its general solution is:

$$x(t) = C_1\, e^{-\eta_1 t} + C_2\, e^{-\eta_2 t} = C_1\, e^{-(\beta-\sigma)t} + C_2\, e^{-(\beta+\sigma)t} \qquad [1.46]$$

$$\dot{x}(t) = C_1(\sigma - \beta)\, e^{-(\beta-\sigma)t} - C_2(\sigma + \beta)\, e^{-(\beta+\sigma)t}\,. \qquad [1.47]$$

The integration constants C_1 and C_2 may be determined from the initial conditions, which are: $x(0) = C_1 + C_2$ and $\dot{x}(0) = C_1(\sigma - \beta) - C_2(\sigma + \beta)$, thus

$$C_1 = (1/2\sigma)\,[(\sigma + \beta)\, x(0) + \dot{x}(0)], \qquad C_2 = (1/2\sigma)\,[(\sigma - \beta)\, x(0) - \dot{x}(0)]. \quad [1.48]$$

For instance, if the body is displaced to a position x_o and released without initial velocity at $t = 0$, the solution is:

$$x = (x_o/2\sigma)\,[(\sigma + \beta)\, e^{-(\beta-\sigma)t} + (\sigma - \beta)\, e^{-(\beta+\sigma)t}],$$
$$\dot{x} = (\omega_o^2 x_o/2\sigma)\,[e^{-(\beta-\sigma)t} - e^{-(\beta+\sigma)t}]. \qquad [1.49]$$

This motion is represented in Figure 1.8a by curve (1) for $\omega_o = 10$ rad/s and $\beta = 15\ \mathrm{s}^{-1}$, and curve (2) for $\omega_o = 10$ rad/s and $\beta = 13\ \mathrm{s}^{-1}$. The solution is the sum of two exponentially decreasing functions of time (due to the negative coefficients of t in the exponentials). In this case, the oscillator returns back to the equilibrium position without oscillation. As $\beta - \sigma < \beta + \sigma$, the first term in [1.49] decreases more slowly than the second term; thus, it dominates the long-term motion. We say that the motion is damped with a *damping coefficient* $\delta = \beta - \sigma$.

B) Critical damping

In the special case where $\beta = \omega_o$, the characteristic equation has two equal roots $\eta_1 = \eta_2 = \beta$ and differential equation [1.43] has the solution $x = F e^{-\beta t}$. As this expression depends on a single constant of integration F, it is not the general solution. Let us write $x = F(t)\, e^{-\beta t}$ and substitute it into equation [1.43] with $\beta = \omega_o$. We find that this equation is identically satisfied if $F(t)$ is a solution of the differential equation $\ddot{F} = 0$; thus, $F = A + Bt$ and the general solution of [1.43] is:

$$x = (A + Bt)\, e^{-\beta t}, \qquad \dot{x} = [B - \beta A - \beta B t\,]\, e^{-\beta t} \qquad [1.50]$$

If the body is displaced to x_o and released without velocity, the solution is:

$$x = x_o(1 + \beta t)\, e^{-\beta t}, \qquad \dot{x} = -x_o \beta^2 t\, e^{-\beta t}\,. \qquad [1.51]$$

This motion is represented by curve (3) in Figure 1.8a. The body returns without oscillation to its equilibrium position with a damping coefficient $\delta = \beta$.

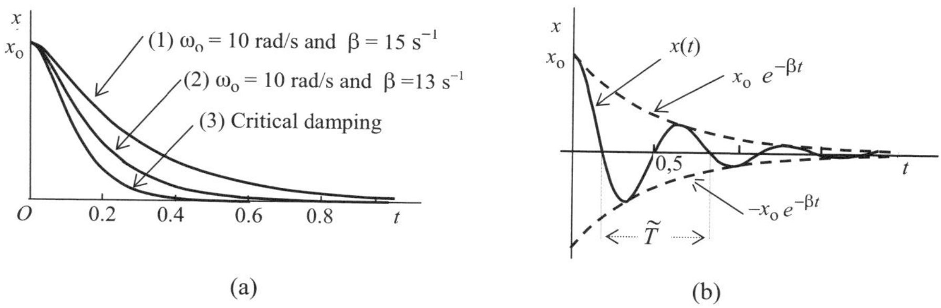

Figure 1.8. *a) Overdamped oscillations (curves (1) and (2)) and critically damped oscillations (curve (3)); b) underdamped oscillations for $\omega_o = 10\ s^{-1}$ and $\beta = 2\ s^{-1}$*

C) Underdamping

If the friction force is weak ($\beta < \omega_o$), we set

$$\hat{\omega} = \sqrt{\omega_o^{\,2} - \beta^2}\ . \qquad [1.52]$$

In this case, the characteristic equation has two roots $\eta_1 = \beta - i\,\hat{\omega}$ and $\eta_2 = \beta + i\,\hat{\omega}$ which are complex conjugate. Thus, the general solution of [1.43] is:

$$x = \underline{C}_1\ e^{-\eta_1 t} + \underline{C}_2\ e^{-\eta_2 t} = e^{-\beta t}\ [\underline{C}_1\ e^{i\hat{\omega} t} + \underline{C}_2\ e^{-i\hat{\omega} t}]. \qquad [1.53]$$

We may write this expression in the following forms:

$$\begin{aligned} x &= A\, e^{-\beta t} \cos(\hat{\omega}\, t + \phi) = A\, e^{-\beta t} \sin(\hat{\omega}\, t + \phi') \\ &= e^{-\beta t}\ [A_1 \cos\ \hat{\omega}\, t + A_2 \sin\ \hat{\omega}\, t] = e^{-\beta t}\ \mathcal{R}e\ [\underline{A}\, e^{i\hat{\omega} t}\ ], \end{aligned} \qquad [1.54]$$

A_1 and A_2 are related to $\underline{C}_1$ and $\underline{C}_2$ by the equations $A_1 = \underline{C}_1 + \underline{C}_2$ and $A_2 = i(\underline{C}_1 - \underline{C}_2)$ and the constants A, ϕ, ϕ', A_1, A_2 are related by equations [1.15] to [1.17].

We may write the real solution x and the corresponding velocity in the form:

$$x = A\, e^{-\beta t} \cos(\hat{\omega}\, t + \phi) \qquad [1.55]$$

$$\begin{aligned} \dot{x} &= -A\, e^{-\beta t}\ [\beta \cos(\hat{\omega}\, t + \phi) + \hat{\omega}\ \sin(\hat{\omega}\, t + \phi)] \\ &= A\omega_o\, e^{-\beta t} \cos(\hat{\omega}\, t + \phi + \pi - \alpha), \quad \text{where } \tan\alpha = \hat{\omega}/\beta \quad (0<\alpha<\pi/2). \end{aligned} \qquad [1.56]$$

For instance, if the system is displaced to x_o and released without initial velocity, the initial conditions are $x_o = A \cos \phi$ and $0 = -A(\beta \cos \phi + \hat{\omega} \sin \phi)$. We deduce that $A = x_o\omega_o/\hat{\omega}$, $\cos \phi = \hat{\omega}/\omega_o$ and $\sin \phi = -\beta/\omega_o$. Thus the solution can be written as:

$$x(t) = (x_o\, \omega_o/\hat{\omega})\; e^{-\beta t} \cos(\hat{\omega} t + \phi). \tag{1.57}$$

The solution $x = A\, e^{-\beta t} \cos(\hat{\omega} t + \phi)$ is shown in Figure 1.8b. It looks like a simple harmonic oscillation, but its amplitude $A\, e^{-\beta t}$ decreases exponentially in time. We say that the oscillations are damped with a damping coefficient β. It is not periodic since x and $\dot{x}$ never simultaneously take the same values again. However, the system passes periodically in the same direction through its equilibrium position; we may say that the oscillations are *quasi-periodic* with a *quasi-frequency* $\hat{\omega}$. The function x has maximums or minimums at t_n such that $\dot{x} = 0$, i.e. $\tan(\hat{\omega} t_n + \phi) = -\beta/\hat{\omega}$; hence $\omega t_n + \phi = \text{Arctan}(-\beta/\hat{\omega}) + n\pi$, where n in an integer. x reaches its maximums periodically with a period $\hat{T} = 2\pi/\hat{\omega}$. The values of these maximums are $x_n = A\, e^{-\beta t_n} \cos(\hat{\omega} t_n + \phi) = \pm\, (A\,\hat{\omega}/\omega_o)\, e^{-\beta t_n}$. After a complete oscillation, the amplitude is reduced by a factor $e^{-\beta \hat{T}}$, that is, $x_{n+2}/x_n = e^{\beta(t_n - t_{n+2})} = e^{-\beta \hat{T}}$. We may also write $\beta \hat{T} = \ln(x_n/x_{n+2})$. The quantity $\beta \hat{T}$ is the *logarithmic decrement.*

In all three cases of damped oscillators, the oscillation amplitude decreases like $e^{-\delta t}$, where the damping coefficient δ is equal to $\beta - \sqrt{\beta^2 - \omega_o^{\,2}}$ in the case of overdamping, and β in the case of underdamping or critical damping. The oscillation dies out more quickly when δ has its maximum value; that is, if the damping is critical ($\beta = \omega_o$). This is preferred for some oscillating systems, like cars, musical instruments and physical instruments to reduce the oscillation time.

The inverse of the damping coefficient $\tau = 1/\delta$ is the *relaxation time.* At $t = 5\tau$, 8τ and 10τ, the values of the damping factor $e^{-t/\tau}$ are 6.7×10^{-3}, 3×10^{-4} and 4.5×10^{-5} respectively. Thus, after an interval of time equal to about $10\,\tau$, the amplitude of the oscillations is practically zero and we may consider that the oscillator is back in equilibrium. The relaxation time is somehow a measure of the duration of oscillation. We may define the *quality factor*

$$f_q = \pi\tau/T_o = \omega_o/2\beta. \tag{1.58}$$

This is approximately the number of oscillations that the system makes before reaching a steady regime or the equilibrium position after excitation. The quality factor is smaller than 0.5 if the oscillator is overdamped, equal to 0.5 if its is critical and larger than 0.5 if it is underdamped. As an order of magnitude, f_q is 250 to 1,500 for earthquakes, 10^3 to 10^4 for musical instruments and 10^8 for excited atoms.

1.7. Dissipation of the energy of a damped oscillator

As in the preceding section, we consider the oscillation of a point-like object on the x-axis. A similar analysis can be made for any mechanical or electromagnetic oscillating system. In the case of underdamping, the general solution [1.43] is in the form [1.55] for x and [1.56] for $\dot{x}$. The total energy of the oscillator is:

$$U_{(T)} = U_{(C)} + U_{(P)} = \tfrac{1}{2}\, mA^2 e^{-2\beta t}\, [\,\tilde{\omega}^2 + 2\beta^2 \cos^2(\tilde{\omega} t+\phi) + \beta\,\tilde{\omega}\, \sin 2(\tilde{\omega} t+\phi)]. \quad [1.59]$$

Its variation between t and $t + dt$ is:

$$dU_{(T)} = -2\beta m A^2 e^{-2\beta t}\, [\,\tilde{\omega} \sin(\tilde{\omega} t+\phi) + \beta \cos(\tilde{\omega} t+\phi)]^2\, dt. \quad [1.60]$$

It is always negative. This means that the total energy decreases. The work of the friction force $F_{(fr)} = -b\,\dot{x}$ in the same interval of time is:

$$dU_{(fr)} = -F_{(fr)}\, dx = -F_{(fr)}\, \dot{x}\, dt = b\,\dot{x}^2\, dt = 2m\beta\, \dot{x}^2\, dt. \quad [1.61]$$

Using expression [1.56] for $\dot{x}$, we verify that $dU_{(fr)} = -\, dU_{(T)}$. This shows that the decrease of the total energy is equal to the energy dissipated by the friction force.

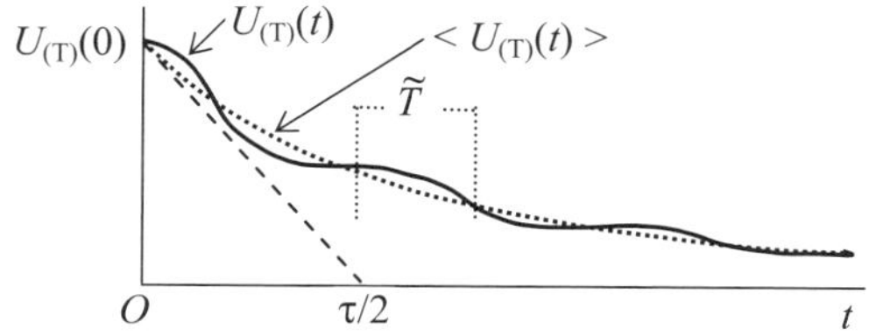

Figure 1.9. *Variation of the energy of an underdamped oscillator (solid curve) and its average over a period (dashed curve)*

While decreasing in time because of the exponential factor $e^{-2\beta t}$, the total energy (see [1.59]) oscillates slightly (Figure 1.9). If the frequency is high, it is sometimes interesting to see how the energy decreases after the elimination of these small oscillations. For this, we calculate the average value of $U_{(T)}$ over a complete oscillation (between $t - \tfrac{1}{2}\tilde{T}$ and $t + \tfrac{1}{2}\tilde{T}$ for instance)

$$< U_{(T)}(t) > = \frac{1}{\tilde{T}} \int_{t-\tilde{T}/2}^{t+\tilde{T}/2} dt' \, U_{(T)}(t'). \qquad [1.62]$$

To simplify the calculation, let us assume that the factor $e^{-2\beta t}$ varies very slowly in this interval of time; this is effectively the case if $\beta T << 1$. In other words, the relaxation time τ is much longer than the quasi-period $\tilde{T}$. Then, we may write:

$$< U_{(T)}(t) > \cong \frac{mA^2}{2\tilde{T}} e^{-2\beta t} \int_{t-\tilde{T}/2}^{t+\tilde{T}/2} dt' \, [\tilde{\omega}^2 + 2\beta^2 \cos^2(\tilde{\omega} t' + \phi) + \beta \tilde{\omega} \sin 2(\tilde{\omega} t' + \phi)]$$
$$= \tfrac{1}{2} KA^2 \, e^{-2\beta t} \equiv U_{(T)}(0) \, e^{-2t/\tau}, \qquad [1.63]$$

where $U_{(T)}(0)$ is the initial energy. Thus, the averaged energy decreases exponentially over time (Figure 1.9).

EXAMPLE 1.4.– Show that the average values of $\sin(n\omega t)$ and $\cos(n\omega t)$ over a period are zero, while the average values of $\cos^2(n\omega t)$ and $\sin^2(n\omega t)$ are ½.

SOLUTION – The average value of $\sin(n\omega t)$ over a period is

$$< \sin(n\omega t) > = (1/T) \int_{t-T/2}^{t+T/2} dt' \sin(n\omega t')$$
$$= (1/nT\omega) \, [\cos(n\omega t - n\omega T/2) - \cos(n\omega t + n\omega T/2)]$$
$$= (1/2n\pi) \, [\cos(n\omega t - n\pi) - \cos(n\omega t + n\pi)] = 0.$$

A similar analysis shows that $< \cos(n\omega t) > = 0$. The interpretation of this result is simple: the functions $\sin(n\omega t)$ and $\cos(n\omega t)$ take opposite values in two consecutive half periods; thus, their average values over a period is zero. As for the squares of these functions, they are always positive. As $\sin^2 n\omega t = ½ - ½ \cos(2n\omega t)$ and $\cos^2 n\omega t = ½ + ½ \cos(2n\omega t)$, the average value of $\cos(2n\omega t)$ being zero, we deduce that the average values of $\sin^2 n\omega t$ and $\cos^2 n\omega t$ are equal to the constant term ½.

1.8. Oscillating LCR circuits

Consider a circuit formed by a capacitor C, an inductor L and a resistor R connected in series. The equilibrium state corresponds to the uncharged capacitor, the current intensity $I = 0$ and all the points of the circuit at the same potential. Assume that the capacitor is initially disconnected and given a charge Q_o (using an auxiliary battery) (Figure 1.10a) then connected to the LR circuit (Figure 1.10b). Its discharge produces a current intensity $I = -dQ/dt$ in the shown direction. The potential drops across the elements are respectively $V_N - V_M = -Q/C$, $V_M - V_P = L\dot{I}$ and $V_P - V_N = RI$. Adding these equations, we get the circuit equation $Q/C + L\dot{I} + RI = 0$, that is,

$$\ddot{Q} + 2\beta\dot{Q} + \omega_o^2 Q = 0, \quad \text{where} \quad \omega_o = 1/\sqrt{LC} \text{ and } \beta = R/2L. \qquad [1.64]$$

This is the standard equation of a damped oscillator with a natural frequency ω_o and a damping constant β.

The analysis of section 1.6 may be repeated. The solution is overdamped if $\beta > \omega_o$ (i.e. $R > 2\sqrt{L/C}$), critically damped if $\beta = \omega_o$ (i.e. $R = 2\sqrt{L/C}$) and underdamped if $\beta < \omega_o$ (i.e. $R < 2\sqrt{L/C}$). In this last case, the quasi-frequency is $\tilde{\omega} = \sqrt{\omega_o^2 - \beta^2} = \sqrt{1/LC - R^2/4L^2}$ and the solution is of the form [1.55] for the charge and [1.56] for the intensity. The electric energy stored in the capacitor is $U_{(E)} = Q^2/2C$ and the magnetic energy stored in the inductor is $U_{(M)} = \frac{1}{2}LI^2$. Thus, the total energy of the circuit is

$$U_{(EM)} = U_{(E)} + U_{(M)} = \tfrac{1}{2}LA^2\, e^{-2\beta t}\, [\tilde{\omega}^2 + 2\beta^2\cos^2(\tilde{\omega}t + \phi) + \beta\tilde{\omega}\, \sin 2(\tilde{\omega}t + \phi)]. \quad [1.65]$$

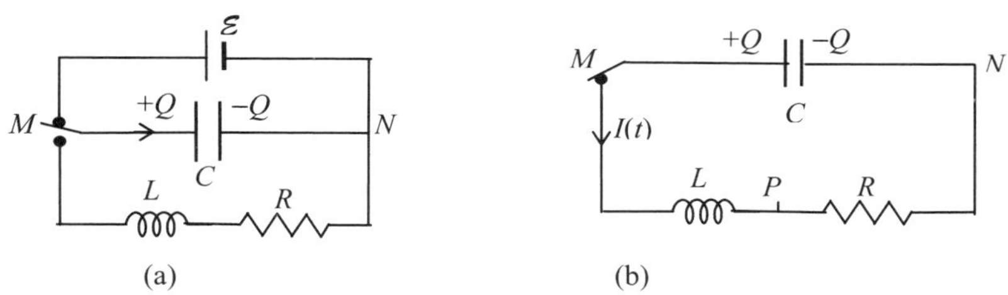

Figure 1.10. *a) Charging a capacitor; and b) oscillating LCR circuit*

The constants A and ϕ may be determined by imposing the initial conditions. We note that the energy of the circuit cannot have discontinuities; thus, Q and I are continuous. For instance, if at $t = 0$, the capacitor has a charge Q_o and is connected to the circuit, we must have $A \cos \phi = Q_o$ and $A\omega_o \sin \phi = 0$. We deduce that $A = Q_o\omega_o/\tilde{\omega}$ and $\tan \phi = -\beta/\tilde{\omega}$ $(-\pi/2 < \phi < 0)$. The total energy decreases from its initial value $Q_o^2/2C$ because of the Joule effect in the resistor. The quality factor of the circuit is $f_q = \omega_o/2\beta = \sqrt{L/CR^2}$.

EXAMPLE 1.5.– A capacitor of 1.0 μF is charged to a voltage of 20 V and then connected to an inductor of 0.20 H and a resistor of 100 Ω in series. **a)** Determine the damping constant, the quasi-frequency, the relaxation time and the quality factor. **b)** Write the expressions of the charge and the current intensity. How long it takes for the amplitude of the intensity to be reduced to 1% of its initial value. How long it takes for the average energy to be dissipated up to 99%.

SOLUTION –

a) The natural frequency ω_o of this circuit is $\omega_o = 1/\sqrt{LC} = 2200$ rad/s. The damping coefficient, the pseudo-frequency, the relaxation time and the quality factor are respectively:

$$\beta = R/2L = 250\ \text{s}^{-1}, \qquad \tilde{\omega} = \sqrt{\omega_o^2 - \beta^2} = 2\,222\ \text{rad/s} \qquad \Rightarrow \tilde{\nu} = 354\ \text{Hz}$$
$$\tau = 1/\beta = 4.00 \times 10^{-3}\ \text{s} \quad \text{and} \quad f_q = \pi\tau/T_o = \tfrac{1}{2}\,\tau\omega_o = 4.45$$

b) The charge and the current intensity are given by the expressions:

$$Q = A\,e^{-\beta t}\cos(\tilde{\omega} t + \phi), \qquad I = A\,e^{-\beta t}\,[-\beta\cos(\tilde{\omega} t + \phi) - \tilde{\omega}\sin(\tilde{\omega} t + \phi)].$$

The initial conditions $Q = Q_o = CV = 20\ \mu$C and $I = 0$ are verified if $A\cos\phi = Q_o$ and $A(-\beta\cos\phi - \tilde{\omega}\sin\phi) = 0$. By taking the amplitude A as positive, the first equation gives $\cos\phi > 0$, thus $-\pi/2 < \phi < \pi/2$ and the second equation gives $\tan\phi = -\beta/\tilde{\omega} = -0.1125$, thus $\phi = -0.112$ rad and $A = 20.1\ \mu$C. The intensity may be written as

$$I = A\,e^{-\beta t}\sqrt{\tilde{\omega}^2 + \beta^2}\,[\sin\phi\cos(\tilde{\omega} t + \phi) - \cos\phi\sin(\tilde{\omega} t + \phi)] = A\omega_o\,e^{-\beta t}\sin(\tilde{\omega} t)$$

The amplitude of the intensity $A\omega_o e^{-\beta t}$ is reduced to 1% of its initial value at time t such that $A\omega_o e^{-\beta t} = 10^{-2} A\omega_o$, thus $e^{-\beta t} = 10^{-2}$ and $t = 18.4$ ms. The average energy decreases according to the equation $\langle U_{EM} \rangle = U_{EM}(0)\,e^{-2\beta t}$. It is dissipated up to 99% in the resistor if $\langle U_{EM} \rangle = 10^{-2}\,U_{EM}(0)$, i.e.

$$e^{-2\beta t} = 10^{-2} \quad \Rightarrow \quad \beta t = \ln 10 \Rightarrow t = 9.21\ \text{ms}.$$

1.9. Small oscillations of a system with one degree of freedom

A physical system has *one degree of freedom* if its state is determined by a single variable or *generalized coordinates* q. A stable equilibrium position q_e always corresponds to a minimum of the potential energy. Instead of q, it is more convenient to use the displacement from equilibrium $u \equiv q - q_e$ as a generalized coordinate and assume that this displacement is small.

The value $u = 0$ corresponds to a stable equilibrium position O of the system if it is a minimum of the potential energy $U_{(P)}$. Indeed, if the system is at rest at O and is subject to no external forces, any displacement u leads to an increase of $U_{(P)}$, and thus a decrease of the kinetic energy $U_{(C)}$ (by energy conservation); this is impossible if the system is initially at rest ($U_{(C)} = 0$) as $U_{(C)}$ is always positive. On the other hand, if the system is slightly displaced from O by an external agent and released, it returns to O with a decreasing $U_{(P)}$ and an increasing $U_{(C)}$. The system continues its displacement beyond O in the opposite direction and so on. Thus, the system oscillates back-and-forth near the stable equilibrium position and, in many cases, it may be modeled as a simple *harmonic oscillator* for small variations of u.

The total energy of the system is a certain function $U_{(T)}(u, \dot{u})$ of the generalized coordinates u and the *generalized velocity* $\dot{u} = \dot{q}$. Assuming that $U_{(T)}$ is continuous and $\dot{u}$ is small, we may write $U_{(T)}$ as a Taylor power series in $\dot{u}$, up to the term $\dot{u}^2$:

$$U_{(T)} = \Phi(u) + \psi(u)\,\dot{u} + \tfrac{1}{2}\,\mu(u)\,\dot{u}^2, \qquad [1.66]$$

where the coefficients Φ, ψ and μ are functions of u such that

$$\Phi(u) = U_{(T)}(u, \dot{u})\Big|_{\dot{u}=0}, \qquad \psi(u) = \partial_{\dot{u}}\, U_{(T)}\Big|_{\dot{u}=0}, \qquad \mu(u) = \partial^2_{\dot{u}\dot{u}}\, U_{(T)}\Big|_{\dot{u}=0}. \qquad [1.67]$$

If $\dot{u} = 0$, the energy of the system is (by definition) wholly potential; thus, $\Phi(u)$ can be identified as the potential energy $U_{(P)}$ and $\psi(u)\,\dot{u} + \tfrac{1}{2}\,\mu(u)\,\dot{u}^2$ as the kinetic energy. If the displacement u is small, we may also write $U_{(P)} = \Phi(u)$ as a Taylor power series in u, up to the term u^2: $U_{(P)}(u) = \Phi_o + \Phi'_o\, u + \tfrac{1}{2}Ku^2$, where Φ_o, Φ'_o and K are respectively the values of $\Phi(u)$, $\partial_u\Phi$ and $\partial^2_{uu}\Phi$ for $u = 0$. The position $u = 0$ is a minimum of $U_{(P)}$ if $\Phi'_o = \partial_u\Phi = 0$ and $K > 0$. On the other hand, we may drop the constant term Φ_o as the potential energy is defined up to the addition of a constant and by convention we take it to be equal to 0 at the equilibrium position.

Consider now the kinetic energy $U_{(C)} = \psi(u)\,\dot{u} + \tfrac{1}{2}\mu(u)\,\dot{u}^2$. As it should be positive for any $\dot{u}$, we must have $\psi(u) = 0$ and $\mu(u) > 0$. If u is small, we may assume that μ is constant (since any u dependence would give a third order term in $U_{(T)}$ (that we have neglected). Thus, we may write the total energy of the system in the form

$$U_{(T)} = U_{(C)} + U_{(P)} = \tfrac{1}{2}\mu\,\dot{u}^2 + \tfrac{1}{2}Ku^2 = \text{Constant}, \qquad [1.68]$$

where $U_{(T)}$ = Constant by energy conservation. Differentiating this equation with respect to time, we obtain

$$\dot{u}\,(\mu\,\ddot{u} + Ku) = 0. \qquad [1.69]$$

This equation may be verified if $\dot{u} = 0$, which means that the system is at rest (at the equilibrium position), or if the motion satisfies the equation:

$$\mu \ddot{u} + Ku = 0, \qquad \text{i.e.} \qquad \ddot{u} + (K/\mu)\, u = 0, \qquad [1.70]$$

This is the equation of a simple harmonic oscillator of natural frequency

$$\omega_o = \sqrt{K/\mu}\,. \qquad [1.71]$$

We designate the constant K as the *restoring force parameter* and the constant μ as an *inertial parameter* near the equilibrium position q_e. If the displacement is not small, the approximations [1.66] and $\Phi(u) = \frac{1}{2} Ku^2$ are not justified and the equation of motion will be nonlinear and more complicated than [1.70]. On the other hand, our analysis assumes that the energy $U_{(T)}$ is a twice differentiable function at $u = 0$ and $\dot{u} = 0$. This condition is not satisfied, for instance, in the case of a ball rolling on the inclines of Figure 1.15 (problem P1.20).

Assume now that the energy of the system is not conserved due to some resistive force or any process, which dissipates energy in the displacement du according to

$$dU_{(fr)} = F_{(fr)}(u, \dot{u})\, du = F_{(fr)}(u, \dot{u})\, \dot{u}\, dt. \qquad [1.72]$$

In the case of small motions, we may write $F_{(fr)}(u, \dot{u}) = a(u) + b(u)\dot{u}$. Very often, $F_{(fr)} = 0$ if $\dot{u} = 0$, thus $a(u) = 0$. If $b(u)$ may be considered as constant for small u, we may write

$$dU_{(fr)} = b\,\dot{u}\, du = b\,\dot{u}^2\, dt, \qquad [1.73]$$

where the constant b is positive. $dU_{(fr)}$ must be equal to the decrease of the total energy $(-dU_{(T)})$ in the interval of time dt; thus, we must have

$$dU_{(T)} + dU_{(fr)} = \dot{u}\,(\mu \ddot{u} + b\dot{u} + K u) = 0. \qquad [1.74]$$

As $\dot{u} = 0$ corresponds to the system in equilibrium, the equation of motion near the equilibrium position is

$$\mu \ddot{u} + b\dot{u} + Ku = 0, \qquad \text{i.e.} \qquad \ddot{u} + (b/\mu)\,\dot{u} + (K/\mu)\, u = 0 \qquad [1.75]$$

This is the equation of a damped oscillator with a damping constant $\beta = b/2\mu$ and a natural frequency $\omega_o = \sqrt{K/\mu}$.

1.10. Nonlinear oscillators

In general, the restoring force near a stable equilibrium position may be any function of u and its time derivative $\dot{u}$. We shall not consider the general case of a force, which depends explicitly on time. If the force is not of the form $-Ku$, it is said to be *nonlinear*. Then the equation of motion has the general form:

$$\ddot{u} = f(u, \dot{u}). \qquad [1.76]$$

If the force does not depend on the velocity $\dot{u}$, the system is said to be *conservative*; then, its equation of motion takes the form:

$$\ddot{u} = f(u) \qquad \text{(nonlinear and conservative systems).} \qquad [1.77]$$

In the following, we assume that $f(u)$ is even in u, that is, $f(-u) = -f(u)$. Then, the motion of the system is periodic but, in general, anharmonic. Assume that the system is initially displaced to a position A and released without initial velocity ($u = A$ and $\dot{u} = 0$ at $t = 0$). It is evident that the system takes a time $T/4$ to return to the equilibrium position $u = 0$. In order to determine this time, we multiply both sides of equation [1.77] by $\dot{u}$ and integrate between $t = 0$ and t; we get

$$\int_0^t dt\ \dot{u}\ \ddot{u} = \int_0^t dt'\ \dot{u}\, f(u) \quad \Rightarrow \quad \dot{u}^2 = 2[F(u) - F(A)] \text{ and } \dot{u} = -\sqrt{2F(u) - 2F(A)}\,,$$

where $F(u)$ is a primitive of $f(u)$ and we have chosen the (−) sign for $\dot{u}$ because u is decreasing during the first quarter period. The expression of the velocity $\dot{u}$ may be cast in the form $dt = -\,du/\sqrt{2F(u) - 2F(A)}$. By integrating once more between $u = A$ (at $t = 0$) and $u = 0$ (at $t = T/4$), we get the period:

$$T = 2\sqrt{2} \int_0^A du / \sqrt{F(u) - F(A)}\,. \qquad [1.78]$$

Thus, in general, T depends on the amplitude of the motion.

1.11. Systems with two degrees of freedom

A system has n degrees of freedom if its state is completely specified by n independent variables q_k ($k = 1, 2, \ldots, n$). In particular, it may be formed by n coupled oscillators. It is *linear* if the n equations of motion are linear as functions of q_k and their time-derivatives $\dot{q}_k$. In this section, we analyze in detail a simple undamped system with two degrees of freedom, namely two masses m_1 and m_2 which are confined to move on the x-axis attached to the ends of two springs with constants K_1 and K_2 and coupled by a spring of constant K_3 (Figure 1.11).

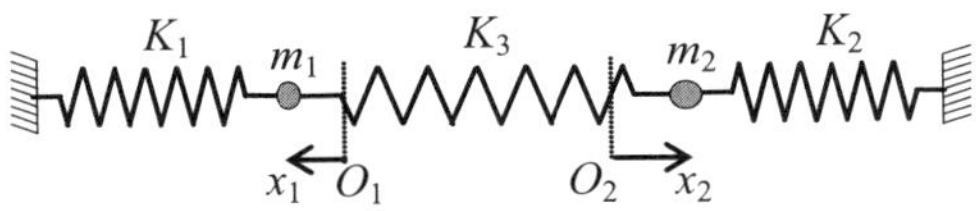

Figure 1.11. *Two masses oscillating at the ends of two springs and coupled by a third spring*

If the coupling spring is removed, the masses oscillate separately with the natural frequencies $\omega_1 = \sqrt{K_1/m_1}$ and $\omega_2 = \sqrt{K_2/m_2}$. Now, if the masses are coupled and displaced by x_1 and x_2 from their respective equilibrium positions O_1 and O_2, the forces which act on the first mass are $-K_1x_1$ due to spring (1), and $-K_3(x_1 - x_2)$ due to the coupling spring. Similarly, the forces, which act on the second mass, are $-K_2x_2$ due to spring (2) and $-K_3(x_2 - x_1)$ due to the coupling spring. Thus, the equations of motion of the masses are:

$$m_1\ddot{x}_1 + K_{11}u_1 + K_{12}u_2 = 0, \qquad m_2\ddot{x}_2 + K_{21}u_1 + K_{22}u_2 = 0, \qquad [1.79]$$

where we have set $K_{11} = K_1+K_3$, $K_{12} = -K_3$, $K_{21} = -K_3$ and $K_{22} = K_2+K_3$. To solve these coupled linear equations, we try the simple harmonic solutions in the complex exponential representation

$$\underline{u}_1 = \underline{C}_1 e^{i\omega t}, \qquad \underline{u}_2 = \underline{C}_2 e^{i\omega' t}. \qquad [1.80]$$

Substituting these expressions into the differential equations of [1.79], we find that they can be identically satisfied (i.e. at any time t) only if the frequencies ω and ω' are equal. This means that the coupled oscillators oscillate with the same frequency ω. Dropping the common exponential $e^{i\omega t}$, the equations of [1.79] become a system of algebraic linear and coupled equations relating ω and the complex amplitudes $\underline{C}_1$ and $\underline{C}_2$:

$$(K_{11} - m_1\omega^2)\,\underline{C}_1 + K_{12}\,\underline{C}_2 = 0, \qquad K_{21}\,\underline{C}_1 + (K_{22} - m_2\,\omega^2)\,\underline{C}_2 = 0.$$

The coefficients of $\underline{C}_i$ in these equations being real, the complex amplitudes $\underline{C}_1$ and $\underline{C}_2$ are in phase or in opposite phase (if ω is real). As a phase of π is equivalent to changing the sign of the amplitude, we may associate a possible common phase ϕ of $\underline{C}_j$ to ωt and write the real part of the solutions directly in the form:

$$x_1 = A_1\cos(\omega t + \phi), \qquad x_2 = A_2\cos(\omega t + \phi), \qquad [1.81]$$

where A_1 and A_2 are independently positive or negative. Substituting these expressions into the equations of [1.79], we find that A_1 and A_2 must satisfy the equations:

$$(K_{11} - \mu_1\omega^2)\, A_1 + K_{12}\, A_2 = 0, \qquad K_{21}\, A_1 + (K_{22} - \mu_2\omega^2)\, A_2 = 0. \qquad [1.82]$$

This system of algebraic, linear and homogeneous equations has a non-trivial solution (i.e. not both A_1 and A_2 are equal to zero) if the determinant of the coefficients is equal to zero. This gives the condition

$$\det\begin{bmatrix} K_{11}-m_1\omega^2 & K_{12} \\ K_{21} & K_{22}-m_2\omega^2 \end{bmatrix} \equiv (K_{11} - \mu_1\omega^2)(K_{22} - \mu_2\omega^2) - K_{21}K_{12} = 0. \quad [1.83]$$

This is the *characteristic equation* of the system of coupled oscillators. In general, it has two solutions $\omega_{(1)}^2$ and $\omega_{(2)}^2$. This determines the frequencies $\omega_{(1)}$ and $\omega_{(2)}$ of the *normal modes*. Substituting one of these normal frequencies for ω into the equations [1.82], we get two equations, which are not independent. They allow us to determine one of the amplitudes A_k in terms of the other in a proportionality relationship. In other words, the *eigenvector* $\mathbf{A}$ of components A_1 and A_2 is determined up to a normalization factor (only its direction is determined).

In the particularly simple symmetric case, $K_1 = K_2 \equiv K$ and $m_1 = m_2 \equiv m$, we get

$$\omega_{(1)} = \sqrt{K/m}\,, \qquad \omega_{(2)} = \sqrt{(K+2K_3)/m}\,. \qquad [1.84]$$

– In normal mode (1), $\omega = \omega_{(1)}$, the equations of [1.82] reduce to a single equation $A_1 - A_2 = 0$. Thus, the displacements in this mode are

$$x_{(1)\,1} = x_{(1)\,2} = A_{(1)} \cos(\omega_{(1)}\, t + \phi_{(1)}). \qquad [1.85]$$

– In normal mode (2), $\omega = \omega_{(2)}$, the equations of [1.82] reduce to a single equation $A_1 + A_2 = 0$. Thus, the displacements in this mode are

$$x_{(2)\,1} = -\, x_{(2)\,2} = A_{(2)} \cos(\omega_{(2)}\, t + \phi_{(2)}) \qquad [1.86]$$

The general solution is a superposition of these modes with arbitrary amplitudes $A_{(1)}$ and $A_{(2)}$ and phases $\phi_{(1)}$ and $\phi_{(2)}$:

$$\begin{aligned} x_1 &= A_{(1)} \cos(\omega_{(1)}t + \phi_{(1)}) + A_{(2)} \cos(\omega_{(2)}t + \phi_{(2)}), \\ x_2 &= A_{(1)} \cos(\omega_{(1)}t + \phi_{(1)}) - A_{(2)} \cos(\omega_{(2)}t + \phi_{(2)}). \end{aligned} \qquad [1.87]$$

The four integration constants $A_{(1)}$, $A_{(2)}$, $\phi_{(1)}$ and $\phi_{(2)}$ may be determined from the initial conditions, i.e. the values of u_1, u_2, $\dot{u}_1$ and $\dot{u}_2$ at $t = 0$.

In the special case where $m_1 = m_2$ and $K_1 = K_2$, the normal frequency $\omega_{(1)}$ is the same as the natural frequency of the oscillators (as if they are uncoupled by removing the coupling spring). Indeed, the coupled masses may oscillate in phase with the same amplitude in such a way that the length of the coupling spring remains unchanged; thus, exerting no variable force on the masses. The system oscillates in mode (1) if, for instance, the masses are displaced by the same distance ($A_{(1)} = A_{(2)}$) and released without initial velocity. In the second normal mode, the distance between the oscillating masses varies and the coupling spring exerts a variable coupling force. The system oscillates effectively in this mode if, for instance, the masses have opposite initial displacements ($A_{(1)} = - A_{(2)}$) and no initial velocity.

To solve the system of differential equations in [1.79] in the special case $K_1 = K_2 \equiv K$ (i.e. $K_{11} = K_{22} = K + K_3$) and $m_1 = m_2 \equiv m$, we may use another procedure. Noting that we may use any two independent variables to analyze the system and guided by the symmetry of the equations, we define two new variables:

$$y_1 = x_1 + x_2, \qquad y_2 = x_1 - x_2. \tag{1.88}$$

Adding the equations of [1.79] and subtracting them, we obtain the equations:

$$m\,\ddot{y}_1 = - K y_1, \qquad m\,\ddot{y}_2 = - (K + 2K_1)\, y_2. \tag{1.89}$$

These new equations of motion are uncoupled; their solutions can be immediately written down:

$$y_1 = a_1 \cos(\omega_{(1)}t + \phi_{(1)}), \qquad y_2 = a_2 \cos(\omega_{(2)}t + \phi_{(2)}), \tag{1.90}$$

where $\omega_{(1)}$ and $\omega_{(2)}$ are just the normal frequencies [1.84]. The variables y_1 and y_2, that uncouple the equations of motion, are the *normal variables* of the system. If we use these variables, the system behaves like two uncoupled oscillators evolving independently. Inversing the relationships of [1.88], we get $x_1 = \frac{1}{2}(y_1 + y_2)$ and $x_2 = \frac{1}{2}(y_1 - y_2)$ and the expressions of the form [1.87].

The energy of this system may have two forms: kinetic for the moving masses and potential which is stored in the springs; thus

$$U_{(C)} = \tfrac{1}{2}\, m\, \dot{x}_1^{\,2} + \tfrac{1}{2}\, m\, \dot{x}_2^{\,2}, \qquad U_{(P)} = \tfrac{1}{2}\, K x_1^{\,2} + \tfrac{1}{2}\, K x_2^{\,2} + \tfrac{1}{2}\, K_3 (x_1 - x_2)^2,$$
$$U_{(T)} = U_{(C)} + U_{(P)} = K A_{(1)}^{\,2} + (K + 2K_3) A_{(2)}^{\,2}. \tag{1.91}$$

The total energy $U_{(T)}$ is conserved and it may be written in the form

$$U_{(T)} = U_1 + U_2 + U_{(int)},$$
$$U_1 = \tfrac{1}{2}\, m\,\dot{x}_1^{\,2} + \tfrac{1}{2}\, Kx_1^2, \qquad U_2 = \tfrac{1}{2}\, m\,\dot{x}_2^{\,2} + \tfrac{1}{2} Kx_2^2,\ \ U_{(int)} = \tfrac{1}{2}\, K_3(x_1 - x_2)^2. \qquad [1.92]$$

U_1 and U_2 are the total energies of each of the oscillators if they are uncoupled and $U_{(int)}$ is the *interaction energy*, which is in this case the potential energy stored in the coupling spring. The energies U_1 and U_2 of the individual oscillators are not constant as the oscillators exchange energy through the coupling spring. Only, the total energy $U_{(T)}$ of the whole isolated system remains constant.

If we use the normal variables of [1.88] to analyze the system, the kinetic energy, potential energy and total energy take the forms:

$$U_{(C)} = \tfrac{1}{4}\, m\, \dot{y}_1^{\,2} + \tfrac{1}{4}\, m\, \dot{y}_2^{\,2}, \qquad U_{(P)} = \tfrac{1}{4}\, Ky_1^{\,2} + \tfrac{1}{4}\,(K + 2K_3)\, y_2^{\,2},$$
$$U_{(T)} = [\tfrac{1}{4}\, m\, \dot{y}_1^{\,2} + \tfrac{1}{4}\, Ky_1^{\,2}] + [\tfrac{1}{4}\, m\, \dot{y}_2^{\,2} + \tfrac{1}{4}\,(K + 2K_3)\, y_2^{\,2}]. \qquad [1.93]$$

$U_{(T)}$ is the same as that of two uncoupled oscillators of energies $U_{(1)}$ and $U_{(2)}$:

$$U_{(T)} = U_{(1)} + U_{(2)}$$
$$U_{(1)} = \tfrac{1}{4}\, m\, \dot{y}_1^{\,2} + \tfrac{1}{4}\, Ky_1^{\,2} \equiv \tfrac{1}{2}\, \mu_{(1)}\, \dot{y}_1^{\,2} + \tfrac{1}{2}\, \mu_{(1)}\, \omega_1^{\,2} y_1^{\,2},$$
$$U_{(2)} = \tfrac{1}{4}\, m\, \dot{y}_2^{\,2} + \tfrac{1}{4}\,(K + 2K_3)\, y_2^{\,2} \equiv \tfrac{1}{2}\, \mu_{(2)}\, \dot{y}_2^{\,2} + \tfrac{1}{4}\, \mu_{(2)}\, \omega_2^{\,2} y_2^{\,2}. \qquad [1.94]$$

Thus, if we use the normal variables, the system is equivalent to two uncoupled oscillators with no interaction energy. Each of them has a constant energy: the oscillation of the first is specified by the variable y_1, its mass is $\mu_{(1)} = m/2$, its force constant is $K_{(1)} = K/2$, its amplitude is $a_1 = 2A_{(1)}$ and its energy is $KA_{(1)}^{\,2}$. The vibration of the second oscillator is specified by the variable y_2, its mass is $\mu_{(2)} = m/2$, its force constant is $K_{(2)} = \tfrac{1}{2}(K + 2K_3)$, its amplitude is $a_2 = 2A_{(2)}$ and its energy is $(K + 2K_3)\, A_{(1)}^{\,2}$. For the same amplitude of oscillation of the masses, the excitation energy $U_{(T)}$ is larger for the mode of higher frequency.

1.12. Generalization to systems with *n* degrees of freedom

The preceding considerations may be generalized to systems with n degrees of freedom and, in particular, to systems of n coupled oscillators. The state of such systems is fully specified by n variables or generalized coordinates q_i. We study in this section the small oscillations near a stable equilibrium position of coordinates $q_{e,i}$. We show in this section that such a system has n *normal modes* characterized by the *normal frequencies* $\omega_{(1)}, \omega_{(2)}, \ldots, \omega_{(n)}$. Some of these frequencies may be equal. If the system evolves in the normal mode (j), all the variables q_k oscillate with the same frequency $\omega_{(j)}$. The general motion of the system is a linear superposition of the normal modes.

Instead of the q_i, it is more convenient to use the "displacements" $u_i = q_i - q_{e,i}$ from the equilibrium position; then, $\dot{q}_i = \dot{u}_i$ and the energy of the system is a certain function $U_{(T)}(u_i, \dot{u}_j)$ of u_i and their first time-derivatives $\dot{u}_j$. As $\dot{u}_j$ is small, we may make a Taylor series expansion of $U_{(T)}$ in powers of $\dot{u}_j$ up to the second order:

$$U_{(T)} = \Phi(u_k) + [\ \Sigma_i\, \psi_i(u_k)\, \dot{u}_i\ + \tfrac{1}{2}\, \Sigma_{ij}\, \mu_{ij}(u_k)\, \dot{u}_i\ \dot{u}_j\,],$$

where $\Phi(u_k) = U_{(T)}\,|_o$, $\psi_i(u_k) = \partial_{\dot{u}_i} U_{(T)}\,|_o$ and $\mu_{ij}(u_i) = \partial^2_{\dot{u}_i \dot{u}_j} U_{(T)}\,|_o$. Here $()|_o$ means that the bracket must be evaluated for all $\dot{u}_i = 0$. The term $\Phi(u_k)$, which is independent on the generalized velocities, is the *potential energy* $U_{(P)}$, and the other terms may be considered to be the *kinetic energy* $U_{(C)}$. As $U_{(C)}$ is independent of the direction of the velocity, there must be no linear term in $\dot{u}_i$; thus, $\psi_i(u_k) = 0$. On the other hand, u_i being small, and neglecting in $U_{(T)}$ powers higher than 2 in u_k and $\dot{u}_i$, we may assume that μ_{ij} are constant; then,

$$U_{(C)} = \tfrac{1}{2}\, \Sigma_{ij}\, \mu_{ij}\ \dot{u}_i\ \dot{u}_j\,. \qquad [1.95]$$

μ_{ij} are the *generalized inertial parameters* of the system near the considered equilibrium position. As they are the second order derivatives of $U_{(T)}$ with respect to $\dot{u}_i$ and $\dot{u}_j$, they are symmetric in *i* and *j* ($\mu_{ij} = \mu_{ji}$). On the other hand, $U_{(C)}$ is always positive. Considering the special case of all the $\dot{u}_i$ equal to zero except two of them, $\dot{u}_r$ and $\dot{u}_s$, the kinetic energy takes the form $U_{(C)} = \tfrac{1}{2}\, \mu_{rr}\, \dot{u}_r^{\,2} + \tfrac{1}{2}\, \mu_{ss}\, \dot{u}_s^{\,2} + \mu_{rs}\, \dot{u}_r \dot{u}_s$. It is positive for any $\dot{u}_r$ and $\dot{u}_s$ if $\mu_{rr} > 0$ and $\mu_{rs}^{\,2} \leq \mu_{rr}\, \mu_{ss}$ for any pair (r, s).

Consider now the potential energy, which may also be written as a power series in u_i up to the second order

$$U_{(P)} \equiv \Phi(u_k) = \Phi_o + \Sigma_i\, \Phi'_{o,i}\, u_i + \tfrac{1}{2}\, \Sigma_{ij}\, K_{ij}\, u_i\, u_j, \qquad [1.96]$$

where we have set $\Phi_o = \Phi(0)$, $\Phi'_{o,i} = \partial_{u_i} \Phi\,|_o$ and $K_{ij} = \partial^2_{u_i u_j} \Phi\,|_o$. Here $()|_o$ means that the bracket must be evaluated for all $u_i = 0$. The constant term Φ_o must be 0 according of our convention of 0 potential energy at the equilibrium position and the coefficients $\Phi'_{o,i}$ must be zero if $U_{(P)}$ is minimum at this position. Thus,

$$U_{(P)} = \tfrac{1}{2} \Sigma_{ij} K_{ij} u_i u_j. \qquad [1.97]$$

K_{ij} are the *generalized restoring force parameters* near the equilibrium position. As they are the second order derivatives of $U_{(P)} = \Phi$ with respect to u_i and u_j, they are symmetric in i and j ($K_{ij} = K_{ji}$) and, in order for the position u_i to be a minimum of the potential energy, Φ must be positive near this position for any u_i. Thus, we must have $K_{ii} > 0$ and $K_{ij}^2 \leq K_{ii} K_{jj}$.

It may be shown that the equations of motion take the form

$$\Sigma_j \mu_{ij} \ddot{u}_j + \Sigma_j K_{ij} u_j = 0 \qquad \text{for } i = 1, 2, ..., n. \qquad [1.98]$$

This is a system of n linear and homogeneous differential equations to determine the n variables u_i. For $i \neq j$ the μ_{ij}, are at the origin of the so-called *inertial coupling* and, for $i \neq j$ the K_{ij}, are at the origin of the so-called *force coupling*.

Let us try solutions of the form $u_j = A_j \cos(\omega_j t + \phi_j)$. Substituting these expressions into [1.9], we find that these equations are satisfied only if the frequencies ω_j are equal and the ϕ_j are equal. To simplify the notation, we introduce the n-dimensional "vector" $\mathbf{u}$ of components u_i, the *inertial tensor* $[\mu]$ and the *restoring force tensor* $[K]$ whose elements are μ_{ij} and K_{ij} respectively. In this section, the second rank tensors $[\mu]$ and $[K]$ are treated as matrices concerning the contracted products $[K]\,\mathbf{u} \equiv \Sigma_{ij}\, \mathbf{e}_i K_{ij} u_j$. Thus, the system of equations [1.98] may be written in the vector form

$$[\mu]\, \ddot{\mathbf{u}} + [K]\, \mathbf{u} = 0 \qquad [1.99]$$

and our trial solution takes the form $\mathbf{u} = \mathbf{A} \cos(\omega t + \phi)$, where $\mathbf{A}$ is an n-dimensional vector whose components are the amplitudes $A_1, A_2, ..., A_n$ of the n variables u_i. Substituting this expression of $\mathbf{u}$ into [1.99], we find that ω and $\mathbf{A}$ must satisfy the equation

$$\{[K] - \omega^2 [\mu]\}\, \mathbf{A} = 0. \qquad [1.100]$$

This equation is a shorthand notation for a system of n linear and homogeneous algebraic equations to determine the n unknown amplitudes A_i. It has a non-trivial solution (i.e. some $A_i \neq 0$) only if the determinant of the coefficient is equal to 0:

$$\det \{[K] - \omega^2 [\mu]\} = 0. \qquad [1.101]$$

This is the *characteristic equation* of the system. Its roots are the normal frequencies. Being algebraic and of degree n in ω^2, it generally has n solutions $\omega_{(k)}^2$. We admit here that these roots are real and positive (see problem P1.33). Thus, in principle, we have, n *normal frequencies* $\omega_{(k)}$ (also called *eigenfrequencies*) which correspond to n *normal modes* of the system. Some of these solutions may be double or triple, etc.; they are then said to be *degenerate* with degeneracies 2, 3, etc. In the case of a non-degenerate normal frequency $\omega_{(k)}$, substituting $\omega_{(k)}$ for ω in equation [1.100] we get $(n - 1)$ independent equations. This allows us to determine $(n - 1)$ components of the vector $\mathbf{A}$ in terms of the n-th component, which remains arbitrary. In other words, $\mathbf{A}$ is determined up to an arbitrary normalization factor. This is the *eigenvector* $\mathbf{A}_{(k)}$ which corresponds to the normal frequency $\omega_{(k)}$. The corresponding solution $\mathbf{u}_{(k)}$ is:

$$\mathbf{u}_{(k)} = \mathbf{A}_{(k)} \cos(\omega_{(k)} t + \phi_{(k)}) \qquad [1.102]$$

or, explicitly,

$$u_{(k)1} = A_{(k)1} \cos(\omega_{(k)} t + \phi_{(k)}), \qquad u_{(k)2} = A_{(k)2} \cos(\omega_{(k)} t + \phi_{(k)}), \ldots,$$
$$u_{(k)n} = A_{(k)n} \cos(\omega_{(k)} t + \phi_{(k)}). \qquad [1.103]$$

If the normal frequency $\omega_{(k)}$ has a degeneracy d, equation [1.100] is equivalent to $(n - d)$ independent equations. Thus, we may determine $(n - d)$ components $A_{(k)i}$ in terms of d components which remain arbitrary and, consequently, d independent eigenvectors $\mathbf{A}_{(k,h)}$ (where $h = 1, 2,\ldots, d$), which correspond to the same normal frequency $\omega_{(k)}$ and, for each one of these vectors, the solution is of the form [1.102]. Any linear combination of the eigenvectors $\mathbf{A}_{(k,h)}$ (for the same k) is an eigenvector with the same normal frequency $\omega_{(k)}$. Therefore, we have a d-dimensional eigenvector space. In conclusion, a system with n degrees of freedom always has n normal modes of oscillation whether they are degenerate or not. The general solution is a superposition of these modes of the form:

$$\mathbf{u} = \Sigma_{k,h}\, \mathbf{A}_{(k,h)} \cos(\omega_{(k)} t + \phi_{(k,h)}), \qquad [1.104]$$

The summation index k labels the different normal frequencies $\omega_{(k)}$ and the index $h = 1, 2, \ldots d$ labels the different degenerate modes that have the same $\omega_{(k)}$. solution [1.104] depends on $2n$ arbitrary constants $\mathbf{A}_{(k,h)}$ and $\phi_{(k,h)}$, which may be determined from the initial conditions.

1.13. Normal variables for systems with *n* degrees of freedom*

The choice of the variables of a physical system with n degrees of freedom is not unique. Some choices are more convenient than others. A set of special variables, called *normal variables*, enables us to consider the system as n independent oscillators. The state of each of these oscillators is specified by a normal variable. These oscillators being independent, they have no interaction energy and the energy of the system is simply the sum of the energies of all the normal oscillators. Thus, the use of normal variables always simplifies the study of the system.

Assume that we choose the variables y_j such that

$$u_i = \Sigma_j S_{ij}\, y_j, \qquad \text{i.e.} \qquad \mathbf{u} = [S]\,\mathbf{y}. \tag{1.105}$$

The new variables y_j are independent if the *transformation matrix* $[S]$, whose elements are S_{ij}, is not singular (i.e. det $[S] \neq 0$). Then, the inverse matrix $[S^{-1}]$ exists and we may write

$$y_i = \Sigma_j S^{-1}{}_{ij}\, u_j, \qquad \text{i.e.} \qquad \mathbf{y} = [S^{-1}]\,\mathbf{u}. \tag{1.106}$$

equation [1.99] becomes an equation for $\mathbf{y}$:

$$[\mu]\,[S]\,\ddot{\mathbf{y}} + [K]\,[S]\,\mathbf{y} = 0. \tag{1.107}$$

By multiplying on the left by the matrix $[\widetilde{S}]$, which is the transposed matrix of $[S]$ (i.e., $\widetilde{S}_{ij} = S_{ji}$), we find

$$[\overline{\mu}]\,\ddot{\mathbf{y}} + [\overline{K}]\,\mathbf{y} = 0, \tag{1.108}$$

where the new tensors $[\overline{\mu}]$ and $[\overline{K}]$ are given by:

$$[\overline{\mu}] = [\widetilde{S}]\,[\mu]\,[S], \qquad [\overline{K}] = [\widetilde{S}]\,[K]\,[S] \tag{1.109}$$

$$[\mu] = [\widetilde{S}^{-1}]\,[\overline{\mu}]\,[S^{-1}], \qquad [K] = [\widetilde{S}^{-1}]\,[\overline{K}]\,[S^{-1}]. \tag{1.110}$$

Thus, the new variables y_j verify a system of linear equations [1.108] similar to the system [1.99] for the variables u_i.

The normal frequencies $\omega_{(k)}$ do not depend on the choice of the generalized variables, which specify the state of the system. Indeed, if $\omega_{(k)}$ and $\mathbf{A}_{(k)}$ satisfy equation [1.100], using the equations of [1.110], we may write equation [1.100] in the form

$$\{[\widetilde{S}^{-1}]\,[\overline{K}]\,[S^{-1}] - \omega_{(k)}{}^2\,[\widetilde{S}^{-1}]\,[\overline{\mu}]\,[S^{-1}]\}\,\mathbf{A}_{(k)} = 0. \tag{1.111}$$

Let us define the vector

$$\mathbf{B}_{(k)} = [S^{-1}]\,\mathbf{A}_{(k)} \tag{1.112}$$

and multiply equation [1.111] on the left by the matrix $[\widetilde{S}]$; we obtain:

$$\{[\overline{K}] - \omega_{(k)}^2 [\overline{\mu}]\} \mathbf{B}_{(k)} = 0 \qquad [1.113]$$

This means that equations [1.99] and [1.108] have the same eigenvalues. The eigenvectors $\mathbf{B}_{(k)}$ and $\mathbf{A}_{(k)}$, which correspond to the same eigenvalue $\omega_{(k)}$, are related by transformation [1.112] or the inverse transformation

$$\mathbf{A}_{(k)} = [S]\, \mathbf{B}_{(k)}, \qquad [1.114]$$

which is the same transformation which transforms the y_i into the u_j. Thus, the solution for the y_i is

$$\mathbf{y}_i = \mathbf{B}_{(i)} \cos(\omega_{(i)}t + \phi_{(i)}). \qquad [1.115]$$

The norm of the eigenvectors is arbitrary. We accept here without proof that it is always possible to choose the norm of the eigenvectors $\mathbf{A}_{(k)}$ so that

$$\Sigma_{ij}\, \mu_{ij}\, A_{(k)i} A_{(k')j} = \delta_{kk'}, \qquad [1.116]$$

where $\delta_{kk'}$ are the Kronecker symbols (i.e. $\delta_{kk'} = 1$ if $k = k'$ and $\delta_{kk'} = 0$ if $k \neq k'$). In the case of a degenerate normal frequency $\omega_{(k)}$, it is always possible to chose the corresponding eigenvectors $\mathbf{A}_{(k,h)}$ to satisfying condition [1.116]. Thus, this equation is satisfied for all the eigenvectors $\mathbf{A}$.

Transformation [1.106] with the matrix $[S]$ having the matrix elements

$$S_{ij} = A_{(j)i} \qquad \text{and} \qquad [S^{-1}] = [\widetilde{S}]\,[\mu] \qquad [1.117]$$

transforms equation [1.99] into equation [1.108] such that $[\overline{\mu}]$ is diagonal with diagonal elements equal to 1, and the new tensor $[\overline{K}]$ is diagonal with diagonal elements given by

$$[\overline{K}]_{ii} = \omega_{(i)}^{\ 2}. \qquad [1.118]$$

Thus, the system of equations for the variables y_i takes the uncoupled form

$$\ddot{y}_i + \omega_{(i)}^{\ 2}\, y_i = 0 \qquad \text{for } i = 1, 2, \ldots, n. \qquad [1.119]$$

This shows that the y_i are the *normal variables* of the system and the solution of the equations [1.119] is of the form $y_i = B_{(i)} \cos(\omega_{(i)} + \phi_{(i)})$.

The energy of the system may be written in the form:

$$U_{(T)} = U_{(C)} + U_{(P)}, \qquad U_{(C)} = ½ \Sigma_i \dot{y}_i^2, \qquad U_{(P)} = ½ \Sigma_i \omega_{(i)}^2 y_i^2. \qquad [1.120]$$

Thus, the system with n degrees of freedom is equivalent to n independent simple harmonic oscillators of normal frequencies $\omega_{(i)}$. Any state of the system is obtained by exciting one or several of these oscillators with appropriate amplitudes and phases. As these oscillators do not interact, the energy of the system is the sum of the energies of these oscillators

$$U_{(T)} = \Sigma_i U_{(i)}, \qquad U_{(i)} = ½ \dot{y}_i^2 + ½ \omega_{(i)}^2 y_i^2. \qquad [1.121]$$

If the non-normal coordinates u_i are used, the energy is given by

$$U_{(T)} = ½ \Sigma_{ij} \mu_{ij} \dot{u}_i \dot{u}_j + ½ \Sigma_{ij} K_{ij} u_i u_j. \qquad [1.122]$$

It may also be written in the form:

$$U_{(T)} = \Sigma_i U_i + \Sigma_{i \neq j} U_{ij}, \qquad [1.123]$$

where U_i has the same expression in u_i and $\dot{u}_i$ as if the oscillator (i) was isolated, and $U_{ij} = U_{ji}$ is the interaction energy of the oscillators (i) and (j). These energies are given by the expressions:

$$U_i = ½ \mu_{ii} \dot{u}_i^2 + ½ K_{ii} u_i^2, \qquad U_{ij} = ½ \mu_{ij} \dot{u}_i \dot{u}_j + ½ K_{ij} u_i u_j. \qquad [1.124]$$

If we divide the normal variable y_i by any constant M_i, the equations of the new variables $z_i = y_i / M_i$ remain uncoupled, in the form [1.119], with a total energy:

$$U_i = ½ {}_i M_i \dot{z}_i^2 + ½ \Sigma_i M_i^2 \omega_i^2 z_i^2. \qquad [1.125]$$

1.14. Summary

Oscillations are motions or changes in the state of physical systems that are repeated in time. $u(t)$ is *periodic* if $u(t) = u(t + T)$. T is the *period*. The frequency is $\tilde{\nu} = 1/T$.

The oscillation equation is $\partial^2_{tt}u + \omega^2 u = 0$. It is *linear* and *homogeneous*, thus, obeying the *superposition principle*. The general solution is the simple harmonic oscillations

$$u(t) = A\cos(\omega t + \phi) = A_1\cos(\omega t) + A_2\sin(\omega t) = \mathcal{Re}\,\underline{u}, \text{ where } \underline{u} = \underline{C}\,e^{i\omega t},$$

where A is the *amplitude*, $\omega = 2\pi\tilde{\nu} = 2\pi/T$ is the *angular frequency* (in rad/s) and ϕ is the *phase* (in rad). $\underline{u}$ is the complex general solution with $\underline{C} = A\,e^{i\phi}$. Its time-derivative and its primitive are:

$$\dot{\underline{u}} = i\omega\,\underline{u}(t) = \omega A\,e^{i(\omega t+\pi/2)}, \qquad \int dt\,\underline{u}(t) = \underline{u}/i\omega = (\underline{A}/\omega)\,e^{i(\omega t-\pi/2)}.$$

The frequency ω is a characteristic quantity of the oscillator while the constants A and ϕ (or $\underline{C}$) depend on the initial excitation, i.e. the initial position and velocity.

If a particle of mass m is subject to a restoring force $F = -Kx$, where K is the *force constant*, it undergoes simple harmonic oscillations $x = A\cos(\omega t + \phi)$ of frequency $\omega = \sqrt{K/m}$. Its *kinetic energy* is $U_{(C)} = \frac{1}{2}m\dot{x}^2$, its *potential energy* is $U_{(P)} = \frac{1}{2}Kx^2$ and its *total energy* is $U_{(T)} = U_{(C)} + U_{(P)} = \frac{1}{2}KA^2$.

If a body is confined to rotate around an axis and it is subject to a restoring moment $\Gamma = -\hat{K}\,\theta$, it undergoes simple harmonic angular oscillations $\theta = A\cos(\omega t + \phi)$ of frequency $\omega = \sqrt{\hat{K}/J}$. Its kinetic energy is $\frac{1}{2}J\dot{\theta}^2$ and its potential energy is $\hat{K}\,\theta^2$. Its total energy is $U_{(T)} = U_{(C)} + U_{(P)} = \frac{1}{2}\hat{K}A^2$.

The equation of motion of a physical pendulum is $\ddot{\theta} + (mgL/J)\sin\theta = 0$. If θ is small, $\sin\theta \cong \theta$ and the oscillation is simple harmonic with an angular frequency $\omega = \sqrt{mgL/J}$. In the case of a simple pendulum of length L, $J = mL^2$ and $\omega = \sqrt{g/L}$.

The equation of damped oscillations is $\ddot{x} + 2\beta\dot{x} + \omega_o^2 x = 0$. It has solutions of the form $x = e^{-\eta t}$, where η is a root of the *characteristic equation* $\eta^2 - 2\beta\eta + \omega_o^2 = 0$.

– In the case of overdamping ($\beta > \omega_o$), the general solution is

$$x(t) = C_1\,e^{-(\beta-\sigma)t} + C_2\,e^{-(\beta+\sigma)t}, \qquad \text{where } \sigma = \sqrt{\beta^2 - \omega_o^2}.$$

The system returns to equilibrium without oscillations. The long-term motion is dominated by the first term which is damped with a damping coefficient $\delta = \beta - \sigma$.

– In the case of critical damping ($\beta = \omega_o$), the general solution is

$$x = (A + Bt)\, e^{-\beta t} .$$

– In the case of underdamping ($\beta < \omega_o$), the general solution is

$$x = e^{-\beta t}\,[\underline{C}_1\, e^{i\hat{\omega} t} + \underline{C}_2\, e^{-i\hat{\omega} t}] = A\, e^{-\beta t} \cos(\hat{\omega}\, t + \phi), \qquad \text{where } \hat{\omega} = \sqrt{\omega_o{}^2 - \beta^2} .$$

This is a quasi-harmonic oscillation of quasi-*frequency* $\hat{\omega}$, damping coefficient $\delta = \beta$, *relaxation time* $\tau = 1/\delta$ and *quality factor* $f_q = \pi\tau/T_o$.

The equation of an *LCR* circuit is $\ddot{Q} + 2\beta\,\dot{Q} + \omega_o{}^2 Q = 0$, where $\omega_o = 1/\sqrt{LC}$. The electromagnetic energy is $U_{(EM)} \equiv U_{(E)} + U_{(M)} \equiv Q^2/2C + \frac{1}{2}LI^2$. It decreases due to the dissipation of energy as Joule heat.

If a system with one degree of freedom whose state is specified by a variable q has small displacements close to a stable equilibrium position q_e, the energy is $U_{(T)} = \frac{1}{2}\mu\,\dot{u}^2 + \frac{1}{2}Ku^2$, where $u = q - q_e$. The conservation of energy leads to the equation $\mu\,\ddot{u} + Ku = 0$. Thus, the system undergoes simple harmonic oscillations of frequency $\omega_o = \sqrt{K/\mu}$. K is the restoring force parameter and μ is the inertial parameter near the equilibrium position q_e. If there is a dissipation of energy of the form $dU_{(fr)} = b\,\dot{u}^2\, dt$, the oscillations obey the equation of motion $\mu\,\ddot{u} + b\,\dot{u} + Ku = 0$.

The energy of a system with n degrees of freedom u_i is $U_{(T)} = U_{(C)} + U_{(P)}$, where $U_{(C)} = \frac{1}{2}\,\Sigma_{ij}\,\mu_{ij}\,\dot{u}_i\,\dot{u}_j$ and $U_{(P)} = \frac{1}{2}\Sigma_{ij}\,K_{ij}\,u_i\,u_j$. Here $[\mu]$ is the *inertial tensor* and $[K]$ is the *restoring force tensor*. If the system is undamped, it has *n normal modes* of the form $u_{(k)} = A_{(k)}\cos(\omega_{(k)}t + \phi_{(k)})$ such $\{[K] - \omega_{(k)}{}^2\,[\mu]\}\mathbf{A} = 0$. The *normal frequencies* $\omega_{(k)}$ are the roots of the *characteristic equation* $\det\{[K] - \omega_{(k)}{}^2[\mu]\} = 0$. The general solution is a linear superposition of the modes. The normal frequencies do not depend on the choice of the variables u_i. It is always possible to find *normal variables* such that $[\,\overline{\mu}\,]$ is diagonal with diagonal elements equal to 1 and $[\,\overline{K}\,]$ is diagonal with the diagonal elements $[\,\overline{K}\,]_{ii} = \omega_{(i)}{}^2$. The system is equivalent to n independent oscillators of frequencies $\omega_{(i)}$, whose states are specified by the normal variables.

In the case of a damped oscillator with n degrees of freedom, the dissipated energy is $dU_{(dis)} = \Sigma_{ij}\, b_{ij}\,\dot{q}_i\, dq_j$. This leads to an equation $[\mu]\,\ddot{\mathbf{u}} + [b]\,\dot{\mathbf{u}} + [K]\mathbf{u} = 0$.

1.15. Problem solving suggestions

1) The SI unit of angles is the *radian* (which is dimensionless) and the unit of angular frequency is radians per second (rad/s). The period is the interval of time that the oscillating system takes to return to the same state; all the physical quantities of the system (position, velocity, energy, etc.) take the same values.

2) A second order differential equation involves derivatives of the second order but no higher order derivatives. Its general solution depends on two arbitrary constants of integration. Conversely, if you find a solution, which depends on two arbitrary and independent constants, you can be sure that this is the general solution.

3) The complex representation is very convenient and, in some cases, is essential to analyze the problem. You must know how to transform real expressions to complex expressions, and conversely, how to add, subtract, multiply, divide complex numbers and expressions etc. The phasor diagram is identical to the Argand diagram for complex numbers. However, only complex representation and complex analysis enable us to make calculations in complex situations.

4) The potential energy of a system may be a complicated function. For instance, in the case of a pendulum, $U_{(P)} = -mgL(1 - \cos\theta)$. By writing the Taylor series for $U_{(P)}$ as a function of u and ignoring the constant term, the lowest order term of the series is the usual expression $\frac{1}{2}Ku^2$ for the simple harmonic oscillator. Higher order terms may be important in some cases. Thus, the oscillation is a simple harmonic, only for small values of u near the equilibrium position. If we have to take into account higher order terms, it is not possible to write the solution of the equation of motion using simple functions. Approximation methods or numerical methods must be used.

5) In some cases, it is possible to guess linear combinations of the coupled equations of motion to find a new set of uncoupled equations for new variables y_i; these are the *normal variables*. The system is equivalent to a set of independent oscillators specified by the variables y_i whose solutions are the normal modes. Otherwise, you may look to these modes in the following general method, which may be used even in the case of damped systems: try an exponential solution of the form $u_i = A_i e^{-\eta t}$. The system of equations for u_i becomes a set of n algebraic linear and homogeneous equations to determine the amplitudes A_i. This set of equations has a non-trivial solution only if the determinant of the coefficients is equal to 0. This condition gives the *characteristic equation*, which is an algebraic equation of degree $2n$ in η. In general, it has $2n$ roots, which come out in pairs. A pair of real and positive roots of the form $\eta_{(k)}^{\pm} = \delta_{(k)} \pm \sigma_{(k)}$ corresponds to a non-oscillatory damped mode. A pair of complex solutions of the form $\eta_{(k)}^{\pm} = \delta_{(k)} \pm i\omega_{(k)}$ corresponds to an

oscillatory damped mode of damping coefficient $\delta_{(k)}$ and pseudo frequency $\omega_{(k)}$. In the case of a conservative system, you may directly try the real solutions $u_i = A_i \cos(\omega t + \phi)$; then, the characteristic equation is algebraic of degree n in ω^2.

6) For each pair of roots of the characteristic equation forming a mode, the set of equations allows us to determine the amplitudes $A_{(k)i}$ of the mode. A pair may be degenerate (i.e. two or more pairs are equal). If the pair is not degenerate, the set of equations for the A_i reduces to $(n - 1)$ independent equations, which allows us to determine $(n - 1)$ amplitudes in terms of the n-th. This comes down to determining the amplitudes $A_{(k)i}$ up to a normalization factor. These are the components of the eigenvector $\mathbf{A}_{(k)}$. If the solution has a degeneracy d, the set of equations reduces to $(n - d)$ independent equations, which allow us to determine $(n - d)$ amplitudes in terms of d arbitrary amplitudes. They correspond to d independent eigenvectors $\mathbf{A}_{(k,h)}$, where $h = 1, 2\ldots, d$. Any linear superposition of these vectors is also an eigenvector.

1.16. Conceptual questions

Q1. Figure 1.12 shows two electric potentials V_1 and V_2 as observed on the screen of an oscilloscope. The time scale is in 10^{-2} s and the voltage scale is in mV. Determine the amplitude and the frequency of V_1 and V_2. What is their phase shift?

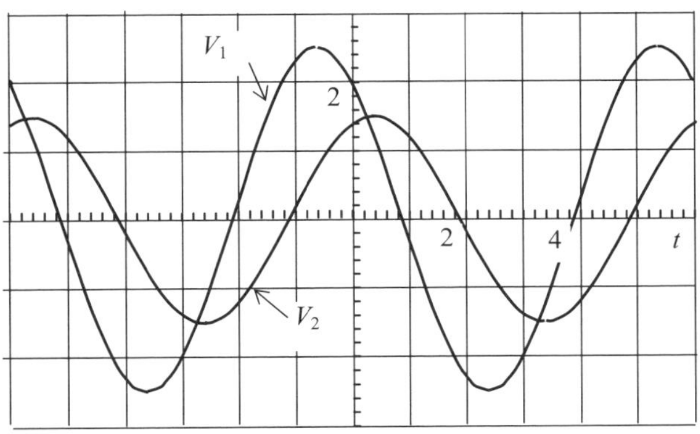

Figure 1.12. *Conceptual question 1*

Q2. A metallic strip oscillates at a frequency of $\tilde{\nu} = 50$ Hz. A *stroboscope* enables us to observe the motion only at predetermined intervals (e.g. by illuminating the object with short flashes of light). Describe what will be observed if the flashing frequency is $\tilde{\nu}_f = 25$ Hz, 50 Hz, 100 Hz, 49 Hz and 51 Hz.

Q3. Specify what is meant by *linearity*. Which of the following algebraic or differential relations are linear?

$$y_1 = 2x + 5, \quad y_2 = x^2, \quad y_3 = 2x, \quad y_4 = (x-1)/(2x+1),$$
$$\dot{u} - 3u = 0, \quad \ddot{u} - 4u = 0, \quad \ddot{u} - 4tu = 0, \quad \ddot{u} + 2\dot{u} - 4u = 0,$$
$$\ddot{u} - 4u^2 = 0, \quad \ddot{u} - 4u = 5, \quad \ddot{u} + t\,\dot{u}^2 - 4u = 0, \quad \ddot{u} + t/u - 4u = 0.$$

Q4. Determine the modulus and the phase of the following complex numbers: 2i, –3i, –5, 1/(2i), $\sqrt{2}$ i, 3 – 4i and (3 + i)/(1 – 3i).

Q5. Draw to scale the phasor diagram for the simple harmonic functions $x_1 = 10\cos(\omega t)$ and $x_2 = 5\cos(\omega t + \pi/6)$. Deduce the expression of their sum.

Q6. Show the expression $U_{(P)} = \frac{1}{2}\hat{K}\theta^2$ for the potential energy of a body subject to a restoring moment $\Gamma_z = -\hat{K}\,\theta$. For this, consider an intermediary position φ and calculate the work required for the variation $d\varphi$. Then, integrate on φ from 0 to θ. What is the total energy of a physical pendulum if it is initially displaced to an angle θ_o and released without initial velocity, and if it is given an initial angular velocity θ'_o from the equilibrium position?

Q7. A point-like body of mass m is placed on the top of a sphere. Is this equilibrium position stable or unstable? What is the horizontal component of the force acting on this body if it is slightly displaced a distance x from this position? What is the expression of its potential energy?

Q8. Show that the equation $A\,e^{-\eta_1 t} + B\,e^{-\eta_2 t} = 0$ can be identically satisfied (at any time t) only if $\eta_1 = \eta_2$ and $A + B = 0$. Show that the equation of the form $A\cos(\omega_1 t + \phi_1) + B\cos(\omega_2 t + \phi_2) = 0$ can be identically satisfied only if $\omega_1 = \omega_2$, $\phi_1 = \phi_2$ and $A = B$ or if $\omega_1 = \omega_2$, $\phi_1 = \phi_2 + \pi$ and $A = -B$.

Q9. Do the normal frequencies depend on the choice of the generalized variables (or coordinates) which specify the state of a system? Can the general solution of a physical system with n degrees of freedom be periodic? On how many independent parameters does it depend?

1.17. Problems

Simple harmonic oscillations: superposition and representations

P1.1. A point-like mass may oscillate on an axis back-and-forth near the origin with a frequency of 20 Hz. **a)** Write down the solution $x_1(t)$ if the mass is displaced a distance $a_1 = 2.0$ cm and given a velocity $v_1 = 1$ m/s; the solution $x_2(t)$ if the mass is displaced a distance $a_2 = -2.0$ cm and given a velocity $v_2 = 3.0$ m/s; and the solution $x_3(t)$ if the mass is displaced a distance $a_3 = 3$ cm and given a velocity $v_3 = 5.0$ m/s. Determine $x_3(t)$ as a superposition of the two motions x_1 and x_2. **b)** Plot x_3 and the velocity $\dot{x}_3$ as functions of time. Determine the maximum displacement ant the maximum velocity.

P1.2. Consider oscillation equation [1.7]. **a)** Verify directly that it has the real general solution $u = A_1 \cos(\omega t) + A_2 \sin(\omega t)$ and the complex general solution $\underline{u} = \underline{C} e^{i\omega t}$. Express A_1 and A_2 in terms of $\underline{C}$ and express $\underline{C}$ in terms of A_1 and A_2. **b)** Draw the phasors of the oscillations $A_1 \cos(\omega t)$ and $A_2 \sin(\omega t)$ as well as their sum. Deduce that this sum may be written in the form $A \cos(\omega t + \phi)$, where A and ϕ are related to A_1 and A_2 by the equations of [1.17].

P1.3. Consider the motion $x = 5 \cos[3\pi(t - 0.25)]$, where x is in cm. What is its frequency and what is its phase? Draw the phasors which represent x, $\dot{x}$ and $\ddot{x}$. Determine the complex amplitude of the complex representation $\underline{x}$.

Point-like mass oscillating on an axis

P1.4. The position of a point-like mass $m = 0.5$ kg on the x-axis is given by the expression $x = 2 \cos[\pi(3t - 1)]$, where x is in cm. **a)** What is the force which produces this motion? **b)** Write down the expressions of the potential energy, the kinetic energy and the total energy as functions of time. Plot these functions against t. **c)** Express the energies as functions of the position x and discuss the form taken by the energy according to the values of x.

P1.5. Consider a spring of natural length L_o, constant K and negligible mass, which hangs vertically. A point-like mass m is attached to its lower end and it is confined to move vertically. Verify that its period of oscillation is $2\pi\sqrt{m/K}$.

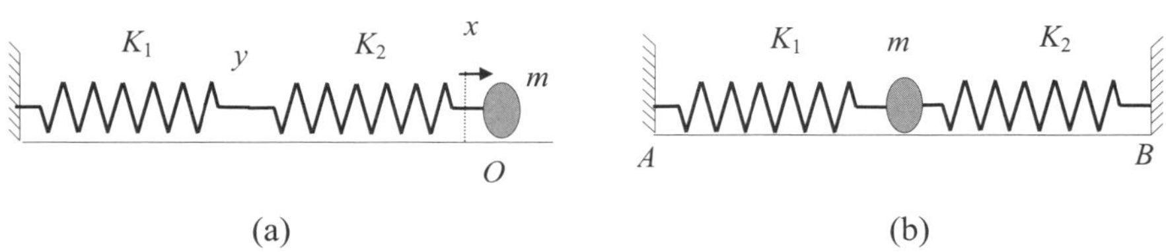

Figure 1.13. *Problem 1.6*

P1.6. a) A point-like mass m is attached to two springs of constants K_1 and K_2 and natural lengths L_1 and L_2 placed in series (Figure 1.13a). The equilibrium position O of the mass is taken as the origin. Calculate the period of small longitudinal oscillations (i.e. in the direction of the springs). **b)** The mass is now attached between the springs (Figure 1.13b) whose other ends are fixed at points A and B separated by a distance L. Determine the equilibrium position of the mass and the period of small longitudinal oscillations.

P1.7. a) A mass m is confined to move on the x-axis and subject to a force $F = -Kx$. Write down the expression of its position as a function of time if it is launched at $t = 0$ from the origin with a velocity v_o in the positive x direction. **b)** Calculate the velocity, the acceleration, the kinetic energy and the potential energy. Plot these quantities against time. **c)** The average value of a quantity $f(t)$ over a period T is defined as $< f > = (1/T) \int_T dt\, f(t)$, where the integration is over a period, between $t - T/2$ and $t + T/2$, for instance. Calculate the mean values of x, $\dot{x}$, x^2, $\dot{x}^2$ and those of the kinetic energy and the potential energy.

P1.8*. A spring of constant K and natural length L has an end fixed at point A of coordinate $-L$ on the horizontal x-axis. A mass M is attached at the other end taken as the origin O at equilibrium. **a)** The mass M is displaced on the x-axis to point B of coordinate x. Calculate the work done if the velocity is negligible in the course of this displacement. Is this work different if the mass is displaced from O to B along an arbitrary curve in the Oxy plane? **b)** Now assume that the spring itself has a uniformly distributed mass m. In order to study the motion of the mass M on the x-axis, we first calculate the total energy of the system. For this, consider a slice of the spring situated between ξ and $\xi + d\xi$. Calculate the mass of this slice, its velocity and its kinetic energy if the mass M has the coordinate x and it is moving with the velocity $\dot{x}$. Deduce the expression of the total kinetic energy of the spring and the mass M. Neglecting all frictions, deduce the equation of motion of the mass M and the period of oscillations. In this problem we assume that the motion is very slow; so all propagation phenomena along the spring are negligible. This means that all the slices of the spring are deformed simultaneously.

Angular oscillations

P1.9. A cylinder is suspended between two vertical wires along its axis. One end of each wire is fixed to the cylinder and the other is fixed to the support. The wires have torsion constants $\hat{K}_1$ and $\hat{K}_2$ respectively and the cylinder has a moment of inertia J about its axis. Calculate the period of angular oscillations of this cylinder.

P1.10*. A simple pendulum consists of a small ball with a mass of 1 kg, suspended to the end of an inextensible and weightless wire of length 2 m whose other end is fixed at O. The ball is initially moved away to an angle of 15° and released without initial velocity. **a)** Considering this amplitude as "small", write down the expressions of the angular displacement, the angular velocity, the angular acceleration, the linear velocity, the centripetal acceleration and the tension of the wire as functions of time. **b)** Using the law of conservation of energy, determine the exact value of the linear velocity when the ball makes an angle of 5° with the vertical, and compare with the expression obtained in question (a). Take $g = 9.81$ m/s^2.

P1.11*. a) A pendulum consists of a ball of radius 5 cm and mass density 5 g/cm^3 suspended on the end of an inextensible and weightless wire of length 1 m. Determine its period if g = 9.81 m/s^2. **b)** What is the value of this period if this pendulum swings in an elevator, which goes up with a constant velocity of 5 m/s and in an elevator, which has an upward acceleration of 1 m/s^2? **c)** What is the period if this pendulum swings in a liquid of mass density 700 kg/m^3 (neglecting the drag force of the liquid)? **d)** The Earth's radius being 6,400 km, what is the variation of the period if the pendulum swings at an altitude of 10 km?

Damped oscillations

P1.12. a) Consider the motion $x = A\,e^{-\beta t}\cos(\hat{\omega} t + \phi)$. Calculate the corresponding velocity and acceleration. Verify that x is a solution of the differential equation: $\ddot{x} + 2\beta\,\dot{x} + \omega_o^{\,2} x = 0$, where $\omega_o^{\,2} = \hat{\omega}^2 + \beta^2$. Write down the expression of x as the sum of two exponential terms of the form $\underline{x} = \underline{C}\,e^{-\eta t}$. **b)** The quasi-period of this oscillator is 2 s. After 10 complete oscillations, its amplitude is reduced to one quarter of its initial value. Determine the damping coefficient and the relaxation time of this oscillator. What is its quality factor?

P1.13. The angle of a pendulum is given by $\theta = 15\,e^{-0.15\,t}\cos(9.42\ t)$ in degrees. **a)** Write down the expression of the angular velocity. Determine its amplitude and its phase. **b)** Determine the amplitude, the period, the initial position and the initial angular velocity. **c)** What is the quality factor of this pendulum? **d)** How long does it take for the amplitude to be reduced to one-tenth of its initial value?

P1.14. A point mass m is confined to move on the x-axis. It is attracted to the origin with a force $F = -Kx$ and it is subject to a friction force $F_{(fr)} = -b\,\dot{x}$.**a)** Write down the expressions of its position and its velocity if it is launched from the origin with a velocity v_o at $t = 0$. **b)** Calculate the total energy of this particle as function of time. **c)** Using the expressions of the friction force and the velocity, calculate the energy which is lost by this particle between $t = 0$ and t. Verify the principle of conservation of energy. **d)** Assuming that m = 2 g, K = 0.050 N/m, v_o = 10 cm/s and b = 0.012 N.s/m, calculate the angular frequency, the damping coefficient and the total energy at $t = 0$ and $t = 5 \times 10^{-2}$ s.

Oscillating circuits

P1.15*. A capacitor of $C = 1\ \mu F$ is charged to a voltage of 100 V and then connected at $t = 0$ to an inductor of $L = 2$ H in series with a resistor of $R = 2\ k\Omega$. **a)** Write the differential equation for the charge Q of the capacitor. What is the solution that corresponds to these initial conditions? **b)** Write down the expressions of the current intensity, the electric energy, the magnetic energy and the total stored energy. **c)** Calculate the dissipated Joule heat between $t = 0$ and t. Compare it with the decrease of the stored energy. **d)** What is the relaxation time of this circuit and what is its quality factor? How long it takes to the amplitude of the charge to be reduced to 1% of its initial value?

P1.16. Consider an LC circuit and assume that, at $t = 0$, the capacitor has a charge Q_o. **a)** Neglecting the resistance of the circuit, write down the expressions of Q and I versus time. What is the energy of the circuit? Show the equation of the circuit from energy conservation. Discuss the form taken by the energy between $t = 0$ and $t = T$. **b)** Now assume that the circuit has a resistance R. Write down the expressions of Q, I and the total energy $U_{(EM)}$. Show that $U_{(EM)}$ decreases always by an amount equal to the energy dissipated as Joule heat. c) Assuming that R is small, show that the dissipated energy ΔU_J in a period is related to the mean stored energy $< U_{(EM)}(t) >$ by the relation $\Delta U_J / < U_{(EM)}(t) > = 2\pi / f_q$, where f_q is the quality factor of the circuit.

*Oscillations near a stable equilibrium position**

P1.17. A cylinder of section S, height h and mass density ρ floats vertically on the surface of a liquid of mass density ρ_o. What is the immersed height? This cylinder is displaced vertically by a distance x_o from its equilibrium position and released without initial velocity. Show that it undergoes harmonic oscillations. What is the period?

P1.18. A tube bent into a U shape contains a liquid. The liquid level in one of the branches is pushed a distance h and then released at $t = 0$. Neglecting all frictions, show that the liquid undergoes oscillations of period $\pi\sqrt{2L/g}$, where L is the total length of the liquid. Analyze this problem, first using the force, and then using the conservation of energy.

P1.19*. A point-like body of mass m slides without friction in a bowl of parabolic shape of equation $z = ax^2$ (Figure 1.14). Study the motion of this body in the vertical plane Oxz. Express the kinetic energy and the potential energy as functions of x and its derivative $\dot{x}$. Deduce the equation of small oscillations back and forth at the bottom O. What is the period?

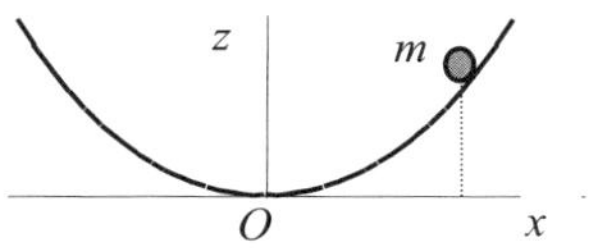

Figure 1.14. *Problem 1.19*

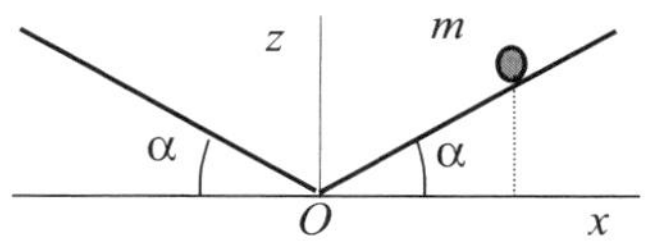

Figure 1.15. *Problem 1.20*

P1.20. Two inclined planes tilted at an angle α with the horizontal plane and set end-to-end along a horizontal axis O (Figure 1.15). A point-like mass m is released without initial velocity on one of the planes at the horizontal distance x_o. Calculate the component of the restoring force in the x direction and the potential energy. Neglecting the friction forces, calculate the period of motion.

P1.21.** A molecule is formed by two atoms of masses m_1 and m_2 whose potential energy of interaction is $U_{(P)} = D\{\exp[-2a(r - r_o)] - 2\exp[-a(r - r_o)]\}$, where D, a and r_o are characteristic constants and r is the distance separating the atoms. $U_{(P)}$ is called the *Morse potential.* We consider in the following only the rectilinear displacements in the x direction. **a)** What is the equilibrium position? Is it stable? **b)** Write down the expression of the total energy of the molecule. Make a Taylor expansion of $U_{(P)}$ near the equilibrium position and show that the relative motion is approximately a simple harmonic. Calculate the frequency of small oscillations near the equilibrium position.

Nonlinear systems *

P1.22.** A particle of mass m moves on the x-axis. Its potential energy depends on x according to the relation $U_{(P)} = \frac{1}{2}Kx^2 - (1/3)\,ax^3$. **a)** What is the force acting on this particle? Write its equation of motion. **b)** This equation cannot be exactly solved. However, if the constant a is very small, the motion is almost a simple harmonic motion. Set $x = A\cos(\omega t + \phi) + u(t)$, where the first term is the simple harmonic vibration due to the harmonic potential energy $\frac{1}{2}Kx^2$ and $u(t)$ is a small correction ($|u| \ll A$) due to the anharmonic term correction $-(1/3)ax^3$. Show that an approximate solution of u is $u(t) = (aA^2/6K)\,[3 - \cos(2\omega t + 2\phi)]$. **c)** Calculate the total energy of the oscillator by keeping only the first order in (aA/K). **d)** Now consider the same problem but a potential energy $U_{(P)} = \frac{1}{2}Kx^2 - \frac{1}{4}ax^4$. Using a similar method as for question (b), show that the approximate solution is $x = A\cos(\omega t + \phi) + u(t)$, where ω is given by $m\omega^2 = K - (3/4)\,aA^2$ and $u = [aA^3/4(K - 9m\omega^2)]\cos[3(\omega t + \phi)]$.

P1.23*. A mass m is attached to two springs of the same constant K (Figure 1.16) and it oscillates in the transverse direction Oy. Calculate the kinetic energy and the potential energy when the mass is at a distance y from its equilibrium position. Make an expansion of the total energy in powers of (y/a) up to the order $(y/a)^4$ and write the law of conservation of energy. Deduce that the period of this motion is $T = 4\sqrt{2}\int_0^A dx/\sqrt{A^4 - x^4} \approx (7.416/A)\sqrt{m/K}$.

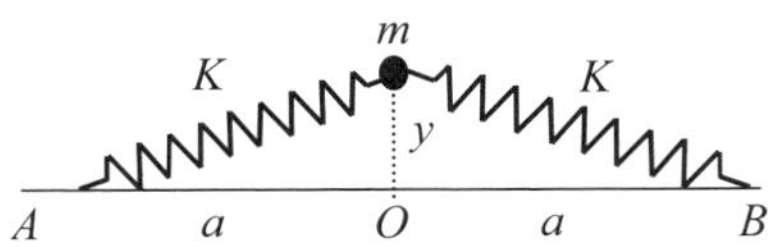

Figure 1.16. *Problem P1.23*

P1.24. a)** Show the equation of motion of a physical pendulum by using the energy conservation. **b)** Writing cos θ as a Taylor series, and keeping only the terms θ^2 and θ^4, show that the potential energy is of the type of question (d) in problem P1.22. Deduce that the period of oscillations for an amplitude θ_o is approximately $T = T_o(1 + \theta_o^2/16)$ and that the angular displacement is:

$$\theta = A\cos(\omega t + \phi) - (1/192)A^3\cos(3\omega t + 3\phi).$$

c) The conservation of energy may be used to find the expression for the period for large amplitudes. For this, assume that the pendulum is displaced to an angle θ_o and released without initial velocity. Let T_o be the period for oscillations of small amplitudes. Show that, in the first half-period of oscillation, the time is related to the angle θ by the equation $t = (T_o/4\pi)\,F(\theta)$, where $F(\theta) = \int_0^\theta d\theta\,[\sin^2(\theta_o/2) - \sin^2(\theta/2)]^{-1/2}$. Deduce that the period is $T = (T_o/4\pi)\,F(\theta_o)$. **d)** To evaluate the integral, set $\alpha = \sin(\frac{1}{2}\theta_o)$ and $\sin(\frac{1}{2}\theta) = \alpha\sin\phi$. Verify that $T = (2T_o/\pi)\int_0^{\pi/2} d\varphi\,(1 - \alpha^2\sin^2\varphi)^{-1/2}$. The integral is called an *elliptic function of the first kind.* As $\alpha < 1$, we may write

$$(1 - \alpha^2\sin^2\varphi)^{-1/2} = 1 + (1/2)\,\alpha^2\sin^2\varphi + (3/8)\,\alpha^4\sin^4\varphi + \ldots$$

Deduce that:

$$T = T_o\left\{1 + \left(\frac{1}{2}\right)^2\sin^2(\tfrac{1}{2}\theta_o) + \left(\frac{1\times 3}{2\times 4}\right)^2\sin^4(\tfrac{1}{2}\theta_o) + \ldots\right\}.$$

Systems of two coupled oscillators

P1.25. Determine the normal modes and the general solution of the circuit of Figure 1.17.

P1.26.** Consider the inertial coupling of two oscillators whose equations are of the form $\mu_{11}\ddot{u}_1 + \mu_{12}\ddot{u}_2 + K_1 u_1 = 0$ and $\mu_{21}\ddot{u}_1 + \mu_{22}\ddot{u}_2 + K_2 u_2 = 0$. **a)** Determine the normal modes and the general solution. **b)** Apply this analysis to study the circuit in Figure 1.18.

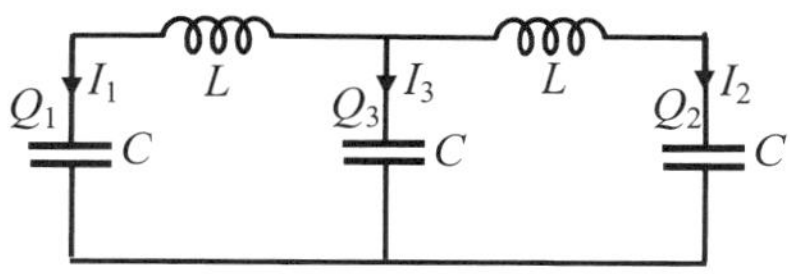

Figure 1.17. *Problem P1.25*

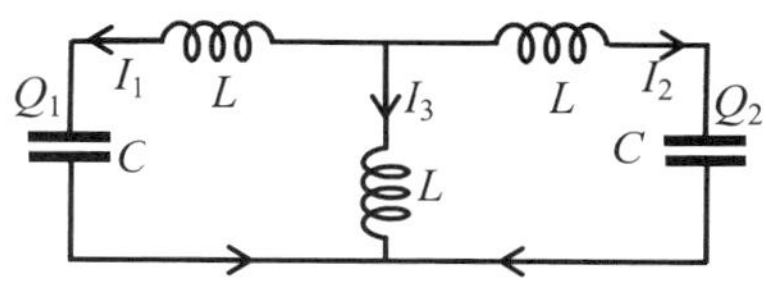

Figure 1.18. *Problem P1.26*

P1.27. Consider the coupled masses in Figure 1.11 in the special case $m_1 = m_2 \equiv m$ and $K_1 = K_2 \equiv K$. The general solution is given by the expressions [1.87]. **a)** Write down the solution which corresponds to the following initial conditions: $x_1(0) = A$, $x_2(0) = 0$ and $\dot{x}_1(0) = \dot{x}_2(0) = 0$. **b)** Calculate the total energy and verify that it is conserved and equal to the work required to displace mass (1) by a distance a, while mass (2) is kept fixed at its equilibrium position. **c)** In the case $K_3 << K$, set $\Delta\omega = ½(\omega_{(2)} - \omega_{(1)})$ and $\omega = ½(\omega_{(1)} + \omega_{(2)})$. Show that a phenomenon of beats appears, such that one of the masses oscillates with a large amplitude while the other oscillates with a small amplitude.

P1.28. a) Write the equations of the circuit in Figure 1.19 and deduce the differential equations for the intensities. **b)** Determine the normal modes. Write down the expressions of the intensities in each mode and the general solution.

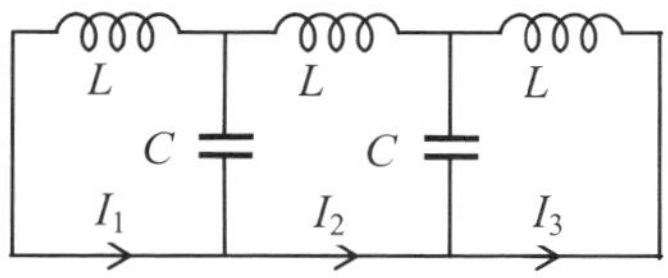

Figure 1.19. *Problem 1.28*

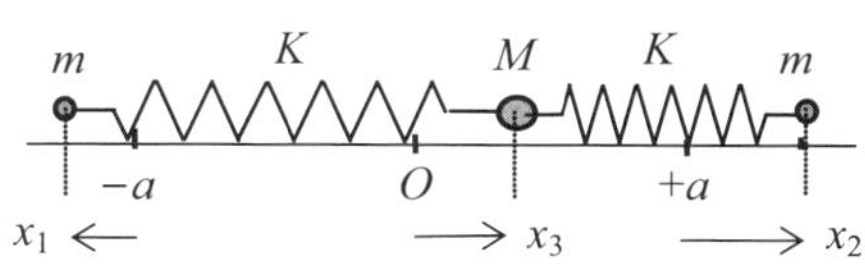

Figure 1.20. *Problem 1.30*

P1.29*. The equations of some mechanical systems are coupled by terms proportional to the velocities (*gyroscopic coupling*). Consider two oscillators whose equations of motion are:

$$\mu_1\ddot{u}_1 + b_{11}\dot{u}_1 + b_{12}\dot{u}_2 + a_{11}u_1 = 0 \quad \text{and} \quad \mu_2\ddot{u}_2 + b_{21}\dot{u}_1 + b_{22}\dot{u}_2 + a_{22}u_2 = 0.$$

a) Write the characteristic equation. Calculate the total energy and its rate of change in time. **b)** Consider the special case $b_{11} = b_{22} = 0$ and $b_{12} = - b_{21}$. Determine the modes and verify that the total energy is constant.

P1.30.** A *triatomic linear molecule* may be modeled as a system of two atoms of mass m situated on both sides of a third atom of mass M on the same x-axis and coupled by two springs of constant K (Figure 1.20). Let $-a$, 0 and a be the equilibrium positions of these atoms and x_1, x_2 and x_3 their displacements from their respective equilibrium positions. **a)** Write down the expressions of the kinetic energy and the potential energy. Deduce the equations of motion, the characteristic equation and the normal modes. **b)** Normalize these modes and determine the normal variables. Write down the expression of the energy as a function of these variables and their derivatives.

P1.31.** Consider the coupled masses in Figure 1.11 in the case $K_1 = K_2 \equiv K$ and $m_1 = m_2 \equiv m$ and assume that these masses are subject, in addition, to the friction forces $F_{(fr)1} = -b\,\dot{x}_1$ and $F_{(fr)2} = -b\,\dot{x}_2$. **a)** Write the equations of motion. **b)** To solve this system of equations, try solutions of the form $x_1 = A_1 e^{-\eta_1 t}$ and $x_2 = A_2 e^{-\eta_2 t}$. Show that η_1 and η_2 must be equal to one of the roots of the characteristic equation $(\eta^2 - 2\beta\eta + \omega_o^2 + \omega_3^2)^2 - \omega_3^4 = 0$, where $\beta = b/2m$, $\omega_o = \sqrt{K/m}$ and $\omega_3 = \sqrt{K_3/m}$. **c)** Discuss whether the modes are oscillatory or not according to the value of b and write the general solution. **d)** Show that $y_1 = x_1 + x_2$ and $y_2 = x_1 - x_2$ are the normal variables. Write the solutions for y_1 and y_2 in the various cases.

*Systems with several degrees of freedom**

P1.32.** Two circuits are formed by two capacitors of equal capacitance C and two coils of negligible resistance and self-inductance L (Figure 1.21). The coils have a mutual inductance coefficient M. Thus, the magnetic energy of these coils is $U_{(M)} = ½\, LI_1^2 + ½\, LI_2^2 + MI_1I_2$. **a)** Write the equations of these circuits and determine the normal modes and the general solution. **b)** Determine the normal variables of this system by using the general method from section 1.13. Write down the expression of the energy as a function of these variables and their derivatives.

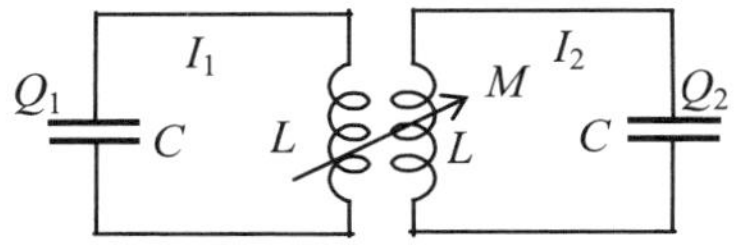

Figure 1.21. *Problem P1.32*

P1.33.** Consider the equation of motion of a system with n degrees of freedom in the vector form $[\mu]\,\ddot{\mathbf{u}} + [K]\mathbf{u} = 0$, where the "vector" $\mathbf{u} \equiv (u_1, u_2, \ldots, u_n)$ represents the displacements from the equilibrium state, $[\mu]$ is the *inertial tensor* and $[K]$ is the *restoring force tensor*. $[\mu]$ and $[K]$ are two real and symmetric tensors. The corresponding *eigenvalue equation* is $\{[K] - \omega^2[\mu]\}\ \mathbf{A} = 0$. **a)**. The kinetic energy $U_{(C)} = \frac{1}{2}\Sigma_{ij}\ \mu_{ij}\,\dot{u}_i\ \dot{u}_j$ is always positive. Deduce that the tensor $[\mu]$ is not singular. **b)** Show that the equation of motion may be written in the *reduced form* $\ddot{\mathbf{u}} + [D]\mathbf{u} = 0$, where $[D]$ is a symmetric tensor, called the *dynamical tensor* of the system. **c)** To solve this equation, write $\mathbf{u} = \mathbf{A}\cos(\omega t + \phi)$. Show that ω^2 is an *eigenvalue* of the tensor $[D]$ and $\mathbf{A}$ is an eigenvector, i.e. $[D]\ \mathbf{A} = \omega^2\mathbf{A}$. Deduce the characteristic equation $\det\{[D] - \omega^2[I]\} = 0$, where $[I]$ is the *identity matrix* in the n dimensional space. We assume in this problem that the roots of the characteristic equation are simple. **d)** ω is acceptable as a frequency of a simple harmonic motion if the eigenvalues of the tensor $[D]$, which are $\lambda \equiv \omega^2$, are real and positive. Show that, because $[D]$ is a symmetric and real tensor and $[D]\mathbf{A} = \lambda\mathbf{A}$, λ is necessarily real. Show that two eigenvectors $\mathbf{A}$ and $\mathbf{A}'$, which correspond to two different eigenvalues λ and λ', are orthogonal. **e)** To show that the eigenvalues are positive, write the equation $[D]\mathbf{A} = \lambda\mathbf{A}$ in the form $[K]\mathbf{A} = \lambda[\mu]\mathbf{A}$. Show that we may write $\lambda = (\Sigma_{ij}\ K_{ij}A_iA_j)/(\Sigma_{ij}\ \mu_{ij}\ A_i\,A_j)$. Deduce that λ is positive. **f)** Show that if $\lambda_{(k)}$ and $\lambda_{(h)}$ are two different eigenvalues, the corresponding eigenvectors $\mathbf{A}_{(k)}$ and $\mathbf{A}_{(h)}$ verify the relationship $\Sigma_{ij}\ \mu_{ij}\ A_{i(k)}A_{j(h)} = 0$. Show that, for a given eigenvalue $\omega_{(k)}$, it is always possible to normalize the eigenvectors $\mathbf{A}_{(k)}$ such that $\Sigma_{ij}\ \mu_{ij}\ A_{(k)i}\ A_{(k)j} = 1$. Then, we may write the general relationship $\Sigma_{ij}\ \mu_{ij}\ A_{(k)i}\ A_{(k')j} = \delta_{kk'}$. **g)** Show that the transformation matrix $[S]$, whose elements are $S_{ij} = A_{i(j)}$, transforms the tensor $[\mu]$ to a tensor $[\bar{\mu}]$ which is diagonal, with the diagonal elements $\bar{\mu}_{ii} = 1$, and it transforms the tensor $[K]$ to a tensor $[\bar{K}]$ which is diagonal with the diagonal elements $\bar{K}_{ii} = \omega_{(i)}{}^2$.

Chapter 2

Superposition of Harmonic Oscillations, Fourier Analysis

The oscillations of a system can never be exactly periodic since they always have a beginning and an end. However, the oscillations may be treated as approximately periodic if they last a very long time, compared to the period of a single oscillation. On the other hand, even if the oscillations of a system are approximately periodic, they are never exactly simple harmonic (i.e. represented by a sinusoidal function with a single frequency). For instance, even if light is a single line of the discrete atomic spectrum or a laser beam, it is always a superposition of monochromatic waves in a more or less wide band. The superposition of oscillations and waves is so real and important that it is often raised to the rank of a principle (called the *superposition principle*). It plays a very important part in the study of interference, diffraction and quantum mechanics. The validity of this principle relies on the linearity of the mechanics and electromagnetism equations.

Our purpose in this chapter is to study the superposition of simple harmonic oscillations. The *Fourier analysis* considers any function as a superposition of simple harmonic functions. We study the case of periodic functions and of non-periodic functions, namely signals of short duration.

2.1. Superposition of two scalar and isochronous simple harmonic oscillations

Consider the oscillations $u_1 = a_1 \cos(\omega t + \phi_1)$ and $u_2 = a_2 \cos(\omega t + \phi_2)$ that are scalar (or vectorial but collinear). Their algebraic sum is

$$u = u_1 + u_2 = a_1 \cos(\omega t + \phi_1) + a_2 \cos(\omega t + \phi_2) = A_1 \cos \omega t - A_2 \sin \omega t, \qquad [2.1]$$

where we have set

$$A_1 = a_1 \cos \phi_1 + a_2 \cos \phi_2, \qquad A_2 = a_1 \sin \phi_1 + a_2 \sin \phi_2. \qquad [2.2]$$

We may write the resulting oscillation u in the simple harmonic form

$$u = A \cos(\omega t + \phi), \qquad [2.3]$$

where the constants A and ϕ are related to A_1, A_2, ϕ_1 and ϕ_2 by the equations

$$A = \sqrt{A_1^2 + A_2^2} = \sqrt{a_1^2 + a_2^2 + 2a_1 a_2 \cos(\phi_1 - \phi_2)} \qquad [2.4]$$

$$\cos \phi = \frac{A_1}{A}, \qquad \sin \phi = \frac{A_2}{A} \qquad \text{and} \qquad \tan \phi = \frac{a_1 \sin \phi_1 + a_2 \sin \phi_2}{a_1 \cos \phi_1 + a_2 \cos \phi_2}. \qquad [2.5]$$

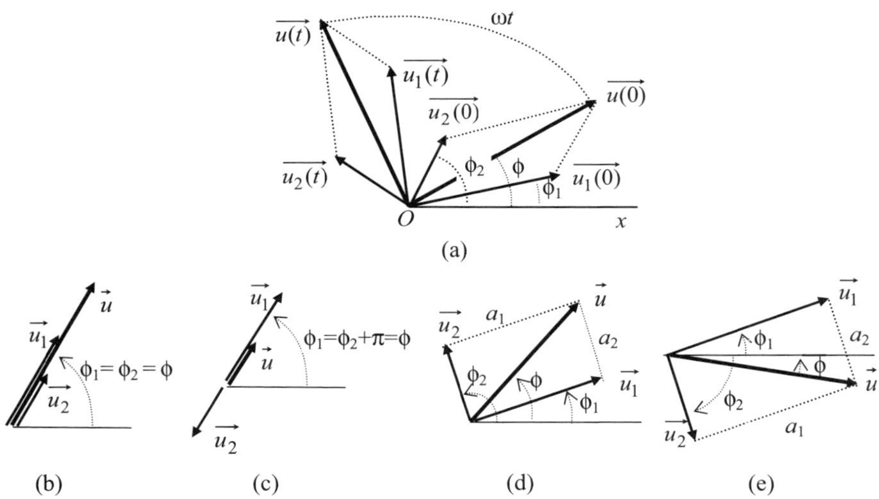

Figure 2.1. *Phasor representation of the superposition of two scalar and isochronous simple harmonic oscillations in the following cases: a) arbitrary phase shift, b) in phase, c) in opposite phase, d) in quadrature ($\phi_2 = \phi_1 + \pi/2$) and e) in quadrature ($\phi_2 = \phi_1 - \pi/2$)*

We may also show this result using the phasor diagram (Figure 2.1a). The algebraic sum $u = u_1 + u_2$ is obtained by projecting, onto Ox, the vector sum $\vec{u}$ of the phasors $\overrightarrow{u_1}$ and $\overrightarrow{u_2}$ representing u_1 and u_2. The length of the vector $\vec{u}$ and the angle that it makes with Ox are obtained by the usual rules of geometry; they are given by [2.4] and [2.5]. Depending on the values of the phase shift between u_1 and u_2, we may consider the following special cases:

– The oscillations u_1 and u_2 are *in phase* ($\phi_1 = \phi_2 + 2n\pi$, Figure 2.1b); then,

$$A = a_1 + a_2, \qquad \phi = \phi_1 = \phi_2. \qquad [2.6]$$

– The oscillations u_1 and u_2 are *in opposite phase* ($\phi_1 = \phi_2 + \pi + 2n\pi$, Figure 2.1c); then,

$$A = |a_1 - a_2|, \qquad \phi = \phi_1 \qquad \text{if } a_1 > a_2 \text{ and } \phi = \phi_2 \text{ if } a_2 > a_1. \qquad [2.7]$$

– The oscillations u_1 and u_2 are *in quadrature* ($\phi_2 = \phi_1 \pm \pi/2$, Figures 2.1d and 2.1e); then,

$$A = \sqrt{a_1^{\,2} + a_2^{\,2}}, \qquad \phi = \phi_1 \pm \text{Arctan}\,(a_2/a_1). \qquad [2.8]$$

EXAMPLE 2.1. Calculate the sum [2.1] using the complex representation.

SOLUTION – Replacing u_1 and u_2 with the complex exponential functions $\underline{u}_1 = \underline{a}_1\, e^{i\omega t}$ and $\underline{u}_2 = \underline{a}_2\, e^{i\omega t}$ with $\underline{a}_1 = a_1\, e^{i\phi_1}$ and $\underline{a}_2 = a_2\, e^{i\phi_2}$, the sum $\underline{u} = \underline{u}_1 + \underline{u}_2$ may be written as $\underline{u} = \underline{A}\, e^{i\omega t}$. Omitting the common factor $e^{i\omega t}$, the calculation of the sum reduces to that of the complex amplitude

$$\underline{A} = \underline{a}_1 + \underline{a}_2 = a_1 \cos\phi_1 + a_2 \cos\phi_2 + i\,(a_1 \sin\phi_1 + a_2 \sin\phi_2).$$

We get $A = A_1 + iA_2$, where the amplitudes A_1 and A_2 are given by [2.2]. We deduce that the amplitude A and the phase ϕ are given by equations [2.4] and [2.5]. Thus, we may write $\underline{u} = A\, e^{i(\omega t+\phi)}$ whose real part is $u = A\cos(\omega t + \phi)$.

2.2. Superposition of two perpendicular and isochronous vector oscillations, polarization

Consider, for instance, a stretched string along the z-axis (Figure 2.2a). A point M_o of the string at equilibrium may be displaced to a position M in any direction, which is perpendicular to the z-axis. Thus, the displacement $\mathbf{u} \equiv \overline{M_0M}$ is a vector quantity. If $\mathbf{u}$ at any point of the string remains parallel to a fixed axis of unit vector $\mathbf{e}_p$, we say that the oscillation is *linearly polarized* in the direction $\mathbf{e}_p$ (or *plane-polarized* in the plane containing Oz and $\mathbf{e}_p$). In particular, the oscillation may be polarized in the direction $\mathbf{e}_x$ or $\mathbf{e}_y$ of the axes of coordinates Ox and Oy. We study in this section the superposition of two isochronous simple harmonic oscillations which are polarized in perpendicular directions, that we may label as $\mathbf{e}_x$ and $\mathbf{e}_y$

$$x(t) = a\cos(\omega t + \phi_1), \qquad y(t) = b\cos(\omega t + \phi_2). \qquad [2.9]$$

In the course of time, the point M of coordinates $x(t)$ and $y(t)$ moves in the Oxy plane. Its trajectory is obtained by eliminating t from the expressions of [2.9]. We obtain[1]

$$x^2/a^2 + y^2/b^2 - 2(xy/ab)\cos(\phi_1 - \phi_2) = \sin^2(\phi_1 - \phi_2). \qquad [2.10]$$

In general, this is an ellipse centered on the origin of coordinates (which coincides with the equilibrium position M_o). It is inscribed within the rectangle $-a < x < a$ and $-b < y < b$ (Figure 2.2b) but its form depends on the phase shift $(\phi_2 - \phi_1)$.

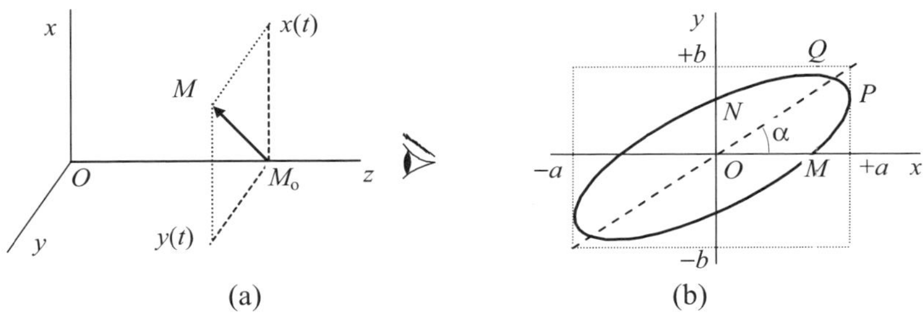

Figure 2.2. *Superposition of two simple harmonic oscillations polarized in perpendicular directions: a) oscillations of a point of a string and b) the corresponding elliptic trajectory*

We may consider the following special cases, which are illustrated in Figure 2.3 as the observer receives the waves:

– If x and y are *in phase* ($\phi_2 - \phi_1 = 2n\pi$), the trajectory is the first diagonal ($y = bx/a$) of the rectangle. Thus, the resulting oscillation of two oscillations x and y in phase is *linearly polarized* in the direction which makes with Ox an angle $\alpha = \text{Arctan}(b/a)$.

1 To obtain the equation of the trajectory [2.10], expand $\cos(\omega t + \phi_1)$ and $\cos(\omega t + \phi_2)$ in terms of $\cos(\omega t)$ and $\sin(\omega t)$, solve the equations of [2.9] to express $\cos(\omega t)$ and $\sin(\omega t)$ in terms of x and y and substitute these expressions in the identity $\cos^2(\omega t) + \sin^2(\omega t) = 1$.

– If x and y are *in phase opposition* ($\phi_2 - \phi_1 = \pi + 2n\pi$), the trajectory is the second diagonal ($y = - bx/a$). Thus, the resulting oscillation of two oscillations x and y in opposite phase is *linearly polarized* in the direction which makes with Ox an angle $\alpha = -\text{Arctan}(b/a)$.

– In the case of an arbitrary phase shift between x and y, the trajectory is elliptic, inscribed within the rectangle. It is clockwise if $0 < \phi_2 - \phi_1 < \pi$ (*right-elliptically polarized motion*) and counterclockwise if $\pi < \phi_2 - \phi_1 < 2\pi$ (*left-elliptically polarized motion*). If x and y are in quadrature ($\phi_2 - \phi_1 = \pm\,\pi/2 + 2n\pi$), the ellipse has its axes in the x and y directions.

– In the special case of x and y having equal amplitudes ($a = b$), the equation of the ellipse takes the form

$$x^2 + y^2 - 2xy\cos(\phi_2 - \phi_1) = a^2 \sin^2(\phi_2 - \phi_1). \qquad [2.11]$$

This is the equation of an ellipse, which is inscribed in a square of sides $2a$. Particularly, if $x(t)$ and $y(t)$ are in quadrature ($\phi_2 - \phi_1 = \pm\,\pi/2 + 2n\pi$), the trajectory is a circle of radius a. If $\phi_2 - \phi_1 = \pi/2 + 2n\pi$, the resulting motion $\mathbf{u}^{(+)}$ is counterclockwise (*right-circularly polarized* motion) and if $\phi_2 - \phi_1 = -\pi/2 + 2n\pi$, the resulting motion $\mathbf{u}^{(-)}$ is clockwise (*left-circularly polarized* motion)

$$\mathbf{u}^{(\pm)} = a\,[\cos(\omega t + \phi_1)\,\mathbf{e}_x + \cos(\omega t + \phi_1 \pm \pi/2)\,\mathbf{e}_y]. \qquad [2.12]$$

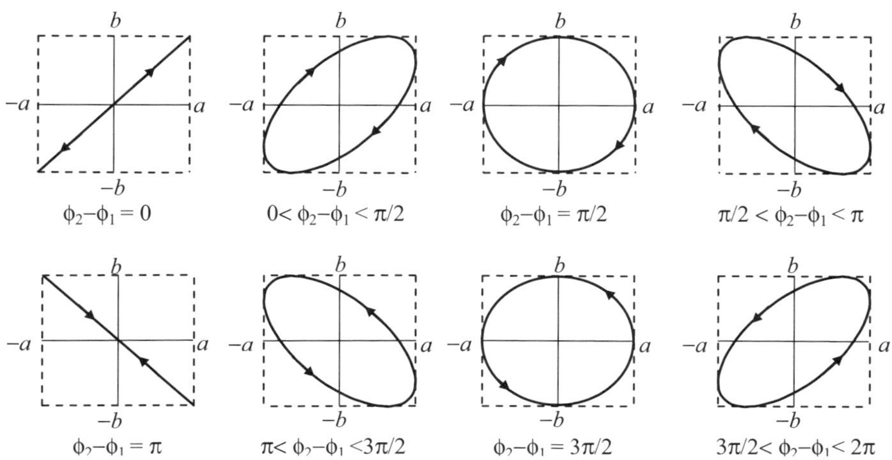

Figure 2.3. *Superposition of two simple harmonic oscillations, which are synchronous and polarized in perpendicular directions for some particular values of the phase shift*

EXAMPLE 2.2. **a)** What is the motion which results from the superposition of the perpendicularly polarized oscillations $x = 2.0\cos(\omega t)$ and $y = 5.0\cos(\omega t)$? **b)** Write the oscillation $u = 4.0\cos(\omega t)$, which is linearly polarized in a direction making an angle of 20° with Ox as a superposition of two linearly polarized oscillations in the directions Ox and Oy.

SOLUTION

a) The motion, which results from the superposition of the oscillations, is

$$\mathbf{u} = 2.0\cos(\omega t)\,\mathbf{e}_x + 5.0\cos(\omega t)\,\mathbf{e}_y = (2.0\,\mathbf{e}_x + 5.0\,\mathbf{e}_y)\cos(\omega t).$$

The vector $(2.0\,\mathbf{e}_x + 5.0\,\mathbf{e}_y)$ has a magnitude $u_o = \sqrt{2{,}0^2 + 5{,}0^2} = 5.4$ and it makes with Ox an angle $\alpha = \text{Arctan}(5.0/2.0) = 1.2$ rad. The resulting oscillation is linearly polarized in this direction and its amplitude is 5.4.

b) The oscillation $u = 4.0\cos(\omega t)$, which is polarized in a direction making 20° with Ox, may be written in the vector form

$$\mathbf{u} = 4.0\cos(\omega t)[\cos 20°\,\mathbf{e}_x + \sin 20°\,\mathbf{e}_y] = 3.8\cos(\omega t)\,\mathbf{e}_x + 1.4\cos(\omega t)\,\mathbf{e}_y.$$

Thus, it is the superposition of two oscillations of amplitudes 3.8 and 1.4, in phase and linearly polarized in the directions Ox and Oy respectively.

EXAMPLE 2.3. **a)** Using the complex exponential representation, write down the expressions of the circularly polarized oscillations. **b)** Write the oscillation, which has an amplitude A and which is polarized in a direction making an angle θ with Ox, as the superposition of two circularly polarized oscillations.

SOLUTION

a) The superposition of the oscillations $x(t) = A\cos(\omega t+\phi_1)$ and $y(t) = A\cos(\omega t+\phi_2)$ such that $\phi_2 = \phi_1 \pm \pi/2 + 2n\pi$, is an oscillation $\mathbf{u}^{(+)}$ and $\mathbf{u}^{(-)}$ respectively right-circularly polarized and left-circularly polarized. Using the complex exponential representation, we may write

$$\begin{aligned}\mathbf{u}^{(\pm)} &= a\,e^{i(\omega t+\phi_1)}\,\mathbf{e}_x + a\,e^{i(\omega t+\phi_2)}\,\mathbf{e}_y = a\,e^{i(\omega t+\phi_1)}\,\mathbf{e}_x + a\,e^{i(\omega t+\phi_1\pm\pi/2)}\,\mathbf{e}_y \\ &= \underline{a}\,e^{i\phi_1}\,e^{i\omega t}(\mathbf{e}_x \pm i\,\mathbf{e}_y) = \underline{A}\,e^{i\omega t}\,\mathbf{e}^{(\pm)},\end{aligned} \qquad [2.13]$$

where we introduced the orthonormal vectors $\mathbf{e}^{(\pm)} = (\mathbf{e}_x \pm \mathrm{i}\mathbf{e}_y)/\sqrt{2}$ and the complex amplitude $\underline{A} = \sqrt{2}\, a\, e^{\mathrm{i}\phi_1}$. Conversely, we have $\mathbf{e}_x = (\mathbf{e}^{(+)} + \mathbf{e}^{(-)})/\sqrt{2}$ and $\mathbf{e}_y = \mathrm{i}(\mathbf{e}^{(-)} - \mathbf{e}^{(+)})/\sqrt{2}$.

b) Using $\mathbf{e}^{(+)}$ and $\mathbf{e}^{(-)}$ as a base, the complex oscillation $\underline{u} = A\ e^{\mathrm{i}\omega t}$, which is polarized in the **e** direction making an angle θ with Ox, may be written in the form

$$\mathbf{u} = A\, e^{\mathrm{i}\omega t}\mathbf{e} = A\, e^{\mathrm{i}\omega t}(\cos\theta\, \mathbf{e}_x + \sin\theta\, \mathbf{e}_y) = \frac{A}{\sqrt{2}} e^{\mathrm{i}\omega t}[\cos\theta\,(\mathbf{e}^{(+)} + \mathbf{e}^{(-)}) + \mathrm{i}\sin\theta\,(\mathbf{e}^{(-)} - \mathbf{e}^{(+)})]$$

$$= \frac{A}{\sqrt{2}} e^{\mathrm{i}\omega t}[e^{-\mathrm{i}\theta}\ \mathbf{e}^{(+)} + e^{\mathrm{i}\theta}\mathbf{e}^{(-)}] = \frac{A}{\sqrt{2}} e^{\mathrm{i}(\omega t - \theta + \phi)}\mathbf{e}^{(+)} + \frac{A}{\sqrt{2}} e^{\mathrm{i}(\omega t + \theta + \phi)}\mathbf{e}^{(-)},$$

where ϕ is the phase of $\underline{A}$. Thus the oscillation **u** is the superposition of a right-circularly polarized oscillation of amplitude $(A/\sqrt{2})$ and phase $\phi - \theta$, and a left-circularly polarized oscillation of amplitude $(A/\sqrt{2})$ and phase $\phi + \theta$.

2.3. Superposition of two perpendicular and non-isochronous oscillations

Consider two non-isochronous oscillations polarized in the Ox and Oy directions respectively

$$x(t) = a\cos(\omega_1 t + \phi_1)\ , \qquad y(t) = b\cos(\omega_2 t + \phi_2). \qquad [2.14]$$

The trajectory may be observed on the screen of an oscilloscope, for instance. We may write $y(t) = b\cos[\omega_1 t + (\Delta\omega\, t + \phi_2)]$, where $\Delta\omega = \omega_2 - \omega_1$. At a given time, the quantity $(\Delta\omega\ t + \phi_2)$ may be considered as the phase of $y(t)$. If the frequency difference is relatively small ($|\Delta\omega| << \omega_1$), the trajectory is approximately an ellipse, which is inscribed within the rectangle of sides $2a$ and $2b$ but it is not exactly closed. In the course of time, the phase shift $(\Delta\omega\ t + \phi_2 - \phi_1)$ varies and the ellipse changes slowly. If $\Delta\omega$ is comparable to ω_1, the deformation is rapid and we cannot really speak of an ellipse. In the special case of frequencies whose ratio is rational: $\omega_1/\omega_2 = T_2/T_1 = n_1/n_2$, where n_1 and n_2 are integers, x and y take back the same values after a time $T = n_1 T_1 = n_2 T_2$. Thus, the trajectory is a closed curve, which is described in a period T. The simpler ratio n_1/n_2 becomes, the simpler the curve becomes. The trajectories, called *Lissajous figures*, are inscribed within the rectangle of sides $x = \pm a$ and $y = \pm b$. They are tangent n_1 times to the sides $x = \pm a$ and n_2 times to the sides $y = \pm b$. Their form also depends on the phase shift $(\phi_2 - \phi_1)$. Some Lissajous figures are shown in Figure 2.4. They can be observed on the screen of an oscilloscope if x and y are electric potentials and may be used to compare the frequencies and phases of the x and y oscillations.

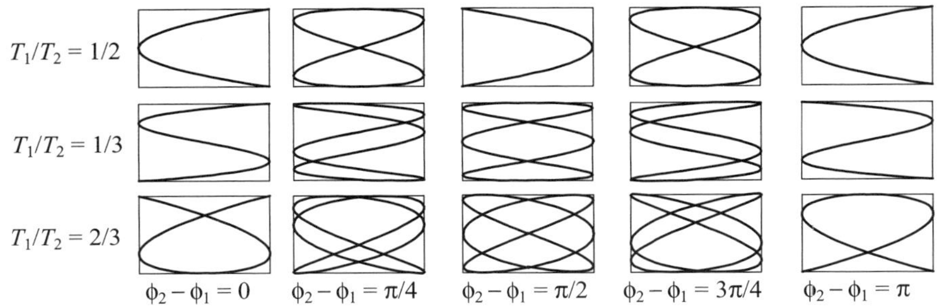

Figure 2.4. *Some Lissajous figures*

The superposition of the perpendicularly polarized simple harmonic oscillations $x(t)$ and $y(t)$ may effectively represent the trajectory of a particle in the Oxy plane if the particle is subject to a force of components $F_x = -K_x x$ and $F_y = -K_y y$. This is the case, for instance, for a small body attached to four springs (as in Figure 2.13, problem P2.4). Such an oscillator may model an atom in a crystal. In the special case of $K_x = K_y$, the force may be written in the vector form $\mathbf{F} = -K\mathbf{r}$ and the bidimensional oscillator is said to be *isotropic*. A point-like mass confined to move frictionlessly near the bottom of a spherical container is an example of a bidimensional system, which are approximately isotropic oscillators (for small oscillations). The trajectories are then illustrated in Figure 2.3. On the other hand, if $K_y/K_x = (n_2/n_1)^2$, then $T_1/T_2 = n_1/n_2$ and the trajectory is a Lissajous figure.

2.4. Superposition of scalar non-synchronous harmonic oscillations, beats

Consider the superposition of two scalar or collinear vector harmonic oscillations

$$u_1(t) = a_1 \cos(\omega_1 t + \phi_1), \qquad u_2(t) = a_2 \cos(\omega_2 t + \phi_2), \qquad [2.15]$$

where the frequencies ω_1 and ω_2 are different ($\omega_1 > \omega_2$). Let $\omega = ½(\omega_1 + \omega_2)$ be their average value and $\Delta\omega = ½\,(\omega_1 - \omega_2)$ their half-difference. Conversely, we may write $\omega_1 = \omega + \Delta\omega$ and $\omega_2 = \omega - \Delta\omega$. The resulting oscillation is

$$u = a_1 \cos(\omega t + \Delta\omega\, t + \phi_1) + a_2 \cos(\omega t - \Delta\omega\, t + \phi_2) = A(t)\cos[\omega t + \phi(t)]. \qquad [2.16]$$

Considering $(\Delta\omega\, t + \phi_1)$ and $(-\Delta\omega\, t + \phi_2)$ as phases and using relations [2.4] and [2.5], we may write:

$$A(t) = \sqrt{a_1^2 + a_2^2 + 2a_1 a_2 \cos(2\Delta\omega\, t + \phi_1 - \phi_2)}\,, \qquad [2.17]$$

$$\phi(t) = \frac{a_1 \sin(\Delta\omega\ t + \phi_1) - a_2 \sin(\Delta\omega\ t - \phi_2)}{a_1 \cos(\Delta\omega\ t + \phi_1) + a_2\ cos(\Delta\omega\ t - \phi_2)}. \quad [2.18]$$

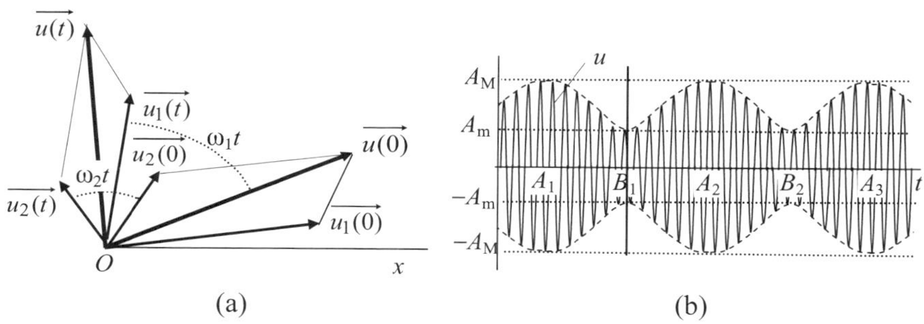

Figure 2.5. *Superposition of two scalar and non-isochronous harmonic oscillations*

We may also obtain this result by using the phasor representation (Figure 2.5a). The phasors $\overrightarrow{u_1(t)}$ and $\overrightarrow{u_2(t)}$, which represent the oscillations u_1 and u_2, have the lengths a_1 and a_2 and they rotate at the angular velocities ω_1 and ω_2. They form at time t a variable angle $(2\Delta\omega\ t + \phi_1 - \phi_2)$. By geometry, we find the amplitude $A(t)$ [2.17] and the phase $\phi(t)$ [2.18]. These are complicated functions of time and the resulting oscillation u is not a simple harmonic function and not even a periodic function in general (Figure 2.5b). However, the amplitude $A(t)$ has minimums $A_m = |a_1 - a_2|$ when $\overrightarrow{u_1(t)}$ and $\overrightarrow{u_2(t)}$ point in opposite directions and maximums $A_M = (a_1 + a_2)$ when $\overrightarrow{u_1(t)}$ and $\overrightarrow{u_2(t)}$ point in the same direction, that is, at $t_n = (2n\pi + \phi_2 - \phi_1)/2\ \Delta\omega$. The amplitude $A(t)$ varies slowly, compared to the function $\cos(\omega t + \phi)$, which oscillates rapidly. Although $A(t)$ is not a sinusoidal function, it has a period $t_{n+1} - t_n = \pi/\Delta\omega$. The function $u(t)$ may be considered as a sinusoidal oscillation of frequency ω with a slowly varying amplitude. We say that $u(t)$ is an *amplitude modulated* simple harmonic oscillation. The ratio $(A_M - A_m)/(A_M + A_m)$ is the *modulation factor* (or *depth of modulation*) and $T_{mod} = \pi/\Delta\omega$ is the period of modulation. The frequency of modulation is $\tilde{\nu}_{mod} = 1/T_{mod} = \tilde{\nu}_1 - \tilde{\nu}_2$.

In the special case of equal amplitudes A_1 and A_2 and nearly equal frequencies, we get

$$u(t) = A(t) \cos[\omega t + \phi(t)]$$
$$A(t) = 2a \cos(\Delta\omega\ t + \phi_1/2 - \phi_2/2),\ \ \phi(t) = ½(\phi_1 + \phi_2). \quad [2.19]$$

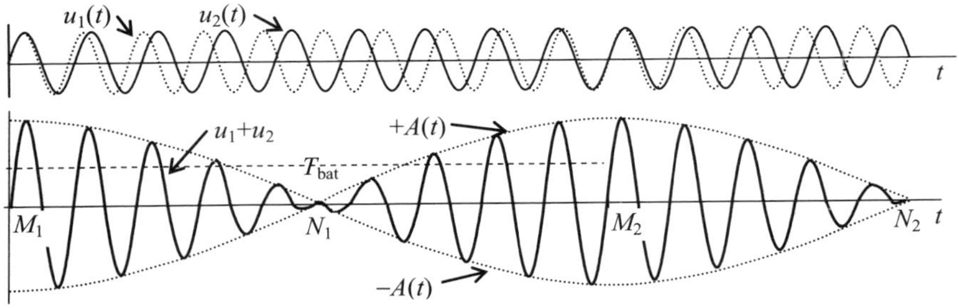

Figure 2.6. *Superposition of two oscillations of the same amplitude and nearly equal frequencies, beats*

The amplitude $A(t)$ is sinusoidal with an angular frequency $\Delta\omega$ (Figure 2.6). The maximums of $|A(t)|$ at M_1, M_2,... are equal to $2a$ and the minimums of $|A(t)|$ at N_1, N_2, ... are equal to 0. This rise and fall of the amplitude is referred to as *beats*. Two consecutive maximums are separated by equal time intervals $T_b = t_n - t_{n-1} = \pi/\Delta\omega$. In the case of sound, the ear is not sensitive to the sign of the oscillation but to the successive maximums of $|A|$ at M_1, M_2,... whose frequency is twice the frequency of $A(t)$. Thus, the beat frequency is

$$\tilde{\nu}_{beat} = 2\,\Delta\omega/\pi = \tilde{\nu}_1 - \tilde{\nu}_2 \,. \qquad [2.20]$$

Beats may be perceived if the playing of two musical instruments differs slightly in frequency. They are more difficult to observe in the case of light because of its high frequency (of the order of 5×10^{14} Hz). Certain light detectors (photoelectric cells, for instance) are only sensitive to light intensity which is proportional to u^2; they are said to be *quadratic detectors*. They enable the observation of light beats if the light waves have a relatively very small frequency difference.

The preceding result may be easily generalized to the superposition of several oscillations whose frequencies lie in a *frequency band* between ω_1 and ω_2. If the band is sufficiently narrow, the resulting oscillation may be considered to be a modulated simple harmonic function of angular frequency ω with slowly varying amplitude and phase.

2.5. Fourier analysis of a periodic function

Consider the superposition of two simple harmonic oscillations whose ratio of frequencies is equal to the ratio of two integers n_1 and n_2:

$$\tilde{\nu}_1 / \tilde{\nu}_2 = \omega_1/\omega_2 = n_1/n_2 \qquad \text{(i.e. } T_2/T_1 = n_1/n_2\text{)}. \qquad [2.21]$$

The oscillations u_1 and u_2 are both repeated after a period $T = n_1T_1 = n_2T_2$. Thus, their sum u is periodic with a period T. Figure 2.7 shows the superposition of two oscillations u_1 and u_2 in the case $T_1/T_2 = 3$. Their sum u is an oscillation which is not simple harmonic but periodic with a period $T = T_1$.

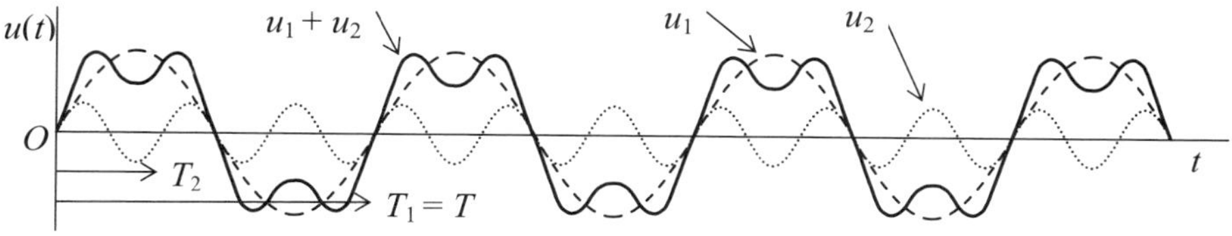

Figure 2.7. *Superposition of two simple harmonic oscillations such that* $T_1 = 3T_2$

The superposition of simple harmonic functions of periods T, $T/2$, $T/3$… is a function of period T. Conversely, any function of period T, may be considered as a superposition of simple harmonic functions of periods T, $T/2$, $T/3$… (i.e. of frequencies ω, 2ω, 3ω, etc.) with appropriate amplitudes and phases. In other words, any real function of period T may be written as a *Fourier series*

$$u(t) = \tfrac{1}{2}\, a_o + \textstyle\sum_{n\geq 1} A_n \cos(n\omega t + \phi_n), \qquad \text{where } \omega = 2\pi/T = 2\pi\,\tilde{\nu}\,. \qquad [2.22]$$

We may also write this series in the form

$$u(t) = \tfrac{1}{2}\, a_o + \textstyle\sum_{n\geq 1} [a_n \cos(n\omega t) + b_n \sin(n\omega t)] \qquad [2.23]$$

with the relations

$$a_n = A_n \cos\phi_n, \qquad b_n = -A_n \sin\phi_n, \qquad A_n = \sqrt{a_n^{\,2} + b_n^{\,2}}, \qquad \tan\phi_n = -b_n/a_n \quad [2.24]$$

The coefficients a_n and b_n in [2.23] are the *Fourier coefficients* for u and the A_n in [2.22] are the *spectral amplitudes* for u. The phases ϕ_n are chosen so that the A_n are positive. The time average of the sinusoidal functions over a period T being 0, we deduce that the average value of u is the constant term $\tfrac{1}{2}a_o$.

The determination of A_n and ϕ_n in expression [2.22] (or the coefficients a_n and b_n in [2.23]) is referred to as the *Fourier analysis* (also called *harmonic analysis* or *spectral analysis*) of the periodic function $u(t)$. The constant term $\tfrac{1}{2}a_o$ has a frequency 0 (infinite period). The $n = 1$ term (of the same period T as u) is the *fundamental* (or first *harmonic*) and the nth terms (of periods T/n) are the nth *harmonics*. In Figure 2.7, for instance, u_1 is the first harmonic and u_2 is the third harmonic of u. The frequencies, which enter the Fourier series of the function u, are the *spectral terms* of u. We say that the spectrum of a periodic function is *discrete*.

If we know a mathematical expression for u, it is more practical to write its Fourier series in the form of [2.23]. To evaluate the coefficients a_n, we multiply both sides of [2.23] by $\cos(p\omega t)$ (where p is an integer) and we integrate over a period and similarly for the b_n by multiplying by $\sin(p\omega t)$. Using the integrals:

$$\int_T dt\,\cos(n\omega t) = \int_T dt\,\sin(n\omega t) = 0, \qquad \int_T dt\,\cos(n\omega t)\,\sin(p\omega t) = 0,$$
$$\int_T dt\,\cos(n\omega t)\,\cos(p\omega t) = \int_T dt\,\cos(n\omega t)\,\cos(p\omega t) = (T/2)\,\delta_{np}, \qquad [2.25]$$

where $\int_T$ stands for the integral over a period T (for instance from $t_o - T/2$ to $t_o + T/2$ for any convenient time t_o) and the *Kronecker symbols* δ_{np} are equal to 1 if $n = p$ and equal to 0 if $n \neq p$, we deduce the expression of the Fourier coefficients

$$a_p = \frac{2}{T}\int_T dt\,\cos(p\omega t)\,u(t),\; b_p = \frac{2}{T}\int_T dt\,\sin(p\omega t)\,u(t). \qquad [2.26]$$

Instead of the *unilateral Fourier series* [2.23], it is possible to write the *bilateral* form

$$u(t) = \tfrac{1}{2}\,\Sigma_{-\infty<n<\infty}\,[a_p\cos(p\omega t) + b_p\sin(p\omega t)], \qquad [2.27]$$

where the coefficients a_p and b_p for negative p are defined by the relations

$$a_{-p} = a_p \qquad \text{and} \qquad b_{-p} = -b_p. \qquad [2.28]$$

Negative frequencies have no direct physical significance but expression [2.27] is sometimes more convenient than [2.23]. The factor ½ takes into account that each harmonic is counted twice ($p\omega$ and $-p\omega$) with equal or opposite coefficients rather than the unilateral Fourier series, where it appears once.

In the case of a periodic complex function $\underline{u}(t)$, we write its real and imaginary parts as Fourier series:

$$\mathcal{Re}\,\underline{u}(t) = \tfrac{1}{2}\,\Sigma_{-\infty<p<\infty}\,[a_p\cos(p\omega t) + b_p\sin(p\omega t)],$$
$$\mathcal{Im}\,\underline{u}(t) = \tfrac{1}{2}\Sigma_{-\infty<n<\infty}\,[a'_p\cos(p\omega t) + b'_p\sin(p\omega t)]$$

and combine these series into a single one with complex coefficients

$$\underline{u}(t) = \Sigma_{-\infty<n<\infty}\, \underline{\mathcal{U}}_{\mathrm{p}}\, e^{\mathrm{i}p\omega t}\,, \qquad [2.29]$$

where the complex coefficients $\underline{\mathcal{U}}_{\mathrm{p}}$ are related to a_{p}, b_{p}, a'_{p} and b'_{p} by the equation

$$\underline{\mathcal{U}}_{\mathrm{p}} = \frac{1}{2}(a_{\mathrm{p}} + b'_{\mathrm{p}}) + \frac{\mathrm{i}}{2}(a'_{\mathrm{p}} - b_{\mathrm{p}}), \qquad [2.30]$$

which is valid for any p. Using expressions [2.26] and [2.30], we deduce that the coefficients $\underline{\mathcal{U}}_{\mathrm{p}}$ are given by the integrals

$$\underline{\mathcal{U}}_{\mathrm{p}} = \frac{1}{T}\int_{\mathrm{T}} dt\; e^{-\mathrm{i}p\omega t}\, \underline{u}(t). \qquad [2.31]$$

If $u(t)$ is real, the complex coefficients $\underline{\mathcal{U}}_{\mathrm{p}}$ verify the symmetry relation

$$\underline{\mathcal{U}}_{\mathrm{p}} = \underline{\mathcal{U}}_{-\mathrm{p}}{}^{*} = \tfrac{1}{2}(a_{\mathrm{p}} - \mathrm{i}b_{\mathrm{p}}) \qquad [2.32]$$

or conversely,

$$a_{\mathrm{o}} = 2\,\underline{\mathcal{U}}_{\mathrm{o}} \quad a_{\mathrm{p}} = 2\mathcal{R}e\,\underline{\mathcal{U}}_{\mathrm{p}}, \quad b_{\mathrm{p}} = -2\mathcal{I}m\,\underline{\mathcal{U}}_{\mathrm{p}}, \quad A_{\mathrm{p}} = 2|\underline{\mathcal{U}}_{\mathrm{p}}| \quad \text{and } \tan\phi_{\mathrm{p}} = \frac{\mathcal{I}m\,\underline{\mathcal{U}}_{\mathrm{p}}}{\mathcal{R}e\,\underline{\mathcal{U}}_{\mathrm{p}}}. \qquad [2.33]$$

The complex exponentials $e^{\mathrm{i}p\omega t}$ are much easier to handle than the trigonometric functions $\cos(p\omega t)$ and $\sin(p\omega t)$. Thus, in the case of a complicated analysis, it is easier to use the complex Fourier series [2.29] and to take, in the end, its real part to represent physical quantities. Some Fourier coefficients may be equal to 0 because of the symmetry of the function u. If $u(t)$ is even, $u(-t) = u(t)$, its Fourier series [2.23] cannot include the terms $\sin(p\omega t)$ which are odd functions of time. Similarly, if $u(t)$ is odd, $u(-t) = -u(t)$, its Fourier series cannot include the even terms $\cos(p\omega t)$. Note also that the Fourier series for the time derivative and the primitive of a function $u(t)$ are not necessarily the derivative and the primitive of its Fourier series.

The Fourier analysis is very useful and it has many applications in acoustics, optics, treatment of signals, telecommunications, etc. Spectroscopy is somehow a Fourier analysis of the wave. It may be used, in particular, to determine the chemical composition of a body, even if it is a very distant star.

EXAMPLE 2.4.– Write the Fourier series for the function $u(t)$ which consists of rectangular pulses of duration τ and is repeated with a period T (Figure 2.8a).

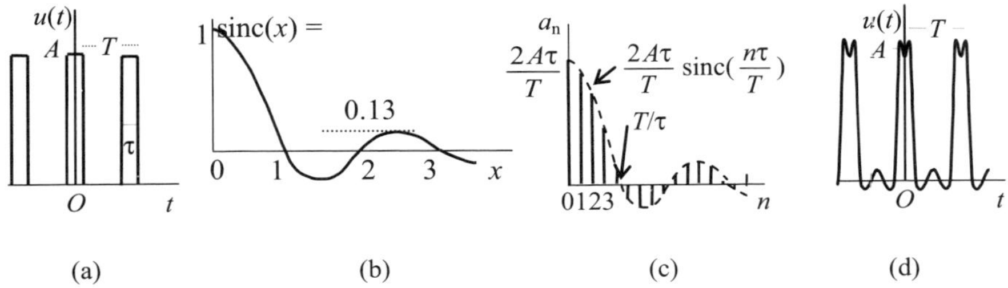

Figure 2.8. *a) Series of rectangular pulses; b) the function sinc(x); c) the Fourier coefficients a_n; d) sum of the first three terms of the Fourier series*

SOLUTION– Taking the origin of time at the mid-point of one of the pulses, the function $u(t)$ is even in t (Figure 2.8a). Thus, all the coefficients b_n are equal to 0. Using [2.26], integrating from $-T/2$ to $+T/2$ and setting $\omega = 2\pi/T$, we obtain

$$a_n = (2/T) \int_{-T/2}^{T/2} dt\ \cos(n\omega t)\, u(t) = (4/T) \int_0^{T/2} dt\ \cos(n\omega t)\, u(t)$$
$$= (4A/T) \int_0^{\tau/2} dt\ \cos(n\omega t) = (2A/n\pi)\sin(n\pi\tau/T).$$

This expression is also valid for a_o by taking the limit $n \to 0$. Thus,

$$a_o = \frac{2A\tau}{T}, \qquad a_n = \frac{2A\tau}{T}\,\mathrm{sinc}(\frac{n\tau}{T}), \qquad b_n = 0, \qquad [2.34]$$

where we have introduced the function

$$\mathrm{sinc}(x) = \frac{\sin(\pi x)}{\pi x} \qquad [2.35]$$

called the *sine cardinal* of x and shown in Figure 2.8b. It is an even function of x with zeros at $x = \pm 1, \pm 2, \ldots$ It has a principal maximum equal to 1 at $x = 0$ and secondary extremums $\cong \pm 1/\pi(n + ½)$ at $x \cong \pm(n + ½)$, where $n = 1, 2, \ldots$ Thus, we may write

$$u(t) = (2A\tau/T)[½ + \Sigma_{n\geq 1}\ \mathrm{sinc}(n\tau/T)\cos(2n\pi t/T)]. \qquad [2.36]$$

The constant term $A\tau/T$ is the average value of $u(t)$. Figure 2.8c shows the Fourier coefficients a_n. In this case, the spectral amplitudes A_n are equal to the coefficients a_n. Figure 2.8d is a plot of the sum of the first three terms of the Fourier series. We note that, if the pulses are not symmetric with respect to the origin O, the Fourier series will contain both $\cos(2n\pi t/T)$ and $\sin(2n\pi t/T)$ terms.

We may also write $u(t)$ as a bilateral series [2.27], using the functions $\cos(n\omega t)$ or the functions $e^{in\omega t}$

$$u(t) = (A\tau/T)\ \Sigma_{-\infty<n<\infty}\ \mathrm{sinc}(n\tau/T)\cos(2n\pi t/T) \qquad [2.37]$$

$$u(t) = (A\tau/T)\ \Sigma_{-\infty<n<\infty}\ \mathrm{sinc}(n\tau/T)\ e^{2in\pi t/T}\ . \qquad [2.38]$$

2.6. Fourier analysis of a non-periodic function

The Fourier analysis may be generalized to non-periodic functions $u(t)$ if they satisfy certain mathematical conditions, such as the convergence of the integral of $|u(t)|$ from $-\infty$ to $+\infty$. For such functions, the Fourier series is replaced by an integral over continuous values of the frequency (*continuous spectrum*).

In the case of a real and non-periodic function of one variable t, the unilateral Fourier series (written in the form of [2.22] or [2.23]) must be replaced by the *unilateral Fourier integral*

$$u(t) = \int_{\Delta\tilde{\nu}} d\tilde{\nu}\ [\,a(\tilde{\nu})\cos(\omega t) + b(\tilde{\nu})\sin(\omega t)] = \int_{\Delta\tilde{\nu}} d\tilde{\nu}\ A(\tilde{\nu})\cos[\omega t + \phi(\tilde{\nu})\,], \qquad [2.39]$$

where $\omega = 2\pi\tilde{\nu}$. The functions $a(\tilde{\nu})$ and $b(\tilde{\nu})$ are, respectively, the *cosine part* and the *sine part* of the spectral function. $A(\tilde{\nu})$ is the *spectral amplitude* and $\phi(\tilde{\nu})$ is the *spectral phase* with the following relationships to $a(\tilde{\nu})$ and $b(\tilde{\nu})$

$$A(\tilde{\nu}) = \sqrt{a(\tilde{\nu})^2 + b(\tilde{\nu})^2} \qquad \text{and} \qquad \tan\phi(\tilde{\nu}) = -\,b(\tilde{\nu})\,/\,a(\tilde{\nu})\,. \qquad [2.40]$$

The integration is over a frequency band $\Delta\tilde{\nu} \equiv [\,\tilde{\nu}_1\,,\tilde{\nu}_2\,]$ which may be formally extended to all frequencies $[0, \infty]$ by taking $a(\tilde{\nu})$ and $b(\tilde{\nu})$ to be equal to 0 outside the band $\Delta\tilde{\nu}$. We may also define a complex spectral function

$$\underline{\mathcal{U}}(\tilde{\nu}) = \tfrac{1}{2}\,[\,a(\tilde{\nu}) - \mathrm{i}\,b(\tilde{\nu})\,] \qquad [2.41]$$

and write

$$u(t) = \int_0^\infty d\tilde{\nu}\ [\,a(\tilde{\nu})\ \cos(\omega t) + b(\tilde{\nu})\sin(\omega t)] = 2\,\mathcal{R}e \int_0^\infty d\tilde{\nu}\ \underline{\mathcal{U}}(\tilde{\nu})\ e^{\mathrm{i}\omega t}\,. \qquad [2.42]$$

In some cases, it is suitable to extend the integration to negative values $\tilde{\nu}$ and to define the spectral function for $\tilde{\nu} < 0$ with the symmetry relation

$$\underline{\mathcal{U}}(-\tilde{\nu}) = \underline{\mathcal{U}}(\tilde{\nu})^* \qquad [2.43]$$

$$\Rightarrow a(-\tilde{\nu}) = a(\tilde{\nu}) \quad \text{and} \quad b(-\tilde{\nu}) = -\,b(\tilde{\nu}) \qquad [\text{if } u(t) \text{ is real}]. \qquad [2.44]$$

Thus, if $u(t)$ is real, the real part of its spectral function $\underline{\mathcal{U}}(\tilde{\nu})$ is even in $\tilde{\nu}$, while its imaginary part is odd. This allows us to write

$$u(t) = \int_{-\infty}^{\infty} d\tilde{\nu}\ \underline{\mathcal{U}}(\tilde{\nu})\ e^{\mathrm{i}\omega t}. \qquad [2.45]$$

Conversely, the spectral function $\underline{\mathcal{U}}(\tilde{\nu})$ is related to $u(t)$ by the equation

$$\underline{\mathcal{U}}(\tilde{\nu}) = \int_{-\infty}^{\infty} dt\ u(t)\ e^{-\mathrm{i}\omega t}. \qquad [2.46]$$

Taking the real part and the imaginary part of this relation, we obtain

$$a(\tilde{\nu}) = 2\int_{-\infty}^{\infty} dt\ u(t)\cos(\omega t), \qquad b(\tilde{\nu}) = 2\int_{-\infty}^{\infty} dt\ u(t)\sin(\omega t)\ [\text{real } u(t)]. \qquad [2.47]$$

Relationships [2.45] and [2.46] may be generalized to complex functions

$$\underline{u}(t) = \int_{-\infty}^{\infty} d\tilde{\nu}\ \underline{\mathcal{U}}(\tilde{\nu})\ e^{\mathrm{i}\omega t} \quad \text{and} \quad \underline{\mathcal{U}}(\tilde{\nu}) = \int_{-\infty}^{\infty} dt\ \underline{u}(t)\ e^{-\mathrm{i}\omega t}\ [\text{complex } u(t)]. \qquad [2.48]$$

The function $\underline{u}(t)$ is the *Fourier transform* of $\underline{\mathcal{U}}(\tilde{\nu})$, and $\underline{\mathcal{U}}(\tilde{\nu})$ is the *inverse Fourier transform* of $\underline{u}(t)$. Note that, in general, the spectral function $\underline{\mathcal{U}}(\tilde{\nu})$ of a complex function $\underline{u}(t)$ does not verify symmetry relationship [2.43].

In some cases, it is suitable to write the Fourier transformation using the angular frequency ω instead of the frequency $\tilde{\nu}$; then, we have

$$u(t) = \frac{1}{2\pi}\int_{-\infty}^{\infty} d\omega\ \underline{\mathcal{U}}(\omega)\ e^{\mathrm{i}\omega t} \quad \text{and} \quad \underline{\mathcal{U}}(\omega) = \int_{-\infty}^{\infty} dt\ u(t)\ e^{-\mathrm{i}\omega t}. \qquad [2.49]$$

Setting $\underline{\hat{u}}(\omega) = \underline{\mathcal{U}}(\tilde{\nu})/\sqrt{2\pi}$, we may cast [2.49] in a symmetric form

$$\underline{u}(t) = \frac{1}{\sqrt{2\pi}}\int_{-\infty}^{\infty} d\omega\ \ \underline{\hat{u}}(\omega)\ e^{\mathrm{i}\omega t}, \qquad \underline{\hat{u}}(\omega) = \frac{1}{\sqrt{2\pi}}\int_{-\infty}^{\infty} dt\ u(t)\ e^{-\mathrm{i}\omega t}. \qquad [2.50]$$

The Fourier transformation may be easily generalized to functions of several variables. For instance, any function $\underline{u}(\mathbf{r})$, where $\mathbf{r}$ is the position of a point of coordinates x, y and z, may be written as the Fourier transform of a function $\underline{\hat{u}}(\mathbf{p})$ and conversely

$$\underline{u}(\mathbf{r}) = (2\pi)^{-3/2}\iiint d^3\mathbf{p}\ \hat{u}(\mathbf{p})\ e^{\mathrm{i}\mathbf{p.r}}, \qquad \hat{u}(\mathbf{p}) = (2\pi)^{-3/2}\iiint d^3\mathbf{r}\ \underline{u}(\mathbf{r})\ e^{-\mathrm{i}\mathbf{p.r}}, \qquad [2.51]$$

where the integrals are over the whole space and we have used symmetric form [2.50] for the three pairs of *conjugate variables* (x, p_x), (y, p_y) and (z, p_z). We note that (x, y, z) may not necessarily be the coordinates of the ordinary space and the number n of these coordinates may be arbitrary; then, we have factors $(2\pi)^{-n/2}$.

2.7. Fourier analysis of a signal, uncertainty relation

A signal is a function, which is negligible but in a narrow interval of time Δt near t_o. The important contribution to the integrals of [2.47], which determine $a(\tilde{\nu})$ and $b(\tilde{\nu})$, comes from $u(t)$ in this interval of time. However; if $\tilde{\nu}$ is such that $\sin(2\pi\tilde{\nu}t)$ and $\cos(2\pi\tilde{\nu}t)$ have many oscillations, as t varies in Δt, the successive positive and negative values of u make the functions $a(\tilde{\nu})$ and $b(\tilde{\nu})$ very small. Thus, these spectral coefficients are important for the frequencies $\tilde{\nu}$, such that the variation of $2\pi\tilde{\nu}t$ is very small when t varies in the interval Δt ($2\pi\tilde{\nu}t << 2\pi$, for instance). This means that the signal is a superposition of simple harmonic oscillations of frequencies varying in a band of width $\Delta\tilde{\nu}$ such that $\Delta\tilde{\nu}\ \Delta t \approx 1$ near a certain frequency $\tilde{\nu}_o$.

Consider, for instance, a *rectangular signal* or *pulse* of width Δt, height u_o and centered at t_o, that is, $u_{sig}(t) = u_o$ for $t_o - ½\,\Delta t < t < t_o + ½\,\Delta t$ (Figure 2.9a). The relationships in [2.47] give the spectral functions

$$a(\tilde{\nu}) = 2\int_{-\infty}^{\infty} dt\ u(t)\cos(\omega t) = 2\,u_o \int_{t_o-\Delta t/2}^{t_o+\Delta t/2} dt\ \cos(\omega t) = F(\tilde{\nu})\ \cos(\omega t_o),$$

$$b(\tilde{\nu}) = 2\int_{-\infty}^{\infty} dt\ u(t)\ \sin(\omega t) = 2u_o \int_{t_o-\Delta t/2}^{t_o+\Delta t/2} dt\ \sin(\omega t) = F(\tilde{\nu})\ \sin(\omega t_o) \qquad [2.52]$$

$$\underline{\mathcal{U}}(\tilde{\nu}) = \int_{-\infty}^{\infty} dt\ u(t)\, e^{-i\omega t} = ½[\,a(\tilde{\nu}) - i\,b(\tilde{\nu})\,] = ½\,F(\tilde{\nu})\ e^{-i\omega t_o}, \qquad [2.53]$$

where we have set $F(\tilde{\nu}) = 2u_o\Delta t\ \mathrm{sinc}\,(\tilde{\nu}\Delta t) = (2u_o/\pi\tilde{\nu})\ \sin(\pi\tilde{\nu}\,\Delta t)$. The functions $\cos(2\pi\tilde{\nu}\ t_o)$, $\sin(2\pi\tilde{\nu}\ t_o)$ and $e^{-2i\pi\nu t_o}$ oscillate rapidly as functions of $\tilde{\nu}$, while the factor $F(\tilde{\nu})$ oscillates slowly with an amplitude which decreases like $1/\tilde{\nu}$. The functions $a(\tilde{\nu})$ and $b(\tilde{\nu})$ are represented in Figures 2.9b and 2.9c. They lie between the curves $\pm F(\tilde{\nu})$. The function $F(\tilde{\nu})$ has a principal maximum equal to $(2u_o\,\Delta t)$ for $\tilde{\nu} = 0$ and it vanishes for $\tilde{\nu} = n/\Delta t$ (where n is an integer that is different from 0). The important part of $a(\tilde{\nu})$ and $b(\tilde{\nu})$ lies between the first zeros $\tilde{\nu} = \pm 1/\Delta t$ of $F(\tilde{\nu})$ situated before and after the origin. We may consider $\Delta\tilde{\nu} \approx 1/\Delta t$ as the bandwidth of the signal. It verifies, with the time duration Δt of the signal, the *uncertainty relation*

$$\Delta\tilde{\nu}\,.\Delta t \approx 1 \qquad \text{or} \qquad \Delta\omega.\Delta t \approx 2\pi. \qquad [2.54]$$

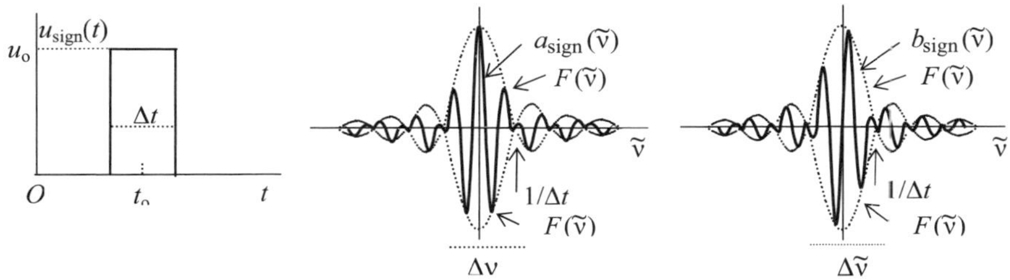

(a) (b) (c)

Figure 2.9. *Fourier analysis of a rectangular signal*

Conversely, consider the superposition of oscillations of frequencies $\tilde{\nu}$ in a band of width $\Delta\tilde{\nu}$ centered on $\tilde{\nu}_o$ with the spectral functions $b(\tilde{\nu}) = 0$ and $a(\tilde{\nu}) = a_o$ for $\tilde{\nu}_o - \frac{1}{2}\Delta\tilde{\nu} < \tilde{\nu} < \tilde{\nu}_o + \frac{1}{2}\Delta\tilde{\nu}$, where a_o is a constant (Figure 2.10a). Equation [2.39] gives the corresponding signal

$$u(t) = \int_{\tilde{\nu}_o - \Delta\tilde{\nu}/2}^{\tilde{\nu}_o + \Delta\tilde{\nu}/2} d\tilde{\nu} \;\; a_o \cos(2\pi\,\tilde{\nu}\, t) = \Delta\tilde{\nu}\, a_o\, \text{sinc}(\Delta\tilde{\nu}\, t) \cos(2\pi\,\tilde{\nu}_o\, t). \qquad [2.55]$$

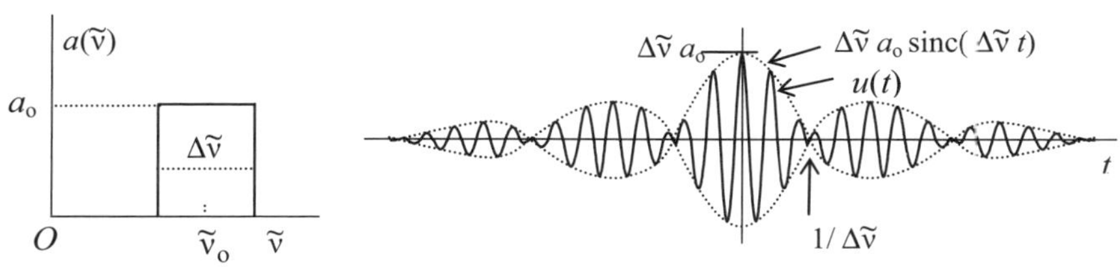

(a) (b)

Figure 2.10. *a) Spectral function* $a(\tilde{\nu}) = a_o$ *constant over* $\Delta\tilde{\nu}$ *; and b) the function u(t)*

The same result may be obtained by using [2.45] with the complex spectral function $\underline{\mathcal{U}}(\tilde{\nu}) = \frac{1}{2}a_o$ over the band $[\tilde{\nu}_o - \frac{1}{2}\Delta\tilde{\nu}\,,\ \tilde{\nu}_o + \frac{1}{2}\Delta\tilde{\nu}\,]$ and the symmetric band $[-\tilde{\nu}_o - \frac{1}{2}\Delta\tilde{\nu}\,,\ -\tilde{\nu}_o + \frac{1}{2}\Delta\tilde{\nu}\,]$, due to relationship [2.44]. Figure 2.10b illustrates the function $u(t)$ in the case $\Delta\tilde{\nu} \ll \tilde{\nu}_o$. Because of the factor $\cos(\tilde{\nu}_o t)$, the function $u(t)$ oscillates rapidly with an amplitude

$$u_m = \Delta\tilde{\nu}\, a_o\, \text{sinc}(\Delta\tilde{\nu}\, t) = (a_o/\pi t)\, |\sin(\pi\, \Delta\tilde{\nu}\, t)|, \qquad [2.56]$$

which decreases in time like $1/t$ and oscillates slowly with a frequency $\frac{1}{2}\Delta\tilde{\nu}$. It is important for $-1/\Delta\tilde{\nu} < t < 1/\Delta\tilde{\nu}$, i.e. between the first zeros of $\sin(\pi\,\Delta\tilde{\nu}\,t)$. Thus, u is a signal of duration $\Delta t \approx 1/\Delta\tilde{\nu}$. This is an example of uncertainty relationship [2.54]. In order to have a narrow signal, the uncertainty relation implies that the frequency band must be wide. Conversely, in order to have a narrow band of frequency, the duration Δt of the signal must be large. In particular, in order to have $\Delta\tilde{\nu} = 0$, i.e. the ideal case of a monochromatic wave, the time duration Δt must be infinite. We shall see that it is also infinite in space: the wave must be permanent and occupy the entire space. This requires infinite energy; thus, exact simple harmonic oscillations and waves are mathematical concepts rather than physical realities.

2.8. Dirac delta-function

A rectangular pulse $u_\tau(t)$ (Figure 2.11b) of width $\Delta t = 2\tau$ and area under the curve equal to 1 may be written as

$$\begin{aligned} &u_\tau(t) = 1/2\tau && \text{if } t_o - \tau < t < t_o + \tau \\ &u_\tau(t) = 0 && \text{if } t < t_o - \tau \quad \text{or } t > t_o + \tau. \end{aligned} \qquad [2.57]$$

In the limit $\tau \to 0$, the signal becomes very narrow and $u(t_o) \to \infty$. This is a representation of the *Dirac delta-function* $\delta(t - t_o)$, centered on t_o and such that

$$\begin{aligned} &\delta(t - t_o) = \infty \quad \text{if } t = t_o \qquad \text{and} \qquad \delta(t - t_o) = 0 \quad \text{if } t \neq t_o, \\ &\int_{-\infty}^{\infty} dt\ \delta(t - t_o) = 1. \end{aligned} \qquad [2.58]$$

Consider a function $f(t)$, which is regular at t_o. Let $F(t)$ be a primitive of $f(t)$. The integral of the product $f(t)\,u_\tau(t)$ in the limit $\tau \to 0$ is given by

$$\lim_{\tau\to\infty} \int_{-\infty}^{\infty} dt\ u_\tau(t)\,f(t) = \lim_{\tau\to\infty} \int_{t_o-\tau}^{t_o+\tau} dt\ \frac{f(t_o)}{2\tau} = \lim_{\tau\to 0} \frac{1}{2\tau}\left[F(t_o+\tau) - F(t_o-\tau)\right] = \left.\frac{dF}{dt}\right|_{t_o} = f(t_o)$$

Thus, the integral of $u_\tau(t)\,f(t)$ is simply $f(t_o)$. Note that the integral over any interval $[t_1, t_2]$ gives the same result provided that t_o is within this interval. In the limit $\tau \to 0$, $u_\tau(t)$ becomes the Dirac delta-function and we may write the useful relationship

$$\int_{-\infty}^{\infty} dt\ \delta(t - t_o)\,f(t) = f(t_o). \qquad [2.59]$$

The inverse Fourier transform of $\delta(t - t_o)$ is

$$\underline{\mathcal{U}}(\tilde{\nu}) = \int_{-\infty}^{\infty} dt\ \delta(t - t_o)\ e^{-i\omega t} = e^{-i\omega t_o}\ , \qquad [2.60]$$

where we used [2.59]. The inverse transformation is

$$\delta(t - t_o) = \int_{-\infty}^{\infty} d\tilde{\nu}\ e^{2i\pi\tilde{\nu}(t-t_o)} = \frac{1}{2\pi}\int_{-\infty}^{\infty} d\omega\ e^{i\omega(t-t_o)}\ . \qquad [2.61]$$

The spectral width of the Dirac delta-function is infinite. Conversely, the Fourier transform of $\delta(\tilde{\nu} - \tilde{\nu}_o)$ is

$$\underline{u}(t) = \int_{-\infty}^{\infty} d\tilde{\nu}\ \delta(\tilde{\nu} - \tilde{\nu}_o)\, e^{i\omega t} = e^{i\omega_o t} \quad (\omega_o = 2\pi\,\tilde{\nu}_o). \qquad [2.62]$$

This is a monochromatic simple harmonic function of frequency $\tilde{\nu}_o$. Thus, it has a spectral width $\Delta\tilde{\nu} = 0$ and an infinite time duration Δt.

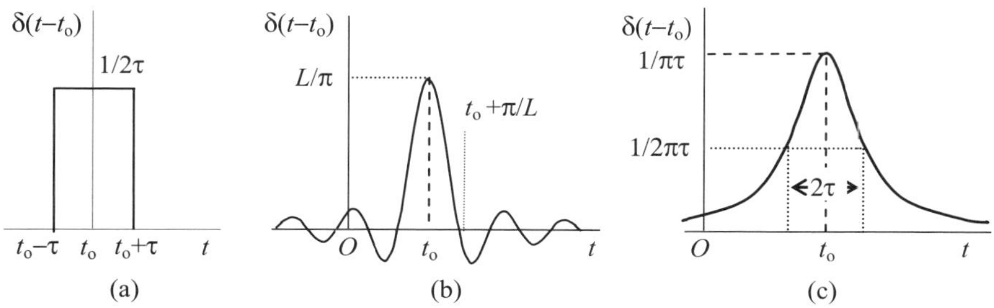

Figure 2.11. *Three representations of the Dirac delta-function $\delta(t - t_o)$: a) the function [2.57], b) the function [2.63] and c) the function [2.64]. The function δ is the limit of the functions (a) and (c) if $\tau \to 0$, and it is the limit of the function (b) if $L \to \infty$*

The integral in equation [2.61] is not finite. Thus, $\delta(t - t_o)$ is not a function in the usual sense; it is infinite for $t = t_o$ and equal to 0 for any other value of t. It is called a *distribution* and it is not well defined except under integrals in the form of [2.59], which may be used as a general definition of the delta-function. It may be considered as the limit of some functions such that

$$\delta(t - t_o) = \lim_{L\to\infty} \frac{\sin[L(t - t_o)]}{\pi(t - t_o)} = \lim_{L\to\infty} \frac{L}{\pi}\ \mathrm{sinc}[\frac{L}{\pi}(t - t_o)] \qquad [2.63]$$

$$\delta(t - t_o) = \lim_{\tau\to 0} \frac{1}{\pi}\frac{\tau}{\tau^2 + (t - t_o)^2}\ . \qquad [2.64]$$

These functions are illustrated in Figures 2.11b and 2.11d. Expression [2.63] is obtained by calculating [2.61] as the limit of the integral from $-L$ to $+L$.

The Dirac delta-function satisfies the following properties that may be shown by making appropriate changes to integration variables:

$$\delta(t) = \delta(-t) \quad [2.65]$$
$$\delta(at) = \delta(t)/|a| \quad [2.66]$$
$$\delta(t^2 - a^2) = [\,\delta(t-a) + \delta(t+a)]/2|a| \quad [2.67]$$
$$\delta[f(t)] = \Sigma_i\, \delta(t - t_i)/|f'(t_i)|, \quad [2.68]$$

where the t_i are the zeros of the equation $f(t) = 0$ and f' is the derivative of f.

The multi-dimensional Dirac delta-function is a simple generalization of the one-dimensional delta-function. For instance, if $\mathbf{r}_o$ is the position of coordinates x_o, y_o and z_o, the three-dimensional Dirac delta-function may be defined as

$$\delta(\mathbf{r} - \mathbf{r}_o) = \delta(x - x_o)\, \delta(y - y_o)\, \delta(z - z_o). \quad [2.69]$$

It has the following properties:

$$\iiint d^3\mathbf{r}\, \delta(\mathbf{r} - \mathbf{r}_o) = 1 \quad [2.70]$$
$$\iiint d^3\mathbf{r}\, f(\mathbf{r})\, \delta(\mathbf{r} - \mathbf{r}_o) = f(\mathbf{r}_o) \quad [2.71]$$
$$\delta(\mathbf{r} - \mathbf{r}_o) = (2\pi)^{-3} \iiint d^3\mathbf{p}\; e^{i\mathbf{p}.\mathbf{r}} \quad [2.72]$$

We note that [2.59] reduces to [2.58] if $f(t) = 1$ and [2.71] reduces to [2.70] if $f(\mathbf{r}) = 1$. Very often, an explicit form of the Dirac delta-function is not required. All that is needed is its property [2.59] or [2.71]. Relationships [2.59] and [2.58] are valid if the domain of integration is finite and contains the point t_o. Similarly, relationships [2.70] and [2.71] are valid if the domain of integration is finite but contains the point $\mathbf{r}_o$. If t_o or $\mathbf{r}_o$ are outside the domain of integration, the right-hand side of [2.58], [2.59], [2.70] and [2.71] must be replaced by 0.

2.9. Summary

The superposition of two scalar harmonic oscillations of equal frequencies is

$$u = a_1 \cos(\omega t + \phi_1) + a_2 \cos(\omega t + \phi_2) = A_1 \cos \omega t - A_2 \sin \omega t = A \cos(\omega t + \phi)$$

with the relationships: $A_1 = a_1 \cos \phi_1 + a_2 \cos \phi_2$ and $A_2 = a_1 \sin \phi_1 + a_2 \sin \phi_2$.

$A = \sqrt{A_1^2 + A_2^2} = \sqrt{a_1^2 + a_2^2 + 2a_1 a_2 \cos(\phi_1 - \phi_2)}$, $\cos\phi = A_1/A$ and $\sin\phi = A_2/A$.

– If u_1 and u_2 are *in phase* ($\phi_1 = \phi_2 + 2n\pi$), $A = a_1 + a_2$.

– If u_1 and u_2 are *in opposite phase* ($\phi_1 = \phi_2 + \pi + 2n\pi$), $A = |a_1 - a_2|$.

– If u_1 and u_2 are *in quadrature* ($\phi_1 = \phi_2 \pm \pi/2 + 2n\pi$), $A = (a_1^2 + a_2^2)^{½}$.

The superposition of two vector harmonic oscillations of equal frequencies and which are perpendicularly polarized is $\mathbf{u} = a\cos(\omega t + \phi_1)\,\mathbf{e}_x + b\cos(\omega t + \phi_2)\,\mathbf{e}_y$. The trajectory of M such that $\overrightarrow{OM} = \mathbf{u}$ is inscribed within the rectangle $-a < x < a$ and $-b < y < b$. Its equation is: $x^2/a^2 + y^2/b^2 - 2xy/ab\cos(\phi_1 - \phi_2) = \sin^2(\phi_1 - \phi_2)$.

– If $\phi_2 - \phi_1 = 2n\pi$ (oscillations in phase), the resulting oscillation is *polarized linearly* in the direction of the first diagonal of the rectangle.

– If $\phi_2 - \phi_1 = \pi + 2n\pi$ (oscillations in phase opposition), the resulting oscillation is *linearly polarized* in the direction of the second diagonal of the rectangle.

– If $a = b$ and $\phi_2 - \phi_1 = \pm\pi/2$ (oscillations in quadrature), the resulting oscillation $\mathbf{u}^{(\pm)}$ is *right* or *left-circularly polarized* respectively:

$$\mathbf{u}^{(\pm)} = a[\cos(\omega t + \phi_1)\,\mathbf{e}_x + \cos(\omega t + \phi_1 \pm \pi/2)\,\mathbf{e}_y].$$

– In the case of arbitrary phase shift, the resulting oscillation is *right-elliptically polarized* if $0 < \phi_2 - \phi_1 < \pi$ and *left-elliptically polarized* if $\pi < \phi_2 - \phi_1 < 2\pi$.

If $x(t) = a\cos(\omega_1 t + \phi_1)$ and $y(t) = b\cos(\omega_2 t + \phi_2)$ with $\omega_1 \neq \omega_2$, the trajectory is a closed curve, if $\omega_2/\omega_1 = n_2/n_1$ (i.e. $T_1/T_2 = n_2/n_1$), where n_1 and n_2 are integers. The motion on this trajectory (called the *Lissajous* figure) has a period of $T = n_1 T_1 = n_2 T_2$.

The resulting oscillation $u = a_1\cos(\omega_1 t + \phi_1) + a_2\cos(\omega_2 t + \phi_2)$ of two scalar simple harmonic oscillations of nearly equal frequencies is an oscillation of frequency $\omega = ½(\omega_1 + \omega_2)$ with a modulating frequency $\Delta\omega = ½(\omega_1 - \omega_2)$. If the amplitudes are also nearly equal, we have *beats* of frequency $\tilde{\nu}_{bat} = \tilde{\nu}_1 - \tilde{\nu}_2$.

A periodic function may be written as a *Fourier series*

$$u(t) = ½\, a_o + \Sigma_{n\geq 1} A_n \cos(n\omega t + \phi_n) = ½\, a_o + \Sigma_{n\geq 1} [a_n \cos(n\omega t) + b_n \sin(n\omega t)],$$

where $\omega = 2\pi/T$. The term of frequency $n\omega$ is the nth *harmonic*. The frequencies $n\omega$ constitute the *spectrum* of u. a_n and b_n are the *Fourier coefficients* and A_n are the *spectral amplitudes* of u. The Fourier coefficients are given by the expressions

$$a_p = (2/T) \int_T dt \cos(p\omega t)\, u(t) \quad \text{and} \quad b_p = (2/T) \int_T dt \sin(p\omega t)\, u(t)\,.$$

We may also write the *bilateral* form

$$u(t) = \tfrac{1}{2} \Sigma_{-\infty<p<\infty} [a_p \cos(p\omega t) + b_p \sin(p\omega t)],$$

where $a_{-p} = a_p$ and $b_{-p} = -b_p$. A periodic complex function may be written as

$$\underline{u}(t) = \Sigma_{-\infty<p<\infty} \underline{\mathcal{U}}_p\, e^{ip\omega t}\,, \qquad \text{where } \underline{\mathcal{U}}_p = (1/T) \int_T dt\; e^{-ip\omega t}\, \underline{u}(t).$$

If $\underline{u}$ is real, we get $\underline{\mathcal{U}}_{-p} = \underline{\mathcal{U}}_p^* = \tfrac{1}{2}(a_p - ib_p)$.

A non-periodic *signal* $u(t)$ may be written as a *Fourier integral*

$$u(t) = \int_{\Delta\tilde{\nu}} d\tilde{\nu}\, [\, a(\tilde{\nu}) \cos(\omega t) + b(\tilde{\nu}) \sin(\omega t)].$$

$a(\tilde{\nu})$ and $b(\tilde{\nu})$ are the *spectral functions*. The spectrum of $u(t)$ is continuous. Introducing the complex spectral function $\underline{\mathcal{U}}(\tilde{\nu}) = \tfrac{1}{2}[\, a(\tilde{\nu}) - i\, b(\tilde{\nu})\,]$, we may write

$$u(t) = \int_0^\infty d\tilde{\nu}\, [\, a(\tilde{\nu}) \cos(\omega t) + b(\tilde{\nu}) \sin(\omega t)] = 2\, \mathcal{R}e \int_0^\infty d\tilde{\nu}\;\; \underline{\mathcal{U}}(\tilde{\nu})\; e^{i\omega t}\,.$$

If we extend the spectral function to negative frequencies such that $\underline{\mathcal{U}}(-\tilde{\nu}) = \underline{\mathcal{U}}(\tilde{\nu})^*$, i.e. $a(-\tilde{\nu}) = a(\tilde{\nu})$ and $b(-\tilde{\nu}) = -b(\tilde{\nu})$, we may write

$$u(t) = \int_{-\infty}^\infty d\tilde{\nu}\;\; \underline{\mathcal{U}}(\tilde{\nu})\; e^{i\omega t} \quad \text{and} \quad \underline{\mathcal{U}}(\tilde{\nu}) = \int_{-\infty}^\infty dt\; u(t)\; e^{-i\omega t}\,.$$

$u(t)$ is the *Fourier transform* of $\underline{\mathcal{U}}(\tilde{\nu})$ and $\underline{\mathcal{U}}(\tilde{\nu})$ is the inverse Fourier *transform* of $u(t)$. $a(\tilde{\nu})$ and $b(\tilde{\nu})$ are given by

$$a(\tilde{\nu}) = 2 \int_{-\infty}^\infty dt\; u(t) \cos(\omega t) \quad \text{and} \quad b(\tilde{\nu}) = 2 \int_{-\infty}^\infty dt\; u(t) \sin(\omega t).$$

If we use the angular frequency ω instead of the frequency $\tilde{\nu}$ and set $\underline{\mathcal{U}}'(\omega) = \underline{\mathcal{U}}(\tilde{\nu}) / \sqrt{2\pi}$, the Fourier transformation may be written in the symmetric form

$$\underline{u}(t) = \frac{1}{\sqrt{2\pi}} \int_{-\infty}^\infty d\omega\, \underline{\mathcal{U}}'(\omega)\, e^{i\omega t}, \qquad \underline{\mathcal{U}}'(\omega) = \frac{1}{\sqrt{2\pi}} \int_{-\infty}^\infty dt\; u(t)\, e^{-i\omega t}.$$

If the function $u(t)$ has a finite duration Δt, the spectral functions are non-negligible in a band of width $\Delta\tilde{\nu}$, obeying the uncertainty relation $\Delta\tilde{\nu}.\Delta t \approx 1$. A monochromatic oscillation corresponds to the limit $\Delta\tilde{\nu} \to 0$ and Δt infinite.

The Dirac delta-function $\delta(t - t_o)$ is equal to 0 for $t \neq t_o$ and infinite for $t = t_o$. For any function $f(t)$, $\delta(t - t_o)$ has the important properties

$$\int_{-\infty}^{\infty} dt\ \delta(t - t_o) = 1 \qquad \text{and} \qquad \int_{-\infty}^{\infty} dt\ \delta(t - t_o)\, f(t) = f(t_o).$$

It may be considered as the limit if $\tau \to 0$ of the impulse function

$$u_{imp}(t) = 1/2\tau \quad \text{if } t_o - \tau < t < t_o + \tau \qquad \text{and} \quad u_{imp}(t) = 0 \text{ otherwise.}$$

We may also write

$$\delta(t - t_o) = \frac{1}{2\pi}\int_{-\infty}^{\infty} d\omega\ e^{i\omega(t-t_o)} = \int_{-\infty}^{\infty} d\tilde{\nu}\ e^{i\omega(t-t_o)}.$$

2.10. Problem solving suggestions

1) The sum of two scalar simple harmonic oscillations of equal frequencies is a simple harmonic function of the same frequency. If $\tilde{\nu}_1/\tilde{\nu}_2 = T_2/T_1 = n_1/n_2$, where n_1 and n_2 are integers, after a time $T = n_1T_1 = n_2T_2$, the first oscillation is repeated n_1 times and the second oscillation n_2 times. Thus, their sum takes the same value. If $\tilde{\nu}_1/\tilde{\nu}_2$ is not equal to the ratio of two integers, $u = u_1 + u_2$ is not periodic.

2) The superposition $\mathbf{u} = \mathbf{u}_1 + \mathbf{u}_2$ of two oscillations of equal frequencies, and polarized linearly in two perpendicular directions, is polarized elliptically for an arbitrary phase shift. $\mathbf{u}$ is polarized linearly in the direction of the first or the second diagonal if $\mathbf{u}_1$ and $\mathbf{u}_2$ are in phase, or in opposite phase, respectively. $\mathbf{u}$ is circularly polarized if $\mathbf{u}_1$ and $\mathbf{u}_2$ have equal amplitudes and are in quadrature.

3) Fourier analysis allows us to write any function $u(t)$ as a superposition of simple harmonic functions. The frequencies of these harmonic functions constitute the spectrum of u. The spectrum of a periodic function u with a period T is discrete with periods T, $T/2$, $T/3$, ... The spectrum of a signal $u(t)$ of duration Δt is continuous in a band of width $\Delta\tilde{\nu}$, such that $\Delta t\ \Delta\tilde{\nu} \approx 1$. The exact value of this product depends on how the width of a curve is defined. However, its value is always of the order of 1.

2.11. Conceptual questions

Q1. Consider the superposition of n oscillations of equal frequencies and equal amplitudes, but a phase shift ϕ between consecutive oscillations. Draw their phasor diagram and determine the amplitude A of the resulting oscillation. Find the same result using the exponential complex representation. What should be the value of ϕ in order to have the maximum amplitude A, and what should be the value of ϕ in order to have $A = 0$?

Q2. Explain what is meant by linearly polarized, right-circularly polarized and left-circularly polarized oscillations. Assume that you have a simple harmonic electric generator, a capacitor C, a variable resistor R and an oscilloscope. Imagine a setup, which enables you to observe such polarized oscillations.

Q3. Is it possible to use Lissajous figures to compare the frequencies of two electric voltages?

Q4. Figures 2.12a and 2.12b show the coordinates $x(t)$ and $y(t)$ of a point M versus time. Draw the trajectory in each case and indicate the time, which corresponds to the particular positions on these curves. Deduce the Lissajous figures in the case of simple harmonic oscillations of similar periods.

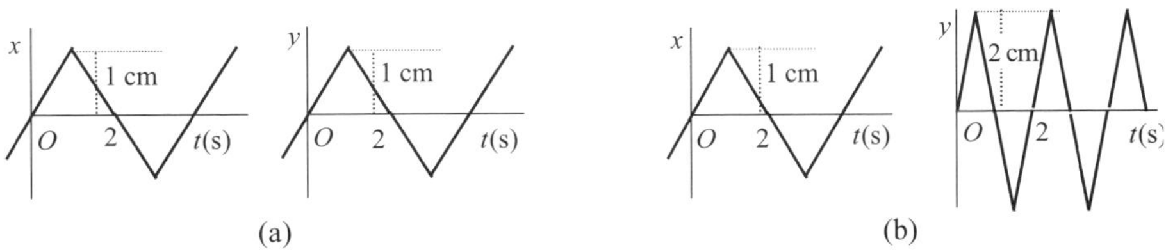

Figure 2.12. *Conceptual question Q5*

Q5. Two electric generators supply voltages of frequencies 100 Hz and 102 Hz. Is it possible to observe beats if they are both connected to an incandescent lamp, a loudspeaker or an oscilloscope?

Q6. Is the superposition of the oscillations $x_1 = a \cos(20\pi t)$ and $x_2 = A \cos(22\pi t)$ periodic? If yes, what is the period? At what times does the amplitude of the resulting oscillation reach its maximum, and at what times does it reach its minimum? Determine the frequency half-bandwidth and the beat frequency. What should be the resulting oscillation if x_2 has a phase ϕ and an amplitude $2A$?

Q7. Show that it is possible to combine the Fourier series for the real part and the imaginary part of $\underline{u}$ in a single Fourier series [2.29], with complex coefficients $\mathcal{U}_p$ related to the coefficients a_p, b_p, a'_p and b'_p by [2.30]. Verify relationship [2.31].

Q8. Assume that an ideal light wave is exactly monochromatic. What is its frequency bandwidth? What is its time duration? Is such a wave physically possible? Assume that a laser beam has an average wavelength of 500 nm and a time duration of 1 ms, what is its average frequency? What are its frequency bandwidth and its wavelength bandwidth?

2.12. Problems

Superposition of two scalar simple harmonic oscillations of equal frequencies

P2.1. Use the trigonometric representation, the phasor diagram and the complex representation to write, in the form $x = A\cos(\omega t + \phi)$, the following oscillations:

a) $x_1 = \cos(\omega t) + 2\sin(\omega t) - 4\cos(\omega t - \pi/4)$,
b) $x_2 = \cos(\omega t + \pi/3) - 2\sin(\omega t)$, and
c) $x_3 = 4\cos(\omega t) + 3\sin(\omega t)$.

Plot against time the two oscillations $4\cos(\omega t)$ and $3\sin(\omega t)$ and their superposition x_3.

P2.2. Consider the motions $x_1 = a_1\cos(\omega t + \phi_1)$ and $x_2 = a_2\cos(\omega t + \phi_2)$ of a point-like mass m on the x-axis subject to two forces $-Kx$ acting simultaneously. Show that the total energy of their superposition $x = x_1 + x_2$ is not equal to the sum of the energies of the two motions, except in some particular cases.

Superposition of two vector oscillations polarized in perpendicular directions

P2.3. A point-like object is subject to two forces, which act in the perpendicular directions Ox and Oy respectively, and produce oscillations of amplitudes A_1 and A_2 and phase shift $\phi \equiv \phi_2 - \phi_1$. What is the resulting motion if: **a)** $A_1 = A_2$ and $\phi = 0$, **b)** $A_1 = A_2$ and $\phi = \pi$, **c)** $A_1 = A_2$ and $\phi = \pi/2$, **d)** $A_1 = A_2$ and $\phi = -\pi/2$, **e)** $A_1 = 2A_2$ and $\phi = 0$,**f)** $A_1 = 2A_2$ and $\phi = \pi/2$, **g)** $A_1 = 2A_2$ and $\phi = \pi/4$?

P2.4*. A point mass M is confined to move in the Oxy plane, attached to four identical springs, as in Figure 2.13. O is the equilibrium position. **a)** Calculate the force acting on the mass if it has the coordinates (x, y) close to the origin O. Write down the equations of motion. **b)** Find the general solutions of these equations. Discuss the form of the trajectory according to the values of the integration constants.

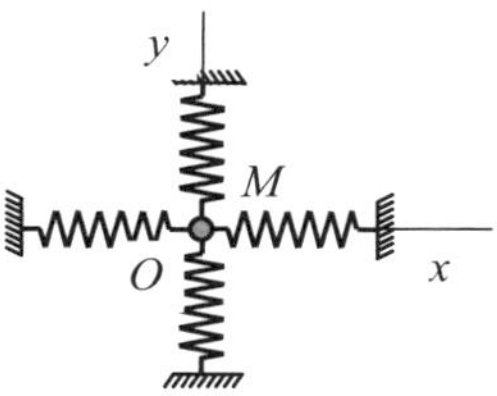

Figure 2.13. *Problem P2.4*

P2.5. a) Consider the oscillations $\mathbf{u}^{(\pm)} = a[\cos(\omega t)\ \mathbf{e}_x + \cos(\omega t \pm \pi/2)\mathbf{e}_y]$, which represent right- and left-circularly polarized motions. Write down the expression of the oscillation $\mathbf{u} = \mathbf{u}^{(+)} + \mathbf{u}^{(-)}$. **b)** How is this result modified if the second oscillation was $\mathbf{u}^{(-)} = a\ [\cos(\omega t + \phi)\ \mathbf{e}_x + \cos(\omega t + \phi - \pi/2)\ \mathbf{e}_y]$?

P2.6. A particle has the coordinates $x = a\cos(\omega t)$ and $y = b\cos(\omega t + \phi)$ on the perpendicular axes Ox and Oy. **a)** Show that the trajectory is elliptical. Considering the derivative $\dot{y}$ for $t = 0$, show that it is right-polarized if $0 < \phi < \pi$ and left-polarized if $\pi < \phi < 2\pi$. **b)** Let Ox' and Oy' be the axes obtained from Ox and Oy by rotation through an angle α. Determine the coordinates with respect to Ox' and Oy' as two simple harmonic oscillations of amplitudes a' and b' respectively with a phase shift ϕ'. **c)** Show that $\frac{1}{2}ab\sin\phi = \frac{1}{2}a'b'\sin\phi'$. Interpret this invariance. **d)** What should α be in order to have the ellipse axes in the directions of Ox' and Oy'?

P2.7. The coordinates of a particle are $x = a\cos(2\pi t)$ and $y = b\sin(3\pi t - \pi/6)$ on the perpendicular axes Ox and Oy. Draw the corresponding Lissajous figure. What is the period of the motion on this trajectory?

Superposition of scalar oscillations of different frequencies, beats

P2.8. Consider the superposition of two scalar oscillations $x_1 = a\cos(\omega_1 t)$ and $x_2 = a\cos(\omega_2 t + \phi)$, where $\omega_1 = \omega + \Delta\omega$ and $\omega_2 = \omega - \Delta\omega$ and $\Delta\omega << \omega$. Using the complex exponential representation, calculate the resulting oscillation.

Fourier analysis of a periodic function

P2.9. Write the Fourier series for the function of Figure 2.14. On the same graph plot the fundamental and the sum of the first two terms of the series and compare with u. Verify that, for $t = 1$, the series gives $u = 1$ exactly.

P2.10. An electric adapter transforms the 220 V and 50 Hz ac into a low dc voltage. It contains a transformer, a capacitor and diodes. A diode allows the current to pass in one direction and blocks it in the other. It is possible to conceive of a circuit, which allows us to have only the positive peaks as an output (Figure 2.15a), or another circuit allowing both positive and negative peaks by changing the direction of the negative peaks (Figure 2.15b). Make the Fourier analysis of these output voltages.

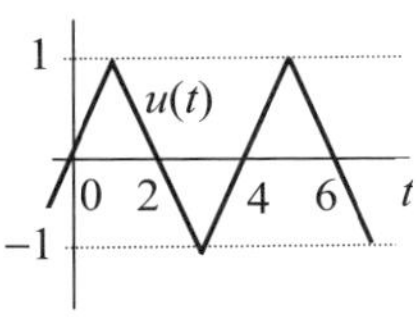

Figure 2.14. *Problem P2.9*

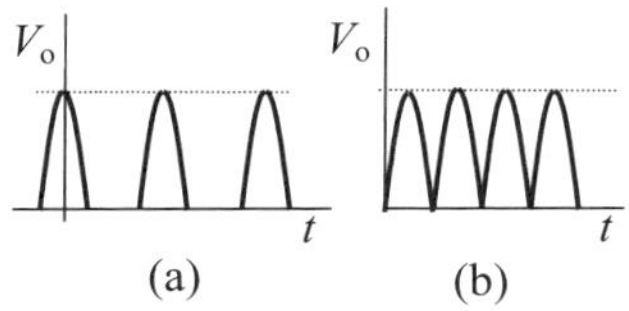

Figure 2.15. *Problem P2.10*

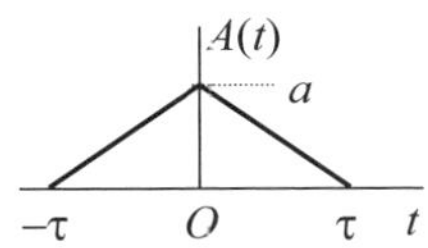

Figure 2.16. *Problem P2.13*

Fourier analysis of a non-periodic signal, uncertainty relations

P2.11*. A signal is represented by the function $u(t) = 1$ if $0 < t < \tau$ and $u(t) = 0$ for $t < 0$ or $t > \tau$. **a)** Show that its spectral functions are $a(\tilde{\nu}) = (1/\pi\tilde{\nu})\sin(2\pi\tilde{\nu}\tau)$ and $b(\tilde{\nu}) = (1/\pi\tilde{\nu})[1\ \cos(2\pi\tilde{\nu}\tau)]$. Verify that, substituting these expressions in [2.42], we find $u(t)$. For this, you may use the integral $\int_0^\infty dx\,(1/x)\sin mx = \pm\frac{1}{2}\pi$, where the sign is (+) if $m > 0$ and (−) if $m < 0$. **b)** Plot these spectral functions and verify that their width increases if the duration τ of the signal decreases.

P2.12. Determine the function $\underline{u}(t)$ which results from the superposition of simple harmonic oscillations, whose frequency lies in a frequency band of width $\Delta\tilde{\nu}$ centered on $\tilde{\nu}_o$ with the constant spectral functions $a(\tilde{\nu}) = a_o$ and $b(\tilde{\nu}) = b_o$.

P2.13*. Consider the oscillation $u = A(t)$, where $A(t)$ is the triangular pulse shown in Figure 2.16. Make the Fourier analysis of this signal. What is the total width $\Delta\omega$ at half-maximum? Verify the uncertainty relation.

P2.14*. Consider the function $u(t) = \cos[\omega_o(t - t_o)]$ for $|t - t_o| < \frac{1}{2}\Delta t$ and $u(t) = 0$ outside this time interval: **a)** Show that $u(t) = \int_0^\infty d\tilde{\nu}\ A(\tilde{\nu})\cos[\omega(t - t_o)]$, where $\omega = 2\pi\tilde{\nu}$, $\omega_o = 2\pi\tilde{\nu}_o$ and $A(\tilde{\nu}) = \Delta t\ \mathrm{sinc}[\Delta t(\tilde{\nu}_o + \tilde{\nu})] + \Delta t\ \mathrm{sinc}[\Delta t(\tilde{\nu}_o - \tilde{\nu})]$. Plot $u(t)$ and $A(\tilde{\nu})$. **b)** Show that, if $\Delta t < 1/\tilde{\nu}_o$, $A(\tilde{\nu})$ is almost constant and equal to $2\,\Delta t$ for $\tilde{\nu} < 1/\Delta t$. **c)** Show that, if $\Delta t > 1/\tilde{\nu}_o$, $A(\tilde{\nu})$ has a significant value only if $\tilde{\nu}$ is very close to $\tilde{\nu}_o$. This analysis applies to light emission by atoms in a discharge tube, for instance. The atom behaves as an oscillator whose relaxation time is of the order of 10^{-8} s, which corresponds to a bandwidth of the order of 10^8 Hz. In fact, the bandwidth is of the order of 10^9 Hz. This may be explained by the behavior of the atom as an oscillator of frequency $\tilde{\nu}_o$ (of the order of 10^{14} Hz) only for a short interval of time $\Delta t \approx 10^{-9}$ s, which is the average time between two collisions of the atom with other atoms. Thus, the bandwidth is of the order of $\Delta\tilde{\nu} \approx 1/\Delta t = 10^9$ Hz.

P2.15.** Consider the rectangular signal $u(t) = a_o$ for $|t - t_o| < ½\Delta t$ and $u(t) = 0$ outside this time interval. **a)** Show that the function u may be written as a Fourier integral $u(t) = \int_0^\infty d\tilde{\nu}\ A(\tilde{\nu}) \cos[\omega(t - t_o)]$, where $A(\tilde{\nu}) = 2a_o\ \Delta t\ \text{sinc}(\tilde{\nu}\ \Delta t)$. Plot $u(t)$ and $A(\tilde{\nu})$. **b)** Define the width $\Delta\tilde{\nu}$ of the function $A(\tilde{\nu})$ as the frequency interval in which $A(\tilde{\nu})$ is larger than half its maximum value. The first roots of the equation $\sin x = x/2$ being $x = \pm 1.8955$, show the uncertainty relationship $\Delta\tilde{\nu}\,.\Delta t = 0.60$. **c)** Now consider an infinite series of rectangular signals repeated with a period T_1. Show that this exact periodic function may be written in the form:

$$u(t) = (a\ \Delta t/T_1)[\,1 + 2\ \Sigma_{n\geq 1}\ (-1)^n\ \text{sinc}(n\ \Delta t/T_1)\cos(2n\pi t/T_1)].$$

Show that the function u may be considered as a superposition of simple harmonic oscillations of discrete spectrum $\tilde{\nu}_1$, $2\tilde{\nu}_1$, $3\tilde{\nu}_1$, ..., where $\tilde{\nu}_1 = 1/T_1$. The important frequencies lie between 0 and $\Delta\tilde{\nu} = 0.60/\Delta t$. **d)** Consider now a function u formed by a finite but large number n of rectangular signals. This function is almost periodic; thus, its spectrum is almost discrete. The frequencies $\tilde{\nu}_1$, $2\tilde{\nu}_1$, $3\tilde{\nu}_1$, etc., now have a width of the order of $\delta\tilde{\nu} = 1/nT_1$ and the most important frequencies lie between 0 and $\Delta\tilde{\nu} = 1/\Delta t$. Plot the spectral function $A(\tilde{\nu})$ in this case.

P2.16*. **a)** Verify that, if we use the angular frequency instead of the frequency, the Fourier transformation and the inverse Fourier transformation take the form:

$$u(t) = \frac{1}{2}\int_{-\infty}^{\infty} d\omega\ \underline{\hat{u}}(\omega)\ e^{i\omega t} \quad \text{and} \quad \underline{\hat{u}}(\omega) = \frac{1}{\pi}\int_{-\infty}^{\infty} dt\ u(t)\ e^{-i\omega t}.$$

b) Let $\hbar$ be any constant. Verify that we may write

$$u(x) = \frac{1}{\sqrt{2\pi\hbar}}\int_{-\infty}^{\infty} dp\ \underline{\hat{u}}(p)\ e^{ipx/\hbar} \quad \text{and} \quad \underline{\hat{u}}(p) = \frac{1}{\sqrt{2\pi\hbar}}\int_{-\infty}^{\infty} dx\ u(x)\ e^{-ipx/\hbar}.$$

c) Show that the Fourier transform of $u(\alpha t)$ is $(1/|\alpha|)\,\mathcal{U}(\tilde{\nu}/\alpha)$.

P2.17*. **a)** Let $\underline{\mathcal{U}}(\tilde{\nu})$ be the inverse Fourier transform of $\underline{u}(t)$:

$$\underline{\mathcal{U}}(\tilde{\nu}) = \int_{-\infty}^{\infty} dt\ \underline{u}(t)\ e^{-i\omega t}.$$

Show that the inverse Fourier transform of $(t - \tau)$ is $e^{-i\omega\tau}\,\underline{\mathcal{U}}(\tilde{\nu})$. This property is known as the *translation theorem*. **b)** Show that the inverse Fourier transform of n repetitions of $\underline{u}(t)$, separated by an interval of time τ, is:

$$\underline{\mathcal{U}}(\tilde{\nu}) = \underline{\mathcal{U}}_o(\tilde{\nu})\ \Sigma_{0\leq p\leq n-1}\ e^{-2ip\pi\nu\tau} = \underline{\mathcal{U}}_o(\tilde{\nu})\ \frac{\sin(n\pi\nu\tau)}{\sin(\pi\nu\tau)}\ e^{-i(n-1)\pi\nu\tau}\ .$$

Dirac delta-function

P2.18*. The Dirac delta-function is defined by the relationship:

$$\int_{-\infty}^{\infty} dx\ \delta(x)\, f(x) = \int_{-A}^{B} dx\ \delta(x)\, f(x) = f(0)$$

provided that $A > 0$ and $B > 0$. Using this definition and making a change of the integration variable if necessary, show the following properties of this function:

$$\int_{-\infty}^{\infty} dx\ \delta(x - x_o)\, f(x) = f(x_o), \qquad \int_{-\infty}^{\infty} dx\ \delta[\alpha(x - x_o)]\, f(x) = \frac{f(x_o)}{|\alpha|}, \qquad \delta[f(x)] = \Sigma_i\, \frac{\delta(x - x_i)}{|f'(x_i)|},$$

where f is any function, x_i are the roots of the equation $f(x) = 0$, which is assumed to be simple, and f' is the derivative of f.

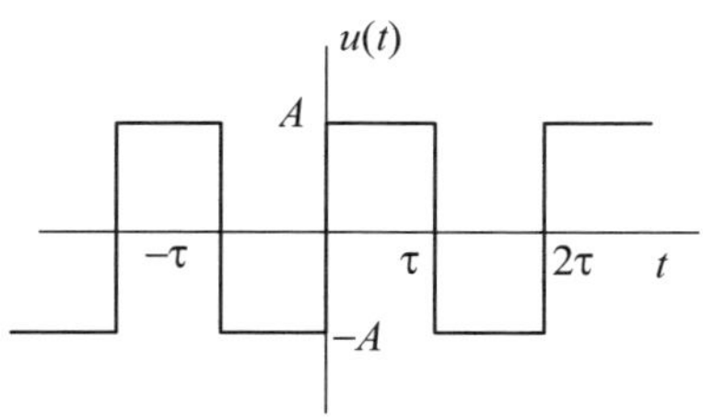

Figure 2.17. *Problem P2.19*

P2.19. a) Using the results from example 2.4, write the unilateral Fourier series and the bilateral Fourier series for the function of Figure 2.17. **b)** Write the spectral function $\underline{\mathcal{U}}(\tilde{\nu})$ of $u(t)$ as a sum of Dirac delta-functions.

P2.20*. a) Consider the oscillation $u = A\cos(\omega t + \phi)$. Verify that the average value of u^2 over a period (or over a long interval of time Δt) is $< u^2 > = ½A^2$. Let

$\underline{u}' = A\,e^{i(\omega t+\phi)}$. Verify that $< |\underline{u}'|^2 > = A^2$. **b)** Let $\underline{u}(t)$ be a signal and $\underline{\mathcal{U}}(\tilde{\nu})$ its spectral function. Verify that $\int_{-\infty}^{\infty} dt\,|u|^2 = \int_{-\infty}^{\infty} d\tilde{\nu}\,|\underline{\mathcal{U}}(\tilde{\nu})|^2$. **c)** Verify that:

$$\int_{-\infty}^{\infty} dt\ \cos(2\pi\tilde{\nu}\,t)\cos(2\pi\tilde{\nu}'\,t) = ½\,[\delta(\tilde{\nu} - \tilde{\nu}') + \delta(\tilde{\nu} + \tilde{\nu}')],$$

$$\int_{-\infty}^{\infty} dt\ \sin(2\pi\tilde{\nu}\,t)\sin(2\pi\tilde{\nu}'\,t) = ½\,[\delta(\tilde{\nu} - \tilde{\nu}') - \delta(\tilde{\nu} + \tilde{\nu}')],$$

$$\int_{-\infty}^{\infty} dt\ \cos(2\pi\tilde{\nu}\,t)\sin(2\pi\tilde{\nu}'\,t) = 0.$$

Let $u_1(t)$ and $u_2(t)$ be two real signals that may be written in the form:

$$\underline{u}_i(t) = \int_0^{\infty} d\tilde{\nu}\ [\,A_i(\tilde{\nu})\cos(\omega t) + B_i(\tilde{\nu})\sin(\omega t)],\ i = 1, 2.$$

Show that $\int_{-\infty}^{\infty} dt\ u_1(t)\,u_2(t) = ½\int_0^{\infty} d\tilde{\nu}\ \{\,A_1(\tilde{\nu})\ A_2(\tilde{\nu}) + B_1(\tilde{\nu})\ B_2(\tilde{\nu})\,\}$.

Chapter 3

Forced Oscillations

Oscillations are said to be *forced* if the oscillator is permanently subject to external forces as well as internal restoring forces and friction forces. In this chapter we assume that the oscillator is *linear*; this is the case for mechanical systems subject to ordinary forces and electric circuits formed by ideal resistors, inductors and capacitors and connected to generators. Then, the superposition principle is valid: if several forces act simultaneously on a linear system, the resulting motion is the superposition of the motions produced by each force acting separately.

In this chapter we study the forced oscillations of a system with one or several degrees of freedom especially in the case of a simple harmonic excitation force. The response of the system is particularly important if the excitation frequency is close to the natural frequency or one of the normal frequencies of the system; this phenomenon is called *resonance*. This study is important because several systems may be modeled as harmonic oscillators, and because the Fourier theory allows us to consider any excitation force as a superposition of harmonic forces.

3.1. Transient regime and steady regime

The equation of an oscillator with one degree of freedom, which is subject to an external force $F(t)$, is $\mu \ddot{u} = -Ku - b\dot{u} + F(t)$ where μ is its inertial parameter, $-Ku$ is the restoring force and $-b\dot{u}$ is the dissipative force. We may also write

$$\ddot{u} + 2\beta\dot{u} + \omega_o^2 u = F(t)/\mu \qquad \text{where } \beta = b/2\mu \quad \text{and } \omega_o = \sqrt{K/\mu} \qquad [3.1]$$

$F(t)$ is a *generalized force*, that is, any cause that may produce or modify the motion or change the state of a system by supplying or extracting energy. If u is the x-coordinate of a particle, the generalized force is the x-component F_x of the ordinary force acting on the particle. If u is the rotation angle about an axis Oz, the generalized force is the z-component of the moment of the forces about O and if u is the charge of a capacitor, the generalized force is the e.m.f. $\mathcal{E}$ of the generator, etc.

Let us assume that u_F is a particular solution of the complete equation, that is,

$$\ddot{u}_F + 2\beta\,\dot{u}_F + \omega_o^2\,u_F = F(t)/\mu \qquad [3.2]$$

Setting $u = u_F + u_o$, equation [3.1] reduces to the homogeneous equation

$$\ddot{u}_o + 2\beta\,\dot{u}_o + \omega_o^2\,u_o = 0. \qquad [3.3]$$

Thus, the solution u of the linear and non-homogeneous differential equation [3.1] is the sum of a particular solution u_F of the complete equation and the general solution of the homogeneous equation (i.e. with $F = 0$). The general solution u depends on two integration constants, which are those of u_o. They may be determined from the initial conditions. Due to damping, u_o dies out exponentially and becomes negligible after an interval of time of the order of 10τ. For this reason, it is called *transient*. After this short transient regime, the oscillator reaches the *steady regime* specified by u_F, which is independent of the initial conditions[1]. Similarly, if the system is in the steady regime and we turn off the force at a time t_o, the system becomes free and evolves according to a transient regime u_o, which is determined by the initial conditions $u(t_o)$ and $\dot{u}(t_o)$. Finally, it returns to the equilibrium state.

If the solution u_o for the damped free oscillator is known (see section 1.6), the study of the forced oscillator reduces to determine the particular solution u_F of the complete equation [3.1]. This solution is often of the same mathematical form as the excitation force $F(t)$. In particular, if $F(t)$ is a simple harmonic function of frequency Ω, it will be the same for the particular solution. In the following, we will only be concerned with the determination of steady regime solution u_F and, for simplicity, we will drop the index (F).

---

1 Let D be the linear operator $d^2/dt^2 + 2\beta(d/dt) + \omega_o^2$. The solution of the equation of the forced oscillator $Du = F/\mu$ is $u(t) = u_F(t) + u_o(t)$ such that $Du_F = F/\mu$ and $Du_o = 0$. Note that u_F is not unique because we may add to it any solution of the homogeneous equation $Du_o = 0$. The oscillator being damped ($\beta \neq 0$), u_o is exponentially decreasing over time. If F is not an exponentially decreasing function, we choose u_F to contain no terms that are decreasing over time. In the special case $\beta = 0$, the oscillator is not damped, the steady regime is never reached and we may add to u_F any solution u_o (see section 3.7).

3.2. Case of a simple harmonic excitation force

If an oscillator is subject to an excitation force $F = F_m \cos(\Omega t)$, its equation of motion is

$$\ddot{u} + 2\beta\,\dot{u} + \omega_o^2\,u = (F_m/\mu)\cos(\Omega t). \qquad [3.4]$$

Let us try a particular solution, which is harmonic of the same frequency Ω

$$u(t) = A\cos(\Omega t - \phi). \qquad [3.5]$$

With a suitable choice of ϕ $(-\pi < \phi \leq \pi)$, it is possible to always have the amplitude A as positive. This is a solution of [3.4] if

$$A(\omega_o^2 - \Omega^2)\cos(\Omega t - \phi) - 2\beta\Omega A\sin(\Omega t - \phi) = (F_m/\mu)\cos(\Omega t). \qquad [3.6]$$

Expanding the trigonometric functions, we obtain

$$\begin{aligned}&A\cos(\Omega t)[(\omega_o^2 - \Omega^2)\cos\phi + 2\beta\Omega\sin\phi]\\&+ A\sin(\Omega t)[(\omega_o^2 - \Omega^2)\sin\phi - 2\beta\Omega\cos\phi] = (F_m/\mu)\cos(\Omega t).\end{aligned} \qquad [3.7]$$

This equation is identically verified if the coefficients of $\cos(\Omega t)$ and $\sin(\Omega t)$ are equal on both sides; hence, the two equations

$$\begin{aligned}&A\,[(\omega_o^2 - \Omega^2)\cos\phi + 2\beta\Omega\sin\phi] = (F_m/\mu),\\&A\,[(\omega_o^2 - \Omega^2)\sin\phi - 2\beta\Omega\cos\phi] = 0.\end{aligned} \qquad [3.8]$$

It is convenient to introduce the *impedance* of the oscillator (see section 3.4)

$$Z = \frac{\mu}{\Omega}\sqrt{(\omega_o^2 - \Omega^2)^2 + 4\beta^2\Omega^2} = \sqrt{b^2 + (K/\Omega - \mu\Omega)^2}\,. \qquad [3.9]$$

The system of equations [3.8] may be solved for $\cos\phi$ and $\sin\phi$:

$$\cos\phi = \frac{\mu F_m}{AZ^2}\,\frac{\omega_o^2 - \Omega^2}{\Omega^2}, \qquad \sin\phi = \frac{\mu F_m}{AZ^2}\,\frac{2\beta}{\Omega}. \qquad [3.10]$$

By choosing the amplitude A to be positive, the second equation in [3.10] shows that $\sin\phi$ is positive; thus ϕ lies between 0 and π and u always has a phase lag over the force F. Substituting the expressions of [3.10] into the identity $\cos^2\phi + \sin^2\phi = 1$, we determine the amplitude A and, by dividing the equations of [3.10], we obtain $\tan\phi$; thus,

$$A = \frac{F_m}{\Omega Z} = \frac{F_m/\mu}{\sqrt{(\Omega^2 - \omega_o^2)^2 + 4\beta^2\Omega^2}}, \qquad \tan\phi = \frac{2\beta\Omega}{\omega_o^2 - \Omega^2} \quad (0 < \phi < \pi). \quad [3.11]$$

This completely determines the particular solution. It should be specified that ϕ is the phase lag of u over the excitation force F. If the force is $F = F_m \cos(\omega t + \alpha)$, the solution would be $u = A\cos(\Omega t + \alpha - \phi)$.

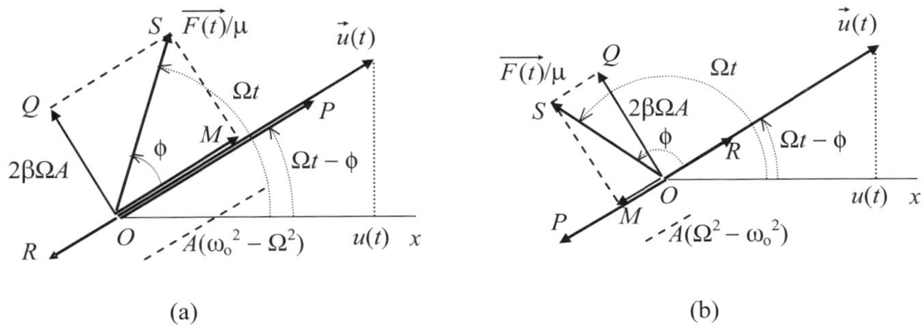

Figure 3.1. *Phasor diagram for a forced oscillator: a) if $\omega_o > \Omega$ and b) if $\omega_o < \Omega$*

The particular solution of equation [3.4] may also be obtained using a phasor diagram (Figure 3.1). For this, let us assume that u is of the form [3.5]. Its first and second time-derivatives are respectively:

$$\dot{u} = -A\Omega\sin(\Omega t - \phi) = A'\cos(\Omega t - \phi') \quad \text{where } A' = A\Omega \text{ and } \phi' = \phi - \pi/2 \quad [3.12]$$

$$\ddot{u} = -A\Omega^2\cos(\Omega t - \phi) = A\Omega^2\cos(\Omega t - \phi + \pi). \quad [3.13]$$

The oscillation $u(t)$ is represented by the phasor $\overrightarrow{u(t)}$ of length A and making an angle $(\Omega t - \phi)$ with the x-axis (which defines the origin of phases). The term $\omega_o^2 u$ in equation [3.4] is represented by the phasor $\overrightarrow{OP}$ of length $\omega_o^2 A$, and pointing in the same direction as $\overrightarrow{u(t)}$. The term $2\beta\dot{u}$ is represented by the phasor $\overrightarrow{OQ}$ of length $2\beta\Omega A$ and in a phase lead of $\pi/2$ over $\overrightarrow{u(t)}$. Finally, the term $\ddot{u}$ is represented by the phasor $\overrightarrow{OR}$ of length $\Omega^2 A$ and in a phase lead of π over $\overrightarrow{u(t)}$. The left-hand side of equation [3.4] is represented by the vector $\overrightarrow{OS}$ sum of the phasors $\overrightarrow{OP}$, $\overrightarrow{OQ}$ and $\overrightarrow{OR}$. The sum of $\overrightarrow{OP}$ and $\overrightarrow{OR}$ is $\overrightarrow{OM}$ of length $|A(\omega_o^2 - \Omega^2)|$ and pointing in the direction of $\overrightarrow{OP}$ or $\overrightarrow{OR}$ depending on whether $\omega_o^2 > \Omega^2$ (Figure 3.1a) or $\omega_o^2 < \Omega^2$ (Figure 3.1b). The vector sum of $\overrightarrow{OM}$ and $\overrightarrow{OQ}$ is $\overrightarrow{OS}$ of length $OS = A\Omega Z/\mu$ and in a phase lead of ϕ over u, given by $\tan\phi = 2\beta\Omega/(\omega_o^2 - \Omega^2)$ (such that $0 < \phi < \pi$). According to [3.4], the phasor $\overrightarrow{OS}$ must be identical to the phasor $\overrightarrow{F(t)}/\mu$. Thus, they must have equal length F_m/μ and equal phase Ωt. The oscillation u must have a phase lag ϕ over the force; hence the expressions of [3.11].

3.3. Resonance

Figure 3.2 shows the variation of the amplitude A and the phase shift ϕ as functions of the excitation frequency Ω for some values of the damping constant β. For a β that is non-zero but is small compared to ω_o, the amplitude A increases from $A_o = F_m/\mu\omega_o^2$ for $\Omega = 0$, reaches a maximum and decreases to zero at high frequency. The phase shift increases from $\phi = 0$ for $\Omega = 0$, reaches the value $\pi/2$ for $\Omega = \omega_o$ and tends asymptotically to π at high frequency.

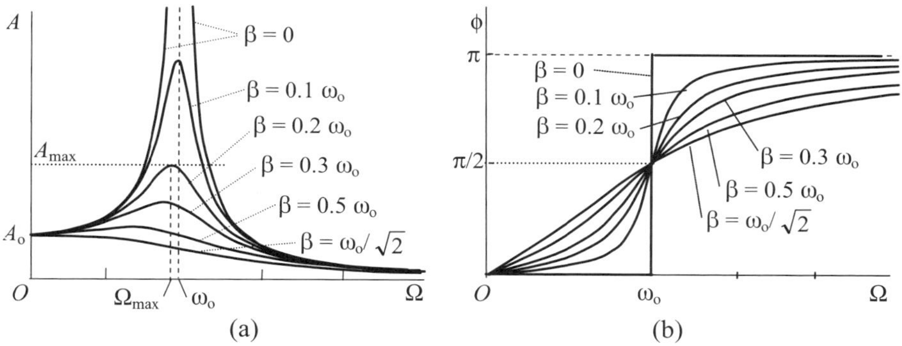

Figure 3.2. *The response of an oscillator versus the excitation frequency Ω for some values of the damping constant β: a) variation of the amplitude, and b) variation of the phase shift*

An excitation frequency equal to 0 is equivalent to a constant external force F. Such force displaces the system from its equilibrium position $u = 0$ to the position $u = A_o$. A phase shift ϕ equal to 0 means that the displacement is in step with the excitation force. For $\Omega = 0$, the slope of curve $A(\Omega)$ is 0. As long as the excitation frequency is low compared to the natural frequency ω_o, the amplitude A is nearly constant and equal to A_o and the phase shift is nearly 0. Noting that $\mu\omega_o^2$ is the restoring force constant K, the oscillation is nearly of the form

$$u \cong A_o \cos(\Omega t) = F(t)/K \qquad [\text{for } \Omega << \omega_o]. \qquad [3.14]$$

Thus, at low excitation frequency, the oscillator behaves like an elastic body: its displacement from equilibrium is small, proportional to the force and nearly independent on the damping constant β.

If $\beta << \omega_o$, the amplitude has a sharp maximum

$$A_{max} = F_m/2\mu\beta\sqrt{\omega_o^2 - \beta^2} \cong F_m/2\mu\beta\omega_o \qquad [3.15]$$

for the excitation frequency $\Omega_{max} = \sqrt{\omega_o^2 - 2\beta^2} \cong \omega_o$. Then, the phase shift is given by $\tan\phi_{max} = \sqrt{\omega_o^2/\beta^2 - 2} \cong \omega_o/\beta >> 1$; thus, ϕ is nearly equal to $\pi/2$. Neglecting the second order in (β/ω_o), we may write the approximate solution for $\Omega \cong \omega_o$

$$u \cong (F_m/2\beta\mu\omega_o)\cos(\Omega t - \pi/2) \qquad [\text{for } \Omega \cong \omega_o]. \qquad [3.16]$$

We say that the oscillator is resonant with the excitation force. *Resonance* occurs at an excitation frequency Ω, close to the natural frequency ω_o of the oscillator[2]. If β increases, A_{max} decreases and Ω_{max} is shifted toward low frequencies. The maximum disappears completely at $\Omega = 0$ for $\beta = \omega_o/\sqrt{2}$. In the special case $\beta = 0$, the amplitude is given by $A = F_m/\mu|\omega_o^2 - \Omega^2|$. It increases indefinitely for $\Omega = \omega_o$. As for the phase shift ϕ, it is equal to 0 for $\Omega < \omega_o$ and π for $\Omega > \omega_o$. Therefore, it has a discontinuity at $\Omega = \omega_o$.

At high frequency ($\Omega >> \omega_o$), the amplitude A of the response decreases asymptotically toward $F_m/\mu\Omega^2$ for all values of β, while the phase shift increases toward π. The oscillation is then of the form

$$u \cong (F_m/\mu\Omega^2)\cos(\Omega t - \pi) \qquad [\text{for } \Omega >> \omega_o]. \qquad [3.17]$$

Thus, it is in opposite phase with the force and its amplitude decreases like $1/\Omega^2$. At high excitation frequency, the response of the oscillator becomes increasingly weaker as Ω increases.

3.4. Impedance and energy of a forced oscillator in the steady regime

The position u and the velocity $\dot{u}$ of an oscillator obeying equation [3.4] are given by equations [3.5] and [3.12]. We define the *impedance* of the oscillator as the ratio of the amplitude of the force to the amplitude of the generalized velocity

$$Z = \frac{F_m}{A'} = \frac{F_m}{\Omega A} = \frac{\mu}{\Omega}\sqrt{(\omega_o^2 - \Omega^2)^2 + 4\beta^2\Omega^2} = \sqrt{b^2 + (K/\Omega - \mu\Omega)^2}\,. \qquad [3.18]$$

---

2 If we consider A and ϕ as functions of ω_o for a given excitation frequency Ω, the curves of A and ϕ are similar to those of Figure 3.2. Resonance occurs exactly at $\omega_o = \Omega$. It corresponds to a maximum of A equal to $F_m/2\mu\beta\Omega$ and a phase shift $\phi = \pi/2$.

The impedance depends only on the characteristics of the oscillator and on the excitation frequency Ω. If the impedance is known, we can determine the amplitude of the velocity $A' = F_m/Z$ as well as the amplitude and the phase shift of the response u of the oscillator to an excitation force $F = F_m \cos(\Omega t)$:

$$A = \frac{F_m}{\Omega Z}, \qquad \cos\phi = \frac{\mu(\omega_o^2 - \Omega^2)}{\Omega Z} = \frac{K/\Omega - \mu\Omega}{Z} \qquad \text{and } \sin\phi = \frac{2\mu\beta}{Z} = \frac{b}{Z}. \qquad [3.19]$$

Thus, the solution in steady regime can be written as

$$u = \text{has } \cos(\Omega t - \phi) = (F_m/\Omega Z)\cos(\Omega t - \phi),$$
$$\dot{u} = A'\cos(\Omega t - \phi') = (F_m/Z)\cos(\Omega t - \phi + \pi/2). \qquad [3.20]$$

The forced oscillator exchanges energy with the generator of the excitation force; this energy may be instantaneously stored or dissipated by the damping forces. In the steady regime, the oscillator returns periodically to the same state. Thus, its average stored energy over a period remains constant, and the net supplied energy by the generator is completely dissipated by the damping forces.

The total energy stored by the oscillator at time t is

$$E_{(T)} = \tfrac{1}{2}\mu\dot{u}^2 + \tfrac{1}{2}Ku^2 = \tfrac{1}{2}\mu A^2[\omega_o^2\cos^2(\Omega t - \phi) + \Omega^2\sin^2(\Omega t - \phi)]. \qquad [3.21]$$

Its rate of variation in time is

$$dE_{(T)}/dt = \mu\Omega A^2(\Omega^2 - \omega_o^2)\cos(\Omega t - \phi)\sin(\Omega t - \phi). \qquad [3.22]$$

The power that is supplied by the generator of the force is

$$P_{(ex)} = F.\dot{u} = -\Omega F_m A\cos(\Omega t)\sin(\Omega t - \phi) = -\Omega^2 A^2 Z\cos(\Omega t)\sin(\Omega t - \phi) \quad [3.23]$$

and the power that is dissipated by the damping forces is

$$P_{(fr)} = b\,\dot{u}\,.\,\dot{u} = 2\mu\beta\,\Omega^2 A^2\sin^2(\Omega t - \phi). \qquad [3.24]$$

Using the relations $Z\sin\phi = 2\mu\beta$ and $Z\cos\phi = (\mu/\Omega)(\omega_o^2 - \Omega^2)$, we easily verify the equation of conservation of energy

$$P_{(ex)} = dE_{(T)}/dt + P_{(fr)}. \qquad [3.25]$$

This relation states that the energy, which is supplied by the generator, is either stored by the oscillator or dissipated by the damping forces. We note that $P_{(ex)}$, as given by equation [3.23], may be momentarily negative; then, the oscillator returns energy back to the generator. To analyze the global energy transfer, regardless of the details of the oscillations, let us evaluate the average value of the oscillator energy over a period (between $t - T/2$ and $t + T/2$); we obtain

$$< E_{(T)} > = \tfrac{1}{4}\mu A^2(\omega_o^2 + \Omega^2) = \tfrac{1}{4}\,\mu(F_m/Z)^2\,(\omega_o^2/\Omega^2 + 1). \qquad [3.26]$$

It remains constant in the course of time. Thus, on average, the supplied energy is positive and completely dissipated by the damping forces. In fact, the average power that is supplied by the generator and the average power that is dissipated by the damping forces are respectively

$$< P_{(ex)} > = -\Omega^2 A^2 Z < \cos(\Omega t)[\sin(\Omega t)\cos\phi - \cos(\Omega t)\sin\phi] > = \tfrac{1}{2}\Omega^2 A^2 Z \sin\phi,$$
$$< P_{(fr)} > = \mu\beta\,\Omega^2 A^2. \qquad [3.27]$$

As $Z \sin\phi = 2\mu\beta$, these powers are equal. Using the definition $Z = F_m/A'$ and the relationships $A' = \Omega A$ and $\phi' = \phi - \pi/2$ for the amplitude and the phase shift of the velocity, we may also write

$$< P_{(ex)} > = < P_{(fr)} > = \frac{F_m^{\ 2}}{2Z}\sin\phi = \frac{1}{2}A'^2 Z\cos\phi' = \frac{1}{2}A' F_m \cos\phi' = \frac{1}{2}b\frac{F_m^{\ 2}}{Z^2}. \qquad [3.28]$$

The quantity $\sin\phi = \cos\phi'$ is called the *power factor* of the oscillator.

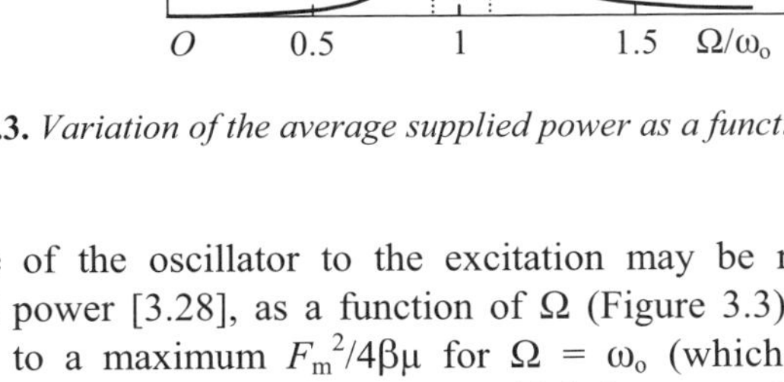

Figure 3.3. *Variation of the average supplied power as a function of* Ω/ω_o

The response of the oscillator to the excitation may be represented by the average supplied power [3.28], as a function of Ω (Figure 3.3). It increases from zero for $\Omega = 0$ to a maximum $F_m^{\ 2}/4\beta\mu$ for $\Omega = \omega_o$ (which corresponds to a minimum of Z) and then it decreases to zero at high frequency like $\beta F_m^{\ 2}/\mu\Omega^2$. The values of Ω, for which the supplied power is equal to half its maximum, are $\Omega^{\pm} = \sqrt{\omega_o^{\ 2} + \beta^2} \pm \beta$. Their difference is the *resonance width*

$$\Gamma \equiv \Omega^+ - \Omega^- = 2\beta = 2/\tau. \qquad [3.29]$$

The response of the oscillator is important if the excitation frequency lies within the frequency band $[\omega_o - \frac{1}{2}\Gamma, \omega_o + \frac{1}{2}\Gamma]$, called the *pass-band* of the oscillator.

The smaller the damping constant β is, the sharper the resonance curve is and the larger the maximum amplitude A_{max} and supplied power $F_m^2/4\beta\mu$ is. The resonance is characterized by the *quality factor* $f_q = \pi\tau/T_o = \frac{1}{2}\omega_o\tau$ (see section 1.6.3). Using the resonance width [3.29], we may also write

$$f_q = \omega_o/\Gamma = \omega_o/2\beta. \qquad [3.30]$$

The ratio of the resonance amplitude to the low frequency amplitude is

$$\frac{A_{max}}{A_o} = \frac{\omega_o^2}{2\beta\sqrt{\omega_o^2 - \beta^2}} \approx \frac{\omega_o}{2\beta} = f_q. \qquad [3.31]$$

If $\beta << \omega_o$, the quality factor is large, the response at resonance is large and the pass band is narrow. In other words, the oscillator is very selective. This should be the case of a radio receiver, for instance. In some other cases, the response of the system must be nearly constant for a wide frequency band in order to avoid signal deformations. In this case, β must be large and f_q small; we cannot then speak of resonance; examples of this case are the eardrum, a microphone or a loud speaker.

EXAMPLE 3.1. A body of mass $m = 0.1$ kg is confined to oscillate on the horizontal x-axis. It is attached to a spring of force constant 40 N/m and subject to a friction force $f = -b\dot{x}$ with $b = 1.2$ kg/s. **a)** Determine the natural frequency and the quality factor. **b)** This body is subject to a force $F = F_m \cos(\Omega t)$ where $F_m = 2$ N. Write down the expression of the displacement as a function of time if the excitation frequency is low? **c)** What should the value of Ω be if the amplitude of oscillation is at its maximum value? What is the value of this maximum? What, then, is the phase of the oscillation? **d)** Assuming that the excitation force has a frequency equal to 4 Hz, calculate the impedance, the amplitude and the phase of the oscillations, the power factor and the required average power to keep this body oscillating.

SOLUTION

a) The damping constant is $\beta = b/2m = 6$ s^{-1}. The natural frequency is $\omega_o = \sqrt{K/m} = 20$ rad/s. The quasi-frequency, if the body is subject to the friction force, is $\omega = \sqrt{\omega_o^2 - \beta^2} = 19.08$ rad/s. The relaxation time is $\tau = 1/\delta = 1/\beta = 0.17$ s. The quality factor is $f_q = \tau\omega_o/2 = 1.7$.

b) At a low excitation frequency, the mass displacement is:

$$x = F/K = (F_m/K)\cos(\omega t) = 5\cos(\omega t) \text{ (in cm).}$$

c) The oscillation amplitude has its maximum value at resonance, that is at $\Omega_{max} = \sqrt{\omega_o^2 - 2\beta^2} = 18.11$ rad/s. Then the amplitude is $A_{max} = F_m/2m\beta\omega = 8.73$ cm and the phase lag of the displacement over the force is given by $\tan\phi_{max} = \sqrt{\omega_o^2/\beta^2 - 2} = 3.02$, i.e. $\phi_{max} = 1.25$ rad.

d) If the excitation frequency is 4 Hz, i.e. $\Omega = 25.13$ rad/s, the impedance is $Z = \sqrt{b^2 + (K/\Omega - m\Omega)^2} = 1.513$ kg/s. The amplitude of the displacement is $A = F_m/\Omega Z = 5.26$ cm. The phase lag of the displacement over the force is given by the relation $\tan\phi = 2\beta\Omega/(\omega_o^2 - \Omega^2) = -1.30$: i.e. $\phi = 2.226$ rad (remember that $0 < \phi < \pi$). Thus, the displacement is $x = A\cos(\Omega t - \phi) = 5.26\cos(\Omega t - 2.226)$. The power factor is $\sin\phi = \sin 2.226$ rad $= 0.793$. The average power required to keep this body oscillating is $\langle P_{gen}\rangle = \frac{1}{2}\Omega^2 A^2 Z \sin\phi = 1.05$ W.

3.5. Complex impedance

To study the response u of a linear oscillator obeying equation [3.4], we may first determine the *complex solution* $\underline{u}$ in the case of the complex force $\underline{F} = F_m e^{i\Omega t}$

$$\ddot{\underline{u}} + 2\beta\dot{\underline{u}} + \omega_o^2\underline{u} = (F_m/\mu)\, e^{i\Omega t}. \qquad [3.32]$$

The solution u of [3.4] is the real part of $\underline{u}$. For equation [3.32] we try a solution $\underline{u}(t) = \underline{A}\, e^{i\Omega t}$, where $\underline{A}$ is generally a *complex amplitude*. Substituting this expression in [3.32], the dependence on t is only in the common factor $e^{i\Omega t}$. The equation becomes an algebraic equation $(\omega_o^2 - \Omega^2 + 2i\beta\Omega)\underline{A} = (F_m/\mu)$. Noting that the modulus of $(\omega_o^2 - \Omega^2 + 2i\beta\Omega)$ is the real impedance Z given by [3.9], we may define the *complex impedance*

$$\underline{Z} = \mu[2\beta + i(\Omega - \omega_o^2/\Omega)] \equiv b + i(\mu\Omega - K/\Omega). \qquad [3.33]$$

We may also write

$$\underline{Z} = Z\, e^{i\phi_z}, \qquad \tan\phi_z = (\Omega^2 - \omega_o^2)/2\beta\Omega \quad (-\pi/2 < \phi_z < \pi/2). \qquad [3.34]$$

The complex amplitude of the forced oscillations may be written in the form

$$\underline{A} = \frac{F_m}{i\Omega\underline{Z}} = \frac{\underline{F}_m/\mu}{\omega_o^2 - \Omega^2 + 2i\beta\Omega} = \frac{\underline{F}_m}{K - \mu\Omega^2 + ib\Omega}. \qquad [3.35]$$

or in the complex exponential form

$$\underline{A} = A\,e^{-i\phi} \qquad \text{where } A = \frac{F_m}{\Omega Z} \qquad \text{and} \quad \tan\phi = \frac{2\beta\Omega}{\omega_o{}^2 - \Omega^2} \quad (0 < \phi < \pi), \quad [3.36]$$

ϕ is the phase of $i\Omega\underline{Z}$. The complex oscillation $\underline{u}$ and the real oscillation u are given by

$$\underline{u} = A\,e^{i(\Omega t - \phi)}, \qquad u = \mathcal{R}e\,\underline{u} = A\cos(\Omega t - \phi). \qquad [3.37]$$

Thus, we again obtain expressions [3.5] and [3.12] with the same amplitude A and the same phase ϕ given by the equations in [3.11].

The complex velocity $\dot{\underline{u}}$ and the real velocity $\dot{u}$ may be written as

$$\dot{\underline{u}} \equiv \underline{A}'\,e^{i\Omega t} = i\Omega\underline{A}\,e^{i\Omega t} = (F_m/Z)\,e^{i\Omega t - \phi'}, \qquad \dot{u} = \mathcal{R}e\;\dot{\underline{u}} = A'\cos(\Omega t - \phi'), \quad [3.38]$$

where $A' = \Omega A = F_m/Z$ and $\phi' = \phi_z = \phi - \pi/2$. The expression of $\dot{u}$ allows us to define the impedance Z as the ratio of the amplitude of the real driving force to the amplitude of the generalized velocity $\dot{u}$ produced by this force. Similarly, the complex impedance $\underline{Z}$ may be defined as the ratio of the complex driving force to the complex generalized velocity $\dot{\underline{u}}$ produced by this force

$$\underline{Z} = \underline{F}\,/\,\dot{\underline{u}} \qquad \text{and} \qquad Z = |\underline{Z}| = F_m/A'. \qquad [3.39]$$

We conclude that, to determine the response of the oscillator, we may first determine the complex impedance $\underline{Z}$, then the velocity $\dot{\underline{u}} = \underline{F}/\underline{Z}$ and $\underline{u}$ by integration with respect to time. Finally, we take the real part of $\underline{u}$ to get the physical response. It should be noted that the impedance Z, sometimes called the *real impedance*, is not the real part of the complex impedance $\underline{Z}$, but its modulus $|\underline{Z}|$. The phase of $\underline{Z}$ is $\phi_z = \phi' = \phi - \pi/2$. This is also the phase lag of the velocity over the force.

The average power supplied by the generator is

$$< P_{(ex)} > = < F\;\dot{u} > = \tfrac{1}{2}\,(F_m{}^2/Z)\cos\phi_z = \mathcal{R}e\,(\tfrac{1}{2}\,\underline{F}\;\dot{\underline{u}}\,). \qquad [3.40]$$

This is the same as [3.28] with the power factor $\cos\phi_z = \sin\phi$.

EXAMPLE 3.2. Calculate the complex impedance of the mass attached to a spring from example 3.1 that is excited by a force of frequency 4 Hz. Using the complex analysis, determine the response if the mass is subject to a force $F = F_m\cos(\Omega t)$ where $F_m = 2$ N, and again find the results of the example 3.1.

SOLUTION – Using the values of b and K from example 3.1, the complex impedance of this oscillator is

$$\underline{Z} = b + \mathrm{i}(m\Omega - K/\Omega) = 1.2 + \mathrm{i}(0.1 \times 8\pi - 40/8\pi) = 1.2 + 0.922\ \mathrm{i}.$$

Calculating the modulus and the phase of $\underline{Z}$, we obtain:

$$\underline{Z} = Z e^{\mathrm{i}\phi_Z} \quad \text{where} \quad Z = (1.2^2 + 0.922^2)^{1/2} = 1.513\ \mathrm{N.s.m^{-1}},$$
$$\tan\phi_Z = 0.922/1.2 = 0.768, \qquad \phi_Z = 0.655\ \mathrm{rad}.$$

If the mass is subject to the complex force $\underline{F} = F_m\, e^{\mathrm{i}\Omega t}$, the complex velocity is

$$\dot{\underline{u}} = \underline{F}/\underline{Z} = (F_m/\underline{Z})\, e^{\mathrm{i}\Omega t} = (F_m/Z)\, e^{\mathrm{i}(\Omega t - \phi_Z)}.$$

Integrating with respect to time, we get the complex displacement

$$\underline{u} = (F_m/\mathrm{i}\Omega Z)\, e^{\mathrm{i}(\Omega t - \phi_Z)} = (F_m/\Omega Z)\, e^{\mathrm{i}(\Omega t - \phi_Z - \pi/2)} \equiv A\, e^{\mathrm{i}(\Omega t - \phi)}$$

where $A = (F_m/\Omega Z) = 5.26$ cm and $\phi = \phi_Z + \pi/2 = 2.226$ rad. If the real force $F = F_m \cos(\Omega t)$ is the real part of $\underline{F}$, the corresponding response is the real part of $\underline{u}$, i.e. $u = A\cos(\Omega t - \phi)$, that we found in example 3.1.

3.6. Sustained electromagnetic oscillations

Consider the LCR circuit in Figure 3.4a connected to the terminals of a generator of electromotive force $\mathcal{E} = \mathcal{E}_m \cos(\Omega t)$. The equation of this circuit as given by the Kirchhoff loop rule is $L\dot{I} + RI + Q/C = \mathcal{E}_m \cos(\Omega t)$, where $I = \dot{Q}$, hence

$$L\ddot{Q} + R\dot{Q} + Q/C = \mathcal{E}_m \cos(\Omega t). \qquad [3.41]$$

This equation is of the form [3.4] with $\mu = L$, $\beta = R/2L$, $\omega_o^2 = 1/LC$ and $F_m = \mathcal{E}_m$.

First we study the equation in the case of a complex e.m.f.

$$L\ddot{\underline{Q}} + R\dot{\underline{Q}} + \underline{Q}/C = \mathcal{E}_m\, e^{\mathrm{i}\Omega t}, \qquad [3.42]$$

which is of the form [3.32]. Equation [3.33] gives the complex impedance

$$\underline{Z} = \underline{\mathcal{E}}/\underline{I} = R + \mathrm{i}\,(L\Omega - 1/\Omega C) = Z e^{\mathrm{i}\phi_Z}. \qquad [3.43]$$

$$Z = \sqrt{R^2 + (L\Omega - 1/C\Omega)^2}\ ,\ \tan\phi_z = (L\Omega - 1/\Omega C)/R\ (-\pi/2 < \phi_z < \pi/2). \quad [3.44]$$

Expression [3.43] agrees with the rule of the addition of complex impedances in series, R for the resistor, $\mathrm{i}L\Omega$ for the inductor and $-\mathrm{i}/C\Omega$ for the capacitor. Conversely, knowing the impedance $\underline{Z}$, we may calculate the complex current intensity and deduce the real current intensity; we obtain:

$$\underline{I} = \frac{\underline{\mathcal{E}}}{\underline{Z}} = \frac{\mathcal{E}_m e^{\mathrm{i}\Omega t}}{Z e^{\mathrm{i}\phi_z}} = \frac{\mathcal{E}_m}{Z}\, e^{\mathrm{i}(\Omega t - \phi_z)}, \qquad I = \mathcal{R}e\,\underline{I} = \frac{\mathcal{E}_m}{Z}\cos(\Omega t - \phi_z). \quad [3.45]$$

The phase ϕ_z of the impedance is equal to the phase lag ϕ' of the current over the e.m.f. $\mathcal{E}_m$. The complex charge $\underline{Q}$ of the capacitor and its real charge are given by

$$\underline{Q} = \frac{\underline{I}}{\mathrm{i}\Omega} = \frac{\mathcal{E}_m}{\Omega Z}\, e^{\mathrm{i}(\Omega t - \phi_z - \pi/2)}, \qquad Q = \mathcal{R}e\,\underline{Q} = \frac{\mathcal{E}_m}{\Omega Z}\cos(\Omega t - \phi_z - \pi/2). \quad [3.46]$$

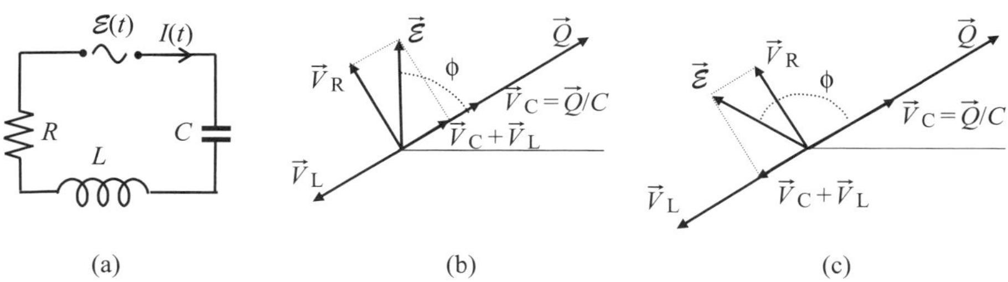

Figure 3.4. *a) Sustained LCR circuit, b) phasor diagram in the case of* $1/C\omega > L\omega$*, c) phasor diagram in the case of* $1/C\omega < L\omega$

The instantaneous power supplied by the generator of the e.m.f. is

$$P_{(ex)} = I\mathcal{E} = I_m\mathcal{E}_m \cos(\Omega t)\cos(\Omega t - \phi_z) \quad [3.47]$$

and its average value over a period is

$$< P_{(ex)} > = \tfrac{1}{2} I_m\mathcal{E}_m \cos\phi_z = \tfrac{1}{2}(RI_m/Z)\, I_m\mathcal{E}_m = \tfrac{1}{2}(R/Z^2)\,\mathcal{E}_m^{\ 2} = \tfrac{1}{2}RI_m^{\ 2}. \quad [3.48]$$

This is also the power that is dissipated in the resistor as Joule heat. The power factor is $\cos\phi_z = R/Z$. We note that the power is not a linear quantity in $\underline{\mathcal{E}}$; thus, we cannot calculate a "complex power" $\underline{\mathcal{E}}(t)\underline{I}(t)$ and take its real part. However, we may write

$$< P_{(ex)} > = \mathcal{R}e\,(\,\tfrac{1}{2}\,\underline{I}^*\,\underline{\mathcal{E}}\,). \quad [3.49]$$

If $R = 0$, the impedance $\underline{Z}$ is imaginary. The circuit is said to be *purely reactive.* Its power factor is equal to 0; thus it consumes, on average, no electric energy. Relationship [3.48] shows that the average power supplied to the circuit by an ac current $I = I_m \cos(\omega t - \phi')$ is the same as the power which is supplied by a dc current of intensity $I_{eff} = I_m/\sqrt{2}$. This is called the *effective intensity* of the alternating current. The instantaneous power supplied by the generator of voltage $V(t) = V_m \cos(\Omega t)$ is

$$P(t) = I(t)V(t) = I_m V_m \cos(\Omega t)\cos(\Omega t - \phi_z). \quad [3.50]$$

It oscillates between $-I_m V_m$ and $I_m V_m$. The product $I_m V_m$ is called the *apparent power*. It is usually expressed in units of *volt-amperes* (VA) to distinguish it from the average power expressed in *watts* (W).

At resonance, we have $1/C\Omega = L\Omega$ and $V_C = -V_L$. The voltage across the resistor becomes simply $V_R = IR = \mathcal{E}R/Z = \mathcal{E}$. The voltage V_L across the inductor and the voltage V_C across the capacitor are opposite and they may be much higher than the e.m.f. $\mathcal{E}$ of the generator. Taking the output at the terminals of the capacitor (or the inductor), the circuit is a *voltage multiplier* with a voltage ratio

$$V_{Cm}/\mathcal{E}_m = Q_m/C\mathcal{E}_m = 1/C\Omega Z. \quad [3.51]$$

This ratio reaches a maximum for $\Omega^2 = \omega_o^2 - 2\beta^2$, that is, at resonance, with a value

$$V_{Cm}/\mathcal{E}_m\,|_{max} = 1/CR\sqrt{\omega_o^2 - \beta^2} \approx \omega_o/2\beta = f_q, \quad [3.52]$$

where we assumed that $\beta << \omega_o$ and f_q is the quality factor of the circuit. The voltage multiplication may damage the capacitor or other electric equipments. The high intensity in the resistor may also cause the circuit to overheat.

3.7. Excitation from a state of equilibrium*

Consider an underdamped oscillator initially in the equilibrium state ($u = 0$ and $\dot{u} = 0$). Assume that it becomes subject to a force $F = F_m \cos(\Omega t)$ starting at $t = 0$. The complete solution in the transient regime is of the form

$$u = u_F + e^{-\beta t}\,[A_1 \cos(\hat{\omega}\,t) + A_2 \sin(\hat{\omega}\,t)], \quad [3.53]$$

where $u_F = A\cos(\omega t - \phi)$ is the steady regime solution. The second term is the general solution of the underdamped free oscillator with $\hat{\omega} = \sqrt{\omega_o^2 - \beta^2}$. The constants A_1 and A_2 may be determined from the initial conditions $u(0) = 0$ and $\dot{u}(0) = 0$; hence, $A_1 = -A \cos\phi$ and $A_2 = -A(\beta/\hat{\omega})\cos\phi - A(\Omega/\hat{\omega})\sin\phi$. In the case of a very small damping ($\beta << \omega_o$, thus $\beta << \hat{\omega}$), we may write

$$u = A\cos\phi\,[\cos(\Omega t) - e^{-\beta t}\cos(\hat{\omega}\,t)] + A\sin\phi\,[\sin(\Omega t) - (\Omega/\hat{\omega})\,e^{-\beta t}\sin(\hat{\omega}\,t)].$$

– if $\Omega = \hat{\omega} \cong \omega_o$, the solution may be written in the approximate form

$$u \cong [1 - e^{-\beta t}]\,u_F. \qquad [3.54]$$

In this case, the solution u in the transient regime is proportional to the steady regime solution. It increases exponentially in time from 0 to u_F, which it reaches theoretically in an infinite time, but practically in a time of the order of 10 τ, where $\tau = 1/\beta$ is the relaxation time.

– If $\beta = 0$, then $\sin\phi = 0$ and the solution takes the form

$$u = A\,[\cos(\Omega t) - \cos(\hat{\omega}\,t)] = 2A\sin(\Delta\omega\,t)\sin(\omega_m t); \qquad [3.55]$$

where we have set $\omega_m = ½(\hat{\omega} + \Omega)$ and $\Delta\omega = ½(\hat{\omega} - \Omega)$. In particular, if $\hat{\omega}$ is close to Ω, a beats phenomenon occurs: the amplitude of the oscillations increases from zero to $2A$, then it decreases to zero and so on. The beat frequency is $\tilde{\nu}_{bat} = \Delta\omega/\pi$. The stored energy also oscillates between 0 and a certain maximum value.

– In the very special case $\beta = 0$ and $\Omega \approx \hat{\omega} = \omega_o$, taking the limits in the right way, we obtain

$$u \approx (F_m\,t/2\mu\Omega)\sin(\Omega t). \qquad [3.56]$$

In this case, the amplitude of u increases indefinitely in time. This result explains why a system, which is subject to oscillating forces, is exposed to serious risks if the excitation frequency is close to the natural frequency of the system.

3.8. Response to an arbitrary force, nonlinear systems*

If an oscillator is subject to an arbitrary excitation force, the Fourier theory allows us to consider this force as a superposition of simple harmonic forces. In the case of a periodic force of period T, it may be written as a Fourier series

$$F(t) = \Sigma_{-\infty < n < \infty}\,\underline{\mathcal{F}}_n\,e^{i\omega_n t} \qquad \text{with } \omega_n = n\omega \equiv 2n\pi/T. \qquad [3.57]$$

A similar expression may be written for a non-periodic force with the summation on the discrete ω_n replaced by an integration on the continuous frequency $\tilde{\nu}$.

In the case of a linear system, its response to the force $F(t)$ is the superposition of the responses to all the spectral simple harmonic forces. Thus, the response is

$$\underline{u} = \Sigma_{-\infty<n<\infty} \underline{\mathcal{U}}_n \, e^{i\omega_n t} . \qquad [3.58]$$

The complex amplitudes $\underline{\mathcal{U}}_n$ are given by [3.35] for each spectral force, i.e. $\underline{\mathcal{U}}_n = \underline{\mathcal{F}}_n / i\Omega \underline{Z}_n$, where $\underline{Z}_n = \mu[2\beta + i(\omega_n - \omega_o^2/\omega_n)]$ is the impedance which corresponds to the excitation frequency ω_n. Using equations [3.5] and [3.11], we may also directly write the force and the response in the form

$$F(t) = \tfrac{1}{2} f_o + \Sigma_{1<n<\infty} f_n \cos(\omega_n + \alpha_n) \qquad [3.59]$$

$$u = \tfrac{1}{2} U_o + \Sigma_{1<n<\infty} U_n \cos(\omega_n + \alpha_n - \phi_n) \qquad [3.60]$$

with

$$U_n = \frac{f_n}{\omega_n Z_n} \text{ and } \phi_n = \text{Arctan} \frac{2\beta\omega_n}{\omega_o^2 - \omega_n^2} . \qquad [3.61]$$

If the oscillator is very damped, its resonance curve is wide and several spectral frequencies of the force may be within the pass band and produce important responses. On the contrary, if the oscillator has little damping, the resonance curve is narrow and very few spectral frequencies are within the pass band and produce important resonance responses. Then, the oscillator is very selective.

In the case of an impulse, the real force and the real response may be written as

$$F(t) = \int_0^\infty d\tilde{\nu} \; f(\tilde{\nu}) \cos[\omega + \alpha(\tilde{\nu})], \qquad u(t) = \int_0^\infty d\tilde{\nu} \; U(\tilde{\nu}) \cos[\omega + \alpha(\tilde{\nu}) - \phi(\tilde{\nu})], \quad [3.62]$$

where $\omega = 2\pi\tilde{\nu}$, $f(\tilde{\nu})$ is important in a frequency band $[\tilde{\nu}_1, \tilde{\nu}_2]$ and

$$U(\tilde{\nu}) = \frac{f(\tilde{\nu})}{\omega Z_n(\omega)}, \qquad \phi(\tilde{\nu}) = \text{Arctan} \frac{2\beta\omega}{\omega_o^2 - \omega^2} . \qquad [3.63]$$

If the natural frequency $\tilde{\nu}_o$ lies within the pass band $[\tilde{\nu}_1, \tilde{\nu}_2]$ of the impulse, the response is large and if it lies outside the pass band, the response is generally small.

If the force has a large amplitude, the effects of a possible nonlinearity of the oscillator may show up. For instance, consider a *nonlinear oscillator*, whose restoring force is $-Kx + \alpha x^3$. Its response to a force of frequency Ω, is the superposition of the response to a simple harmonic force of frequency Ω and its harmonics $p\Omega$. If the oscillator is subject to two forces of frequencies Ω_1 and Ω_2 (for instance, a nonlinear microphone receiving two sounds of different frequencies), the response is the superposition of oscillations whose frequencies are Ω_1, Ω_2, $\Omega_1 \pm \Omega_2$, $\Omega_1 \pm 2\Omega_2$, $2\Omega_1 \pm \Omega_2$, etc.

After the advent of computers in the 1970s, numerical solutions for nonlinear equations become possible. These numerical solutions have shown that nonlinear systems may have some properties that are completely different from those of linear systems. In particular, they may have chaotic behavior; this phenomenon shows up in many branches of physics (mechanical systems, fluid flow, lasers, etc.) and certainly in other domains, like population growth, economy etc. This means that, although the system obeys deterministic equations of motion (such as Newton laws), its real response to excitation may be unpredictable and even chaotic.

Consider, for instance, a pendulum subject to an excitation force of frequency Ω and a friction force. If the excitation force is weak ($\ll mg$), the angle θ of the pendulum remains small; the approximation $\sin\theta \cong \theta$ in the expression of the restoring moment of the force remains valid and the pendulum is a linear system. Its oscillation rapidly reaches the steady regime $\theta = \theta_m \cos(\Omega t - \phi)$ independently of the initial conditions. If the excitation force is increased, the angle θ may become large. The approximation $\sin\theta \cong \theta$ is no longer valid and the pendulum is nonlinear. The numerical solution shows that the nth harmonics (of frequencies $n\Omega$) become increasingly important in steady regime. For a more important force, sub-harmonics of frequencies Ω/n appear and, finally, the motion becomes completely chaotic. The important property of this chaotic motion is its extreme sensitivity to initial conditions. For instance, if two pendulums have a very small difference in their initial conditions, they end up with completely different motions. This is equivalent to saying that their motion is unpredictable.

3.9. Excitation of a system of coupled oscillators*

Consider a system of coupled oscillators with a force $F_m \cos(\Omega t)$ acting on one of them. This oscillator starts to oscillate and, because of the coupling, the other oscillators oscillate in turn. Here we neglect the propagation time of the oscillations from one oscillator to another. The resulting oscillation of the system is the superposition of the particular harmonic solution of frequency Ω and the general solution of free oscillations, which is itself an arbitrary superposition of the normal modes. Each mode of frequency $\omega_{(k)}$ is characterized by a damping coefficient $\delta_{(k)}$, i.e. a relaxation time $\tau_{(k)} = 1/\delta_{(k)}$ such that the oscillation of this mode becomes negligible after a time of the order of $10\tau_k$. Thus, after a certain time, the system reaches its steady regime, oscillating according to the particular solution, which has the same frequency as the excitation force.

A resonance phenomenon occurs if Ω is close to the frequency of one of the normal modes of the system; the amplitudes of certain variables u_i of the system are large and the generator of the force supplies a large average power. The resonance bandwidth, which corresponds to mode (k) is $\Gamma_{(k)} = 2/\tau_{(k)} = 2\delta_{(k)}$.

Consider two simple pendulums of equal mass m and equal length L coupled by a spring of constant K (Figure 3.5a). We assume that these masses are subject to two friction forces $-2m\beta\,\dot{u}_1$ and $-2m\beta\,\dot{u}_2$ respectively, and that a force $F = F_m \cos(\Omega t)$ acts on the first mass. The equations of motion are

$$\ddot{u}_1 + 2\beta\,\dot{u}_1 + \omega_o^2\,u_1 + \omega_r^2\,(u_1 - u_2) = F/m,$$
$$\ddot{u}_2 + 2\beta\,\dot{u}_2 + \omega_o^2 u_2 + \omega_r^2(u_2 - u_1) = 0, \qquad [3.64]$$

where we have set $\omega_o = \sqrt{g/L}$ and $\omega_r = \sqrt{K/m}$. Let us first determine the normal frequencies of this system (i.e. its frequencies of oscillation if it is free, $F = 0$). Setting $u_1 = C_1\,e^{-\eta t}$ and $u_2 = C_2\,e^{-\eta t}$ in the equations of [3.64] with $F = 0$, we obtain the characteristic equation

$$(\eta^2 - 2\beta\eta + \omega_o^2)(\eta^2 - 2\beta\eta + \omega_o^2 + 2\omega_r^2) = 0. \qquad [3.65]$$

The roots of this equation are $\beta \pm i\omega_{(1)}$ and $\beta \pm i\omega_{(2)}$, where $\omega_{(1)} = (\omega_o^2 - \beta^2)^{1/2}$ and $\omega_{(2)} = (\omega_o^2 + 2\omega_r^2 - \beta^2)^{1/2}$. Thus we have two normal modes of frequencies $\omega_{(1)}$ and $\omega_{(2)}$ with an equal damping coefficient $\delta = \beta$.

It is convenient to solve the system of linear equations [3.64] for the complex force $\underline{F} = F_m\,e^{i\Omega t}$ first, and then take the real part of the solution. Let us determine a solution of the form $\underline{u}_i = \underline{A}_i\,e^{i\Omega t}$. Substituting this expression in the equations [3.64] (with a complex force $\underline{F}$), we obtain a system of algebraic linear equations for the complex amplitudes $\underline{A}_i$

$$(\omega_o^2 + \omega^2 - \Omega^2 + 2i\beta\Omega)\underline{A}_1 - \omega_r^2\underline{A}_2 = F_m/m,$$
$$-\,\omega^2\underline{A}_1 + (\omega_o^2 + \omega_r^2 - \Omega^2 + 2i\beta\Omega)\,\underline{A}_2 = 0. \qquad [3.66]$$

The solution of these equations is

$$\underline{A}_1 = (F_m/m\Delta_1\Delta_2)(\omega_r^2 + \omega_o^2 - \Omega^2 + 2i\beta\Omega), \qquad \underline{A}_2 = (F_m/m\Delta_1\Delta_2)\,\omega_r^2 \qquad [3.67]$$

where $\Delta_1 = \omega_o^2 - \Omega^2 + 2i\beta\Omega$ and $\Delta_2 = 2\omega_r^2 + \omega_o^2 - \Omega^2 + 2i\beta\Omega$. We may also write $\underline{A}_1 = A_1 e^{i\phi_1}$ and $\underline{A}_2 = A_2 e^{i\phi_2}$ with the real amplitudes A_i given by

$$A_1 = \frac{F_m}{m|\Delta_1||\Delta_2|}\sqrt{(\omega_o^2 + \omega_r^2 - \Omega^2)^2 + 4\beta^2\Omega^2}\,,\; A_2 = \frac{F_m}{m|\Delta_1||\Delta_2|}\omega_r^2 \qquad [3.68]$$

where $|\Delta_1| = \sqrt{(\omega_o^2 - \Omega^2)^2 + 4\beta^2\Omega^2}$ and $\Delta_2 = \sqrt{(\omega_o^2 + 2\omega_r^2 - \Omega^2)^2 + 4\beta^2\Omega^2}$ while the phases ϕ_i are given by the equations: $\phi_1 = \alpha_3 - \alpha_1 - \alpha_2$ and $\phi_2 = -\alpha_1 - \alpha_2$, with the α_i lying between 0 and π and given by the equations:

$$\tan\alpha_1 = \frac{2\beta\Omega}{\omega_o^2 - \Omega^2}, \qquad \tan\alpha_2 = \frac{2\beta\Omega}{2\omega_r^2 + \omega_o^2 - \Omega^2}, \qquad \tan\alpha_3 = \frac{2\beta\Omega}{\omega_r^2 + \omega_o^2 - \Omega^2}. \; [3.69]$$

If Ω is close to one of the normal frequencies $\omega_{(1)}$ and $\omega_{(2)}$, the product $|\Delta_1||\Delta_2|$ is small and the amplitudes A_1 and A_2 are large. Thus, the coupled pendulums resonate if the excitation frequency is close to one of their normal frequencies $\omega_{(k)}$. The resonance pass band is $\Gamma_{(k)} = 2\delta_{(k)}$ where $\delta_{(k)}$ is the damping coefficient of mode (k). Figure 3.5b shows the variation of the ratio A_2/A_1 as a function of Ω.

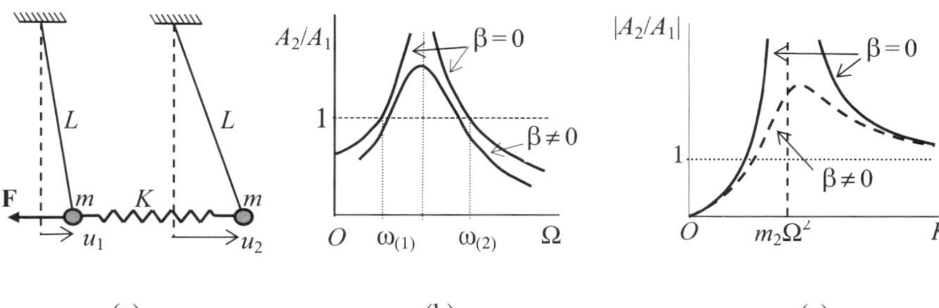

Figure 3.5. *a) Two coupled pendulums; b) variation of the ratio A_2/A_1 of the amplitudes as a function of the excitation frequency Ω; c) the ratio $|A_2/A_1|$ for two masses, which are coupled by a spring as a function of the spring constant*

It is possible to use the normal coordinates $y_1 = u_1 + u_2$ and $y_2 = u_1 - u_2$, which obey the uncoupled differential equations:

$$\ddot{y}_1 + 2\beta\,\dot{y}_1 + \omega_o^2 y_1 = F/m, \qquad \ddot{y}_2 + 2\,\beta\,\dot{y}_2 + (\omega_o^2 + 2\omega_r^2)y_1 = F/m. \qquad [3.70]$$

The system of coupled pendulums is thus equivalent to two uncoupled oscillators of impedances $\underline{Z}_1 = -im\Delta_1/\Omega$ and $\underline{Z}_2 = -im\Delta_2/\Omega$. Each is excited by a force F. The solution may be written in the form

$$y_1 = C\cos(\Omega t - \phi_1), \qquad y_2 = D\cos(\Omega t - \phi_2), \qquad [3.71]$$

where, using [3.11],

$$C = F_m/\Omega Z_1 = F_m/m|\Delta_1|, \qquad D = F_m/\Omega Z_2 = F_m/m|\Delta_2|,$$
$$\tan\phi_1 = 2\beta\Omega/(\omega_o^2 - \Omega^2) \qquad \tan\phi_2 = 2\beta\Omega/(\omega_o^2 + 2\,\omega_r^2 - \Omega^2) \quad (0 < \phi_i < \pi).$$

If Ω is close to one of the normal frequencies $\omega_{(1)}$ or $\omega_{(2)}$, the corresponding normal oscillator resonates, its variable y has a large amplitude and u_1 and u_2 are also large. The power, which is supplied by the excitation force then has a maximum.

This result may easily be generalized to systems with several degrees of freedom. The amplitude of the variables, which specify the system, and the power transfer to the system have peaks if the excitation frequency is close to one of the normal frequencies.

As an application, consider two coupled masses m_1 and m_2. We would like to greatly reduce the oscillation of one of them if a force $F = F_m e^{i\Omega t}$ acts on the first mass. For this, we couple the masses by a spring of constant K. If the coupling is too weak ($K \cong 0$), m_2 is not affected by the oscillation of m_1. On the contrary, if the coupling is too rigid, the oscillations of m_1 are completely transmitted to m_2 and the system behaves as a single body. Neglecting the friction forces, the masses' equations of motion are

$$m_1\ddot{x}_1 + K(x_1 - x_2) = F_m e^{i\Omega t}, \qquad m_2\ddot{x}_2 + K(x_2 - x_1) = 0. \qquad [3.72]$$

The solution of these equations in the steady regime is of the form $x_i = A_i e^{i\Omega t}$ where $A_1 = (F_m/\Delta)(K - m_2\Omega^2)$ and $A_2 = F_m K/\Delta$ with $\Delta = m_1 m_2\Omega^4 - K\Omega^2(m_1 + m_2)$. Figure 3.5c illustrates the variation of the transmission ratio $|A_2/A_1| = K/|K - m_2\Omega^2|$ as a function of K. It increases from 0 for $K = 0$ (no coupling at all) to infinity for $K = m_2\Omega^2$, then it decreases to 1 if K is very large ($K >> M\Omega^2$, i.e. a very rigid coupling). In reality, $|A_2/A_1|$ cannot be infinite as the friction is never completely negligible. $|A_2/A_1|$ has a peak larger than 1 for a certain value of K close to $m_2\Omega^2$; this corresponds to a small oscillation of m_1. On the contrary, the oscillation of m_2 is small if the spring constant is small ($K << m_2\Omega^2$). For instance, if $K = 0.1\,m_2\Omega^2$ we find $A_2/A_1 \approx -0.11$.

3.10. Generalization of the concepts of external force and impedance*

In section 3.1.9, we generalized the concept of mass μ and the concept of restoring force constant K for any oscillator, by considering the expression of the total energy

$$E_{(T)} = \tfrac{1}{2}\,\mu\,\dot{u}^2 + \tfrac{1}{2}\,Ku^2, \qquad [3.73]$$

where u is a small displacement from the equilibrium position. We have also generalized the concept of the friction coefficient by considering the dissipated energy. We now generalize the concept of external force F as the ratio of the supplied energy $dE_{(ex)}$ to the variation du of the variable of the system

$$dE_{(ex)} = F\,du. \qquad [3.74]$$

We note that, if the system is subject to no internal forces or if its internal variables are constant, its internal potential energy is constant. Then, the conservation of energy implies that the energy [3.74], which is supplied by the external force F, is transformed into kinetic energy: $dE_{(ex)} = F\,\dot{u}\,dt = d(\tfrac{1}{2}\mu\,\dot{u}^2) = \mu\,\dot{u}\ddot{u}\,dt$. We deduce the relation $F = \mu\,\ddot{u}$, which generalizes Newton's second law for a point-particle. If a rigid body is free to rotate about the z-axis, the energy that this body receives by rotating by an angle $d\theta$ is $dE_{(ex)} = \Gamma_z\,d\theta$. The generalized force is thus the moment of the external forces Γ_z about Oz. In the case of a circuit connected to the terminals of a generator of e.m.f. $\mathcal{E}$, the energy supplied by the generator in the time interval dt is $dE_{(ex)} = \mathcal{E}I\,dt = \mathcal{E}\,dq$. The generalized force in this case is the e.m.f.

If an oscillator with one degree of freedom is subject to a dissipative force $-b\,\dot{u}$ its equation of motion may be obtained from the conservation of energy: the work of external force [3.74], in the time interval dt, is equal to the sum of the variation of total energy [3.73] and the dissipated energy in this interval of time; hence, the equation $dE_{(T)} + b\,\dot{u}^2\,dt = F\,\dot{u}\,dt$, that is,

$$\mu\,\ddot{u} + b\,\dot{u} + Ku = F \qquad [3.75]$$

In the case of a system with several degrees of freedom specified by the displacements u_i from a stable equilibrium position, if the transferred energy in the displacements du_i is $dE_{(ex)} = \Sigma_i F_i\,du_i$, then F_i is the *generalized force* associated with the variable u_i. We may show that the equations of motion of the system are

$$\Sigma_j\,\mu_{ij}\,\ddot{u}_j + \Sigma_j\,b_{ij}\,\dot{u}_j + \Sigma_j\,K_{ij}\,u_j = F_i \qquad \text{with } i = 1, 2, \ldots, n. \qquad [3.76]$$

3.11. Some applications

Forced oscillations and resonance phenomena are important both in theoretical physics and in applied physics. Here we discuss some applications.

– A musical instrument is essentially an oscillator (a violin string, for instance) coupled to a resonator (resonant cavity or an organ pipe), whose oscillation modes correspond to the standing waves of the instrument. The excitation force may be a superposition of harmonic forces of various frequencies. Only the frequencies, which are close to the normal frequencies of the instrument, are excited with large amplitude by going into resonance.

– If the damping coefficient is large, the resonance curve is wide and the oscillator responds to a wide band of excitation frequency. This is effectively the case of detectors, such as the eardrum and microphones.

– At resonance, the system receives a large amount of power. This may have harmful effects, and sometimes catastrophic consequences: in buildings if their natural frequency is close to that of an earthquake, in ships if their rolling frequency is close to that of the sea waves, etc. As famous examples, we can mention the collapse of the Tacoma Narrows Bridge in 1940 caused by a 65 km/h wind, and the collapsing bridges at Manchester (1831) and Angers (1849) when a platoon of soldiers marched across them in step with a frequency, which happened to be close to the natural frequency of the bridges.

– Atoms and molecules may be in states of characteristic energy E_i. They are excited with a high probability if they are exposed to an electromagnetic radiation of frequency $\tilde{\nu}_{ij} = (E_i - E_j)/h$, where h is the Planck constant. The *absorption spectroscopy* uses this resonance property. It is used, for instance, to determine the chemical composition of a substance by exposing it to a radiation with a wide frequency band and analyzing the absorbed frequencies.

– Atomic nuclei have a magnetic moment $\mathcal{M} = g\mu\mathbf{S}$, where $\mathbf{S}$ is the spin (intrinsic angular momentum in unit of $\hbar = h/2\pi$), g is the *gyromagnetic ratio* and $\mu = e\hbar/2m_p$ is the *nuclear magneton*. If the nucleus is exposed to a magnetic field $\mathbf{B}$, it obtains a magnetic energy $U_{(M)} = -g\mu BS_z$. The component S_z of the spin in the direction Oz is quantized and may change by one unit. The radiation is also quantized as photons of energy $E = h\tilde{\nu}$. Thus the nucleus may absorb a photon and re-emit it with a high probability, if $h\tilde{\nu}$ is equal to the difference between two possible values of the energy (*magnetic resonance*); hence, the relation $h\tilde{\nu} = U_{(M)}(S_z) - U_{(M)}(S_z + 1) = g\mu B$, that is, $\tilde{\nu} = g\mu B/h$. The emitted photons may be detected by appropriate electronic devices. The *nuclear magnetic resonance* (*NMR*) and the *electronic spin resonance* are examples of this type of resonance with several practical applications.

– Relationship [3.29] may be used to measure the relaxation time τ if the resonance width is easier to measure as in nuclear physics and particle physics.

3.12. Summary

The equation of a forced oscillator is $\ddot{u} + 2\beta\dot{u} + \omega_o^2 u = F(t)/\mu$ where $\beta = b/2\mu$ and $\omega_o = \sqrt{K/\mu}$. Its solution is the sum of the general solution u_o for the free equation ($F = 0$) and the particular solution u_F of the complete equation. u_o is only important in the *transient regime* (which lasts a time of the order of $10\tau = 10/\beta$ if the oscillator is underdamped). Thus, the solution in the *steady regime* is u_F.

If $F = F_m \cos(\Omega t)$, the steady regime solution is $u(t) = A\cos(\Omega t - \phi)$ where:

$$A = \frac{F_m}{\Omega Z} \quad \text{and} \quad \tan\phi = \frac{2\beta\Omega}{\omega_o^2 - \Omega^2} \quad (0 < \phi < \pi), \qquad Z = \frac{\mu}{\Omega}\sqrt{(\omega_o^2 - \Omega^2)^2 + 4\beta^2\Omega^2}$$

Z is the impedance of the oscillator. This is the ratio of the amplitude F_m of the excitation force to the amplitude ΩA of the velocity.

As a function of Ω, the amplitude A increases from $A_o = F_m/\mu\omega_o^2$ for $\Omega = 0$ to a maximum $A_{max} = F_m/2\beta\mu(\omega_o^2 - \beta^2)^{1/2}$ for a frequency $\Omega_{max} = (\omega_o^2 - 2\beta^2)^{1/2}$; then, it decreases to 0 as $\Omega \to \infty$. ϕ increases from 0 at $\Omega = 0$ to $\pi/2$ for $\Omega = \omega_o$ and approaches π as $\Omega \to \infty$. The oscillator goes into resonance if the angular frequency Ω of the excitation force is close to ω_o; then, $A = A_{max}$ and u has a phase lag of $\pi/2$ over the force. The smaller the damping coefficient β is, the shaper the resonance peak is. For $\beta = 0$, A_{max} is infinite and, as β is increased, the peak becomes smaller. It disappears completely at $\Omega = 0$ if $\beta = \omega_o/\sqrt{2}$.

The energy of an oscillator is $E_{(T)} = \frac{1}{2}\mu\dot{u}^2 + \frac{1}{2}Ku^2$. The instantaneous power $P_{(ex)} = F\dot{u}$ that is supplied by the generator is equal to the sum of the rate of variation of $E_{(T)}$ in time, and the dissipated power by the friction force $P_{(fr)} = b\dot{u}^2$. Thus, $P_{(ex)} = dE_{(T)}/dt + P_{(fr)}$. The average value of $E_{(T)}$ over a period is $< E_{(T)} > = \frac{1}{4}\mu A^2(\omega_o^2 + \Omega^2)$. It remains constant in the course of time; thus,

$$< P_{(ex)} > = < P_{(fr)} > = \tfrac{1}{2}\,\Omega^2 A^2 Z \sin\phi = \mu\beta\Omega^2 A^2 = (F_m^2/2Z)\sin\phi = \tfrac{1}{2}\,A' F_m \sin\phi$$

The average transferred power to the oscillator $< P_{(ex)} >$ increases from zero for $\Omega = 0$ to a maximum $P_{max} = F_m^2/4\beta\mu$ for $\Omega = \omega_o$; then, it decreases to zero at high frequency. The *resonance width* is $\Gamma \equiv \Omega^+ - \Omega^- = 2\beta = 2/\tau$ where $\Omega^\pm$ correspond to $<P_{(ex)}> = ½ P_{max}$. The *quality factor* is $f_q = \pi\tau/T_o = ½\, \omega_o\tau = \omega_o/\Gamma \approx A_{max}/A_o$.

A complex force $\underline{F} = F_m e^{i\Omega t}$ produces a complex oscillation $\underline{u}(t) = \underline{A}\, e^{i\Omega t}$ where

$$\underline{A} = A\, e^{-i\phi},\ A = F_m/\Omega Z \text{ and } \tan\phi = 2\beta\Omega(\omega_o^2 - \Omega^2).$$

The complex impedance is:

$$\underline{Z} = \frac{\underline{F}}{\dot{\underline{u}}} = \frac{F_m}{i\Omega\underline{A}} = \mu[2\beta + i(\Omega - \omega_o^2/\Omega)] \equiv b + i(\mu\Omega - \frac{K}{\Omega}) = Z e^{i(\phi - \pi/2)} = Z e^{i\phi_Z}.$$

The modulus of $\underline{Z}$ is the real impedance Z and the phase of $\underline{Z}$ is $\phi_Z = \phi - \pi/2$ which is equal to the phase lag of the velocity $\dot{u}$ over the force.

The complex impedance of an LCR circuit is $\underline{Z} = R + i(L\Omega - 1/\Omega C)$. The power supplied by the e.m.f. generator is $P_{(ex)} = I\mathcal{E}$, and its average value over a period is $<P_{(ex)}> = ½ I_m\mathcal{E}_m \cos\phi_Z = (R/2Z^2)\mathcal{E}_m^2 = ½ R I_m^2$. The power factor is $\cos\phi_Z = R/Z$.

Any excitation force $F(t)$ may be written as a superposition of harmonic forces. If the oscillator is linear, its response to $F(t)$ is the superposition of the responses to the spectral terms of $F(t)$. The response is large if a spectral frequency of F is close to the natural frequency of the oscillator (resonance).

A system of coupled oscillators, which is subject to a simple harmonic force $F_m \cos(\Omega t)$ goes into resonance if Ω is close to one of the normal frequencies of the system. The power supplied by the excitation force then has a peak.

An external generalized force is defined by the power that it supplies to the system $P_{(ex)} = F\dot{u}$. Thus, the equation of the oscillator is $\mu\ddot{u} + b\dot{u} + Ku = F$.

3.13. Problem solving suggestions

1) The resonance frequency depends on how it is defined: the amplitude of u as a function of Ω for a fixed value of ω_o has a maximum for $\Omega_{max} = (\omega_o^2 - 2\beta^2)^{½}$. The amplitude of u, the amplitude of the velocity $\dot{u}$ and the average power, which is supplied by the generator as functions of ω_o for fixed Ω, have maximums for $\Omega = \omega_o$ exactly. The frequencies Ω_{max} and ω_o differ very little if the damping coefficient is small ($\beta << \omega_o$). The forced oscillations always have a phase lag ϕ ($0 < \phi < \pi$) over the force. At resonance ($\Omega = \omega_o$), this phase lag is $\pi/2$.

2) The complex impedance $\underline{Z} = b + i(\mu\Omega - K/\Omega)$ is a useful concept for studying the response of a linear oscillator. The real impedance is $Z = |\underline{Z}|$ (not $\mathcal{R}e\,\underline{Z}$):

– The equivalent complex impedance of circuit elements, of impedances $\underline{Z}_i$ connected in series, is $\underline{Z} = \Sigma_i\, \underline{Z}_i$; and the equivalent complex impedance of elements connected in parallel is given by $1/\underline{Z} = \Sigma_i\, 1/\underline{Z}_i$. These rules are not valid if we use their real impedances.

– If the impedance $\underline{Z}$ of an oscillator is known, the complex amplitude of the generalized velocity $\underline{\dot{u}}$ is $\underline{F}_m/\underline{Z}$ and that of $\underline{u}$ is $\underline{F}_m/i\Omega\underline{Z}$. The real amplitude of $\dot{u}$ is F_m/Z and that of u is $F_m/\Omega Z$.

– The phase lag of $\dot{u}$ over the force is equal to ϕ_z ($-\pi/2 < \phi_z < \pi/2$) and that of u is $\phi = \phi_z + \pi/2$ such that $0 < \phi < \pi$ and $\tan\phi = 2\beta\Omega/(\omega_o^2 - \Omega^2)$.

– The average power, which is supplied by the generator of the force is $\langle P_{(ex)}\rangle = (F_m^2/2Z)\sin\phi = (F_m^2/2Z)\cos\phi_z$, where $\sin\phi = \cos\phi_z = b/Z$.

3) Do not confuse *linearity* and *proportionality*. The relation of the voltage across an electric element to the intensity may be a proportionality ($V = RI$), a differentiation $[V = L\ (dI/dt)]$ or an integration $[V = (1/C)\ \int dt\ I]$. All these relationships are linear. In the case of a sinusoidal current, the ratio of the amplitude of V to that of I is Z, but V and I are not always in phase. In complex representation, the ratio of $\underline{V}$ to $\underline{I}$ is $\underline{Z}$, but this is not valid for currents if they are not simple harmonic, such as saw-tooth or repeated pulses.

3.14. Conceptual questions

Q1. Recall the definitions and expressions of the natural frequency ω_o, the damping coefficient β and the pseudo-angular frequency $\hat{\omega}$, as functions of the mass, the friction constant b and the restoring force constant K. Define the relaxation time and the quality factor. Earthquakes have a quality factor, which varies between 250 and 1,500; what does that means?

Q2. Draw the phasor diagram for the response $u(t)$ of an oscillator subject to a simple harmonic force of angular frequency Ω. Draw a similar diagram for the response at resonance. Can you draw such a diagram for the response to a force, which is not a simple harmonic, such as a constant force, a saw-tooth and a periodic series of rectangular pulses? Verify that the quality factor is equal to the ratio A_{max}/A_o where A_{max} is the amplitude at resonance, and A_o is the amplitude at low frequency.

Q3. If you move a piece of rigid cardboard alternatively to the right and to the left, it follows your hand without a phase lag. If you do the same for a thin piece of paper, it follows your hand with a phase lag of π. Can you explain why?

Q4. To analyze a sustained *LCR* circuit, it is convenient to study the current intensity I instead of the charge Q of the capacitor. Plot the current amplitude as a function of the excitation angular frequency Ω, and verify that it has a maximum for $\Omega = \omega_o = 1/\sqrt{LC}$ exactly. You get the same result if you study the supplied power of the generator, while the amplitude of Q has a maximum at the frequency $\Omega = \sqrt{1/LC - R^2/2L^2}$.

Q5. Draw the phasor diagram for the impedance of an *LCR* circuit sustained with an e.m.f. of angular frequency Ω. Draw this diagram at resonance itself. Deduce the phase ϕ_z and the real impedance.

Q6. If you observe the light spectrum of a gas, formed by well-separated molecules, you find spectral lines corresponding to distinct frequencies; but if you observe the spectrum of the emitted light by an incandescent solid body, you find that it is continuous. If a beam of light, emitted by an incandescent lamp, passes through sodium vapor, we observe dark lines at the same emission frequencies of a sodium lamp: explain why.

3.15. Problems

Transient and steady regimes

P3.1. The equation of motion of a particle of mass m confined to move on an axis is $\ddot{u} + 2\beta\dot{u} + \omega_o^2 u = 0$. Assume that it is initially at rest and a constant force F_o is exerted on it starting at $t = 0$. What is the solution in steady regime? What is the transient solution? What is the final position x_{lim} of the particle? How long does it take to reach 0.99 x_{lim} if $F = -m\omega_o^2$, $m = 1$ kg, $\beta = 0.2$ s^{-1} and $\omega_o = 1$ rad/s? What is the work of the force required to reach the final position?

P3.2. What is the angular frequency corresponding to the maximum amplitude of a forced oscillator, and what is the value of this maximum? What is the angular frequency corresponding to the maximum of the average power, which is supplied by the generator of the force, and what is the value of this maximum? Determine the angular frequencies corresponding to a power equal to half the maximum power. Deduce the resonance width.

P3.3. a) Verify that the response of an oscillator may be written in the form $u = A_{\text{el}} \cos(\Omega t) + A_{\text{ab}} \sin(\Omega t)$. Calculate A_{el} and A_{ab}. **b)** Show that the average value of the power transferred to the oscillator is $< P_{(ex)} > = ½\, A_{ab}\Omega F_m = \beta\mu F_m{}^2/Z^2$. **c)** Analyze the variations of the *absorption amplitude* A_{ab} and the *elastic amplitude* A_{el} as functions of the angular frequency Ω.

P3.4. A point-like body of mass $m = 0.2$ kg is suspended from a spring of constant $K = 80$ N/m, and it is subject to a friction force $-b\dot{x}$. **a)** It was found that the ratio of the damped frequency to the undamped frequency is 0.97. What are the values of the friction coefficient b and the attenuation constant δ? By what factor is the amplitude reduced after 10 complete oscillations? **b)** A simple harmonic force $F = F_m \cos(\Omega t)$, such that $F_m = 2$ N and $\Omega = 30$ rad/s is now exerted on this body. Calculate the amplitude of the resulting motion and its phase lag over the force. Calculate the average power supplied by the generator of the force. Calculate the impedance and the power factor.

P3.5. A point-like body of mass m is confined to move on the horizontal x-axis. **a)** Assuming that it is subject only to a friction force $F_{(fr)} = -b\dot{x}$, write its equation of motion, and its solution corresponding to the initial position x_o and the initial velocity x'_o. Analyze the special case $x_o \neq 0$ and $x'_o = 0$ and the special case $x_o = 0$ and $x'_o \neq 0$. **b)** Assuming that the body is subject to a small friction force $F_{(fr)} = -b\dot{x}$ and a restoring force $F = -Kx$, write its equation of motion and its solution. Calculate the period if $m = 250$ g, $b = 5$ N.s/m and $K = 100$ N/m. **c)** Assume that the restoring force is turned off and the body is subject to an external force $F = F_m \cos(\Omega t)$. Write the solution, which corresponds to the initial conditions x_o and x'_o. Deduce the expression of steady motion. **d)** Now assume that the body is subject to the excitation force $F = F_m \cos(\Omega t)$ with $F_m = 3$ N and $\Omega = 45$ rad/s, the restoring force and the friction force of question (b). Calculate the amplitude of the steady oscillations and the phase lag of the displacement over the force F. Calculate the kinetic energy, the potential energy and the total energy as functions of time. Calculate the mean values of these quantities over a period. Calculate the average energy supplied by the force generator, the impedance and the power factor.

Sustained electromagnetic oscillations

P3.6. An LCR circuit is connected to the terminals of a generator of constant e.m.f. $\mathcal{E}_o$. **a)** Write the circuit equation for the charge Q of the capacitor. Assume that the capacitor charge is Q_o at $t = 0$, when the generator is turned on. Write the expression of the charge Q as a function of t in the case of small resistance, and in the case of large resistance. Plot Q and the current intensity I against t in both cases. **b)** Write down the expressions for the electric, magnetic and total energies. **c)** Now assume that the LCR circuit with small resistance is connected to a generator of a simple harmonic e.m.f. $\mathcal{E} = \mathcal{E}_m \cos(\Omega t)$. Write down the expression of the current intensity. Assuming that L, C and Ω are maintained constant, for which value of R is the dissipated power in the resistor maximum? Again assuming that L, C and R are maintained constant, for which value of Ω is the dissipated power maximum?

P3.7. A generator of e.m.f. $\mathcal{E} = \mathcal{E}_m \cos(\Omega t)$ is connected to a capacitor of capacitance C in series with a coil of self-inductance L and internal resistance R. **a)** Write down the expressions for the current intensity and the charge of the capacitor using the phasor diagram. **b)** Calculate the ratio of the voltage across the coil to $\mathcal{E}_m$ at resonance if $C = 1\ \mu\text{F}$, $L = 0.2$ H and $R = 100\ \Omega$.

P3.8. An LCR circuit is connected to a generator of e.m.f. $\mathcal{E} = \mathcal{E}_m \cos(\Omega t)$. **a)** Write down the expressions of the capacitor charge Q and the current intensity I as functions of time in the steady regime. **b)** Calculate the electromagnetic energy of the circuit. **c)** Calculate the instantaneous power supplied by the generator, and verify that it is equal to the sum of the variation rate of the stored energy and the dissipated power in the resistor. **d)** Calculate the average power supplied by the generator and verify that it is equal to the average power dissipated in the resistor. **e)** Determine the value of Ω corresponding to the maximum supplied power, and the value P_{max} of this maximum. Determine the values of Ω, which correspond to a supplied power equal to ½ P_{max}.

P3.9. A generator of e.m.f. $\mathcal{E} = \mathcal{E}_m \cos(\Omega t)$ is connected to a capacitor of capacity C in parallel with a coil of resistance R and self-inductance L (Figure 3.6). Calculate the current intensity and the average power supplied by the generator. Assuming that R is small, show that the intensity has a minimum (*anti-resonance*) for an angular excitation frequency close to $\omega_o = 1/\sqrt{LC}$.

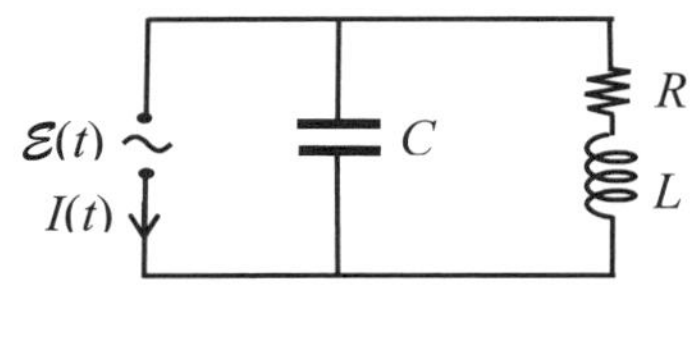

Figure 3.6. *Problem P3.9*

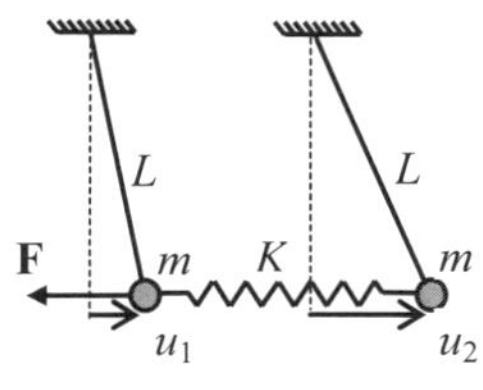

Figure 3.7. *Problem P3.10*

Coupled oscillators, response to an arbitrary force

P3.10. Two pendulums without friction ($\beta_i = 0$) are coupled as in Figure 3.7. The first pendulum is subject to a force $F = F_m e^{i\Omega t}$. **a)** Write the steady solution and plot the ratio A_2/A_1 of the amplitudes as a function of Ω. Verify that, if Ω is in the band $[\omega_{(1)}, \omega_{(2)}]$ between the normal frequencies of the system, the ratio $|A_2/A_1|$ is larger than 1. Determine the pass-band. **b)** What is the pass-band if the length of the pendulums is very large? **c)** What is the pass-band if the spring coupling is very rigid (the spring is replaced by a rigid rod, for instance)?

P3.11*. A force has a constant magnitude F_m but it changes direction periodically with a period τ. **a)** Plot the force against time. Create its Fourier analysis and show that it may be written as the Fourier series

$$F(t) = \frac{4F_m}{\pi} \Sigma_{p\geq 0} \frac{1}{2p+1} \sin[(2p+1)\Omega t] \text{ where } \Omega = \pi/\tau.$$

Make a separate plot for the fundamental and for the sum of the first two terms of this series and compare them with $F(t)$. **b)** Using this Fourier series, study the motion of a point-like mass m subject only to the force $F(t)$. Plot the position $x(t)$ if we keep the $p = 0$ and $p = 1$ harmonics. **c)** A mass m is attracted to the origin with a force $-m\omega_o^2 x$ and is subject, in addition, to a friction force $-2m\beta\dot{x}$ and the force $F(t)$. Calculate the amplitude and the phase of the motion due to the $(2p + 1)$ harmonic. Compare the amplitudes of the harmonics $p = 0, 1, \ldots, 5$ if $\beta = 1\ \text{s}^{-1}$, ω_o= 8 rad/s and $\Omega = 4$ rad/s. What conclusion can you draw?

P3.12. Two LC circuits are coupled by a capacitor (Figure 3.8). **a)** Assuming that there is no e.m.f., write the differential equations for the intensities I_1 and I_2 and determine the normal modes and the general solution. **b)** Assume now that the first circuit contains a generator of e.m.f. $\underline{\mathcal{E}} = \mathcal{E}_m e^{i\Omega t}$. Determine the steady intensities.

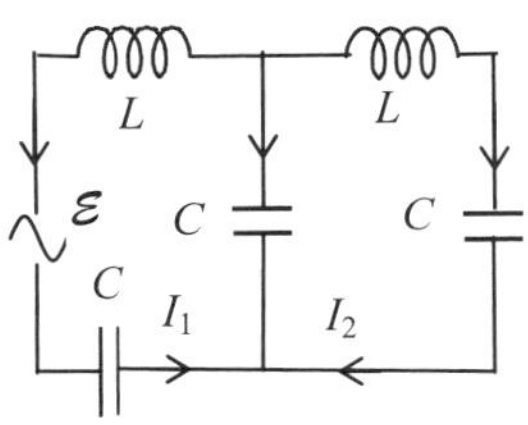

Figure 3.8. *Problem P3.12*

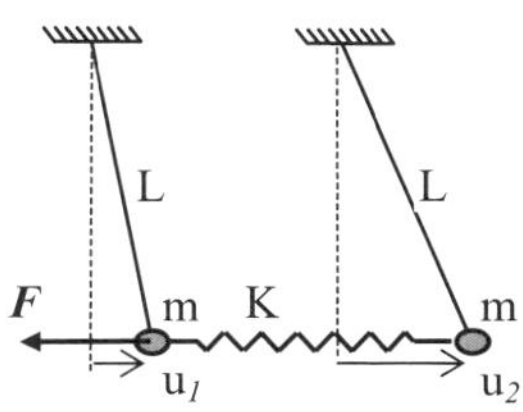

Figure 3.9. *Problem P3.13*

P3.13.** Three simple pendulums are coupled by two springs of equal constant K (Figure 3.9). Assume that the masses are equal and that the first and third masses are subject to small friction forces $-2m\beta\dot{x}_1$ and $-2m\beta\dot{x}_3$ respectively, and a force $F = F_m \cos(\Omega t)$ acts on the first mass. **a)** Write the equations of motion for the masses. For this, set $\omega_o^2 = g/L$ and $\omega_r^2 = K/m$. **b)** Assuming that β is small enough to be able to neglect the β^2 terms, compared to ω_o^2 and ω_r^2, show that the normal angular frequencies are ω_o, $\sqrt{\omega_o^2 + \omega_r^2}$ and $\sqrt{\omega_o^2 + 3\omega_r^2}$, with the damping coefficients $2\beta/3$, β and $\beta/3$ respectively. **c)** To determine the steady motion in the case of a force $F_m \cos(\Omega t)$, it is convenient to solve the equations for a complex force $F_m e^{i\Omega t}$ first. Show that the response to the force $F_m \cos(\Omega t)$ is the real part of the response to the complex force $F_m e^{i\Omega t}$. Try complex solutions $x_i = \underline{A}_i e^{i\Omega t}$ and determine the complex amplitudes $\underline{A}_i$. Deduce the amplitudes A_i and the phases ϕ_i of the real motions.

P3.14*. It is possible to sustain the oscillations of some systems even though the source is non-oscillatory. These are the so-called *self-excited oscillations* or *relaxation oscillations.* We consider an electromagnetic example. A neon lamp glows brightly if the voltage across its terminals exceeds an ignition value V_i, and it remains so as long as the voltage is higher than an extinction voltage V_e ($V_e < V_i$). It may be considered as having an infinite resistance when it is extinct, and a finite resistance if it is bright. It is connected in parallel with a capacitor C to the terminals of an e.m.f. $\mathcal{E}$ generator (Figure 3.10). The generator is switched on at $t = 0$, the charge of the capacitor is then 0. Study the voltage u across the lamp terminals as a function of time. Show that the period of the self-excited oscillations is

$$T = \tau_2 \ln[(V_i - \mathcal{E}')/(V_e - \mathcal{E}')] + \tau_1 \ln[(V_e - \mathcal{E})/V_i - \mathcal{E})],$$

where we have set: $\mathcal{E}' = \mathcal{E}/(1 + R/R_o)$, $\tau_1 = CR$ and $\tau_2 = CR/(1 + R/R_o)$.

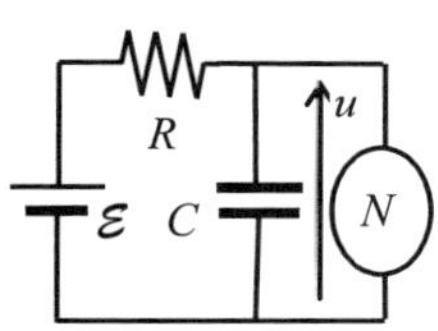

Figure 3.10. *Problem P3.14*

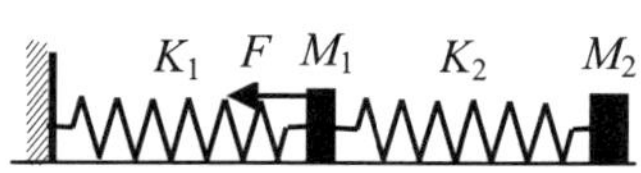

Figure 3.11. *Problem P3.15*

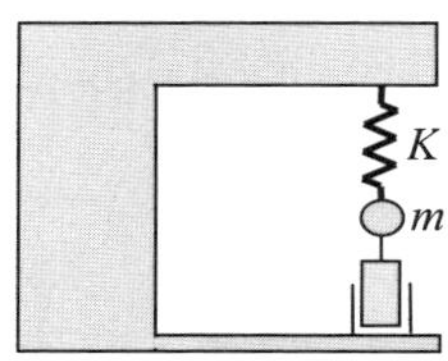

Figure 3.12. *Problem P3.17*

P3.15. Two masses M_1 and M_2 are confined to oscillate on the horizontal x-axis subject to the friction forces $-b_1 \dot{u}_1$ and $-b_2 \dot{u}_2$. They are coupled by two springs, as in Figure 3.11. The mass M_1 is subject to an external force $F = F_m \cos(\Omega t)$. Write the equations of motion. Using complex analysis and designating by $\underline{z}_1$ and $\underline{z}_2$ the impedances of the masses M_1 and M_2 subject respectively to the springs (1) and (2) and to the friction forces, write the steady solution in the case of the complex force $\underline{F} = F_m e^{i\Omega t}$. What is the complex entry impedance, $\underline{Z}_1 = \underline{F}_m / \dot{u}_{1,m}$?

P3.16. Two masses m_1 and m_2 are confined to frictionless movement on the horizontal x-axis. They are coupled by a spring of constant K. Assume that the mass

m_1 is subject to a force $F = F_m e^{i\Omega t}$. **a)** Determine the motions u_1 and u_2 of the masses. Show that the ratio of their amplitudes is $A_2/A_1 = K/(K - m_2\Omega^2)$. **b)** Is it possible to choose the spring constant in order to have a small amplitude A_1 for a given angular frequency Ω? Explain how to reduce the oscillations of m_1 by coupling it to the mass m_2. **c)** Is it possible to choose the spring in order to have a very small A_2? Assume that a sensitive instrument is placed, without friction, on an air cushion, and it must be fixed horizontally to the laboratory walls. The horizontal oscillations of the walls may disturb the instrument if it is coupled to the walls by rigid rods. Explain how to use the results of question (b) to reduce the effect of these oscillations. **d)** The vertical oscillations are uncomfortable in cars, train wagons, etc. To reduce their effects, the chassis are mounted on springs (shock absorbers). Consider for instance a mass m_1 that is separated from the floor by a spring. How much must this mass contract the spring in order to reduce the floor oscillations of frequency 30 Hz by a factor 100?

P3.17. Figure 3.12 shows a simple seismograph, formed by a mass m suspended from the bottom of a spring of constant K, which hangs from a stand fixed to the floor. The mass m is also subject to a friction force $F_{(fr)} = -2m\beta\dot{u}$. Calculate the amplitude of the oscillations of m if the earthquake has an angular frequency Ω and an amplitude A. What should the spring constant be in order to have a large amplitude for a given frequency Ω?

Chapter 4

Propagation in Infinite Media

A wave is a disturbance, which propagates in a medium carrying energy and other physical quantities. The wave is characterized by its speed, direction of propagation and profile. If it is a simple harmonic its profile is specified by its amplitude, angular frequency and phase. Certain properties are common to waves and free particles: rectilinear propagation in homogeneous media, reflection, transfer of energy, etc. However, some properties are specifically characteristic of waves: space and time extension, interference, diffraction, etc.

In this chapter we study the propagation of waves in infinite media. We write the wave equation in one, two and three dimensions. We consider the harmonic waves and we introduce the Fourier analysis of waves. Finally, we discuss the energy density and its propagation.

In order to simplify the notation, we designate the partial derivatives by $\partial_t u \equiv \dot{u} \equiv \partial u/\partial t$, $\partial_x u \equiv \partial u/\partial x$ and $\partial^2_{xy} u \equiv \partial^2 u/\partial x \partial y$, etc. A summation over the components will be written Σ_i, where $i = x$, y and z. For instance, $\Sigma_i \, \partial_i u \, \mathbf{e}_i$ means $\partial_x u \, \mathbf{e}_x + \partial_y u \, \mathbf{e}_y + \partial_z u \, \mathbf{e}_z$.

4.1. Propagation of one-dimensional waves

Everyday experience shows that a disturbance, which is produced in a medium, propagates from one point to another. The propagation of mechanical waves, such as elastic waves and sound waves, is due to the interaction of the particles of the medium; and the propagation of electromagnetic waves is due to the electromagnetic induction and the interaction of the particles of the medium with the wave.

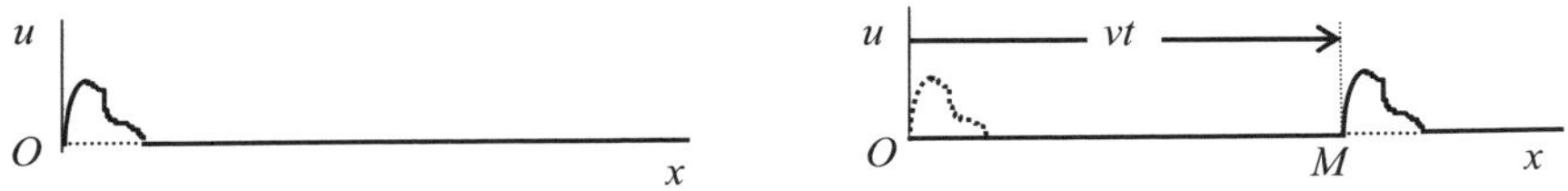

Figure 4.1. *Propagation of a pulse on a taut string*

Consider, for instance, a taut elastic string. A pulse that is produced at a point O may be specified by the transverse displacement of O as a function of time $u_O = f(t)$. Due to the tension, a displaced element of the string acts on the following element and the disturbance propagates with a certain speed v producing a visible deformation of the string (Figure 4.1). After a time t, the deformation reaches the point M of coordinate $x = vt$. In the ideal case, the speed depends only on the characteristics of the string and the pulse propagates without any change in shape. The displacement at M is then the same as at O, but with a time delay x/v, that is,

$$u = f(t - x/v), \qquad [4.1]$$

where f is the same function as at O. A disturbance, which propagates indefinitely in an infinite medium, is a *progressive wave*.

The physical quantity u, which represents the disturbance, is the *wave function*. We consider in this book *scalar wave functions* (such as the pressure in a gas), and *vector wave functions* (such as the displacement of the points of a string or the electric field in the three-dimensional space). Note that the same wave might be represented by different physical quantities. For instance, a sound wave may be represented by the local variation of pressure, the mean displacement of molecules, etc. In the absence of disturbance, the molecules of the medium are not at rest but are in thermal agitation, especially in the case of fluids. In the case of a gas, for instance, they are moving at random in all directions with a velocity, which may be very high, but the average velocity is equal to zero. If a mechanical wave propagates in the medium, it produces a small average local displacement $\mathbf{u}$ of the molecules, and a small mean velocity $\dot{\mathbf{u}}$. This velocity should not be confused with the speed of propagation v of the disturbance. In the case of a simple harmonic wave, $\mathbf{u}$ and $\dot{\mathbf{u}}$ are oscillatory; thus, their time averages are equal to zero while the speed of propagation v of the wave, from one point of the medium to another, is certainly not equal to zero. The wave propagation is not usually accompanied by any transfer of matter. However, in the case of a strong disturbance (such as an explosion), there may be some transfer of matter.

In the case of a one-dimensional medium, such as a string, the wave function u depends on time and a single space variable x. Setting $u = f(s)$ where $s \equiv t - x/v$ and designating the derivatives of f by $f' = df/ds$ and $f'' = d^2f/ds^2$, we get the partial derivatives of u:

$$\partial_t u = (df/ds)(\partial s/\partial t) = f', \qquad \partial^2_{tt}u = (df'/ds)(\partial s/\partial t) = f'',$$

$$\partial_x u = (df/ds)(\partial s/\partial x) = -f'/v, \qquad \partial^2_{xx}u = -(1/v)(df'/ds)(\partial s/\partial t) = f''/v^2.$$

Comparing these expressions, we may write the *one-dimensional wave equation*

$$\partial^2_{tt}u - v^2\,\partial^2_{xx}u = 0. \qquad [4.2]$$

This equation has several types of solutions, in particular,

$$u_+ = f(t - x/v) \qquad \text{and} \qquad u_- = g(t + x/v), \qquad [4.3]$$

where f and g are arbitrary functions. These solutions represent *progressive waves*. u_+ propagates toward the positive x while u_- propagates toward the negative x. The wave equation being linear, the propagation equation also has a solution, which is a superposition of u_+ and u_-, that is,

$$u = f(t - x/v) + g(t + x/v). \qquad [4.4]$$

4.2. Propagation of two- and three-dimensional waves

Consider waves propagating in a two-dimensional isotropic medium such as the transverse deformation of a taut elastic membrane, a wave on the surface of water, etc. Assume that the same disturbance, $u_O = f(t)$, is produced at all the points of a straight line D_O passing through the origin O. Due to the translation symmetry parallel to D_O, the wave will be the same at all points of any straight-line D, parallel to D_O (Figure 4.2). Thus, we have a *rectilinear wave* and the lines D are its *wave fronts*. Let **e** be the unit vector, which is perpendicular to the wave fronts and α and β the components of **e** (with the relation $\alpha^2 + \beta^2 = 1$). **e** is the *direction of propagation* of the wave. All the points of D have the same disturbance as the points of D_O, but with a time delay d/v where v is the speed of propagation and d is the perpendicular distance separating D from D_O. If a point M of D has the coordinates x and y, the distance d is ($\overrightarrow{OM}.\mathbf{e}$) and the disturbance at M is

$$u = f(t - \overrightarrow{OM}.\mathbf{e}/v) = f[t - (\alpha x + \beta y)/v]. \qquad [4.5]$$

Setting $s = t - (\alpha x + \beta y)/v$ and $f'' = d^2u/ds^2$, we get the partial derivatives:

$$\partial^2_{xx}u = (\alpha^2/v^2)f'', \qquad \partial^2_{yy}u = (\beta^2/v^2)f'' \qquad \text{and} \qquad \partial^2_{tt}u = f''. \qquad [4.6]$$

As $\alpha^2 + \beta^2 = 1$, we may combine these relations to get the two-dimensional *wave equation*

$$\partial^2_{tt}u - v^2 (\partial^2_{xx} u + \partial^2_{yy}u) = 0. \qquad [4.7]$$

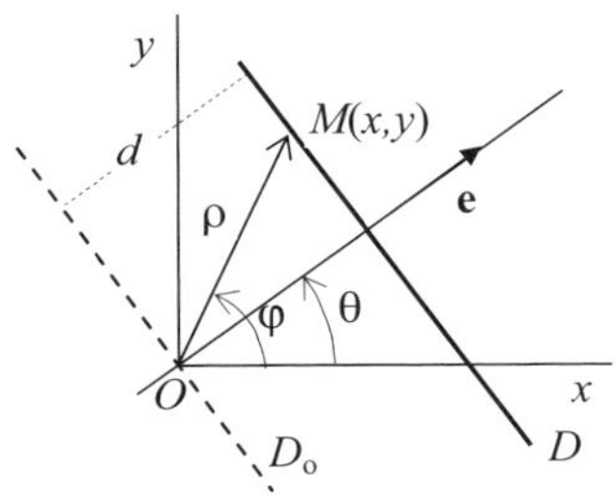

Figure 4.2. *Rectilinear surface wave*

We may use the polar coordinates ρ and φ where ρ is the distance OM and φ is the angle of $\overrightarrow{OM}$, measured anticlockwise from the x-axis ($0 < \varphi < 2\pi$) with the relations:

$$x = \rho \cos \varphi, \quad y = \rho \sin \varphi, \quad \rho = \sqrt{x^2+y^2}, \quad \cos \varphi = x/\rho, \quad \sin \varphi = y/\rho. \quad [4.8]$$

Wave equation [4.7] takes the form

$$\partial^2_{tt}u - v^2\left(\frac{\partial^2 u}{\partial \rho^2} + \frac{1}{\rho}\frac{\partial u}{\partial \rho} + \frac{1}{\rho^2}\frac{\partial^2 u}{\partial \varphi^2}\right) = 0. \qquad [4.9]$$

In particular, if the wave is symmetrical about the point O, u does not depend on φ and the wave equation reduces to

$$\partial^2_{tt}u - v^2\left(\frac{\partial^2 u}{\partial \rho^2} + \frac{1}{\rho}\frac{\partial u}{\partial \rho}\right) = 0 \qquad \text{[symmetrical wave about } O\text{]}. \qquad [4.10]$$

Now consider the waves $u(\mathbf{r}, t)$ propagating in a three-dimensional isotropic medium. Assume that the same disturbance $u_O = f(t)$ is produced at all points of a plane P_O passing through the origin. Because of the translation symmetry parallel to P_O, the disturbance will be the same at all points of any plane P parallel to P_O (Figure 4.3a). The unit vector **e**, perpendicular to the wave fronts P, determines the *direction of propagation*. The disturbance u at all the points M of the wave front P is the same as u_O on the wave front P_O but with a time delay d/v: where d is the perpendicular distance separating P from P_O and v is the speed of propagation. Thus, the wave at M of coordinates x, y, z and at time t is

$$u = f(t - \overrightarrow{OM}.\mathbf{e}/v) = f[t - (\alpha x + \beta y + \gamma z)/v], \qquad [4.11]$$

where α, β and γ are the components of $\mathbf{e}$. The second derivatives of u are:

$$\partial^2_{xx} u = (\alpha^2/v^2)f'', \quad \partial^2_{yy} u = (\beta^2/v^2)f'', \quad \partial^2_{zz} u = (\gamma^2/v^2)f'', \quad \partial^2_{tt} u = f'', \qquad [4.12]$$

where $f'' = d^2u/ds^2$ and $s \equiv t - (\alpha x + \beta y + \gamma z)/v$. As $\alpha^2 + \beta^2 + \gamma^2 = 1$, we may combine these relations to get the three-dimensional wave equation

$$\partial^2_{tt} u - v^2 \Delta u = 0, \qquad [4.13]$$

called the *d'Alembert equation*. Δ is the *Laplacian* operator defined by

$$\Delta = \nabla^2 = \partial^2_{xx} + \partial^2_{yy} + \partial^2_{zz}, \qquad [4.14]$$

where ∇ is the *differential vector operator* (known as *del* or *nabla*) whose components are ∂_x, ∂_y and ∂_z. Wave equation [4.13] does not depend on the direction of propagation. It has expression [4.11] as a solution, which represents a progressive plane wave propagating in the direction $\mathbf{e}$. It has other types of solutions.

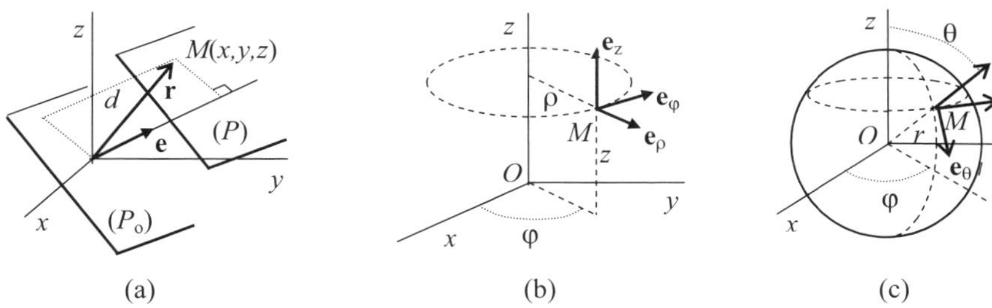

Figure 4.3. *a) Plane wave, b) cylindrical coordinates, and c) spherical coordinates*

Sometimes, it is convenient to use the *cylindrical coordinates* (ρ, φ and z) about a z-axis, where ρ and φ are related to x and y by the same equations of [4.8] (Figure 4.3b). The Laplacian in these coordinates is provided in Appendix A and the wave equation takes the form

$$\partial^2_{tt} u - v^2\left(\frac{\partial^2 u}{\partial \rho^2} + \frac{1}{\rho}\frac{\partial u}{\partial \rho} + \frac{1}{\rho^2}\frac{\partial^2 u}{\partial \varphi^2} + \frac{\partial^2 u}{\partial z^2}\right) = 0. \qquad [4.15]$$

In particular, if the wave is symmetric about the z-axis, the wave function u does not depend on φ and the wave equation reduces to

$$\partial^2_{tt} u - v^2\left(\frac{\partial^2 u}{\partial \rho^2} + \frac{1}{\rho}\frac{\partial u}{\partial \rho} + \frac{\partial^2 u}{\partial z^2}\right) = 0 \qquad \text{[cylindrical symmetry].} \qquad [4.16]$$

Sometimes, we use the *spherical coordinates* (r, θ and φ) about the origin O, such that $r = OM$, θ is the angle of $\overrightarrow{OM}$ measured from Oz, and φ is the angle that the MOz plane makes with the Oxz plane with a sign given by the right-hand rule (Figure 4.3c). These coordinates are related to x, y and z by the equations:

$$x = r \sin\theta \cos\varphi, \qquad y = r \sin\theta \sin\varphi, \qquad z = r\cos\theta,$$

$$r = \sqrt{x^2+y^2+z^2}, \qquad \tan\theta = \sqrt{x^2+y^2}\,/z \qquad (0 < \theta < \pi),$$

$$\cos\varphi = x/r, \qquad \sin\varphi = y/r \qquad (0 < \varphi < 2\pi). \qquad [4.17]$$

In these coordinates, wave equation [4.13] takes the form

$$\partial^2_{tt} u - v^2\left[\frac{1}{r^2}\frac{\partial}{\partial r}\left(r^2\frac{\partial u}{\partial r}\right) + \frac{1}{r^2 \sin\theta}\frac{\partial}{\partial \theta}\left(\sin\theta\frac{\partial u}{\partial \theta}\right) + \frac{1}{r^2 \sin^2\theta}\frac{\partial^2 u}{\partial \varphi^2}\right] = 0. \qquad [4.18]$$

In particular, if the wave is symmetric about the origin O, the wave function u does not depend on θ and φ. The wave equation reduces to

$$\partial^2_{tt} u - v^2\left[\frac{\partial^2 u}{\partial r^2} + \frac{2}{r}\frac{\partial u}{\partial r}\right] = 0 \qquad \text{[spherical symmetry about } O\text{].} \qquad [4.19]$$

It is easy to verify that this equation has solutions of the form

$$u_\pm = \frac{1}{r} f\left(t \mp \frac{r}{v}\right). \qquad [4.20]$$

In this case, the wave function is the same at all the points M, which are at the same distance r from the origin O. Thus, $u_\pm$ represent *spherical waves*. Comparing $u_\pm(t, r)$ at r to $u_S(t)$ on a sphere S of radius r_o, we may write

$$u_\pm(t, r) = \frac{r_o}{r}\, u_S\left(t \mp \frac{r - r_o}{v}\right). \qquad [4.21]$$

The wave at r has the same profile as the wave at S but it is multiplied by a factor r_o/r, and it has a time delay of $(r - r_o)/v$ in the case of u_+ and a time lead $(r - r_o)/v$ in the case of u_-. Thus, u_+ is a *spherical wave*, which *diverges* from O; and u_- is a *spherical wave*, which *converges* at O. We note that the wave at the origin is theoretically infinite; thus, O is a singular point of the wave. The waves u_+ may be effectively emitted by spherical sources of radius r_o and the waves u_- are produced by systems, which focus the wave in O. Equation [4.19] being linear, it also admits the superposition of these two types of spherical waves as solutions

$$u = u_+(t, r) + u_-(t, r). \qquad [4.22]$$

It is not necessary for a medium to effectively be a surface in order for a wave to be two-dimensional. It is sufficient that the wave, which propagates in a three-dimensional medium, does not depend on one of the coordinates. For instance, if the wave does not depend on z, wave equation [4.13] reduces to wave equation [4.7] and the wave equation in cylindrical coordinates [4.15] reduces to wave equation [4.9] in polar coordinates. Similarly, if a wave propagates in a three-dimensional space but does not depend on two coordinates y and z, it may be considered as a one-dimensional wave in x. This is the case for a sound wave propagating in a pipe parallel to the x-axis if it has the same value on any normal section of the pipe (thus, it is independent on y and z).

4.3. Propagation of a vector wave

If a wave is represented by a vector quantity $\mathbf{u}(\mathbf{r}, t)$, its equation of propagation is

$$\partial^2_{tt}\mathbf{u} - v^2\,\Delta\mathbf{u} = 0. \qquad [4.23]$$

In Cartesian coordinates, this equation represents three equations of the form [4.13] for each of the components u_x, u_y and u_z. It has special solutions of the form

$$\mathbf{u} = \mathbf{u}_o f(\mathbf{r}, t), \qquad [4.24]$$

where $\mathbf{u}_o$ is a constant vector. Substituting this expression for $\mathbf{u}$ in [4.23], we find that f must verify the wave equation

$$\partial^2_{tt}f - v^2\,\Delta f = 0. \qquad [4.25]$$

In particular, if $\mathbf{e}$ is a unit vector, this equation has solutions of the form

$$f = f(t - \mathbf{e}.\mathbf{r}/v). \qquad [4.26]$$

Expression [4.24] represents a plane wave, which propagates in the direction $\mathbf{e}$ (Figure 4.4a). At each point $\mathbf{r}$, the vector $\mathbf{u}$ is in the direction of $\mathbf{u}_o$. If we use axes of coordinates, in such a way that Oz is in the direction of $\mathbf{e}$, f depends only on z and equation [4.25] is equivalent to a one-dimensional propagation equation

$$\partial^2_{tt} f - v^2\, \partial^2_{zz} f = 0. \qquad [4.27]$$

In the general case, **u** does not have the same direction everywhere and at any time. If **u** is always in the direction of propagation **e**, the wave is said to be *longitudinal* and, if **u** is always perpendicular to **e**, the wave is said to be *transverse*. Some waves are a combination of longitudinal waves and transverse waves. This is the case of the waves on the surface of a liquid and earthquakes, for instance.

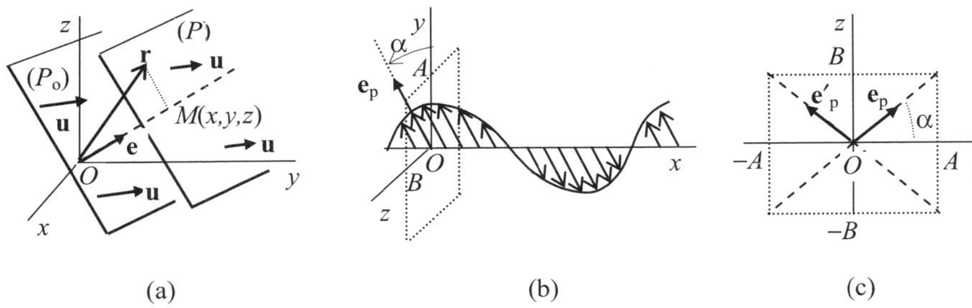

Figure 4.4. *a) Plane vector wave* $\mathbf{u} = \mathbf{u}_0 f(t - \mathbf{e}.\mathbf{r}/v)$, *b) transverse vector wave propagating in the direction Ox and polarized in the direction* $\mathbf{e}_p$, *and c) superposition of two waves in phase and polarized linearly in the directions* $\mathbf{e}_y$ *and* $\mathbf{e}_z$

Consider the case of a spherical vector wave. Ideally, it is emitted by a point source O, or it converges in a focal point O that we take as origin. It is evident that, in this case, spherical coordinates are better adapted than the Cartesian coordinates. The function **u** cannot have the same direction at all space points M (the wave would not have the spherical symmetry). The study of the propagation of spherical transverse vector waves is quite involved and we will consider only the case of longitudinal waves. In this case, **u** is in the direction of $\mathbf{r} = \overrightarrow{OM}$ (i.e. in the direction of the radial unit vector $\mathbf{e} = \mathbf{r}/r$). Thus, we write

$$\mathbf{u}(r, t) = \frac{\mathbf{r}}{r}\, F(r, t) \qquad \text{[longitudinal wave]}. \qquad [4.28]$$

Spherical symmetry requires F not to depend on the direction of **r**. substituting expression [4.28] in equation [4.23], we obtain the equation

$$\partial^2_{tt} F - v^2 \left(\frac{\partial^2 F}{\partial r^2} + \frac{2}{r}\frac{\partial F}{\partial r} - \frac{2}{r^2} F \right) = 0 \qquad \text{[spherical symmetry]}. \qquad [4.29]$$

It may be verified that this equation has the diverging wave solutions

$$F_+ = \frac{v}{r^2}\, f(t - \frac{r}{v}) + \frac{1}{r}\, f'(t - \frac{r}{v}), \qquad [4.30]$$

where f is an arbitrary function of $s \equiv t - r/v$ and f' is its derivative with respect to s. At long distance, the $(1/r)$ term dominates

$$F_+ \to \frac{1}{r} f'(t - \frac{r}{v}) \qquad [\text{if } r \to \infty]. \qquad [4.31]$$

Thus, the magnitude of the wave decreases like $1/r$. We shall see that the intensity of a simple harmonic wave is always proportional to its squared amplitude. Thus, the intensity of the wave decreases like $1/r^2$, which is what we expect from energy conservation, as the wave power is distributed uniformly on a sphere or radius r. However, at short distance, the $(1/r^2)$ term dominates

$$F_+ \to \frac{v}{r^2} f(t - \frac{r}{v}) \qquad [\text{if } r \to 0]. \qquad [4.32]$$

Equation [4.29] also has converging wave solutions of the form

$$F_- = \frac{v}{r^2} f(t + \frac{r}{v}) - \frac{1}{r} f'(t + \frac{r}{v}). \qquad [4.33]$$

They behave like $(v/r^2)\, f(t + r/v)$ at short distance, and like $(-1/r)f'(t + r/v)$ at long distance. $\mathbf{u}_+ = (\mathbf{r}/r)F_+$ is a longitudinal wave which is emitted by a point source at O and $\mathbf{u}_- = (\mathbf{r}/r)F_-$ is a wave which converges in a focal point O. In particular, the longitudinal sinusoidal waves of angular frequency ω are of the form

$$\mathbf{u}_\pm(\mathbf{r}, t) = \frac{\mathbf{r}}{r} \{\frac{v}{r^2} u_o \sin(t \mp \frac{r}{v}) \pm \frac{\omega}{r} u_o \cos[\omega(t \mp \frac{r}{v})]\}. \qquad [4.34]$$

4.4. Polarization of a transverse vector wave

In this section we consider the displacement $\mathbf{u}$ of the points of a taut string. Of course, the analysis holds for any type of transverse vector wave (mechanical or electromagnetic). If the emitter oscillates permanently in a fixed direction $\mathbf{e}_p$ perpendicular to the string, the displacement $\mathbf{u}(x, t)$ at any point of the string and at any time will be parallel to $\mathbf{e}_p$. This wave is said to be *linearly polarized* in the direction of $\mathbf{e}_p$, or *plane polarized* in the plane $(Ox, \mathbf{e}_p)$. A transverse wave, which propagates in the direction Ox, may be linearly polarized in any direction of the plane Oyz, that is, polarized in any plane about Ox (Figure 4.4b). In particular, we may consider two independent states of polarization in the directions Oy and Oz:

$$\mathbf{u}_y = A \cos(\omega t - kx)\, \mathbf{e}_y, \qquad \mathbf{u}_z = B \cos(\omega t - kx + \phi)\, \mathbf{e}_z. \qquad [4.35]$$

Due to the linearity, the wave may be a superposition of these waves with any amplitudes and any phase shift ϕ. We may consider the following special cases:

– If the waves $\mathbf{u}_y$ and $\mathbf{u}_z$ are in phase ($\phi = 0$), their superposition is the wave

$$\mathbf{u} = \cos(\omega t - kx)(A\ \mathbf{e}_y + B\ \mathbf{e}_z) = u_m \cos(\omega t - kx)\ \mathbf{e}_p, , \qquad [4.36]$$

where we have set $u_m = \sqrt{A^2+B^2}$ and $\mathbf{e}_p = (A\mathbf{e}_y + B\mathbf{e}_z)/u_m$. This wave has an amplitude u_m and it is linearly polarized in the direction of the unit vector $\mathbf{e}_p$ of the first diagonal of the rectangle of sides $2A$ and $2B$, thus making an angle $\alpha = \text{Arctan}(B/A)$ with Oy (Figures 4.4b and c). If $\mathbf{u}_y$ and $\mathbf{u}_z$ are in opposite phase ($\phi = \pi$), their superposition is $\mathbf{u} = \cos(\omega t - kx)(A\ \mathbf{e}_y - B\ \mathbf{e}_z) = u_m \cos(\omega t - kx)\ \mathbf{e}'_p$, where $\mathbf{e}'_p = (A\mathbf{e}_y - B\mathbf{e}_z)/u_m$. This is a wave of amplitude u_m and it is linearly polarized in the direction of the unit vector $\mathbf{e}'_p$ of the second diagonal of the rectangle, thus making with Oy an angle $\alpha' = -\text{Arctan}(B/A)$ (Figures 4.4b and c).

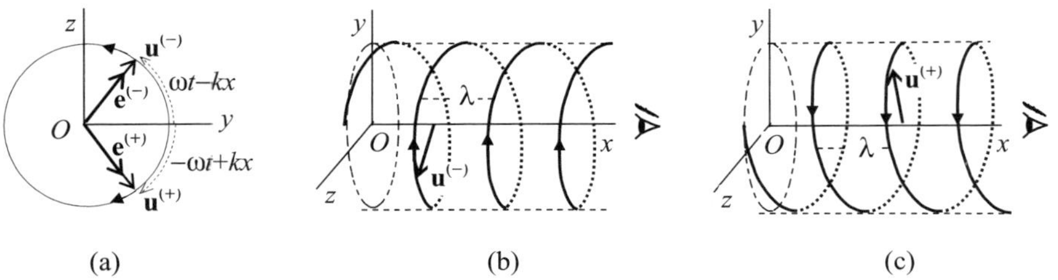

Figure 4.5. *Circularly polarized waves, a) rotation of $\mathbf{u}^{(-)}$ and $\mathbf{u}^{(+)}$ in time at a given point, b) profile of the string in the case of the left-circularly polarized wave $\mathbf{u}^{(-)}$, c) profile of the string in the case of the right-circularly polarized wave $\mathbf{u}^{(+)}$*

– If the waves $\mathbf{u}_y$ and $\mathbf{u}_z$ have equal amplitudes but $\mathbf{u}_z$ has a phase lag of $\pi/2$ over $\mathbf{u}_y$ ($A = B$ and $\phi = -\pi/2$), the resulting wave is

$$\begin{aligned}\mathbf{u}^{(-)} &= A\ [\cos(\omega t - kx)\ \mathbf{e}_y + \cos(\omega t - kx - \pi/2)\ \mathbf{e}_z] \\ &= A\ [\cos(\omega t - kx)\ \mathbf{e}_y + \sin(\omega t - kx)\ \mathbf{e}_z] = A\ \mathbf{e}^{(-)},\end{aligned} \qquad [4.37]$$

where $\mathbf{e}^{(-)}$ is the unit vector in the plane Oyz which makes an angle $\omega t - kx$ with Oy. At a given point x of the string, the amplitude of $\mathbf{u}^{(-)}$ conserves the same magnitude A in the course of time, but its direction (determined by the vector $\mathbf{e}^{(-)}$) rotates about the axis of propagation in the positive direction (counterclockwise). An observer who receives the wave, sees every point of the string describing a circle of radius A in the positive direction (Figure 4.5a). At a given time, the oscillation at x has a phase lag equal to $kx = 2\pi x/\lambda$ with respect to the origin. The points, which are separated by an integer multiple of the wavelength λ, have the same displacement and velocity. The string forms a left-handed spiral of radius A (i.e. in the direction of the left-hand fingers if the thumb points in the direction of propagation Ox) (Figure 4.5b). We say that the wave is *left-handed circularly polarized.*

– If the waves $\mathbf{u}_y$ and $\mathbf{u}_z$ have equal amplitudes but $\mathbf{u}_z$ has a phase lead of $\pi/2$ over $\mathbf{u}_y$ ($A = B$ and $\phi = +\pi/2$), the resulting wave is

$$\begin{aligned}\mathbf{u}^{(+)} &= A\,[\cos(\omega t - kx)\,\mathbf{e}_y + \cos(\omega t - kx + \pi/2)\,\mathbf{e}_z] \\ &= A\,[\cos(-\omega t + kx)\,\mathbf{e}_y + \sin(-\omega t + kx)\,\mathbf{e}_z] = A\,\mathbf{e}^{(+)},\end{aligned} \qquad [4.38]$$

where $\mathbf{e}^{(+)}$ is the unit vector in the plane Oyz which makes an angle $(-\omega t + kx)$ with Oy. Any point x of the string describes a circle of radius A in the negative direction (clockwise, Figure 4.5a). At a given time, the string forms a right-handed spiral of radius A (i.e. in the direction of the right-hand fingers if the thumb points in the direction of propagation Ox) (Figure 4.5c). The wave is said to be *right-handed circularly polarized.*

– In the more general case of the superposition of two waves $\mathbf{u}_y$ and $\mathbf{u}_z$ of fixed arbitrary amplitudes A and B and fixed arbitrary phase shift ϕ, the projection of $\mathbf{u}$ onto the plane Oyz is an ellipse which is inscribed within a rectangle of sides $y = \pm A$ and $z = \pm B$. The various points of the string describe identical ellipses in the positive direction if $-\pi < \phi < 0$ (*left-handed elliptically polarized wave*) or in the negative direction if $0 < \phi < \pi$ (*right-handed elliptically polarized wave*).

– If the direction of oscillation changes arbitrarily in time, or if the wave is a superposition of waves, which are polarized in random directions, the wave is *non-polarized.* Such a wave may be considered to be the superposition of two linearly polarized waves [4.35] with variable amplitudes A and B, or variable phase shift ϕ, or a superposition of a right-handed circularly polarized wave and a left-handed circularly polarized wave with variable amplitudes or variable phase-shift. It is possible to produce a polarized wave from a non-polarized wave using a *polarizer.* This is a device, which absorbs one of the components $\mathbf{u}_y$ and $\mathbf{u}_z$. In the case of a string, it may simply be a transverse slit, through which the string passes.

4.5. Monochromatic wave, wave vector and wavelength

Consider first the case of a one-dimensional wave, such as the wave on a string. Let us assume that the origin O is forced by an oscillator to have a permanent simple harmonic oscillation of the form $f(t) = A\cos(\omega t)$, and that the wave propagates with a speed $v_{(p)}$. The oscillation at point M of coordinate x is the same as at O, but with a time delay $x/v_{(p)}$

$$u = A\cos[\omega(t - x/v_{(p)})] \equiv A\cos(\omega t - kx). \quad [4.39]$$

k is the *wave number*. Using the period $T = 2\pi/\omega$, we may write

$$u = A\cos[2\pi(t/T - x/\lambda)], \quad [4.40]$$

where λ is the *wavelength*. ω, $v_{(p)}$, k and λ are related by the equations:

$$\lambda = v_{(p)}T = v_{(p)}/\tilde{\nu}, \quad [4.41]$$

$$k = \omega/v_{(p)} = 2\pi/\lambda. \quad [4.42]$$

Thus, λ is the distance travelled by the simple harmonic wave in a time equal to the period T. Sometimes, the number of wavelengths per unit of length $1/\lambda = k/2\pi$ is called the *wave number*.

Expression [4.40] may be interpreted in two ways:

– At a given point x, u represents its oscillation as a function of time. It is a simple harmonic with a period T, and it is the same as the oscillation of the entry ($x = 0$), but with a time delay $x/v_{(p)}$, that is, a phase lag $\phi = kx = 2\pi x/\lambda$ (Figure 4.6a).

– At a given time t, u represents the oscillation as a function of x, that is, the profile of the string at this time. The string returns back to the same configuration after a time equal to the period T. The points of the string, which are separated by a distance λ, have the same state of oscillation (that is, the same u, $\partial_t u$, etc.). The wave function u is a simple harmonic function of x of period λ (Figure 4.6b).

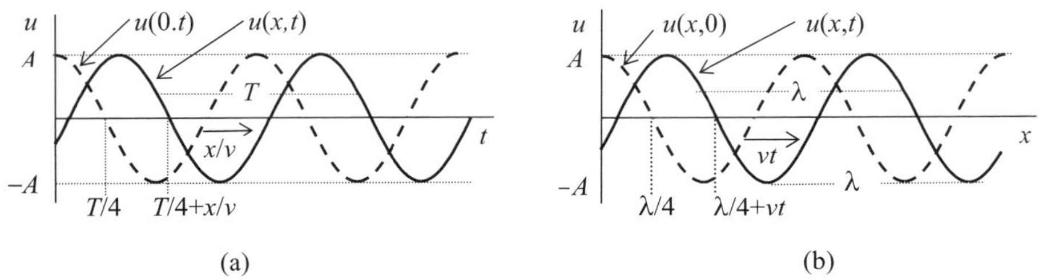

Figure 4.6. *a) The wave at a point of a string versus t, b) profile of the string at a given time t*

– Expression [4.39] shows that a point, where the wave has a phase equal to a given value α, has the coordinate $x = (\omega/k)t - \alpha/k$. This point moves with the velocity $v_{(p)} = \omega/k$. For this reason, $v_{(p)}$ is called the *phase velocity*.

Consider now a three-dimensional plane wave of the form [4.26], which propagates in the direction of the unit vector **e**. If it is a simple harmonic wave (that is, monochromatic) with angular frequency ω and phase velocity $v_{(p)}$, it may be written in the form

$$u = A\cos[\omega(t - \mathbf{e.r}/v_{(p)})] \equiv A\cos(\omega t - \mathbf{k.r}), \qquad [4.43]$$

$$\mathbf{k} = (\omega/v_{(p)})\,\mathbf{e}, \qquad k = 2\pi/\lambda = \omega/v_{(p)}. \qquad [4.44]$$

k is the *wave vector*. It points in the direction of propagation **e**; thus, it is perpendicular to the wave fronts $\omega t - \mathbf{k.r} = \alpha$. The magnitude of **k** is related to the wavelength λ by relation [4.42], which is the same as for one-dimensional waves. Thus, a simple harmonic wave is characterized by its amplitude A, its angular frequency ω and its wave vector **k**. Instead of **k**, we may use the direction of propagation **e** and k and, instead of k, we may use the wavelength $\lambda = 2\pi/k$ or the phase velocity $v_{(p)} = \omega/k = \tilde{\nu}\,\lambda = \omega\lambda/2\pi$.

EXAMPLE 4.1.– Waves are produced on the surface of a pool by moving up and down a long horizontal rod, which slightly touches the water surface. Assume that the vertical position of the rod is $u = 3\sin(\omega t)$, where u is in mm and the frequency is 10 Hz. The observed ripples travel in the direction perpendicular to the rod, with crests 5 cm apart. Determine the wave number and speed of propagation. Write down the expressions of the vertical displacement and velocity at a point situated at a perpendicular distance x from the rod.

SOLUTION – The wavelength is λ = 5 cm. The wave number is $k = 2\pi/\lambda = 126\ \text{m}^{-1}$. The speed of propagation is $v_{(p)} = \lambda/T = \lambda\tilde{\nu} = 0.5$ m/s. The vertical displacement (in m) and the velocity (in m/s) at a distance x to the rod are:

$$u(x,t) = u(0.\ t - x/v_{(p)}) = 3\times10^{-3}\sin[2\pi\tilde{\nu}\,(t - x/v_{(p)})] = 3\times10^{-3}\sin[62.83(t - x/0.5)]$$

$$\dot{u}(x.t) = 3\times 0{,}06283\cos[62.83(t - x/0.5)] = 0{,}1885\cos[62.83(t - x/0.5)].$$

Notice the large difference between the horizontal speed of propagation of the wave and the vertical velocity of water.

4.6. Dispersion

To write the wave equations and their solutions, we have assumed that the *speed of propagation v* depends on the physical conditions of the medium but not on the characteristics of the wave (its amplitude, its frequency and its direction of propagation). In reality, the propagation of waves is more complicated for two reasons:

– The propagation of a wave is a complex phenomenon of displacement or a change of state of the atoms or molecules of the medium under the influence of the wave and, sometimes, the emission of secondary waves. The superposition of all these waves constitutes the wave in the medium. In general, the interaction of the particles of the medium with the wave depends on its frequency (then, the medium is said to be *dispersive*). It may also depend on the direction of propagation of the wave (then, the medium is said to be *anisotropic*). In such media, it is not possible to write a wave equation with a speed of propagation v independent of the frequency and the direction of propagation.

– A material medium of propagation is always bound by surfaces and the wave must obey the boundary conditions on these surfaces. This is the case, for instance, for a sound wave in a pipe and light in an optical fiber. Then, the speed of propagation depends on the dimensions of the medium. It is equal to v only if the wavelength λ is much smaller than the dimensions of the medium.

It is only in the case of an infinite, isotropic and non-dispersive medium that we can write d'Alembert equation [4.13] with a constant speed of propagation v. A plane wave $u = A\cos(\omega t - \mathbf{k}.\mathbf{r})$ verifies this equation if $v_{(\mathrm{P})} = \omega/k = v$ independently of the wave characteristics. In general, the phase velocity $v_{(\mathrm{P})} = \omega/k$ of a simple harmonic wave depends on the angular frequency of the wave, its direction of propagation and polarization (if it is a transverse vector wave). $v_{(\mathrm{P})}$ also depends on the transverse dimensions of the medium if they are finite. In a *dispersive* but *isotropic* medium, the relation of ω to k is not a simple proportionality relation $\omega = vk$, but a more complicated relation

$$\omega = \omega(k), \qquad [4.45]$$

called the *dispersion relation*. Then, the phase velocity may be written in the form

$$v_{(\mathrm{P})} = \frac{\omega(k)}{k}, \qquad [4.46]$$

which determines $v_{(\mathrm{P})}$ as a function of k (or the wavelength $\lambda = 2\pi/k$). If dispersion relation [4.45] may be solved to express k as a function of ω, the phase velocity $v_{(\mathrm{P})}$ may be written as a function of ω

$$v_{(\mathrm{P})} = \frac{\omega}{k(\omega)}. \qquad [4.47]$$

If the propagation medium is anisotropic, the phase velocity depends on the direction of propagation, its frequency and its polarization – if any. Then, it is better to characterize the wave using the wave vector $\mathbf{k}$, and write the dispersion relation in the form

$$\omega = \omega(\mathbf{k}) \equiv \omega(k_x, k_y, k_z). \qquad [4.48]$$

In spherical coordinates, ω is a function $\omega = \omega(k, \theta_k, \varphi_k)$ of the magnitude of $\mathbf{k}$ and the angles θ_k and φ_k, which determine the direction (and possibly the direction of polarization). The phase velocity is then

$$v_{(p)}(k, \theta_k, \varphi_k) = \frac{\omega(k, \theta_k, \varphi_k)}{k}. \qquad [4.49]$$

This relation determines the phase velocity as a function of the direction of propagation and the wave number (or the wavelength via the relation $k = 2\pi/\lambda$).

In the case of electromagnetic waves (and light in particular), the vacuum is non-dispersive. The speed of light in vacuum is now fixed at

$$c = 2.997\,924\,58 \times 10^8 \text{ m/s} \qquad [4.50]$$

for any frequency (that is, any wavelength). Any other medium is dispersive. The *index of refraction* of the medium is by definition $n = c/v_{(p)}$; thus, it depends on the frequency of the wave.

4.7. Group velocity

The concept of a simple harmonic (i.e. *monochromatic*) wave is a mathematical idealization. However, Fourier's theory allows us to write any disturbance in a medium as an appropriate superposition of simple harmonic waves. To analyze the consequences of dispersion, let us first consider two simple harmonic waves of equal amplitudes, angular frequencies ω_1 and ω_2 and which propagate in the same direction Ox. Their wave functions are respectively $u_1 = A\cos(\omega_1 t - k_1 x)$ and $u_2 = A\cos(\omega_2 t - k_2 x)$, where k_1 and k_2 are related by the dispersion relation to the angular frequencies ω_1 and ω_2. We set $\omega_1 = \omega - \Delta\omega$, $k_1 = k - \Delta k$, $\omega_2 = \omega + \Delta\omega$ and $k_2 = k + \Delta k$. In the case of scalar waves, longitudinal waves or vector waves that are polarized in the same direction, the resulting wave is their algebraic sum

$$u = u_1 + u_2 = 2A\cos(\Delta\omega\, t - \Delta k\, x)\cos(\omega t - kx). \qquad [4.51]$$

This expression represents a wave of angular frequency $\omega = \frac{1}{2}(\omega_1 + \omega_2)$ and wave number $k = \frac{1}{2}(k_1 + k_2)$, which propagates in the direction Ox with a phase velocity $v_{(p)} = \omega/k = (\omega_1 + \omega_2)/(k_1 + k_2)$ and a variable amplitude $2A\cos(\Delta\omega\, t - \Delta k\, x)$. The wave is said to be *amplitude-modulated* (Figure 4.7). It is enveloped by the simple harmonic functions $\pm\, 2A\cos(\Delta\omega\, t - \Delta k\, x)$, called a *modulating wave*, of angular frequency $\Delta\omega$ and wave number $\Delta k = \frac{1}{2}(k_2 - k_1)$. The *modulation velocity* (i.e. the phase velocity of the modulating wave) is

$$v_M = \frac{\Delta\omega_1}{\Delta k} = \frac{\omega_2 - \omega_1}{k_2 - k_1}. \quad [4.52]$$

If the angular frequencies ω_1 and ω_2 are very close, the modulation velocity v_M becomes the *group velocity*

$$v_{(g)} = d\omega/dk. \quad [4.53]$$

If the medium is non-dispersive, ω is related to k by the equation $\omega = vk$. Thus, the group velocity and the phase velocity are both equal to v. On the contrary, if the medium is dispersive, $v_{(p)}$ and $v_{(g)}$ are different from v and the wave profile is deformed while propagating.

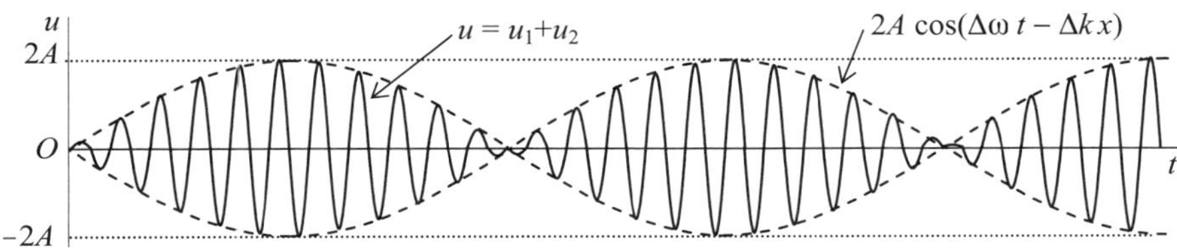

Figure 4.7. *Superposition of two waves of equal amplitudes and close frequencies*

4.8. Fourier analysis for waves*

Let us assume that at the entry the emitter produces a permanent oscillation $f(t)$ of period T. The Fourier theory allows us to consider $f(t)$ as a superposition of simple harmonic oscillations of angular frequencies ω, 2ω, 3ω, etc., where $\omega = 2\pi/T$:

$$\underline{u}(t)\,|_{x=0} \equiv f(t) = \Sigma_{-\infty<p<\infty}\, \underline{\mathcal{U}}_p\, e^{ip\omega t}. \quad [4.54]$$

If the medium is non-dispersive, all the spectral harmonic oscillations propagate with the same speed v and the oscillation at a distance x to the entry may be written as

$$\underline{u}(x, t) = \Sigma_{-\infty<p<\infty}\, \underline{\mathcal{U}}_p\, e^{2i\pi p(t/T - x/\lambda)} = \Sigma_{-\infty<p<\infty}\, \underline{\mathcal{U}}_p\, e^{ip\omega(t - x/v)}. \quad [4.55]$$

The first expression means that any wave of period T is a superposition of simple harmonic waves of periods T, $T/2$, $T/3$, etc., and wavelengths λ, $\lambda/2$, $\lambda/3$, etc. These are, respectively, the fundamental (or first harmonic), the second harmonic, the third harmonic, etc. The second expression means that the oscillation at x is the same as at the entry, but with a time delay x/v,

$$\Delta\omega.\Delta t \approx 2\pi \qquad \text{and} \qquad \Delta k.\Delta x \approx 2\pi. \qquad [4.56]$$

In the case of an aperiodic oscillation at the entry, the Fourier theory allows us to consider it as a superposition of simple harmonic oscillations with a continuous spectrum of the form

$$\underline{u}(t)\,|_{x=0} \equiv f(t) = \int_{-\infty}^{\infty} d\tilde{\nu}\ \underline{\mathcal{U}}(\tilde{\nu})\ e^{i\omega t}. \qquad [4.57]$$

In particular, a signal of short duration Δt at the entry has a frequency band width $\Delta\tilde{\nu}$ such that

$$\Delta\tilde{\nu}\,.\Delta t \approx 1. \qquad [4.58]$$

where the 1 on the right-hand side is an order of magnitude. The exact value depends on how $\Delta\tilde{\nu}$ and Δt are defined. Thus, the integration in [4.57] is essentially over a band $[\tilde{\nu}_1, \tilde{\nu}_2] \approx \Delta\tilde{\nu}$. If the medium is non-dispersive, all the spectral components of the oscillation propagate with the same speed v and the wave at a distance x may be written as

$$\underline{u}(x, t) = \int_{-\infty}^{\infty} d\tilde{\nu}\ \underline{\mathcal{U}}(\tilde{\nu})\ e^{i\omega(t-x/v)} = f(t - x/v). \qquad [4.59]$$

The oscillation at point x is the same as at the entry, but with a time delay x/v.

Equations [4.56] and [4.59] express that any disturbance (whether it is periodic or aperiodic) propagates in a non-dispersive medium without any change in the profile. On the contrary, if the medium is dispersive, each wave of the series or the Fourier integral propagates with the corresponding speed. Some harmonics arrive before others to the point of observation and the profile of the wave changes as it propagates. Then, equations [4.55] and [4.59] may be written as

$$\underline{u}(x, t) = \Sigma_p\, \underline{\mathcal{U}}_p\, e^{i(p\omega t - k_p x)} \qquad \text{or} \qquad \underline{u}(x, t) = \int_{-\infty}^{\infty} d\tilde{\nu}\ \underline{\mathcal{U}}(\tilde{\nu})\ e^{i(\omega t - kx)}, \qquad [4.60]$$

where k_p is related to $p\omega$ and k is related to ω by the dispersion relation of the medium.

Consider a signal [4.60] of narrow band width $\Delta\omega$ close to ω_o. If x and t are such that $(\omega t - kx)$ has a large variation in the band $\Delta\omega$, the real part and the imaginary part of $e^{i(\omega t - kx)}$ oscillate rapidly between -1 and $+1$ and the integral is expected to be small. On the contrary, the wave function $u(x, t)$ is large if the phase $(\omega t - kx)$ is almost constant, that is, $(d/d\omega)(\omega t - kx) = t - x\,(dk/d\omega)\big|_{\omega=\omega_o} \approx 0$. We say that we

have a *wave packet* (or a *signal*) whose center is at the point $x = t\ (d\omega/dk)\big|_{\omega=\omega_o}$.
This center moves with the *group velocity*

$$v_{(g)} = (d\omega/dk)\big|_{\omega=\omega_o}. \qquad [4.61]$$

The band width of the wave number k is

$$\Delta k \approx (dk/d\omega)\,\Delta\omega = \Delta\omega/v_{(g)}. \qquad [4.62]$$

As the disturbance has a time duration Δt and propagates with the group velocity $v_{(g)}$, the extension of the wave packet in space is $\Delta x = v_{(g)}\,\Delta t = (d\omega/dk)\,\Delta t$. Thus, we have the *uncertainty relations*

$$\Delta k.\Delta x = \Delta\omega.\Delta t \approx 2\pi. \qquad [4.63]$$

If we reduce the frequency band width $\Delta\omega$, Δk decreases, the duration Δt of the signal and its space extension Δx increase. In the ideal case of a monochromatic wave ($\Delta\omega \longrightarrow 0$), its space extension Δx and its time duration Δt are infinite. This is a plane wave which occupies all the space, and which has no beginning and no end.

As the angular frequency and the wave number are related by the dispersion relation of the medium, we may take k instead of ω to characterize the band and write

$$\underline{u}(x,t) = \int_{-\infty}^{\infty} dk\ \ \hat{\mathcal{U}}(k)\ e^{i(\omega t - kx)}, \qquad [4.64]$$

where the spectral function $\hat{\mathcal{U}}(k) = (d\omega/dk)\,\underline{\mathcal{U}}(\tilde{\nu})\,/2\pi$ is important in the band Δk.

To study the superposition of waves, which propagate in the three-dimensional space, we must specify the angular frequency, the direction of propagation and the propagation velocity. All these quantities are completely determined by the wave vector **k** and the dispersion relation of the medium (which determines the angular frequency ω). In general, a three-dimensional wave packet is a superposition of waves of various frequencies and directions. Thus, its most general expression is

$$\underline{u}(\mathbf{r},t) = \iiint d^3k\ \hat{\mathcal{U}}(\mathbf{k})\ e^{i(\omega t - \mathbf{k}.\mathbf{r})}, \qquad [4.65]$$

where $d^3k = dk_x\ dk_y\ dk_z$ and the integration is formally from $-\infty$ to ∞ for all the components of **k**. However, $\hat{\mathcal{U}}(\mathbf{k})$ is important only in a small domain of **k** close to a certain wave vector $\mathbf{k}_o$. The angular frequency ω is related to **k** by the dispersion relation and the central value $\mathbf{k}_o$ corresponds to ω_o.

The wave function $\underline{u}(\mathbf{r},t)$ has a large value if the phase $(\omega t - \mathbf{k}.\mathbf{r})$ is almost constant for **k** close to $\mathbf{k}_o$. That is, near the point **r** and at time t, which verify the relations:

$$\frac{\partial}{\partial k_x}(\omega t - \mathbf{k}.\mathbf{r}) = 0, \qquad \frac{\partial}{\partial k_y}(\omega t - \mathbf{k}.\mathbf{r}) = 0, \qquad \frac{\partial}{\partial k_z}(\omega t - \mathbf{k}.\mathbf{r}) = 0, \qquad [4.66]$$

This gives, $x = (\partial\omega/\partial k_x)t$, $y = (\partial\omega/\partial k_y)t$ and $z = (\partial\omega/\partial k_z)t$. Thus, the wave packet moves with a velocity equal to the group velocity

$$\mathbf{v}_{(g)} = \nabla_k \,\omega|_{\mathbf{k}_o} \equiv (\partial\omega/\partial k_x)|_{\mathbf{k}_o}\, \mathbf{e}_x + (\partial\omega/\partial k_y)|_{\mathbf{k}_o}\, \mathbf{e}_y + (\partial\omega/\partial k_x)|_{\mathbf{k}_o}\, \mathbf{e}_z. \qquad [4.67]$$

If the medium is isotropic, ω is only a function of k (independent of the direction of **k**) and the group velocity is

$$\mathbf{v}_{(g)} = (d\omega/dk)|_{\mathbf{k}_o}\, \mathbf{e}_k \qquad \text{where} \qquad \mathbf{e}_k = \mathbf{k}_o/k_o. \qquad [4.68]$$

$\mathbf{e}_k$ is the unit vector, which points in the direction of propagation of the wave packet.

If k_x, k_y and k_z vary in the band widths Δk_x, Δk_y and Δk_z respectively, the wave packet occupies a region of extension space Δx, Δy and Δz, which verify the uncertainty relations:

$$\Delta k_x.\Delta x \approx 2\pi, \qquad \Delta k_y.\Delta y \approx 2\pi, \qquad \Delta k_z.\Delta z \approx 2\pi. \qquad [4.69]$$

If the medium is dispersive, it may be shown that signals widen as they propagate.

4.9. Modulation*

If a signal is repeated indefinitely and identically, the only information that it transmits is the existence of the emitter or an obstacle, which reflects the wave. A useful information (sound, image, etc.) may be transmitted, recorded or treated only as a *modulated wave*, that is, whose amplitude, frequency or phase vary according to the information. All systems of telecommunication consist of an *emitter* and a *receiver*. The emitter generates an electromagnetic wave of angular frequency ω called a *carrier wave*. The transmitted signal is converted to a *modulating wave* by an *input transducer* (a microphone, an electric photocell, etc.) and it is impressed upon the carrier wave producing a variation of its characteristics. This modulated wave propagates to the receiver where an *output transducer* (a speaker, for instance) extracts the modulating wave and converts it back to a non-electrical signal.

The *amplitude modulation* consists of varying the amplitude of the carrier wave according to the signal to be transmitted. In the absence of a signal, the emitter generates the carrier simple harmonic wave $u_o = A\cos(2\pi\tilde{\nu}_o t + \phi)$ of constant amplitude A (Figure 4.8a). To transmit a sound, light or another signal $s(t)$ (Figure 4.8b), the input transducer (a microphone in the case of sound) modifies the amplitude of the emitted wave to become $u = A[1 + \alpha\, s(t)]\cos(2\pi\tilde{\nu}_o t + \phi)$ where α is a constant called the *modulation ratio* (Figure 4.8c). Then, the spectral function of u is

$$
\begin{aligned}
\underline{\mathcal{U}}(\tilde{\nu}) &= \int_{-\infty}^{\infty} dt\; e^{-2i\pi\tilde{\nu}t}\; \underline{u}(t) \\
&= \tfrac{1}{2}A \int_{-\infty}^{\infty} dt\; e^{-2i\pi\tilde{\nu}t}\, \{1 + \alpha \int_{-\infty}^{\infty} d\tilde{\nu}'\, e^{2i\pi\tilde{\nu}'t}\, \underline{\mathcal{S}}(\tilde{\nu}')\, \}\, [\, e^{i(2\pi\tilde{\nu}_o t+\phi)} + e^{-i(2\pi\tilde{\nu}_o t+\phi)}\,] \\
&= \tfrac{1}{2}A e^{i\phi} \{ \int_{-\infty}^{\infty} dt\, e^{2i\pi(\tilde{\nu}_o-\tilde{\nu})t} + \alpha \int_{-\infty}^{\infty} d\tilde{\nu}' \underline{\mathcal{S}}(\tilde{\nu}') \int_{-\infty}^{\infty} dt\; e^{2i\pi(\nu'+\tilde{\nu}_o-\tilde{\nu})t} \} \\
&\qquad\qquad + (\phi \to -\phi \text{ and } \tilde{\nu}_o \to -\tilde{\nu}_o) \\
&= \tfrac{1}{2}A\, e^{i\phi}\, \{ \delta(\tilde{\nu}_o-\tilde{\nu}) + \alpha \int_{-\infty}^{\infty} d\tilde{\nu}'\; \underline{\mathcal{S}}(\tilde{\nu}')\, \delta(\tilde{\nu}' - \tilde{\nu} + \tilde{\nu}_o)\} + (\phi > -\phi \text{ and } \tilde{\nu}_o \to -\tilde{\nu}_o) \\
&= \tfrac{1}{2}A\, [e^{i\phi}\, \delta(\tilde{\nu}_o-\tilde{\nu}) + e^{-i\phi}\, \delta(\tilde{\nu}_o+\tilde{\nu})\,] + \tfrac{1}{2}\alpha A[e^{i\phi}\; \underline{\mathcal{S}}(\tilde{\nu}-\tilde{\nu}_o) + e^{-i\phi}\; \underline{\mathcal{S}}(\tilde{\nu}+\tilde{\nu}_o)\,],
\end{aligned}
$$

where, in the second line, we have introduced the spectral function $\underline{\mathcal{S}}(\tilde{\nu}')$ of $s(t)$, in the third line, we used expression [2.61] of the Dirac delta-function and, in the fourth line, we used property [2.59] of the Dirac delta-function to integrate over $\tilde{\nu}'$.

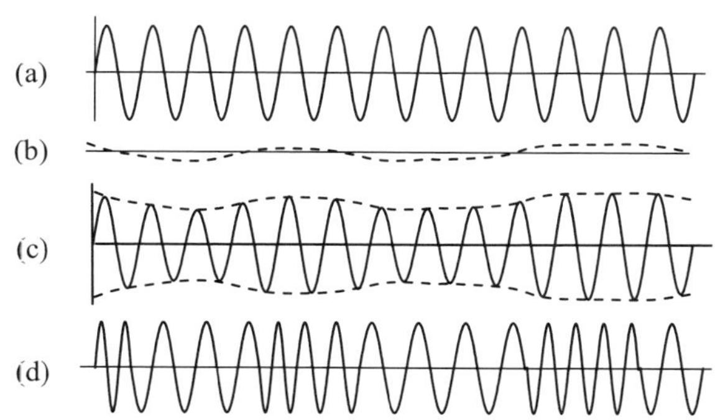

Figure 4.8. *Modulation: a) the carrier wave, b) the signal to be transmitted, c) the amplitude-modulated wave, and d) a frequency-modulated wave*

If $s(t)$ is a real function, we have $\underline{S}(-\tilde{\nu}) = \underline{S}(\tilde{\nu})^*$ and we may write

$$\underline{\mathcal{U}}(\tilde{\nu}) = \frac{A}{2}\,[\,e^{i\phi}\delta(\tilde{\nu}_o - \tilde{\nu}) + e^{-i\phi}\delta(\tilde{\nu}_o + \tilde{\nu})\,] + \frac{\alpha A}{2}\,[\,e^{i\phi}\underline{S}(\tilde{\nu} - \tilde{\nu}_o)^* + e^{-i\phi}\underline{S}(\tilde{\nu} + \tilde{\nu}_o)\,]. \quad [4.70]$$

This is the sum of four terms: the first two correspond to an oscillation of frequency $\tilde{\nu}_o$ equal to the carrier frequency. If the signal $s(t)$ has a spectrum in the band $[\,\tilde{\nu}_1, \tilde{\nu}_2\,]$, the term $\underline{S}(\tilde{\nu}_o - \tilde{\nu})^*$ has a frequency band $[\,\tilde{\nu}_o - \tilde{\nu}_2, \tilde{\nu}_o - \tilde{\nu}_1\,]$, called the *lower band*, and the term $\underline{S}(\tilde{\nu}_o + \tilde{\nu})$ has a frequency band $[\,\tilde{\nu}_o + \tilde{\nu}_1, \tilde{\nu}_o + \tilde{\nu}_2\,]$ called the *higher band*.

In order to have a high-quality transmission, reproducing the fine details of the signal, the period of the carrier wave must be much shorter than the characteristic time of the signal to be transmitted. Thus, the frequency of the carrier wave must be much higher than the frequency of the modulating wave.

The major failing of the amplitude modulation in radio emissions is that the receiver picks up too much extraneous noise, which is generated by all electric sources, such as lightning and nearby electric installations (electric motors, computers etc.). To avoid such signals, we use *frequency modulation* with a high carrier frequency (88 MHz $< \tilde{\nu} <$ 108 MHz, i.e. 3.4 m$< \lambda <$ 2.8 m). Frequency modulation consists of varying the wave frequency in a narrow band on both sides of the carrier wave according to the signal to be transmitted (Figure 4.8d). We may also use *phase modulation*, which consists of varying the phase of the carrier wave. Phase modulation is reserved to some special pieces of equipment because it requires a very wide pass-band.

4.10. Energy of waves

Like any oscillation, a certain energy density is associated to the wave in the region that it occupies. This energy has two forms, it may be transformed from one form to the other, and it moves from one region to another with the group velocity. To simplify, we assume in the following that the medium is non-dispersive (then, $v_{(g)} = v_{(p)} = v$). In the case of a one-dimensional wave, we define the *energy density per unit length* $E_l(x,t)$. We also define the *energy flux* $P(x,t)$ as the power that is carried by the wave at the point x and at time t.

Consider a segment situated between x and $x + \Delta x$ in the direction of propagation of the wave (Figure 4.9a). The energy of the wave in this segment is $E_l\,\Delta x$ and its variation in the interval of time dt is $(\partial_t E_l)\,\Delta x\,dt$. During this same interval of time, the energy which enters at x is $P(x, t)\,dt$, and the energy which leaves the segment at

$x + \Delta x$ is $P(x + \Delta x, t)\, dt \approx P(x, t) + \partial_x P\, \Delta x$. The conservation of energy requires that $P(x, t)\, dt = P(x + \Delta x, t)\, dt + \partial_t E_l\, \Delta x\, dt$; hence, the relation

$$\partial_t E_l + \partial_x P = 0. \qquad [4.71]$$

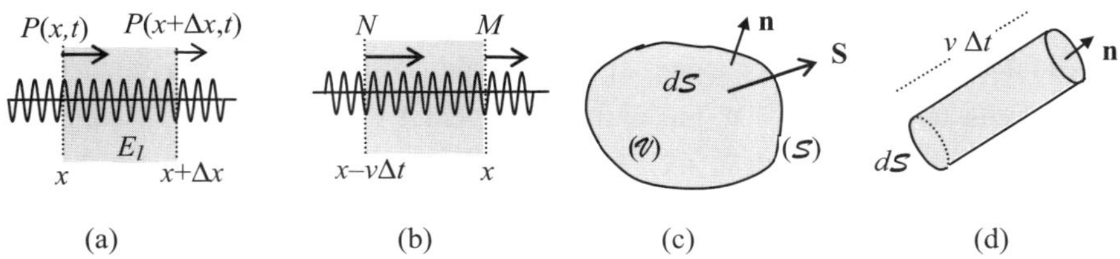

Figure 4.9. *Energy of a progressive wave: a) energy density and energy flux of a one-dimensional wave; b) interpretation of the relationship $P = vE_l$; c) energy density and energy flux density of a three-dimensional wave, d) interpretation of the relationship $\mathbf{S} = vE_v\, \mathbf{e}$*

All physical quantities associated with a progressive wave are carried at the same velocity as the wave. Thus, we may write a simple equation relating the energy flux and the energy density (Figure 4.9b). The energy, which crosses point M of coordinate x between t and $t + \Delta t$, is $P(x, t)\, \Delta t$. As the wave propagates at the speed v, this energy is contained between the points N and M such that $NM = v\, \Delta t$, that is $E_l(x, t)\, v\, \Delta t$. Comparing the two expressions, we deduce that

$$P(x, t) = v\, E_l(x, t) \qquad \text{[progressive waves]}. \qquad [4.72]$$

In the case of a three-dimensional wave, we have at each space point an *energy density* $E_v(\mathbf{r}, t)$ per unit volume and the energy propagates in the direction of propagation of the wave. We define the *vector energy flux density* $\mathbf{S}(\mathbf{r}, t)$ per unit area. The flux of $\mathbf{S}$ across an element of area $d\mathcal{S}$ is $(\mathbf{S}.\mathbf{n})\, d\mathcal{S}$ where $\mathbf{n}$ is the unit vector, which is normal to $d\mathcal{S}$. This is the power of the wave, which is intercepted by $d\mathcal{S}$. The power intercepted by a finite surface $\mathcal{S}$, is

$$P_{\mathcal{S}}(t) = \iint_{\mathcal{S}} d\mathcal{S}\, (\mathbf{n}.\mathbf{S}). \qquad [4.73]$$

Consider a closed surface $\mathcal{S}$ enclosing a volume $\mathcal{V}$ (Figure 4.9c). The total energy of the wave stored in the volume $\mathcal{V}$ is

$$E_{\mathcal{V}}(t) = \iiint_{\mathcal{V}} d\mathcal{V}\, E_v(\mathbf{r}, t). \qquad [4.74]$$

The variation of $E_{\mathcal{V}}$ in the interval of time dt is $dE_{\mathcal{V}} = dt \iiint_{\mathcal{V}} d\mathcal{V}\, \partial_t E_v$, and the energy, which gets out of the surface $\mathcal{S}$ in the same interval of time, is

$$dE_{\mathcal{S}} = dt\, P_{\mathcal{S}}(t) = dt \iint_{\mathcal{S}} d\mathcal{S}\,(\mathbf{n}.\mathbf{S}) = dt \iiint_{\mathcal{V}} d\mathcal{V}\,(\nabla.\mathbf{S}), \quad [4.75]$$

where $\mathbf{n}$ is the outward unit vector normal to $\mathcal{S}$ and we used the Gauss-Ostrogradski theorem to transform the surface integral to a volume integral (see section A1.6 of Appendix 1). The conservation of energy requires that $dE_{\mathcal{S}} + dE_{\mathcal{V}} = 0$ for any volume $\mathcal{V}$ if it contains no source or dissipation of energy. We deduce that

$$\partial_t E_v + (\nabla.\mathbf{S}) = 0. \quad [4.76]$$

Consider a progressive wave, which propagates in the direction $\mathbf{e}$ in an isotropic medium, and an element of area $\delta\mathcal{S}$ perpendicular to $\mathbf{e}$ (Figure 4.9d). The energy, which is intercepted by $\delta\mathcal{S}$ during dt, is $(\mathbf{S}.\mathbf{n})\,\delta\mathcal{S}\,dt = S\,\delta\mathcal{S}\,dt$. This energy is stored in a cylinder of length $v\,dt$ parallel to $\mathbf{e}$. Its volume is $v\,dt\,\delta\mathcal{S}$ and the energy that it contains is $E_v(\mathbf{r}, t)\,v\,dt\,\delta\mathcal{S}$. Comparing the two expressions, we get

$$S(\mathbf{r}, t) = v\,E_v(\mathbf{r}, t), \qquad \text{i.e. } \mathbf{S}(\mathbf{r}, t) = v\,E_v(\mathbf{r}, t)\,\mathbf{e} \quad \text{[progressive waves]}. [4.77]$$

Equations [4.71] and [4.76] are the local expressions of the law of conservation of energy. They hold for any type of waves, whether they are progressive or standing. Relationships [4.72] and [4.77] are valid only in the case of progressive waves. Standing waves, for instance, have energy density but no transfer of energy for distances larger than the wavelength. Note also that the energy, which is carried by a wave packet, is concentrated in the region, which is occupied by the wave packet. As the center of the wave packet moves with the group velocity $\mathbf{v}_{(g)}$, the energy is carried with this velocity. This is also true for all physical quantities of the wave. If the medium is not isotropic, the energy flux density $\mathbf{S}$ is not necessarily in the direction of propagation of the wave $\mathbf{e}$.

In the case of a periodic wave, we define the *wave intensity* as the average power that is intercepted by the unit area perpendicular to the direction of propagation (i.e. the energy flux density $\mathbf{S}$). This is the average of the magnitude of $\mathbf{S}$ over a period

$$\mathcal{I}(\mathbf{r}, t) = \langle S(\mathbf{r}, t) \rangle = \frac{1}{T}\int_t^{t+T} dt\;\; S(\mathbf{r}, t). \quad [4.78]$$

4.11. Other unattenuated wave equations, conserved quantities*

The Lagrangian formalism and the symmetries of the wave allow us to establish the general form of the wave equation. We do not use this formalism in this text. In some cases, the medium may be considered as the limit of a system of coupled oscillators. For instance, a chain of masses on a taut and massless string (see section 5.1) or coupled by springs of negligible mass (see section 5.3) leads to the d'Alembert equation for massive strings, springs and elastic rods in the limit of distributed masses; this is the simplest, linear and second order wave equation. A model of an infinite set of coupled pendulums (see problem P4.18) allows us to establish, in the limit of a distributed system, a more general wave equation called the *Klein-Gordon equation*

$$\mu\, \partial^2_{tt}u - B_o\, \Delta u + \mu\omega_o^{\,2}\, u = 0 \quad \text{or} \quad \partial^2_{tt}u - v^2\, \Delta u + \omega_o^{\,2}\, u = 0, \quad \text{where } v = \sqrt{B_o/\mu} \qquad [4.79]$$

μ is an inertial parameter and $K = \mu\omega_o^{\,2}$ is a restoring force parameter for each oscillator, while B_o is an interaction parameter for two adjacent oscillators.

In the case of a vector wave propagating in an isotropic medium, a possible wave equation is

$$\mu\, \ddot{\mathbf{u}} - B_o\Delta\mathbf{u} + \mu\omega_o^{\,2}\, \mathbf{u} + C_o\, \nabla(\nabla.\mathbf{u}) = 0. \qquad [4.80]$$

We will encounter the supplementary term $\mu\omega_o^{\,2}\mathbf{u}$ in our study of propagation in *reactive media*, such as the plasmas (see section 6.3) and the term $C_o\nabla(\nabla.\mathbf{u})$ in the study of elastic waves (see section 5.6).

The Lagrangian formalism also allows us to write down the expressions of some conserved quantities, which play a very important part in physics. The *Noether's theorem* associates a conservation law with each invariance, which is expressed as a continuous transformation of the variables of the system. The invariance in the translation of time leads to the existence of an *energy density* and an *energy flux*. In the case of a one-dimensional wave, which obeys the Klein-Gordon equation $\partial^2_{tt}u - v^2\, \partial^2_{xx}u + \omega_o^{\,2}u = 0$, we find an *energy density per unit length* $E_l(x, t)$ and an *energy flux* $P(x, t)$, which verify the local conservation of energy equation [4.71]:

$$E_l(x, t) = \tfrac{1}{2}\mu\, \dot{u}^2 + \tfrac{1}{2}\mu\omega_o^{\,2}u^2 + \tfrac{1}{2}B_o(\partial_x u)^2, \qquad P(x, t) = -\, B_o\, \dot{u}\; \partial_x u. \qquad [4.81]$$

In the case of a three-dimensional wave, which obeys equation [4.80], the invariance in the translation of time and the translations in space lead to an *energy density per unit volume* $E_v(\mathbf{r}, t)$ and an *energy flux density* $\mathbf{S}(\mathbf{r}, t)$, which verify the equation of conservation of energy [4.76] and also a momentum density $\mathbf{P}_v$ given by

$$E_v = \tfrac{1}{2}\{\mu\, \dot{\mathbf{u}}^2 + \mu\omega_o^{\,2}\mathbf{u}^2 + B_o\Sigma_{ij}\,(\partial_i u_j)(\partial_i u_j) + (D_o - C_o)\Sigma_{ij}(\partial_i u_j)(\partial_j u_i) - D_o(\nabla\mathbf{u})^2\}$$

$$S^i = -\, B_o\, \Sigma_j\, \dot{u}_j\,(\partial_i u_j) + (C_o - D_o)\, \Sigma_j\, \dot{u}_j\,(\partial_j u_i) + D_o\, \dot{u}_i\, (\nabla.\mathbf{u})$$

$$\mathbf{P}_v^{\,i} = -\, \Sigma_j\, (\partial_i u_j)[\mu\, \dot{u}_j + \tfrac{1}{2}\, F_o u_j]. \qquad [4.82]$$

D_o and F_o are constants which may appear in the Lagrangian but not in the wave equation. Thus, the wave has a momentum density and it carries momentum. If the wave is intercepted by the surface of a body, which is perpendicular to the direction of propagation, this momentum transfer is equivalent to a *radiation pressure*

$$p_r = v\,|\mathbf{P}_v| \qquad [4.83]$$

if the body totally absorbs the wave. We will consider this question in our study of sound waves (section 5.8) and electromagnetic waves (section 6.6).

The invariance of the laws of physics in the rotations of the reference frame shows that circularly or elliptically polarized transverse waves have an intrinsic angular momentum density and carry intrinsic angular momentum (in addition to orbital angular momentum). A scalar wave, a longitudinal vector wave or a linearly polarized transverse vector wave have no intrinsic angular momentum.

4.12. Impedance of a medium*

The concept of impedance that we introduced for oscillators in section 3.4, may be generalized to waves. Assume that a generator acts on the entry surface $\mathcal{S}$ of a medium, emitting a vector wave $\mathbf{u} = \mathbf{A}\cos(\omega t - \mathbf{k}.\mathbf{r})$ in the normal direction to $\mathcal{S}$. The wave $\mathbf{u}$ is a solution to equation [4.80] if $(\mu\omega^2 - \mu\omega_o{}^2 - B_o k^2)\mathbf{A} = C_o \mathbf{k}\,(\mathbf{k}.\mathbf{A})$. This leads to the dispersion relation in the form $\mu\omega^2 = \mu\omega_o{}^2 + B_o k^2$ in the case of a transverse wave ($\mathbf{k}.\mathbf{A} = 0$), and in the form $\mu\omega^2 = \mu\omega_o{}^2 + (B_o + C_o)\,k^2$ in the case of a longitudinal wave (then $\mathbf{A} = A\mathbf{k}/k$). Using the equations of [4.82], we obtain the energy flux density

$$S^i = [B_o\,\omega\,k_i\,\mathbf{A}^2 - C_o\,\omega\,A_i\,(\mathbf{k}.\mathbf{A})]\,\sin^2(\omega t - \mathbf{k}.\mathbf{r}). \qquad [4.84]$$

As $\mathcal{S}$ is normal to the direction of propagation $\mathbf{e} = \mathbf{k}/k$, it transmits a power

$$P = \mathcal{S}\,(\mathbf{S}.\mathbf{e}) = \mathcal{S}\,[B_o\omega k\,\mathbf{A}^2 - C_o\,(\omega/k)\,(\mathbf{k}.\mathbf{A})^2]\,\sin^2(\omega t - \mathbf{k}.\mathbf{r})$$
$$= (\mathcal{S}/k\omega)\,[B_o\,k^2\,\dot{\mathbf{u}}^2 - C_o\,(\mathbf{k}.\dot{\mathbf{u}})^2]. \qquad [4.85]$$

This is also the power that the generator emits. In the case of a mechanical wave, this power is the same as if the medium opposes a force $-\,Z_s\,\dot{\mathbf{u}}$ per unit area of the entry. Conversely, the generator exerts, on the medium, a force per unit area

$$F_s = Z_s\,\dot{u}\,. \qquad [4.86]$$

Z_s is the *impedance of the medium per unit area*. The power that the generator must supply to sustain the wave is $P = SZ_s\dot{u}^2$. In the case of a transverse wave ($\mathbf{k}.\dot{\mathbf{u}} = 0$) and in the case of a longitudinal wave ($\mathbf{k}.\dot{\mathbf{u}} = k\dot{u}$), we obtain, respectively,

$$Z_s = B_o\frac{k}{\omega} = \sqrt{\mu B_o(1-\omega_o^2/\omega^2)} \qquad \text{[transverse wave]}$$

$$Z_s = (B_o - C_o)\frac{k}{\omega} = (B_o - C_o)\sqrt{\frac{\mu}{B_o + C_o}(1-\frac{\omega_o^2}{\omega^2})} \quad \text{[longitudinal wave]}. \qquad [4.87]$$

In particular, if the wave obeys the d'Alembert wave equation ($\omega_o = 0$ and $C_o = 0$) or if it is a scalar wave, we get the impedance

$$Z_s = B_o k/\omega = B_o/v = \sqrt{\mu B_o}\ . \qquad [4.88]$$

The concept of impedance plays an important part in the study of the reflection and transmission of waves (Chapter 7).

4.13. Attenuated waves

With the remarkable exception of electromagnetic waves in vacuum, a plane wave cannot propagate indefinitely in a medium without attenuation. The first reason for attenuation is that, in certain media, the wave obeys the Klein-Gordon equation [4.79] instead of the d'Alembert equation. A wave of the form $u(x, t) = A\,e^{i(\omega t - kx)}$ must obey the dispersion relation

$$\omega^2 = v^2k^2 + \omega_o^2. \qquad [4.89]$$

If $\omega > \omega_o$, setting $\gamma = \sqrt{1-\omega_o^2/\omega^2}$, the wave number $k = \omega\gamma/v$ is real and the wave is progressive with a phase velocity and a group velocity:

$$v_p = \omega/k = v/\gamma > v, \qquad v_g = d\omega/dk = v^2k/\omega = v^2/v_p = \gamma v < v. \qquad [4.90]$$

We say that the medium is *dispersive* for waves of frequency $\omega > \omega_o$; on the other hand, if $\omega < \omega_o$, k is imaginary, of the form $k = \pm i\delta$, where $\delta = (1/v)\sqrt{\omega_o^2 - \omega^2}$. In this case, the wave is exponential, of the form $u(x, t) = A\,e^{-\delta x}e^{i\omega t}$ or $A\,e^{\delta x}e^{i\omega t}$ (or a superposition of these solutions). If the medium is infinite toward positive x, the solution $A\,e^{\delta x}e^{i\omega t}$ is physically impossible, as it requires infinite energy. The exponentially decreasing solution $A\,e^{-\delta x}e^{i\omega t}$ is not really a wave. It represents oscillations without any transfer of energy or other physical quantities. If a wave of angular frequency $\omega < \omega_o$ falls on this medium, all the energy is reflected as if by an ideal mirror. We say that the medium is *reactive* for waves of frequency $\omega > \omega_o$. We note that, the attenuation of the wave is not due to a dissipation of energy in the medium but an interaction which blocks the wave.

The attenuation of the wave may be due to the absorption of energy as heat or other processes; in this case we say that the medium is *resistive*, and the one-dimensional wave equation may be written as

$$\ddot{u} + 2\beta\,\dot{u} - v^2\,\Delta u + \omega_o^{\,2}\,u = 0. \qquad [4.91]$$

If $\beta = 0$, we find the non-attenuated Klein-Gordon equation and, if $\beta = 0$ and $\omega_o = 0$, we find the d'Alembert equation. If $\omega_o = 0$, the equation is called the *equation of telegraphy*, because it is encountered when studying the propagation of electromagnetic signals along a transmission line and in conductors. Equation [4.91] has the plane wave solutions $\underline{u} = A\,e^{\mathrm{i}(\omega t - q\mathbf{e}.\mathbf{r})}$ if q verifies the equation $v^2q^2 = \omega^2 - \omega_o^{\,2} - 2\mathrm{i}\beta\omega$. This equation shows that q is always complex in the form $k - \mathrm{i}\delta$. Thus, the wave is always attenuated with an attenuation coefficient δ. In the special case where $\omega_o = 0$, the solution may be written in the form

$$\underline{u} = A\,e^{-\delta x}\;e^{\mathrm{i}(\omega t - kx)}$$

$$k = \frac{\omega}{v\sqrt{2}}\,\{\sqrt{1 + 4\beta^2/\omega^2} + 1\}^{½}, \qquad \delta = \frac{\omega}{v\sqrt{2}}\,\{\sqrt{1 + 4\beta^2/\omega^2} - 1\}^{½}. \qquad [4.92]$$

In the case of a dispersive medium, the intensity $\mathcal{I}$ of a plane wave is the same at any time and at any point. In the case of a reactive medium, the intensity $\mathcal{I}$ equals zero. In the case of a resistive medium, because of the dissipation of energy, the wave intensity decreases as it propagates.

In an absorbing medium, regardless of the equation of propagation, the intensity of a plane wave (or a parallel beam of particles) obeys a general exponential decrease of intensity law. Consider a beam (of radiation or particles) which enters an absorbing medium at $x = 0$ with an intensity $\mathcal{I}_o$ and propagates in the direction Ox (Figure 4.10a). The intensity at x is $\mathcal{I}(x)$. It is evident that the variation $d\mathcal{I}(x)$ over a distance dx is proportional to dx (to the first order in dx) and, in most physical processes, the dissipation of energy is proportional to the wave intensity. Thus, we may write $d\mathcal{I}(x) = -\mu\;\mathcal{I}(x)\;dx$, where the (–) sign is due to the decrease of the wave intensity and μ is the *absorption coefficient*, which depends on the specific process of energy absorption and on the nature of the medium and its physical conditions. This equation may be integrated in the form

$$\ln(\mathcal{I}/\mathcal{I}_o) = -\mu x \qquad \text{or} \qquad \mathcal{I}(x) = \mathcal{I}_o\, e^{-\mu x}, \qquad [4.93]$$

where the condition $\mathcal{I} = \mathcal{I}_o$ for $x = 0$ is used to determine the integration constant. This is a generalization of the Bouguer-Lambert-Beer law in optics. Noting that the intensity of a wave is always proportional to its amplitude squared, the amplitude decreases exponentially according to the law

$$A(x) = A(0)\, e^{-\delta x} = A(x_o)\, e^{-x/l}. \qquad [4.94]$$

$\delta = \mu/2$ is the *attenuation constant* (or *attenuation coefficient*) and $l = 2/\mu = 1/\delta$ is the *penetration depth* of the wave. The variation of $\mathcal{I}$ as a function of x is illustrated in Figure 4.10b, while that of $\ln \mathcal{I}$ is illustrated in Figure 4.10c. $1/\mu$ is the distance corresponding to the reduction of the intensity by a factor $1/e = 0.368$. We sometimes introduce the *half-value thickness*, which corresponds to the reduction of the intensity by ½

$$x_{1/2} = \frac{\ln 2}{\mu} = \frac{0.693}{\mu} = 1.386\, l. \qquad [4.95]$$

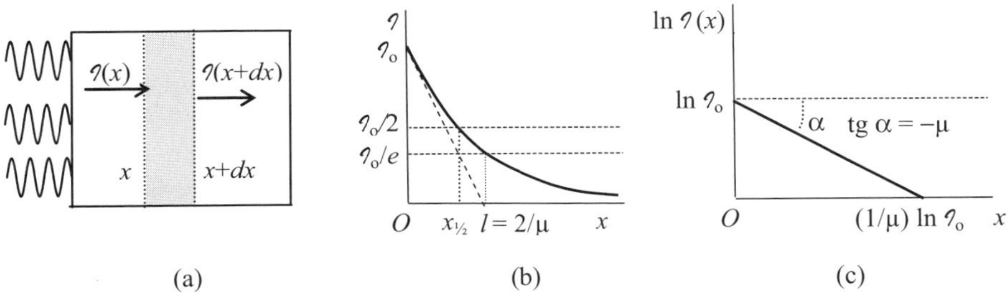

Figure 4.10. *Absorption of a wave in a medium: a) the intensity of the wave decreases exponentially, b) variation of $\mathcal{I}$ versus x, c) variation of* $\ln \mathcal{I}$ *versus x*

We frequently use a logarithmic scale for the intensity level and define the attenuation in decibels for a propagation distance Δx as

$$(\Delta\mathcal{I})_{dB} = 10 \log[\mathcal{I}(x+\Delta x)/\mathcal{I}(x)] = -10\,(\log e)\,\mu\,\Delta x = -4.3429\,\mu\,\Delta x$$
$$= -8.6858\,\Delta x/l = -10(\log 2)\,\Delta x/x_{1/2} = -3.01\,\Delta x/x_{1/2}. \qquad [4.96]$$

Thus, the intensity level decreases by 3.01 dB for each propagation of half-value thickness $x_{1/2}$ and 8.68 dB for each penetration depth l.

If several linear processes contribute to the attenuation of the wave, each process (i) contributes a variation $d\mathcal{I}_i(x) = -\mu_i\,\mathcal{I}(x)\,dx$ with a characteristic absorption coefficient μ_i. The total decrease of the intensity is

$$d\mathcal{I}(x) = \Sigma_i\, d\mathcal{I}_i(x) = -\,(\mu_1 + \mu_2 + \mu_2 + ...)\,\mathcal{I}(x)\,dx. \quad [4.97]$$

Thus, the intensity decreases according to [4.93] with an absorption coefficient

$$\mu = \mu_1 + \mu_2 + \mu_2 + ... \quad [4.98]$$

This is the case for the attenuation of sound waves due to the viscosity, the absorption etc., and for electromagnetic waves, which are attenuated because of their diffusion, absorption, etc. The absorption coefficient μ_i depends on the type of wave, the physical conditions of the medium and on the wave frequency. A resonance phenomenon may occur for some frequencies, producing a peak for one of the coefficients μ_i, thus for μ.

4.14. Sources and observers in motion, the Doppler effect and shock waves

In this section we consider the effects of the motion of the source and observer on the characteristics of the wave: its velocity, its direction of propagation and its frequency. We also consider shock waves. For this, let us consider two inertial reference frames, R and R', such that R' moves with a velocity $\mathbf{v}_o$ with respect to R. We assume that the axes of coordinates are parallel and coincide at $t = 0$ and that the velocity $\mathbf{v}_o$ points in the z-direction. The coordinates of the same space point in these frames are related by the *Galilean transformation*:

$$x = x', \qquad y = y', \qquad z = z' + v_o t. \quad [4.99]$$

Differentiating with respect to time (assumed to be the same in both frames), we obtain the *Galilean transformation of the velocity*, relating the velocity $\mathbf{v}$ of a body in R to its velocity $\mathbf{v}'$ in R'

$$\mathbf{v} = \mathbf{v}' + \mathbf{v}_o. \quad [4.100]$$

If we use spherical coordinates around the origin, we specify the velocity by its magnitude v, its angle θ with Oz and its azimuthal angle φ around Oz. The vector relation [4.100] is equivalent to the relations:

$$v = \sqrt{v'^2 + v_o^2 + 2\,v' v_o \cos\theta'} \quad [4.101]$$

$$\tan\theta = \sin\theta' / [\cos\theta' + v_o/v'], \qquad \varphi = \varphi'. \quad [4.102]$$

Consider a monochromatic plane wave, which may be written in R and R' as:

$$u(\mathbf{r}, t) = u_m \cos[\omega(t - \mathbf{e}.\mathbf{r}/ v_{(P)})] \quad [4.103]$$

$$u'(\mathbf{r}', t') = u'_m \cos[\omega'(t' - \mathbf{e}'.\mathbf{r}'/v'_{(P)})]. \quad [4.104]$$

The phase at each point M of space at a given time t must be the same in both frames. Indeed, if the wave equals zero, reaches a maximum or a minimum in R, this must also be the case in R'. Thus, we must have

$$\omega(t - \mathbf{e}.\mathbf{r}/v_{(P)}) = \omega'(t' - \mathbf{e}'.\mathbf{r}'/v'_{(P)}). \quad [4.105]$$

The direction of propagation $\mathbf{e}$ is determined by its angle θ with $\mathbf{v}_o$ and its azimuthal angle φ about $\mathbf{v}_o$. Expressing x, y and z as functions of x', y' and z' using the Galilean transformation [4.99], relationship [4.105] may be written in the form

$$\omega'\{ t - [x' \sin\theta' \cos\varphi' + y' \sin\theta' \cos\varphi' + z' \cos\theta']/v'_{(P)} \}$$
$$= \omega\{ t - [x' \sin\theta \cos\varphi + y' \sin\theta \cos\varphi + (z' + v_o t) \cos\theta]/v_{(P)} \}. \quad [4.106]$$

This relation is identically verified, that is, for any x', y', z' and t, if the coefficients of these quantities on both sides are equal, hence:

$$\theta' = \theta, \qquad \varphi' = \varphi \quad [4.107]$$

$$v'_{(P)} = v_{(P)} - v_o \cos\theta \quad [4.108]$$

$$\omega' = \omega[1 - (v_o/v_{(P)}) \cos\theta]. \quad [4.109]$$

The inverse transformations are obtained by changing v_o to $-v_o$, that is,

$$\theta = \theta', \qquad \varphi = \varphi' \quad [4.110]$$

$$v_{(P)} = v'_{(P)} + v_o \cos\theta' \quad [4.111]$$

$$\omega = \omega' [1 + (v_o/v'_{(P)}) \cos\theta']. \quad [4.112]$$

The equations [4.110] state that *the direction of propagation of a wave is the same in all reference frames.* The aberration of light, discovered by Bradley in 1727, confirms the transformation property for the direction of the velocity of light in corpuscular model [4.102], but not for the direction of propagation of a wave [4.110]. Equation [4.111] is the law of transformation of the phase velocity (identical to the group velocity in the case of propagation of electromagnetic waves in vacuum). It is not the same as the transformation law of the velocity of a particle [4.101] and both disagree with the Michelson experiment, which in 1881 showed that the speed of light in vacuum is independent on the reference frame.

Equation [4.109] is the transformation law of frequency. It agrees with the *Doppler effect* discovered in 1842 for sound waves, and was later generalized by Fizeau to light waves. Relations [4.108] and [4.109] imply that the wavelength $\lambda = 2\pi v_{(p)}/\omega$ is the same in both frames. This result agrees with the invariance of distances in classical physics and the interpretation of the wavelength as the distance separating two crests and measured in interference experiments (see Chapter 8). We specify that, in the case of electromagnetic waves and especially light waves in vacuum, the wavelength means $\lambda = 2\pi c/\omega$; thus it depends on the frame.

To analyze the Doppler-Fizeau effect, we designate the proper frames of the medium of propagation, the observer and the source by R_M, R_O and R_S respectively. The velocities of the source and of the observer with respect to the medium are designated $\mathbf{v}_S$ and $\mathbf{v}_O$. The angles of these velocities with the direction of propagation $\mathbf{e}$ of the wave (Figure 4.11a) are designated by θ_O and θ_S. The physical quantities are measured in R_O, the speed of propagation $v_{(P)M}$ in R_M is known, and in R_S the proper angular frequency ω_S of the source is known. Using equation [4.109] to transform from R_M to R_O and then to transform from R_M to R_S, we obtain

$$\omega_O = \omega_M\,[1 - \frac{v_O}{v_{p(M)}}\cos\theta_O] = \omega_S\,\frac{1-(v_O/v_{p(M)})\cos\theta_O}{1-(v_S/v_{p(M)})\cos\theta_S} = \omega_S\,\frac{1-\mathbf{e}.v_O/v_{p(M)}}{1-\mathbf{e}.v_S/v_{p(M)}}\,. \quad [4.113]$$

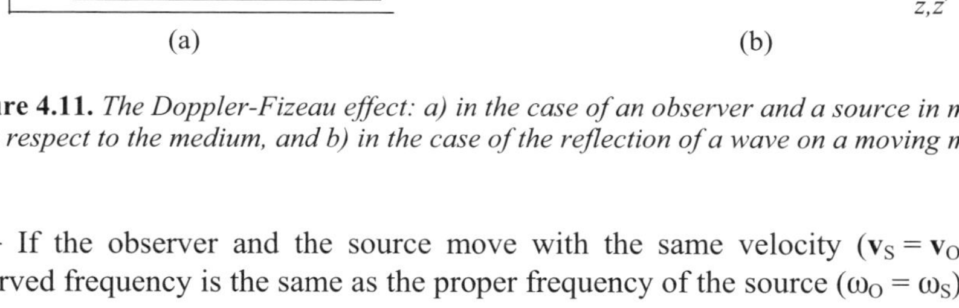

Figure 4.11. *The Doppler-Fizeau effect: a) in the case of an observer and a source in motion with respect to the medium, and b) in the case of the reflection of a wave on a moving mirror*

– If the observer and the source move with the same velocity ($\mathbf{v}_S = \mathbf{v}_O$), the observed frequency is the same as the proper frequency of the source ($\omega_O = \omega_S$).

– If the observer is at rest in the medium ($R_O \equiv R_M$ and $\mathbf{v}_O = 0$) and the source is moving with a velocity $\mathbf{v}_S$, we obtain

$$\omega_O = \frac{\omega_S}{1-(v_S/v_{p(M)})\cos\theta_O} = \frac{\omega_S}{1-(\mathbf{v}_S.\mathbf{e})/v_{p(M)}}, \quad [4.114]$$

where we used the equality of θ_O and θ_S to express ω_O in terms of θ_O, which is effectively the angle, which is measured by the observer in his proper frame R_O. If the source moves away from the observer the angle θ_O is obtuse and $\omega_O < \omega_S$, whereas if the source approaches the observer the angle θ_O is acute and $\omega_O > \omega_S$.

– If the source is at rest in the medium ($R_M = R_S$ and $\mathbf{v}_S = 0$) and the observer is moving with a velocity $\mathbf{v}_O$, we obtain

$$\omega_O = \omega_S \left[1 - \frac{v_O}{v_{p(M)}} \cos\theta_O\right] = \omega_S \left[1 - \frac{\mathbf{e}.\mathbf{v}_O}{v_{p(M)}}\right]. \qquad [4.115]$$

If the observer is moving away from the source θ_O is acute and $\omega_O < \omega_S$, and if the source moves toward the source θ_O is obtuse and $\omega_O > \omega_S$. We note that the Doppler shift does not have the same expression if the source is moving or the observer is moving with the same relative velocity.

The wave also undergoes a Doppler-Fizeau effect if it is reflected on a moving body. Consider a wave that is emitted by a source *S*, and received by an observer after its reflection off a distant "mirror". We assume that the mirror is moving with a velocity **v** toward the source and the observer, which are at rest in the propagation medium (Figure 4.11b). The incident wave (1) propagates in the opposite direction to **v**, while the reflected wave (2) propagates in the same direction as **v**. We associate a frame of reference *R* with the medium and a frame *R′* to the mirror. *R* is also the proper frame of the source and the observer. Equation [4.115] gives the frequency of wave (1) if it is observed in *R′* by an observer moving with the same velocity as the mirror

$$\omega'_1 = \omega_S\, [1 - (v/v_{(p)}) \cos\theta\,] = \omega_S\, [1 + v/v_{(p)}].$$

The mirror reflects the wave without changing its frequency (see Chapter 7). Thus, the angular frequency of wave (2) is $\omega'_2 = \omega'_1$, if it is measured in the frame *R′*. The angular frequency of wave (2), as measured by the observer at *O*, is obtained by using equation [4.114]

$$\omega_2 = \frac{\omega'_2}{1-(v/v_{(p)})\cos\theta} = \frac{\omega'_1}{1-(v/v_{(p)})} = \omega_S \frac{1+(v/v_{(p)})}{1-(v/v_{(p)})}. \qquad [4.116]$$

Note that ω_2 is not equal to the observed angular frequency $\omega_S(1 + 2v_S/v_{(p)})$ of the image of the source in the mirror (which moves towards the observer with speed $2v$).

The Doppler effect can be used to measure the velocity of moving objects. For this, a wave of known frequency is emitted toward the object. A measurement of the frequency of the reflected wave allows us to determine the component of the velocity of the object in the direction of observation: thus, it is used to measure the velocity of aeroplanes using radar, detect underwater obstacles using sonar, satellites, cars, etc. In medicine, the Doppler shift is used to monitor the flow of blood circulation (this technique uses 8 MHz ultrasound).

The radiation spectrum emitted by an atom is characteristic of that atom, independently of the physical or chemical conditions,. By analyzing light coming from far away galaxies, we find the same spectra as those of the elements on Earth, except that they are all "red-shifted". This shows that these galaxies were formed from the same elements as the Earth and that they move away. In other words, the Universe is expanding.

There is another effect of moving sources which arises if their velocity v_S exceeds the speed of propagation v of the wave in the medium. Figure 4.12 shows the waves, which are emitted by a source S and received by an observer O, if the source is respectively at rest (Figure 4.12a), moving toward the observer with a velocity $v_S < v$ (Figure 4.12b) and moving with a velocity $v_S = v$ (Figure 4.12c). In the last case, the source and all the waves that it has emitted while travelling on its trajectory reach the observer at the same time, forming a *shock wave* – so-called because it has a high intensity. Figure 4.12d represents the case of a source whose velocity v_S exceeds v. Consider the positions S_o, S_1, S_2 and S_3 of the source at times 0, τ, 2τ and 3τ, such that $S_oS_1 = S_1S_2 = S_2S_3 = v_S\tau$. At time 3τ, the waves which were emitted by the source at times 0, τ, 2τ and 3τ are spheres of radii $3v\tau$, $2v\tau$, $v\tau$ and 0 centered on S_o, S_1, S_2 and S_3 respectively. These spherical waves, as well as all the intermediary waves, are enveloped by a shock wave which has the form of a cone, whose vertex coincides with the position S_3 of the source, and which is symmetric about the source trajectory with a half angle α_M given by

$$\sin \alpha_M = v/v_s = 1/M. \qquad [4.117]$$

α_M is called the *Mach angle* and the ratio $M = v_s/v$ is called the *Mach number*. This conic wave front of high intensity moves in the direction **e** making an angle $90 - \alpha_M$ with the source trajectory. Outside the Mach cone, there is no wave at all. On the conic shock wave, the physical quantities of the wave have large discontinuities. An acoustic shock wave is produced if a projectile or an airplane travels at supersonic velocity (v_S exceeding the sound velocity) and a light shock wave is produced if a charged particle moves with a velocity exceeding the speed of light c/n in the medium (*Tcherenkov effect*). According to the special theory of relativity, the velocity of a body never reaches or exceeds the velocity of light in vacuum, so there is no Tcherenkov effect in vacuum.

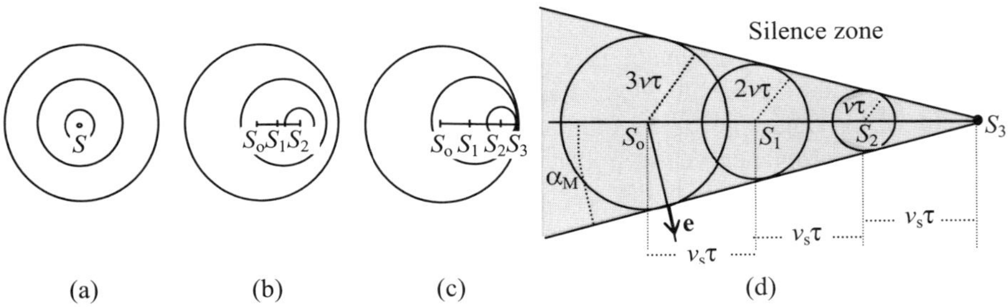

Figure 4.12. *Profile of the waves that are emitted by a source: a) if it is at rest, b) if its velocity v_S is less than the propagation velocity v, c) if $v_S = v$ and d) if $v_S > v$*

4.15. Summary

A wave is characterized by its speed of propagation, direction of propagation, amplitude and frequency (if it is periodic). Waves differ from particles by their large extension in space and time, interference, diffraction and, sometimes, polarization.

The one-dimensional d'Alembert wave equation and its progressive solution are

$$\partial^2_{tt}u - v^2\,\partial^2_{xx}u = 0, \qquad u = f(t - x/v) + g(t + x/v).$$

f is a wave, which propagates in the positive x direction, and g propagates in the negative x direction.

The three-dimensional d'Alembert wave equation and its progressive solution are

$$\partial^2_{tt}u - v^2\,\Delta u = 0 \qquad \text{and} \qquad u = f(t - \mathbf{r}.\mathbf{e}/v).$$

If the wave is symmetric around the origin O, the wave equation in spherical coordinates is $\partial^2_{tt}u - v^2[\partial^2 u/\partial r^2 + (2/r)(\partial u/\partial r)] = 0$. This equation has solutions of the form $u^{(\pm)}(r, t) = (1/r)\,f(t \mp r/v)$ where $u^{(+)}$ is a diverging spherical wave and $u^{(-)}$ is a converging spherical wave.

The wave equation for a vector wave is $\partial^2_{tt}\mathbf{u} - v^2\,\Delta\mathbf{u} = 0$. It has special solutions of the form $\mathbf{u} = \mathbf{u}_o\,f(\mathbf{r}, t)$. The wave is *longitudinal* if $\mathbf{u}$ points in the direction of propagation $\mathbf{e}$ and it is *transverse* if $\mathbf{u}$ is orthogonal to $\mathbf{e}$. In the last case, the direction of $\mathbf{u}$ determines the *polarization* of the wave. A spherical symmetric longitudinal vector wave is of the form $\mathbf{u} = (\mathbf{r}/r)F(r, t)$, where F is a solution to the wave equation

$$\partial^2_{tt}F - v^2[\partial^2_{rr}F + (2/r)\,\partial_r F - 2F/r^2] = 0.$$

This equation has the solutions $F_\pm = (v/r^2)\, f(t \mp r/v) \pm (1/r)\, f'(t \mp r/v)$. In the *wave zone* $[r \to \infty]$, $F_\pm \to (\pm 1/r)\, f'(t \pm r/v)$.

A simple harmonic scalar wave may be written in the form

$$u^\pm(x, t) = A \sin[\omega(t \pm x/v_{(p)}) \equiv A \sin(\omega t \pm kx) = A \sin[2\pi(t/T \pm x/\lambda)].$$

$k = \omega/v_{(p)} = 2\pi/\lambda$ is the *wave number* and $\lambda = v_{(p)}T = 2\pi/k$ is the *wavelength*.

A medium is *non-dispersive* if the phase velocity $v_{(p)} = \omega/k$ does not depend on ω. In the case of a dispersive medium, the *dispersion relation* may be written in the form $\omega = \omega(\mathbf{k})$, the phase *velocity* is $v_{(p)} = \omega(\mathbf{k})/k$ and the group velocity is $\mathbf{v}_{(g)} = \nabla_k\omega$.

A wave of period T is a superposition of simple harmonic waves of periods T, $T/2$, $T/3$, etc. (or angular frequencies ω, 2ω, 3ω, etc., with $\omega = 2\pi/T$). A signal is a superposition of simple harmonic waves with a continuous spectrum in a band $\Delta\omega$:

$$u(x, t) = \tfrac{1}{2}\, a_o + \Sigma_{n\geq 0}\{\, a_n \cos[n\omega(t - x/v)] + b_n \sin[n\omega(t - x/v)]\,\}$$
$$u(x, t) = \int_{\Delta\omega} d\omega\, \{A(\omega)\cos[\omega(t - x/v)] + B(\omega)\sin[\omega(t - x/v)]\}.$$

A wave packet may be written in the form $\underline{u}(\mathbf{r}, t) = \int d^3k\ \hat{\mathcal{U}}(\mathbf{k})\ e^{i(\omega t - \mathbf{k}.\mathbf{r})}$. Its center moves with the *group velocity* $\mathbf{v}_{(g)} = \nabla_k\omega$. It verifies the *uncertainty relations*

$$\Delta k_x.\Delta x \approx 2\pi, \qquad \Delta k_y.\Delta y \approx 2\pi, \qquad \Delta k_z.\Delta z \approx 2\pi.$$

A wave is amplitude-modulated if its amplitude varies slowly in time according to $A(t) = A\,[1 + \alpha s(t)]$ and it is frequency-modulated if its frequency varies slowly in time according to $\tilde{\nu}(t) = \tilde{\nu}_o + \alpha s(t)$.

In the case of a one-dimensional wave, the *energy density per unit length* $E_l(x, t)$ and the *energy flux* $P(x, t)$ verify the *energy conservation equation* $\partial_t E_l + \partial_x P = 0$. If the wave is progressive of the form $u = f(t - x/v)$, the two forms of energy have the same density and $P(x, t) = v\, E_l(x, t)$.

In the case of a three-dimensional wave, the *energy density per unit volume* $E_v(\mathbf{r}, t)$ and the vector *energy flux density* $\mathbf{S}(\mathbf{r}, t)$ verify *the equation of energy conservation* $\partial_t E_v + \nabla.\mathbf{S} = 0$. If the wave is progressive, the two forms of energy have the same density and $\mathbf{S}(\mathbf{r}, t) = vE_v(\mathbf{r}, t)\,\mathbf{e}$. The *intensity of the wave* is

$$\mathcal{I}(\mathbf{r}, t) = < S(\mathbf{r}, t) > = (1/T) \int_t^{t+T} dt\ S(\mathbf{r}, t).$$

If the wave equation in a medium is $\partial^2_{tt}u - v^2\Delta u + \omega_o^2 u = 0$ (*Klein-Gordon equation*) and $\omega < \omega_o$, the wave is attenuated with an attenuation constant $\delta = \sqrt{\omega_o^2 - \omega^2}\ /v$ but there is no dissipation of energy. Then the medium is said to be *reactive*.

In a resistive medium, the attenuation is due to a dissipation of energy. This is the case if the wave equation is $\ddot{u} + 2\beta\,\dot{u} - v^2(\partial^2_{xx}u) = 0$. An attenuated simple harmonic wave has the form $\underline{u} = A\, e^{-\delta x}\ e^{i(\omega t - kx)}$, where

$$k = \frac{\omega}{v\sqrt{2}}\ \{\sqrt{1+\beta^2/\omega^2} + 1\}^{1/2}, \qquad \delta = \frac{\omega}{v\sqrt{2}}\ \{\sqrt{1+\beta^2/\omega^2} - 1\}^{1/2}.$$

In general, the intensity varies according to the law $\mathcal{I}(x) = \mathcal{I}_o\, e^{-\mu x}$ with $\mu = \Sigma_i\ \mu_i$, where μ_i correspond to the various attenuation processes.

The *Doppler-Fizeau effect* is the frequency shift due to the motion of the source, the motion of the observer or the reflection of the wave on a moving body. If the source and the observer move with the velocities $\mathbf{v}_S$ and $\mathbf{v}_O$,

$$\omega_O = \omega_S\ \frac{1 - (v_O/v_{p(M)})\cos\theta_O}{1 - (v_S/v_{p(M)})\cos\theta_S} = \omega_S\ \frac{1 - (\mathbf{e}.\mathbf{v}_O)/v_{p(M)}}{1 - (\mathbf{e}.\mathbf{v}_S)/v_{p(M)}}\ .$$

If the observer or the source move towards one another $\omega_O > \omega_S$, otherwise $\omega_O < \omega_S$.

The shock wave arises if the source moves with a velocity v_S that is greater than the phase velocity $v_{(p)}$ in the medium. It has the profile of a cone, which is symmetric about the source trajectory, its vertex coincides with the source and its half angle α is given by $\sin\alpha = 1/M = v_{(p)}/v_S$.

4.16. Problem solving suggestions

1) In this chapter we have considered propagation in infinite media. A wave propagating on a string toward positive x, for instance, has the form $u_+ = f(t - x/v)$. If the string is fixed at the other end, the wave will be reflected to become $u_- = g(t + x/v)$, which propagates toward negative x. The total wave on the string is then $u = u_+ + u_-$. Note that the speed of propagation v is positive in u_+ and u_-.

2) Two-dimensional and three-dimensional progressive plane waves always have the form $u = f(t - \mathbf{r}.\mathbf{e}/v)$. We do not consider solutions of the form $u = f(t + \mathbf{r}.\mathbf{e}/v)$, because they simply correspond to a direction of propagation $\mathbf{e}' = -\mathbf{e}$. On the other hand, if we consider spherical symmetric three-dimensional waves, we must consider the two types of waves $u_{\pm} = (1/r)\, f(t \mp r/v)$. Here also, v is always positive.

3) A wave is symmetric around the z-axis if a rotation about Oz does not change the wave. In this case, it is more convenient to use cylindrical coordinates ρ, φ and z about Oz or spherical coordinates r, θ and φ about O. If the wave is scalar, it is represented by a function $u(\rho, z, t)$ or $u(r, \theta, t)$, which does not depend on φ. If the wave $\mathbf{u}$ is vectorial, the symmetry about Oz implies that its components u_ρ, u_φ and u_z or u_r, u_θ and u_φ do not depend on φ. Similarly, if a wave has rotation symmetry about a point O taken as the origin, it is convenient to use the spherical coordinates about O. A scalar wave or the components u_r, u_θ and u_φ of a vector wave do not depend on θ and φ.

4) Any wave has two forms of energy density. We also define the energy flux P in the case of a one-dimensional wave, and vector energy flux density $\mathbf{S}$ in the case of a three-dimensional wave. The flux of $\mathbf{S}$ across an element of area $d\mathcal{S}$ is the power dP, which is intercepted by $d\mathcal{S}$. The law of conservation of energy may be written in the form $\partial_t E_l + \partial_x P = 0$ and $\partial_t E_v + \nabla.\mathbf{S} = 0$ where E_l is the total energy density per unit length and E_v is the total energy density per unit volume. If the wave is progressive, we may write $P = vE_l$ and $\mathbf{S} = vE_v\,\mathbf{e}$ ($\mathbf{e}$ is the unit vector in the direction of propagation). A sound wave in a pipe of cross-section $\mathcal{S}$, for instance, is a three-dimensional wave. However, if it is uniform over the cross-section, it may be considered to be a one-dimensional wave with $E_l = \mathcal{S}E_v$ and $P = \mathcal{S}|\mathbf{S}|$.

5) All the physical quantities, which are associated with a wave, propagate with the group velocity (not the phase velocity $v_{(p)}$ or the speed v, which appears in the wave equation). If the medium is bounded in a transverse direction, the wave is successively reflected on the boundary surfaces (for instance, in a waveguide). Then, the phase velocity $v_{(p)}$ and the group velocity $v_{(g)}$ depend on v, the transverse dimensions of the medium and the frequency of the wave.

6) If a reactive or a resistive medium is infinite in the direction of propagation of the wave, the wave cannot be exponentially increasing, for this requires infinite energy. On the other hand, if the medium is bounded in the direction of propagation, the wave may be a superposition of an exponentially increasing wave and an exponentially decreasing wave. This is required for the wave to verify the boundary conditions at the boundary of the medium.

7) The expression of the Doppler-Fizeau effect is not the same, whether the source is moving and the observer is at rest in the propagation medium, or whether

the source is at rest and the observer is moving with the opposite velocity. If we analyze the Doppler effect by associating reference frames with the medium, the source and the observer, these frames must correspond to the considered situation.

4.17. Conceptual questions

Q1. Consider the vector wave $\underline{\mathbf{u}} = \underline{\mathbf{u}}_{\mathrm{m}}\, e^{\mathrm{i}(\omega t - k\mathbf{e}.\mathbf{r})}$. Specify the following concepts: longitudinal wave, transverse wave, linearly-polarized wave, circularly and elliptically-polarized wave and non-polarized wave. Evaluate $\nabla.\underline{\mathbf{u}}$, then $\Delta\underline{\mathbf{u}}$ and show that $\underline{\mathbf{u}}$ is a solution to the d'Alembert equation if $k = \omega/v$.

Q2. Consider a plane wave. Define what is meant by a progressive wave, standing wave, monochromatic wave, speed of propagation, wavefront, direction of propagation, phase velocity, wave number, wave vector and wavelength.

Q3. What are the causes of dispersion? Give examples. Is there any difference in the dispersion relationship for an isotropic medium and a non-isotropic medium? Write down the expression of the phase velocity and that of the group velocity in both cases. If you measure the speed of light by measuring its travel time, do you measure the phase velocity or the group velocity? If you use a long pipe to transmit words, they are slightly deformed. Explain the reason.

Q4. Explain what is meant by a wave packet and its velocity. Write the relationships giving its frequency band width, its time duration and its space extension in terms of the wave vector band width.

Q5. Explain what is meant by a carrier wave, amplitude modulation, modulating wave and frequency modulation. Providing an approximate answer, what should the band width for a radio emission be? If the band width for TV emissions is 10 MHz, how many channels can be transmitted in a fiber using the entire visible light spectrum (400 nm $< \lambda <$ 700 nm)?

Q6. What are the high-frequency limits of the telegraphy equation and the corresponding phase velocity and attenuation constant?

Q7. Verify that the intensity of a wave may be written in the form $\mathcal{I} = \mathcal{I}_o 2^{-x/x_{1/2}}$. Interpret $x_{½}$. What would $\mathcal{I}/\mathcal{I}_o$ be if the wave travels a distance equal to $10\, x_{½}$?

4.18. Problems

Propagation of waves

P4.1. Consider an infinite string taut in the direction *Ox*. Its origin is excited by an oscillator, which has a frequency of 5 Hz and an amplitude of 1 cm. It starts from the origin at $t = 0$ moving towards the positive *y*-axis. The speed of propagation is 10 m/s. Write down the expressions of the displacement at the points $x = 0$ and $x = 25$ cm of the string. What is the expression of the velocity as a function of time at the point $x = 25$ cm? Draw the profile of the string at $t = 0$, $t = 0.05$ and $t = 0.10$ s.

P4.2. The origin *O* of a string is displaced up and down according to the equation $u(0, t) = A \sin(2\pi t/T)$ only between $t = 0$ and $t = T/2$. Determine the displacement of the string at the point *x* and the profile of the string at time *t*.

P4.3. Consider the wave equation $\partial^2_{tt}u - v^2\partial^2_{xx}u = 0$ for one-dimensional waves. Introduce the variables $\xi = t - x/v$ and $\eta = t + x/v$. **a)** Show that the wave equation becomes $\partial^2_{\xi\eta}u = 0$. **b)** Deduce that the solution of this equation may be written in the form $u = f(\xi) + g(\eta)$ where *f* and *g* are two arbitrary functions.

P4.4*. Verify that wave equation [4.7] may be written in the form of [4.9] in polar coordinates. Verify that wave equation [4.13] may be written in the form of [4.15] in cylindrical coordinates and [4.18] in spherical coordinates.

P4.5. a) A wave of amplitude 4 mm and frequency 10 Hz propagates on the surface of the water in the direction of the first bisector of the axes *Ox* and *Oy*, with a speed of 0.5 m/s. Assume that the origin starts to vibrate upward at $t = 0$, write the wave function at point *M* of coordinates *x* and *y*. **b)** A plane wave propagates with a speed $v = 300$ m/s in a direction **e** which makes an angle of 45° with *Oz*. The plane formed by **e** and *Oz* makes an angle of 60° with the plane *Oxz*. The oscillation has a frequency of 400 Hz and reaches its maximum at the origin at $t = 0$. Write down the expression of this wave.

P4.6. Verify that the three-dimensional wave equation for a scalar wave has solutions of the form $u^{\pm} = (1/r)\, f(t \pm r/v)$ in the case of spherical symmetric wave.

P4.7. a) Show that a one-dimensional progressive simple harmonic wave may be represented by the following mathematical expressions:

$$u = A\cos[\omega(t - x/v)] = A\cos[k(x - vt)] = A\cos 2\pi(\tilde{\nu}\, t - x/\lambda) = A\cos 2\pi(t/T - x/\lambda)$$

b) Consider a wave $u = A\cos(\omega t - mx + ny - pz)$. Determine its direction of propagation, its wavelength and its phase velocity.

P4.8*. Consider a longitudinal vector wave, which passes by a focal point O. Writing the expressions of the wave before and after O, show that it undergoes a change of sign; that is, a phase shift of π. This analysis may be undertaken for the case of a transverse wave, but the analysis is more involved. The Meslin experiment verified this phase shift in the case of light waves.

Dispersion, modulation velocity and group velocity

P4.9. a) Knowing the expression of the phase velocity as a function of the wavelength λ, show that the group velocity may be written in the form $v_{(g)} = v_{(p)} - \lambda(dv_{(p)}/d\lambda)$. **b)** In optics, the index of refraction n is usually expressed as a function of the wavelength. Show that the group velocity may be written in the form $v_{(g)} = c/n + (c\lambda/n^2)(dn/d\lambda)$. **c)** Knowing the expression of the phase velocity as a function of the frequency, show that the group velocity is given by the equation $1/v_{(g)} = 1/v_{(p)} - (\tilde{\nu}/v_{(p)}^2)(dv_{(p)}/d\tilde{\nu}$. **d)** The refraction index of glass depends on the light wavelength in vacuum, according to Cauchy's empirical formula $n = A + B/\lambda^2$. Calculate the phase velocity and the group velocity in a glass, whose parameters are $A = 1.584$ and $B = 1.270 \times 10^4$ nm^2 for the wavelength $\lambda = 600$ nm (in vacuum).

P4.10. Consider an isotropic medium whose dispersion relation is $\omega = \omega(k)$. Show that the group velocity is $\mathbf{v}_{(g)} = \nabla\omega = (\partial\omega/\partial k)\,\mathbf{k}/k$.

Fourier analysis for waves

P4.11.** A vibrator acts on the origin of a string producing a rectangular signal such that the displacement is $u = 1$ mm for $0 < t < \tau$ and $u = 0$ for $t < 0$ or $t > \tau$. **a)** Create the Fourier analysis of this signal. **b)** Assume that the speed of propagation does not depend on the frequency, write down the expression of the signal at time t. **c)** Now assume that the speed varies with the frequency according to the relation $v(\omega) = v_o\omega/(\alpha + \omega)$. What is the group velocity? **d)** What is the expression of the signal at time t? Draw the profile of this signal.

P4.12. a)** The spectral amplitude of a signal is $\hat{\mathcal{U}}(k) = A_m\, e^{-(k-k_o)^2/4\kappa^2}$ where κ is a constant (called a *Gaussian signal*). Study its propagation by expanding the dispersion relation $\omega = \omega(k)$ as a power series in $(k - k_o)$ and neglecting powers higher than 2. **b)** A three-dimensional Gaussian wave packet has the spectral amplitude $\underline{A}(\mathbf{k}) = A\; e^{-(k_x-k_{ox})^2/\kappa_x^2}\; e^{-(k_y-k_{oy})^2/\kappa_y^2}\; e^{-(k_z-k_{oz})^2/\kappa_z^2}$. Calculate its wave function by neglecting powers of $(k - k_o)$ that are higher than 1. Verify the uncertainty relations [4.69].

P4.13.** Consider a wave packet in the form of [4.64], with an amplitude $\hat{\mathcal{U}}(k) = 1$ if $|k - k_o| < k_1$ and $\hat{\mathcal{U}}(k) = 0$ if $|k - k_o| > k_1$ where k_o and k_1 are constants. Retaining only the first two terms of the Taylor series of ω as a function of $(k - k_o)$, show that the wave may be written as

$$u(x,t) = \frac{2}{x - v_g t} e^{i(\omega_o t - k_o x)} \sin k_1 (x - v_{(g)} t)$$

where $v_{(g)}$ and ω_o correspond to k_o. Draw the profile of this wave. Determine the center of the wave packet, its velocity and its width. Verify the uncertainty relation.

Modulation

P4.14. Two waves $y_1 = A \sin(50\,x - 200\,t)$ and $y_2 = A \sin(54\,x - 230\,t)$ superpose on a string. **a)** Write down the expression of the resulting wave. **b)** Calculate the modulation velocity. **c)** What is the distance between two points where the amplitude equals zero at a given time t? **d)** What would be your answer to question (c) if the second wave is $y_2 = A \cos(54\,x - 230\,t)$?

P4.15*. **a)** The amplitude of a wave $u = A \cos(\omega t - kx)$ is modulated according to the relation $A = a + b \cos(\Delta\omega\, t) + c \cos^2(\Delta\omega\, t)$. Show that this wave may be considered as a superposition of simple harmonic waves with angular frequencies ω, $\omega \pm \Delta\omega$ and $\omega \pm 2\Delta\omega$. What are their respective amplitudes? **b)** The frequency of the wave $u = A \cos(\omega t - kx)$ is modulated according to $\omega = \omega_o + a \cos(\Delta\omega\, t)$ where ω_o, a and $\Delta\omega$ are constants. Assume that $a^2 << 1$, show that, to the first order in a, this wave may be considered as a superposition of a wave of angular frequency ω_o and amplitude A and four waves of angular frequencies $\omega_o \pm \Delta\omega$ and $\omega_o + a \pm \Delta\omega$ and amplitudes $A/2$. **c)** The phase of a wave $u = A \cos(\omega_o t - kx + \phi)$ is modulated according to $\phi = a + b \cos(\Delta\omega\, t)$ where a, b and $\Delta\omega$ are constants. Show that, to the first order in b, this wave may be considered as a superposition of a wave with angular frequency ω_o and amplitude A, and two waves with angular frequencies $\omega_o \pm \Delta\omega$ and amplitudes $Ab/2$.

P4.16*. **a)** It is possible to produce an amplitude-modulated voltage as follows: consider a resistance, which varies in time according to $R = R_o[1 + a_m \cos(\omega_m t)]$, where ω_m is the modulating angular frequency. What is the voltage across this resistance if it carries a current $I = I_o \cos(\omega t)$, where ω is the carrier angular frequency? Write this voltage as a superposition of simple harmonic voltages of angular frequencies ω (the carrier), $\omega + \omega_m$ (the upper lateral band) and $\omega - \omega_m$ (the lower lateral band). The *carbon microphone* uses a resistance of this type: the sound wave of angular frequency ω_m acts on the diaphragm and compresses the carbon grains, thus, producing a variation of the resistance as indicated above. **b)** Suppose that we have at our disposal a nonlinear amplifier, that is, whose output voltage V_s is related to the input voltage V_e by an nonlinear equation $V_s = \alpha V_e + \beta V_e^2$. Assume that

V_e is the superposition of $V_1 = A\cos(\omega t)$ and $V_2 = a\cos(\omega_m t)$. Show that V_s is a superposition of a modulated wave, other waves of angular frequencies ω_m, $2\omega_m$ and 2ω and a constant term. Is it possible to isolate the output modulated signal and eliminate the others using an appropriate band-pass filter? **c)** Consider the amplitude-modulated voltage $V = [A + a\cos(\omega_m t)]\cos(\omega t)$. To retain a voltage of frequency equal to the modulating frequency ω_m, we use the voltage V as an input of a quadratic amplifier. Analyze the output spectrum. What band-pass filter should we choose to isolate the voltage of angular frequency ω_m?

P4.17*. A voltage $V = V_o\cos(\omega t + \phi)$ is frequency-modulated if ω varies slowly according to $\omega = \omega_o[1 + a\cos(\omega_m t)]$ where ω_o is the carrier frequency and ω_m is the modulating frequency. To transmit a sound of angular frequency ω_m, we use a *condenser microphone*. The sound wave of angular frequency ω_m acts on a diaphragm, which is attached to the plate of a capacitor. Thus, the capacitance varies according to $C = C_o + c_o\cos(\omega_m t)$ with $c_o << C_o$. **a)** Show that an LC circuit, using this capacitor, has a natural angular frequency $\omega = \omega_o[1 + a\cos(\omega_m t)]$ with $\omega_o = 1/\sqrt{LC_o}$ and $a = -c_o/2C_o$. Deduce that the voltage across the capacitor is frequency-modulated with a carrier frequency ω and a modulating frequency ω_m. **b)** Show that a frequency-modulated voltage may be considered as a superposition of simple harmonic voltages of angular frequencies ω_o, $\omega_o \pm \omega_m$ and $\omega_o + a\omega_o \pm \omega_m$. Calculate their relative amplitudes. How can we use a nonlinear amplifier and filters to retain a voltage with the modulating frequency ω_m?

Attenuated waves

P4.18.** Consider a chain of simple pendulums of natural angular frequency ω_o. Consecutive pendulums are separated by a distance d and coupled by springs of constant $K \equiv m\omega_1^2$ (Figure 4.13). **a)** Show that the equation of motion for the nth mass is $\partial^2_{tt}u + \omega_o^2 u_n + \omega_1^2(2u_n - u_{n-1} - u_{n+1}) = 0$. The first pendulum is subject to a force $\underline{F} = \underline{F}_m e^{i\omega t}$. If $\omega_o^2 < \omega^2 < \omega_o^2 + 4\omega_1^2$, verify that the steady solution may be written in the form $\underline{u}_n = \underline{A}\, e^{i(\omega t - nkd)}$ with the dispersion relation $\sin^2(kd/2) = \gamma$ where $\gamma = (\omega^2 - \omega_o^2)/4\omega_1^2$. **b)** Verify that the solution is of the form $\underline{u}_n = \underline{A}\, e^{-n\delta d}\, e^{i\omega t}$ if $\omega^2 < \omega_o^2$ and $\underline{u}_n = (-1)^n \underline{A}\, e^{-n\delta d}\, e^{i\omega t}$ if $\omega^2 > \omega_o^2 + 4\omega_1^2$ with $\delta d = \ln[1 - 2\gamma + 2(\gamma^2 - \gamma)^{1/2}]$. Interpret these solutions. **c)** What will the pass-band be if the system of pendulums becomes distributed ($d \to 0$ and $m/d \to \mu$)? Verify that we obtain the Klein-Gordon equation. Find the solutions of this equation directly. Write the dispersion relation and the expressions of the phase velocity, the group velocity and the damping coefficient according to the values of ω. **d)** Assume that the pendulums are subject to the friction forces $-2\beta\,\partial_t u_n$. Write the wave equation in the limit of a distributed system. Verify that the wave of angular frequency ω propagates with a wave number and an attenuation coefficient δ given by

$$k = (\omega^2 - \omega_o{}^2 + \alpha^2)^{½}/v\sqrt{2} \qquad \text{and} \qquad \delta = (\omega_o{}^2 - \omega^2 + \omega_1{}^2)^{½}/v\sqrt{2}$$

where $\alpha^4 = (\omega^2 - \omega_o{}^2)^2 + 4\beta^2\omega^2$. Plot k and δ versus ω for the cases $\beta < \omega_o$ and $\beta > \omega_o$.

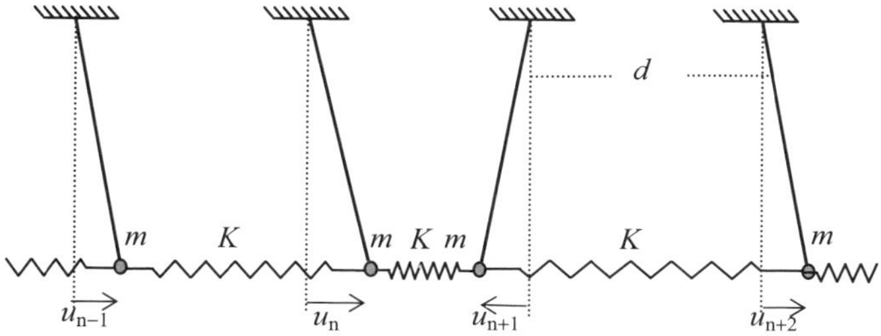

Figure 4.13. *Problem P4.18*

Sources and observers in motion, the Doppler effect and shock waves

P4.19. A sound source has a proper frequency $\tilde{\nu}_S = 500$ Hz. The speed of sound in air is $v_s = 330$ m/s. **a)** Assume that the source moves toward an observer with a velocity $v_o = 40$ m/s. Determine the frequency and the wavelength as measured by the observer at rest. **b)** Assume that the source is at rest but the observer moves toward it with a velocity $v_o = 40$ m/s. Determine the frequency and the wavelength measured by the observer.

P4.20. a) An ambulance moves toward an observer with a velocity of 100 km/h and it emits a sound at the frequency 3,000 Hz. What is the frequency that the observer perceives? **b)** What will be the perceived frequency after the ambulance passes the observer and moves away with the same velocity? **c)** The observer also hears the sound reflected off a distant wall. What is the frequency of the reflected sound? Which effect results from the superposition of the direct and the reflected sounds?

P4.21. A satellite passes at the zenith of a radar station at $t = 0$ with a velocity of 3×10^4 km/h. **a)** Study the variation of the reflected wave frequency as function of time. **b)** Sending a signal at $t = 0$, it takes 0.5 ms to receive the echo. What is the altitude of the satellite?

P4.22. The Sun has a radius of 6.96×10^8 m and its period of rotation is 24.7 (terrestrial) days. What is the Doppler shift for the 550 nm light emitted by the periphery of the solar disk? Assuming the invariance of the velocity of light in vacuum, what is the shift in the wavelength of this radiation?

P4.23. A supersonic aircraft flies at mach-2 (velocity twice the speed of sound) at an altitude of 1,000 m and passes at the zenith of an observer O at $t = 0$. **a)** Draw a figure showing the shock wave of this aircraft. **b)** How does the shock wave propagate and what is its velocity toward the observer? **c)** At what time does the observer perceive the wave? What is the position of the aircraft at this time?

Chapter 5

Mechanical Waves

In this chapter we study mechanical waves in solids and fluids. Any disturbance displaces the molecules or atoms of the medium and manifests itself as local displacements or deformations (called *strains*) and internal forces (called *stress*). If the disturbance depends on time, it propagates as an *elastic wave* in solids, *sound wave* in fluids or a *surface wave* in liquids. We study the elastic waves in the case of simple systems: strings, rods and membranes. The study of sound, also called *acoustics*, includes *physical acoustics* (which we introduce in this chapter), *physiological acoustics, and architectural acoustics.* Sound waves include audible sounds as well as *infrasounds* and *ultrasounds*, which are inaudible. We also briefly consider surface waves.

5.1. Transverse waves on a taut string

Consider a line of masses m joined by an ideally massless string taut to a tension F (Figure 5.1a). Let Ox be the axis of equilibrium of the string, and d the distance between consecutive masses. Due to the string tension, a displacement u_n of the mass (n) in the transverse y-direction propagates to the mass (n + 1) producing a displacement u_{n+1} and so on. Each mass (n) is subject to two forces of equal magnitude F exerted by the string toward the masses (n –1) and (n + 1). Their resultant is generally not in the transverse direction Oy. However, if the oscillations are small, the longitudinal component of this resultant is of the second order in u_n, u_{n-1} or u_{n+1} and the transverse component may be written in the form

$$F_n = (F/d)(u_{n-1} - u_n) + (F/d)(u_{n+1} - u_n). \qquad [5.1]$$

Thus, the equation of motion of the nth mass is

$$\partial^2_{tt}u_n + \omega_o^2 (2u_n - u_{n+1} - u_{n-1}) = 0, \qquad [5.2]$$

where we have set $\omega_o^2 = F/md$. If the first mass is excited by a simple harmonic force $F = F_o e^{i\omega t}$, the masses have simple harmonic displacements of the same angular frequency ω, of the form $\underline{u}_n = A_n e^{i\omega t}$. Substituting such expressions for u_n and $u_{n\pm 1}$ in equation [5.2], we find that the amplitudes A_n must verify the recurrence relation

$$A_{n+1} + A_{n-1} = (2 - \omega^2/\omega_o^2)\, A_n. \qquad [5.3]$$

Inspired by a wave $u(x, t) = A\, e^{i(\omega t - kx)}$ where $x = nd$ is the coordinate of the mass (n), we try a solution of the form $A_n = A\, e^{-inkd}$. We find that this is effectively a solution of [5.3] if $e^{ikd} + e^{-ikd} = (2 - \omega^2/\omega_o^2)$. This condition on k may be written in the form $\sin^2(½\, kd) = \omega^2/4\omega_o^2$ or

$$\omega = 2\omega_o\, |\sin(½kd)|. \qquad [5.4]$$

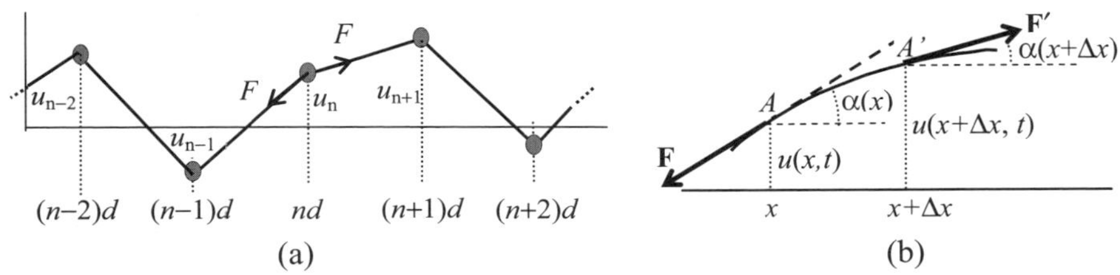

Figure 5.1. *a) Transverse oscillations of a line of masses connected by a taut massless string and b) transverse wave on a taut massive string*

This equation determines the "*wave number*" k if $\omega < 2\omega_o$. The *cut-off angular frequency* $2\omega_o$ corresponds to $kd = (2q + 1)\pi$ where q is an integer number. We note that ω and the u_n are invariant if kd changes by 2π. Two wave numbers k and $k + 2q\pi/d$ correspond, exactly, to the same displacement of the masses and cannot be distinguished in any physical way. On the other hand, the wave numbers $k_1 < \pi/d$ and $k_2 = 2\pi/d - k_1$ correspond to identical waves which propagate toward the positive x and the negative x respectively. Thus we may consider only the values of k which lie between 0 and π/d (*Brillouin zone*). Then, the displacement of the mass (n) may be written in the form

$$u_n = A\, e^{i(\omega t - ndk)} = A\, e^{i\omega(t - ndk/\omega)}. \qquad [5.5]$$

The mass (n) oscillates like the mass ($n - 1$), but with a time delay kd/ω. Therefore, if $\omega < 2\omega_o$, a *progressive wave* propagates along the chain. The energy is then transferred from the mass (n) to the mass ($n + 1$). Equation [5.4], which relates the angular frequency ω to the wave number k, is the *dispersion relation* of the chain. The wave propagates with the phase velocity and the group velocity

$$v_{(p)} = \omega/k = (2\omega_o/k)\,|\sin(kd/2)|, \qquad [5.6]$$

$$v_{(g)} = d\omega/dk = d\,\omega_o \cos(\tfrac{1}{2}\,kd) = \tfrac{1}{2}\,d\,\sqrt{4\omega_o^2 - \omega^2}\,. \qquad [5.7]$$

On the other hand, if the angular frequency ω is higher than the cut-off frequency $2\omega_o$, the equation $e^{ikd} + e^{-ikd} = (2 - \omega^2/\omega_o^2)$ cannot be satisfied by a real value of k, instead it is satisfied by an imaginary $k = -i\delta$ such that

$$\delta d = \ln\,[1 - \omega^2/2\omega_o^2 + (\omega/2\omega_o^2)\sqrt{\omega^2 - 4\omega_o^2}\,]. \qquad [5.8]$$

In this case, solution [5.5] may be written in the form

$$u_n = (-1)^n A\, e^{-nd\delta}\, e^{i\omega t}\,. \qquad [5.9]$$

Therefore, the mass (n) has a simple harmonic oscillation but with an amplitude reduced by a factor $e^{-d\delta}$, compared to the amplitude of the preceding mass ($n - 1$) and in opposite phase. We say in this case that the chain is *reactive*. In other words, the chain constitutes a *low-pass* filter with a *cut-off angular frequency* $2\omega_o$.

Let us assume now that the distance d and the mass m are increasingly reduced with the ratio m/d constant and equal to m_l. At the limit, we get a massive string of mass density m_l per unit length. As $\omega_o^2 \rightarrow F/m_l d^2$, the cut-off frequency $2\omega_o$ becomes infinite and the waves of all frequencies propagate on this string. Dispersion relation [5.4] becomes the proportionality relation $\omega = v\,k$, where

$$v = \sqrt{F/m_l}\,. \qquad [5.10]$$

The phase velocity and the group velocity become equal to v. The position nd of mass (n) must now be considered as a continuous coordinate x, and the displacement $u_n(t)$ of the mass (n) becomes a wave function $u(x, t)$. Making the Taylor expansion $u_{n\pm1}(t) \rightarrow u(x \pm d, t) \approx u(x, t) \pm (\partial_x u)\,d + \tfrac{1}{2}(\partial^2_{xx}u)d^2$, the equation of motion [5.2] becomes the d'Alembert wave equation for elastic transverse waves on the string

$$\partial^2_{tt}u - v^2(\partial^2_{xx}u) = 0. \qquad [5.11]$$

It is possible to directly study the propagation of the *elastic transverse wave* on a massive string (Figure 5.1b). A small element AA' of the string situated between x and $x + \Delta x$ is displaced by $u(x, t)$. If the displacement u of the string is small, the angle α between Ox and the tangent to the string (oriented toward the positive x) is small. The slope of the string at x is $\tan \alpha \cong \alpha = \partial_x u$. The element AA' is subject to two forces $\mathbf{F}$ and $\mathbf{F}'$ exerted at A and A' by the other parts of the string. These forces have magnitudes equal to the tension F of the string and they are tangent to the string at A and A'. They point toward the negative x and the positive x respectively. Thus, they make with Ox the respective angles

$$\alpha(x) - \pi \cong \partial_x u(x, t) - \pi \quad \text{and} \quad \alpha(x + \Delta x) \cong \partial_x u(x + \Delta x, t) \cong \partial_x u + \Delta x\, \partial^2_{xx} u.$$

The x-component of the resultant force acting on the element AA' is

$$F_x + F'_x = F \cos[\partial_x u - \pi] + F \cos[\partial_x u + \Delta x\, \partial^2_{xx} u] \cong 0$$

up to terms of the second order in u or its derivatives. The y-component of $\mathbf{F} + \mathbf{F}'$ is

$$\Delta F_y = F_y + F'_y = F \sin[\partial_x u - \pi] + F \sin[\partial_x u + \Delta x\, \partial^2_{xx} u] \cong F\, \Delta x\, \partial^2_{xx} u.$$

The equation of motion of this element $\Delta m\, \partial^2_{tt} u = \Delta F_y$ can then be written in the form [5.11] with $v = \sqrt{F/m_l}$.

In fact, the general displacement of a point of a taut string is a vector $\mathbf{u}(x, t)$ which may be oriented in any direction perpendicular to the string (transverse wave). The direction of $\mathbf{u}$ determines the *polarization of the wave* (see section 4.4).

5.2. Strain and stress in elastic solids

A solid body is never completely rigid; it deforms if it is subject to an external force. If, after the removal of the external force, the body returns rapidly to its original configuration, we say that the material is *elastic*; otherwise, it is *plastic*. The basic notions in the study of elastic properties are strain and stress.

Strain is related to the deformation of the object (that is, a modification of its dimensions, its volume and its shape). The particles of the body in a small volume element near the point $\mathbf{r}$ undergo a *displacement* $\mathbf{u}(\mathbf{r})$ from their equilibrium position. If $\mathbf{u}$ is independent of $\mathbf{r}$, the displacement of the body is a uniform translation, there is no deformation and, consequently, no internal forces will result in the body. If, for instance, two rods of length 2 m and 5 m have one end fixed and the other compressed by 1 cm, it is evident that the first rod is more deformed than the second. This can be understood if we consider the interatomic distance, which varies more in the first rod. A good measure of the *strain* in this case is the relative variation of length $\hat{\varepsilon} \equiv \delta x/x$. We may consider other deformations; in all cases, the strain is defined as a dimensionless ratio, such as the relative variation of length $\delta x/x$, the relative variation of volume $\delta\mathcal{V}/\mathcal{V}$ or an angle of torsion. Although the displacement **u** may be large (in the case of one-dimensional objects like a strip or a wire undergoing bending), the strain $\hat{\varepsilon}$ is always small.

To produce a certain strain $\hat{\varepsilon}$ (of any type), a distributed force must be exerted on a part $\mathcal{S}$ of the surface of the body (or on the entire surface). This force δF may be normal or tangential to $\mathcal{S}$. The experiment shows that the relevant physical quantity is the *stress*

$$\Pi = \delta F/\mathcal{S}, \qquad [5.12]$$

which may be very large. It has the dimensions of a pressure. Its *SI* unit is the *pascal* (1 Pa = 1 N.m^{-2}). The *megapascal* (1 MPa = 10^6 Pa) and the *gigapascal* (1 GPa = 10^9 Pa) are frequently used.

Figure 5.2 shows the variation of the stress Π as a function of the strain $\hat{\varepsilon}$. If $\hat{\varepsilon}$ is increased but it remains small, the experiment shows that the stress remains proportional to the strain (the segment *OA* of the graph). If the stress is suppressed at any point *M* of this segment, the body returns to its initial configuration *O* following the opposite path *MO*. This law, called *Hooke's law*, is valid as long as the stress is less than the *elastic limit* Π_l or *yield strength* (the point *A* of the graph). Beyond Π_l, the relationship between Π and $\hat{\varepsilon}$ is no longer linear. If the stress is gradually reduced starting from the point *N*, the body describes the path *NN′*. At *N′*, the stress Π is equal to zero but with a certain remnant strain. The body is permanently deformed. If the stress reaches a certain value Π_t called the *tensile strength* (or *ultimate strength*) corresponding to point *B*, the solid will start to fracture (it looses its cohesion) and continue to deform even with decreasing stress until the *breaking strength* Π_b (point *C*) which corresponds to the rupture of the sample.

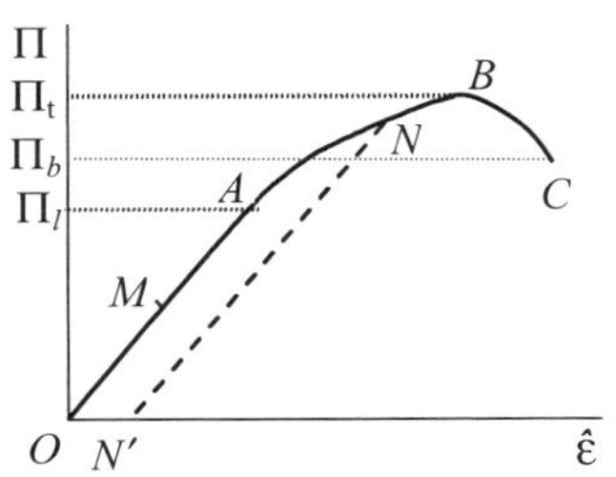

Figure 5.2. *Stress-strain curve*

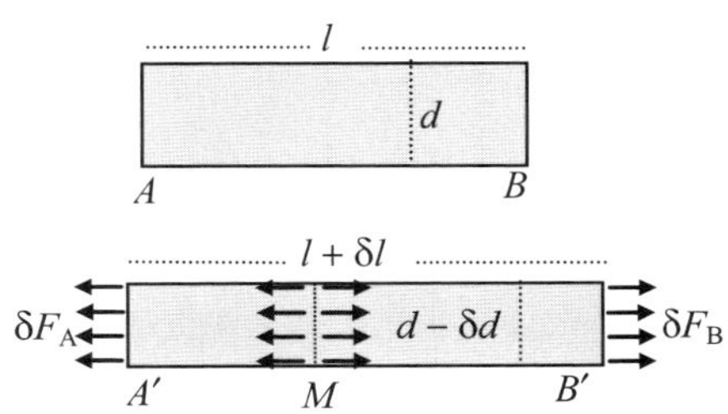

Figure 5.3. *Stress and strain in a rod*

For each type of stress and strain, we define the *elasticity modulus* as the ratio $\Pi/\hat{\varepsilon}$ of the stress to the strain. These are characteristic quantities of the material. Thus, they do not depend on the shape or the size of the sample.

a) Consider again a uniform rod of length l and section $\mathcal{S}$ (Figure 5.3) and assume that its end sections are submitted to two longitudinal and opposite (algebraic) forces δF_A and δF_B. We adopt the convention that the normal stress and displacement are positive if they point toward the exterior of the body. Thus, in the case of *traction*, $\delta F_B = -\delta F_A > 0$, the stress is $\Pi \equiv \delta F_B/\mathcal{S} > 0$, the rod undergoes an elongation ($\delta l \equiv \delta x_B - \delta x_A > 0$) and the strain is $\hat{\varepsilon} = \delta l/l > 0$ while, in the case of a *compression*, all quantities have opposite signs. The strain produces internal forces in the rod. If we imagine a section at M, the rod being in equilibrium, the part $A'M$ of the rod acts on the part MB' with a force equal to δF_A and MB' acts on $A'M$ with a force equal to δF_B. Thus all the parts (or the particles) of the body are subject to opposite forces and this is due to a modification of the interatomic distance everywhere within the body. On the other hand, the experiment shows that each of the transverse dimensions d of the rod undergoes a relative variation $\delta d/d = -\sigma\hat{\varepsilon}$, where σ is the *Poisson ratio*. Thus, d decreases if l increases or vice versa. In this case, Hooke's law may be written in the form $\Pi = Y\hat{\varepsilon}$, where Y is the *elasticity modulus* (or *Young's modulus*). Thus, we have for both traction and compression:

$$\frac{\delta l}{l} = \frac{1}{Y}\Pi = \frac{1}{Y}\frac{\delta F}{\mathcal{S}}, \qquad \frac{\delta d}{d} = -\sigma\frac{\delta l}{l} = -\frac{\sigma}{Y}\Pi. \qquad [5.13]$$

The values of Y and σ for some common materials are given in Table 5.1. We note the similarity of the variation of the length of a rod and that of an elastic spring with the same length l. A force δF, acting on the spring of force constant K, produces a variation of length $\delta l = \delta F/K$. Thus, according to [5.13], the rod behaves like a spring of constant $K = Y\mathcal{S}/l$.

b) If a uniform stress Π acts on the entire surface of a body (whether it is solid or fluid), its volume $\mathcal{V}$ undergoes a relative variation

$$\delta\mathcal{V}/\mathcal{V} = \Pi/B, \qquad [5.14]$$

where B is the *bulk modulus* (or *compression modulus*) of the material. In the case of an elastic solid, B may be expressed in terms of Y and σ. Consider a rod with a rectangular section of sides a and b (Figure 5.4a). The stress Π, applied only on its opposite faces $AEHD$ and $KCGF$ deforms it length AK by $\delta l = (l/Y)\Pi$ and its transverse dimensions AD and AE by $\delta d_1 = -(\sigma\, d_1/Y)\Pi$ and $\delta d_2 = -(\sigma\, d_2/Y)\Pi$. The relative variation of the volume $\mathcal{V} = l d_1 d_2$ is

$$\frac{\delta \mathcal{V}}{\mathcal{V}} = \frac{\delta l}{l} + \frac{\delta d_1}{d_1} + \frac{\delta d_2}{d_2} = \frac{1-2\sigma}{Y}\Pi. \qquad [5.15]$$

$\delta\mathcal{V}/\mathcal{V}$ is independent of the dimensions of the body. Thus, if the same stress Π acts on the three pairs of opposite faces, the relative variation of the volume is multiplied by 3 and we get the bulk modulus such that

$$\Pi = B\,(\delta\mathcal{V}/\mathcal{V}), \qquad B = Y/3(1-2\sigma). \qquad [5.16]$$

The inverse of the bulk modulus $\kappa = 1/B$ is the *compressibility* of the material. For any material as the volume of a compressed body decreases ($B > 0$) we must have $0 < \sigma < 0.5$.

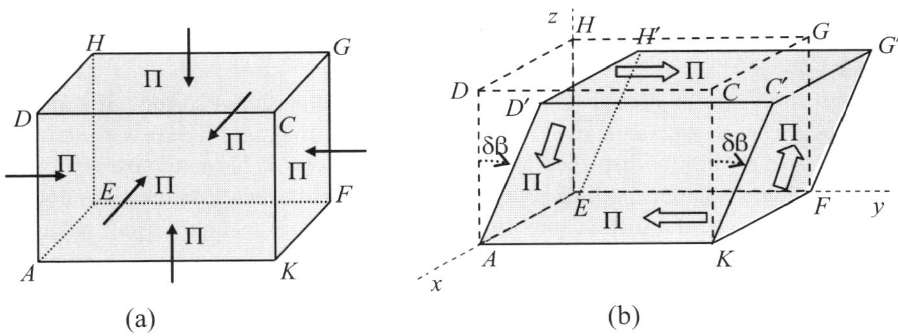

Figure 5.4. *a) Variation of the volume of a rectangular block under a uniform compressive stress, b) shear strain of a rectangular block*

Material	m_v (kg/m^3)	Y (10^{10} Pa)	σ	$B =$ (10^{10} Pa)	μ (10^{10} Pa)
Lead	11,340	1.6	0.4	0.8	0.5
Aluminum	2,700	5.6-7.7	0.33	7.2	2.4-3
Copper	8,920	11-12.5	0.34	14	4.2-5
Steel	7,500	21	0.29	7	8.5
Iron (pure)	7,860	20.6	0.27	6	5-8
Nickel	8,800	21.5	0.35	26	7.7
Tungsten	19,300	36	0.2	20	15
Glass	2,400-2,800	5.5-7	0.25	3.7-4	0.8-3

Table 5.1. *Elastic characteristics of some common solid materials*

c) A solid may also be under *shear stress*. Consider, for instance, a rectangular block, which is twisted by acting on two pairs of opposite faces with tangential and oppositely directed forces Π per unit area as in Figure 5.4b. The block is twisted, without changing volume, through an angle δβ proportional to Π:

$$\Pi = \mu\, \delta\beta. \qquad [5.17]$$

μ is the *shear modulus* (*or modulus of rigidity*). The general theory of elasticity enables us to express μ in terms of Y and σ

$$\mu = \frac{Y}{2(1+\sigma)}\,. \qquad [5.18]$$

5.3. Elastic waves in massive springs and rods

Consider a line of identical masses m attached by massless springs of constant K and located at $x = d, 2d, ..., nd, ...$ at equilibrium (Figure 5.5a). We set $K \equiv m\omega_o^2$. If the first mass is slightly displaced longitudinally in a simple harmonic oscillation of angular frequency ω, the coupling spring transmits this disturbance to the second mass and so on. Let u_n be the displacement of the nth mass from its equilibrium position. This mass is subject to the force $m\omega_o^2(u_{n-1} - u_n)$ exerted by the preceding spring, and the force $m\omega_o^2(u_{n+1} - u_n)$ exerted by the next spring. Its equation of motion has the form [5.2]. Thus, its solution is $u_n = A_n e^{i(\omega t - nkd)}$ with the same properties of propagation as the masses joined by the massless taut string (see section 5.1). Figure 5.5b shows the dispersion relation and the *Brillouin zone* ($0 < k < \pi/d$).

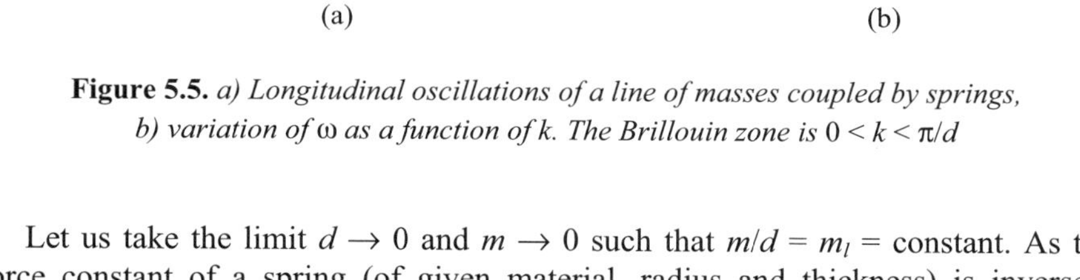

Figure 5.5. *a) Longitudinal oscillations of a line of masses coupled by springs, b) variation of* ω *as a function of k. The Brillouin zone is* $0 < k < \pi/d$

Let us take the limit $d \to 0$ and $m \to 0$ such that $m/d = m_l$ = constant. As the force constant of a spring (of given material, radius and thickness) is inversely proportional to its length, we also take $K \to K_l/d$ where K_l is the force constant for a spring of unit natural length. The line of coupled masses becomes a massive spring of mass m_l and force constant K_l per unit length. This limit also models the coupled atoms in a crystalline solid. The functions $u_n(t)$ become the wave function $u(x, t)$ obeying the wave equation

$$\ddot{u} - v^2(\partial^2_{xx}u) = 0 \qquad \text{with} \qquad v = \sqrt{K_l/m_l}\,. \qquad [5.19]$$

Now consider a cylindrical rod of section $\mathcal{S}$ and Young's modulus Y. If two time-independent and opposite forces F are exerted on its end sections, its length varies (Figure 5.6). A thin slice AB situated between x and $x + \Delta x$ at equilibrium is displaced to $A'B'$ such that

$$AA' = u(x), \qquad BB' = u(x + \Delta x) \cong u(x) + \partial_x u\, \Delta x.$$

Because of the stress in the rod, this slice is subject to two opposite forces F and F' due to the action of that part of the rod which are situated to the left of A' and the right of B' respectively. According to [5.13], these forces are given by

$$F' = -F = Y\mathcal{S}\,\frac{A'B' - AB}{AB} = Y\mathcal{S}\frac{du}{dx}\,. \qquad [5.20]$$

With the convention that the forces are positive if they point outside the slice.

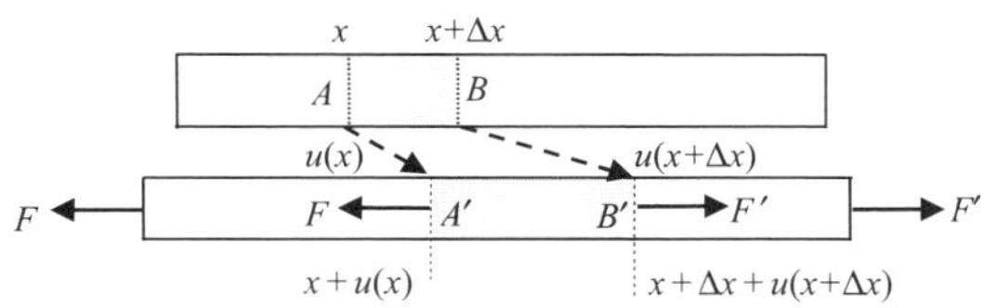

Figure 5.6. *Longitudinal wave in a rod*

If the origin of the rod undergoes a time-dependent displacement $f(t)$ (produced by a vibrator, for instance), the entry slice ($x = 0$) has a displacement $u(0, t) = f(t)$. It exerts a force on the next slice, which moves in turn, and so on. Thus every point of coordinate x in the rod undergoes a displacement $u(x, t)$ measured from its equilibrium position x and producing a *longitudinal elastic wave* in the rod. In this case, the forces F and F' acting on a slice AB are not exactly opposite (otherwise, the slice will not oscillate). The resultant force on the slice is

$$\Delta F = F'_x(x+\Delta x) + F_x(x) = Y\mathcal{S}\,\partial_x u(x + \Delta x) - Y\mathcal{S}\,\partial_x u(x) = Y\mathcal{S}\,\Delta x\, \partial^2_{xx}u.$$

If m_v is the rod mass density per unit volume, the mass of the slice AB or $A'B'$ is $\Delta m = m_v \mathcal{S} \Delta x$ and its position is $u(x, t)$. Its equation of motion $\Delta m \ddot{u} = \Delta F$ may be written as

$$\partial^2_{xx} u - v^2 \partial^2_{xx} u = 0 \qquad \text{where} \qquad v = \sqrt{Y/m_V}\,. \qquad [5.21]$$

This is the propagation equation for longitudinal waves in the elastic rod. Note the similarity of expressions [5.19] and [5.21] for the propagation velocity. Indeed, the elastic rod of length l behaves like a spring of constant $K = Y\mathcal{S}/l$. Thus, we may write $K_l = Kl = Y\mathcal{S}$. Using the relationship $m_l = m_v \mathcal{S}$ of the mass per unit length, the expression $v = \sqrt{K_l/m_l}$ becomes $v = \sqrt{Y/m_V}$.

5.4. Propagation of sound in a pipe

Sonic or acoustic waves are weak disturbances (displacement and variation of the physical quantities) in fluids. At equilibrium, each volume element $\mathcal{V}$ of the fluid contains a very large number of molecules in thermal agitation. The velocity of the molecules is randomly oriented; thus, the fluid has no average velocity. The state of equilibrium is characterized by a pressure $\overline{p}$, temperature $\overline{T}$, mass density $\overline{m_v}$, etc. A disturbance of the fluid produces an orderly displacement $\mathbf{u}(\mathbf{r}, t)$ of the molecules. The volume $\mathcal{V}$ of the fluid is displaced to $\mathcal{V}'$; thus, undergoing a change $\delta\mathcal{V} = \mathcal{V}' - \mathcal{V}$. According to [5.14], this volume variation corresponds to a pressure variation

$$\delta p = -B\,(\delta\mathcal{V}/\mathcal{V}), \qquad [5.22]$$

where the (–) sign is due to the usual definition of the *pressure* in a fluid as the normal force per unit area that the gas exerts on its surrounding while the stress Π is the normal force per unit area exerted on the medium. B is the *bulk modulus* (related to the *compressibility* by $\kappa = 1/B$).

In the case of a sound wave, the displacement of the fluid $\mathbf{u}$ from the equilibrium position is oscillatory with a very small amplitude. The variation of the volume is very small and B is approximately constant. If $p(\mathbf{r}, t)$ is the pressure at point $\mathbf{r}$ and time t, the pressure in excess of the equilibrium value, $p_a(\mathbf{r}, t) = p(\mathbf{r}, t) - \overline{p}$, is called the *acoustic pressure*; it is usually very small ($|p_a| << \overline{p}$). Due to the variation of pressure, the element with displaced volume acts on the next element, which moves in turn and so on. The disturbance propagates as a *sonic* or *acoustic* wave with a speed, which depends on the nature of the medium and its physical conditions. In the case of a gas, these conditions are modified by the wave itself. The study of the propagation is considerably simplified if the amplitude of the wave is small (as in the case of acoustic waves) and the fluid is ideal (that is, of negligible viscosity). In this case, there is no transfer of matter from one region to another and the speed of propagation does not depend on the amplitude of the wave. On the contrary to elastic waves in solids, which may be transverse or longitudinal, sound waves are always longitudinal because of the absence of shear forces in fluids.

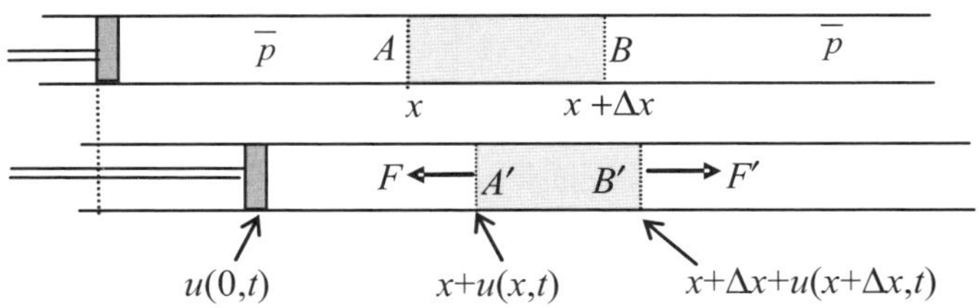

Figure 5.7. *Sound wave in a pipe*

To study the propagation of sound, we need the relationship between the displacement **u** and the acoustic pressure p_a. In this section we consider the propagation in a long and uniform pipe filled with a gas at the equilibrium pressure $\overline{p}$ (Figure 5.7). One of the modes of propagation corresponds to the sections of the pipe as wavefronts and the displacement oriented in the direction of the axis and depending only on x and t exactly, as in the case of a one-dimensional propagation. A volume element situated at equilibrium between the sections at A and B of coordinates x and $x + \Delta x$ becomes, after the displacement, situated between A' and B' of coordinates $x + u(x, t)$ and $x + \Delta x + u(x + \Delta x, t)$ respectively. The variation of its volume is $\delta\mathcal{V} \equiv \mathcal{V}' - \mathcal{V} = \mathcal{S}\,\Delta x\,(\partial_x u)$. As $\mathcal{S}\,\Delta x = \mathcal{V}$, we deduce that

$$\delta\mathcal{V} = \mathcal{V}\,(1 + \partial_x u) \qquad \text{and} \qquad \delta\mathcal{V}/\mathcal{V} = \partial_x u. \qquad [5.23]$$

Relationship [5.22] gives the variation of pressure, which is the acoustic pressure

$$p_a(x, t) \equiv p(x, t) - \overline{p} = -B\,\partial_x u. \qquad [5.24]$$

Thus, the variation of pressure is due to the gradient of the displacement u not on u itself. Indeed, if all the points of a gas undergo the same translational displacement, there will be no volume variation and no pressure variation.

To write the equation of propagation, assume that u decreases as a function of x like in Figure 5.7, for instance. According to [5.24], the volume $\mathcal{V}$ is at a relatively higher pressure. Thus, it is subject at A' to a force $-p(x, t)\,\mathcal{S}$ (toward the negative x)

and at B' to a force $p(x + \Delta x, t)\,\mathcal{S}$ (toward the positive x). The resultant force acting on this volume element is

$$\Delta F = \mathcal{S}\,p(x, t) - \mathcal{S}\,p(x + \Delta x, t) = -\,\Delta x\,\mathcal{S}\,\partial_x p = -\Delta x\,\mathcal{S}\,\partial_x\,p_a.$$

Using equation [5.24], we may also write

$$\Delta F = B\,\Delta x\,\mathcal{S}\,\partial^2{}_{xx}u.$$

The mass of the volume $\mathcal{V}$ of gas being $\Delta m = m_v\mathcal{V} = m_v\mathcal{S}\,\Delta x$, its equation of motion $\Delta m\,\partial^2{}_{tt}u = \Delta F$ has the form of the d'Alembert wave equation

$$\partial^2{}_{tt}u - v_s{}^2\,\partial^2{}_{xx}u = 0 \qquad \text{where } v_s = \sqrt{B/m_v}\,. \qquad [5.25]$$

Differentiating this equation with respect to x and using equation [5.24], we obtain the wave equation of the acoustic pressure

$$\partial^2{}_{tt}p_a - v_s{}^2\,\partial^2{}_{xx}\,p_a = 0. \qquad [5.26]$$

To determine the speed of sound, we need the bulk modulus B. In the case of a gas, we may determine it if we know the nature of the thermodynamic process. If the gas is ideal and the sound propagates in the gas at constant temperature (*isothermal process*), the gas obeys the Boyle-Mariotte law $p\mathcal{V}$ = Constant, and equation [5.22] gives $dp = -p(d\mathcal{V}/\mathcal{V})$. This corresponds to $B = p$ and a speed of sound $v_s|_{T=\text{Const}} = \sqrt{p/m_v}$. In the case of air at standard temperature and pressure (STP) (p = 1 atm = 1.01×10^5 Pa and $m_v = 1.29$ kg/m^3) we find a speed of 280 m/s. This model, initially proposed by Newton, does not agree with the experimental value v_s = 331.6 m/s at 0°C and 344 m/s at 20°C. On the other hand, if the propagation is *adiabatic* (without heat exchange within the gas), the gas obeys the law $p\mathcal{V}^\gamma$ = constant where γ is the *adiabatic constant* (equal to the ratio of the specific heat capacities in isobaric and isovolumic processes $\gamma = c_P/c_v$), we obtain $dp = -\,\gamma p\,(d\mathcal{V}/\mathcal{V})$. This leads to $B = \gamma p$ and

$$v_s = \sqrt{\gamma p/m_v} \qquad \text{[adiabatic]}. \qquad [5.27]$$

This result is known as the *Laplace formula*. In the case of air at STP (γ = 1.4), we get v_s = 332.

Using equation [5.27], it is possible to study the variation of the speed of sound with temperature if the equation of state of the gas is known. For instance, the equation of state of an ideal gas is $p\mathcal{V} = (m/M)RT$ where R = 8.3145 joule.K^{-1}.mole^{-1}

is the *gas constant*, m is the mass of the gas, M is its molar mass, T is its absolute temperature, $\mathcal{V}$ is its volume and p is its pressure. Thus, we have $m_v = pM/RT$ and the speed of sound may be written as

$$v_s = \sqrt{\gamma RT/M} \ . \qquad [5.28]$$

Contrary to gases, there is no equation of state for liquids. The bulk modulus B and the mass density m_v must be determined experimentally. Liquids are much denser than gases and much less compressible (κ small, i.e. B large). For instance, in the case of water at 20°C, $B = 2.2$ GPa and $m_v = 1{,}000$ kg/m^3. We obtain a speed of sound $v_s = 1{,}500$ m/s, compared to the experimental value of 1,435 m/s for fresh water at 8°C and 1,475 m/s for seawater. Table 5.2 lists the speed of sound in some fluids and the speed of longitudinal and transverse elastic waves in some solids.

Gas or vapor	**Air**	**Steam (134°)**	**H_2**	**O_2**	**N_2**	**He_2**	**CO_2**
v_s ($m.s^{-1}$)	331.45	494	1 284	316	334	965	259
Z_s (N.s.m^{-3})	430	397	115	452	418	172	512

Liquid	**Fresh water**	**Seawater**	**Benzene**	**Mercury**	**Glycerin**	**Alcohol**
v_s (10^3 $m.s^{-1}$)	1.498	1.531	1.295	1.45	1.904	1.21
Z_s (10^6 N.s.m^{-3})	1.50	1.57	1.13	19.6	2.40	0.953

Solid	**Aluminum**	**Iron**	**Copper**	**Steel**	**Granite**	**Quartz**
v_l (m.s^{-1})	6,420	5,950	3,560	5,800-6,100	≈ 6,000	≈ 5,600
v_t (m.s^{-1})	3,040	3,240	2,270	3,100-3,270	...	...
m_v (kg.m^{-3})	2,700	7,900	8,900	7,800	2,700	2,650

Table 5.2. *Speed of sound and acoustic impedance for some gases at 0°C and some liquids at 20°C and speed of longitudinal and transverse elastic waves in some solids at 20°C.*

EXAMPLE 5.1. **a)** Evaluate the speed of sound in a mixture of two ideal gases of masses m_1 and m_2 occupying the unit volume. Their molar masses are M_1 and M_2 and their adiabatic constants are γ_1 and γ_2. **b)** Calculate the speed of sound at 20°C in air (21% of O_2 and 79% of N_2 in volume). $M_O = 32$ and $M_N = 28$.

SOLUTION

a) The density of this mixture is $m_v = (m_1 + m_2)$. The unit volume contains $n_1 = m_1/M_1$ moles of the first gas and $n_2 = m_2/M_2$ moles of the second. Using the equation $p\mathcal{V} = nRT$ for each one, we obtain the partial pressures $p_1 = m_1RT/M_1$ and $p_2 = m_2RT/M_2$. According to [5.22], if a volume $\mathcal{V}$ of this gas varies by $\delta\mathcal{V}$, the variation of the total pressure is

$$\delta p = \delta p_1 + \delta p_2 = -\gamma_1 p_1\, \delta\mathcal{V}/\mathcal{V} - \gamma_2 p_2\, \delta\mathcal{V}/\mathcal{V} = -RT\,(\gamma_1 m_1/M_1 + \gamma_2 m_2/M_2)\, \delta\mathcal{V}/\mathcal{V}.$$

Thus, the global bulk modulus is $B = RT(\gamma_1 m_1/M_1 + \gamma_2 m_2/M_2)$ and

$$v_s = \sqrt{B/m_v} = \sqrt{RT(\gamma_1 m_1/M_1 + \gamma_2 m_2/M_2)/(m_1 + m_2)}\,.$$

b) The unit volume of air contains 0.21 m^3 of oxygen and 0.79 m^3 of nitrogen. Their masses are (in kg/m^3):

$$m_O = 0.032 \times (0.21/0.0224) = 0.300 \text{ and } m_N = 0.028 \times (0.79/0.0224) = 0.988.$$

As $\gamma_O = \gamma_N = 1.4$, the speed of sound in air (STP) is

$$v_s = [RT(\gamma_O m_O/M_O + \gamma_N m_N/M_N)/(m_O + m_N)]^{1/2} = 344 \text{ m/s}.$$

5.5. Transverse waves on elastic membranes

Consider a rectangular elastic membrane with Oxy as its plane of equilibrium. We assume that it has very large dimensions (in order to neglect reflections on its boundaries). It is taut along the sides parallel to Ox with a force per unit length F_y in the direction of Oy and taut along the sides parallel to Oy with a force per unit length F_x in the direction of Ox (Figure 5.8). We study its small transverse vibrations, that is, in the normal direction Oz. Let $u(x, y, t)$ be the displacement at the point of coordinates x and y of the membrane. Consider a rectangular element of area $d\mathcal{S}$ of sides dx and dy centered on the point (x, y). A similar analysis to that of the taut string, shows that the force acting on $d\mathcal{S}$ has negligible x and y components, while its z component is

$$dF_z = F_y\, dx[\partial_y u(x, y+\tfrac{1}{2}dy) - \partial_y u(x, y-\tfrac{1}{2}dy)] + F_x\, dy[\partial_x u(x+\tfrac{1}{2}dx, y) - \partial_x u(x-\tfrac{1}{2}dx, y)]$$
$$= dx\, dy\, (F_x\, \partial^2_{xx}u + F_y\, \partial^2_{yy}u) = d\mathcal{S}\, (F_x\, \partial^2_{xx}u + F_y\, \partial^2_{yy}u).$$

The equation of motion of this area element is $m_s\, dS\, \partial^2_{tt}u = dF_z$ where m_s is the mass per unit area. The wave equation can then be written as

$$m_s\, \partial^2_{tt}u - F_x\, \partial^2_{xx}u - F_y\, \partial^2_{yy}u = 0. \qquad [5.29]$$

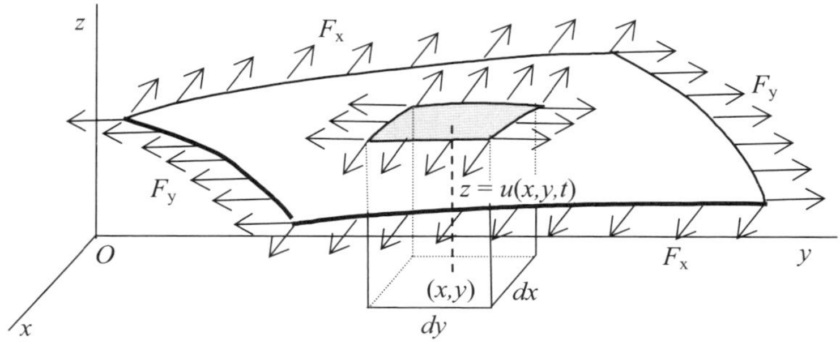

Figure 5.8. *Wave on an elastic membrane*

The taut membrane is a simple example of a non-isotropic medium. If a rectilinear wave propagates in the direction Ox, u is independent of y and the equation becomes

$$m_s\, \partial^2_{tt}u - F_x\, \partial^2_{xx}u = 0. \qquad [5.30]$$

Thus, the speed of propagation is $v_x = (F_x/m_s)^{1/2}$. Similarly, the wave propagates in the direction Oy with a speed $v_y = (F_y/m_s)^{1/2}$. In general, wave equation [5.29] has solutions of the form $u = u_m\, e^{i(\omega t - xk_x - yk_y)}$ if the angular frequency ω and the wave vector $\mathbf{k} = k_x\, \mathbf{e}_x + k_y\, \mathbf{e}_y$ verify the dispersion relation

$$\omega^2 = v_x^2 k_x^2 + v_y^2 k_y^2. \qquad [5.31]$$

If the wave propagates in a direction $\mathbf{e}$, which makes an angle θ with Ox, its wave vector is $\mathbf{k} = k\ \mathbf{e} = k\ (\cos\theta\ \mathbf{e}_x + \sin\theta\ \mathbf{e}_y)$ and dispersion relation [5.31] may be written in the form

$$\omega^2 = k^2\, [v_x^2 \cos^2\theta + v_y^2 \sin^2\theta]. \qquad [5.32]$$

Thus, the phase velocity and the group velocity are

$$v_{(p)} = \omega/k = \sqrt{v_x^2 \cos^2\theta + v_y^2 \sin^2\theta}\ , \qquad [5.33]$$

$$\mathbf{v}_{(g)} = \Sigma_{i = x,y}\, (\partial\omega/\partial k_i)\, \mathbf{e}_i = (v_x^2 k_x/\omega)\, \mathbf{e}_x + (v_y^2 k_y/\omega)\, \mathbf{e}_y. \qquad [5.34]$$

If the tension per unit length is the same in the directions Ox and Oy ($F_x = F_y \equiv F$), the wave equation is isotropic

$$\partial^2_{tt}u - v^2(\partial^2_{xx}u + \partial^2_{yy}u) = 0 \qquad \text{with} \quad v = \sqrt{F/m_s}\,. \qquad [5.35]$$

In this case, the phase velocity and the group velocity are equal to v for any propagation direction.

5.6. Mechanical waves in three dimensions

In the case of a sound wave, which propagates in a three-dimensional medium, the displacement $\mathbf{u}(\mathbf{r}, t)$ and the velocity $\dot{\mathbf{u}}(\mathbf{r}, t)$ of the molecules are vectors. The acoustic pressure is given by the equation $p_a(\mathbf{r}, t) = -B(\nabla.\mathbf{u})$, which is a generalization of [5.24]. These quantities obey the d'Alembert propagation equations:

$$\partial^2_{tt}\mathbf{u} - v_s^2\,\Delta\mathbf{u} = 0, \qquad \text{and} \qquad \partial^2_{tt}p_a - v_s^2\,\Delta p_a = 0. \qquad [5.36]$$

When studying this wave, it is convenient to use a scalar function rather than a vector function. The acoustic pressure p_a is not a good choice because the relation $p_a = -B(\nabla.\mathbf{u})$ does not enable us to determine $\mathbf{u}$. If the motion of the fluid is irrotational (non-turbulent), that is, the circulation of the velocity $\dot{\mathbf{u}}$ on any closed path is 0 (in other words $\nabla\times\dot{\mathbf{u}} = 0$), it is possible to define a *velocity potential* Ψ such that

$$\partial_t\mathbf{u}(\mathbf{r}, t) = -\nabla\Psi(\mathbf{r}, t). \qquad [5.37]$$

$\Psi(\mathbf{r}, t)$ obeys the d'Alembert propagation equation

$$\partial^2_{tt}\Psi - v_s^2\,\Delta\Psi = 0. \qquad [5.38]$$

Knowing Ψ, equation [5.37] enables us to determine the velocity $\dot{\mathbf{u}}(\mathbf{r}, t)$. Integrating with respect to time, we determine $\mathbf{u}(\mathbf{r}, t)$ and the relationship $p_a(\mathbf{r}, t) = -B(\nabla.\mathbf{u})$ enables us to determine the acoustic pressure.

Equation [5.38] has simple harmonic plane wave solutions of the form

$$\Psi(\mathbf{r}, t) = \Psi_m\, e^{i(\omega t - \mathbf{k}.\mathbf{r})} \qquad \text{if} \quad \mathbf{k} = k\mathbf{e} = (\omega/v_s)\,\mathbf{e}. \qquad [5.39]$$

This wave propagates in the direction of **e** (Figure 5.9a). The corresponding velocity, displacement and acoustic pressure of the fluid are:

$$\partial_t\mathbf{u}(\mathbf{r},t) = -\nabla\Psi = \mathrm{i}\mathbf{k}\Psi = \mathbf{e}\,k\Psi_m\,e^{\mathrm{i}(\omega t-\mathbf{k.r}+\pi/2)}, \qquad \mathbf{u} = \partial_t\mathbf{u}/\mathrm{i}\omega = \mathbf{e}(\Psi_m/v_s)\,e^{\mathrm{i}(\omega t-\mathbf{k.r})}$$

$$p_a(\mathbf{r},t) = -B(\nabla.\mathbf{u}) = (iBk^2/\omega)\Psi_m\,e^{\mathrm{i}(\omega t-\mathbf{k.r})} = m_v\omega\Psi_m\,e^{\mathrm{i}(\omega t-\mathbf{k.r}+\pi/2)}. \qquad [5.40]$$

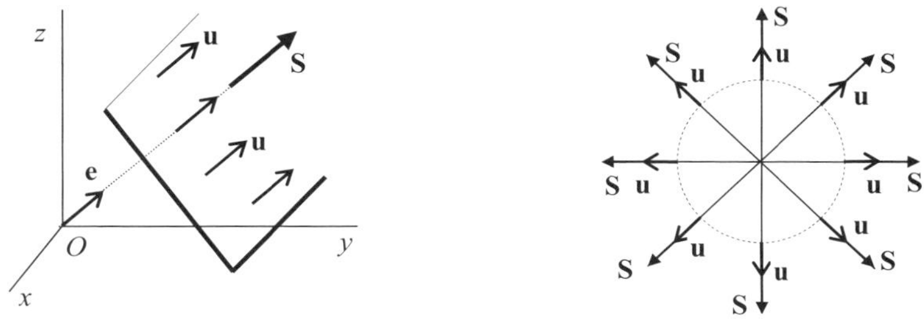

(a) (b)

Figure 5.9. *a) Plane sound wave, and b) diverging spherical sound wave*

In the case of spherical symmetry (see section 4.2), wave equation [5.38] has solutions of the form

$$\Psi(r,t) = \frac{R}{r}\,\Psi_m\cos(\omega t \pm kr), \qquad [5.41]$$

which represent spherical waves diverging from the origin O (sign –) (Figure 5.9b) or converging at O (sign +). R is the radius of a sphere on which the amplitude of the velocity potential is Ψ_m. The corresponding velocity, displacement and acoustic pressure are respectively:

$$\partial_t\mathbf{u}(\mathbf{r},t) = R\Psi_m[\,(1/r^2)\cos(\omega t \pm kr) \pm (k/r)\sin(\omega t \pm kr)]\,\mathbf{e}_r,$$

$$\mathbf{u}(\mathbf{r},t) = (R\Psi_m/\omega)[(1/r^2)\sin(\omega t \pm kr) \mp (k/r)\cos(\omega t \pm kr)]\,\mathbf{e}_r,$$

$$p_a(\mathbf{r},t) = (Bk^2R\Psi_m/\omega r)\cos(\omega t \pm kr + \pi/2), \qquad [5.42]$$

where $\mathbf{e}_r = \mathbf{r}/r$ is the radial unit vector. At a short distance from the emitter, the dominant terms of $\partial_t\mathbf{u}$ and $\mathbf{u}$ are proportional to $1/r^2$ and, at large distance from the emitter (the wave zone), the dominant terms are proportional to $1/r$.

Now we consider elastic waves in a three-dimensional medium. Wave equation [5.21] does not take into account the deformation of the transverse dimensions of the rod. The theory of elasticity of isotropic media enables us to establish the wave equation for three-dimensional waves

$$\ddot{\mathbf{u}} - v_T^2\,\Delta\mathbf{u} + (v_T^2 - v_L^2)\,\nabla(\nabla.\mathbf{u}) = 0$$

$$\text{where } v_L^2 = \frac{Y(1-\sigma)}{m_v(1+\sigma)(1-2\sigma)}, \qquad v_T^2 = \frac{Y}{2m_v(1+\sigma)}. \qquad [5.43]$$

For instance, a plane wave, which propagates in the direction Ox, may be written as $\mathbf{u} = \mathbf{u}_L + \mathbf{u}_T$, where $\mathbf{u}_L = u_L(x, t)\,\mathbf{e}_x$ is the longitudinal component and $\mathbf{u}_T(x, t)$ is the transverse component. As $\nabla.\mathbf{u} = \partial_x u_x$, wave equation [5.43] may be written as

$$\ddot{\mathbf{u}}_L - v_L^2\,\partial^2_{xx}\mathbf{u}_L = 0, \qquad \ddot{\mathbf{u}}_T - v_T^2\,\partial^2_{xx}\mathbf{u}_T = 0. \qquad [5.44]$$

This shows that v_L is the speed of longitudinal waves and v_T is the speed of transverse waves. As $0 < \sigma < ½$, we always have $v_L > \sqrt{4/3}\,v_T$. For instance, in the case of steel ($m_v = 7.5 \times 10^3$ kg/m^3, $Y = 21 \times 10^{10}$ Pa and $\sigma = 0.3$), we obtain the velocities $v_L = 6\,140$ m/s and $v_T = 3\,280$ m/s. These values may be compared to the expression $v = \sqrt{Y/m_v} = 5290$ m/s, which does not take into account transverse deformations. A seismic wave may be longitudinal (then, it is called *wave P* or *primary* as it propagates with the higher speed $v_L \approx 6$ to 10 km/s depending on the nature of the rock and the depth) or transverse (then, it is called *wave S* or *secondary* as it propagates at a lower speed $v_T \approx 5$ to 6 km/s).

5.7. Energy of mechanical waves

a) One-dimensional waves

The energy of a mechanical wave may be kinetic, associated with the velocity of the masses, or potential, due to the deformations under stress. For instance, in the case of a string of mass per unit length m_l and tension F, a small element of the string situated between x and $x + \Delta x$ has a mass $m_l\,\Delta x$, a velocity $\dot{u}$ (Figure 5.10a) and a kinetic energy $\Delta E_{(C)} = ½\,m_l\,\Delta x\,\dot{u}^2$. On the other hand, because of the displacement, the length of the element becomes

$$\Delta x' = \{\Delta x^2 + [u(x + \Delta x, t) - u(x, t)]^2\}^{½} \approx \Delta x[1 + ½(\partial_x u)^2].$$

The work required for the lengthening from Δx to $\Delta x'$ is the potential energy $\Delta E_{(P)} = F(\Delta x' - \Delta x) \approx ½\,\Delta x\,F(\partial_x u)^2$. Thus, the total energy per unit length is

$$E_l = ½\,m_l\dot{u}^2 + ½\,F\,(\partial_x u)^2 \qquad \text{[taut string]}. \qquad [5.45]$$

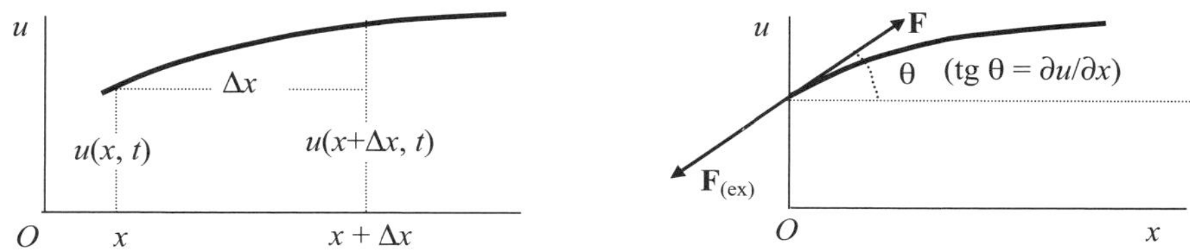

Figure 5.10. *Energy of a wave on a taut string: a) deformation of an element Δx of the string and b) force $\mathbf{F}_{(ex)}$ exerted by the generator of the wave*

In the case of an elastic membrane, the energy density per unit area is

$$E_s = E_{(C)\,s} + E_{(P)\,s} = \tfrac{1}{2}\, m_s\, \dot{u}^2 + \tfrac{1}{2}\, F_x\, (\partial_x u)^2 + \tfrac{1}{2}\, F_y\, (\partial_y u)^2 \quad \text{[membrane]}. \quad [5.46]$$

The energy of a wave propagates from one region to another. In the case of a one-dimensional wave, the energy flux $P(x, t)$ is related to the energy density by the conservation of energy equation (see section 4.10)

$$\partial_t E_l + \partial_x P = 0. \quad [5.47]$$

Using the expression E_l and the wave equation, and integrating this equation, we obtain the expression $P(x, t)$. For instance, in the case of a wave on a taut string, using expression [5.45] of E_l, we obtain

$$\partial_x P = -\, m_l\, \dot{u}\, \ddot{u} \; - F\, (\partial_x u)\, (\partial^2_{xt} u) = -\, v_s^2\, m_l\, \dot{u}\, (\partial^2_{xx} u) - F\, (\partial_x u)\, (\partial^2_{xt} u)$$

$$= -\, F\, [\, \dot{u}\, (\partial^2_{xx} u) + (\partial_x u)(\partial^2_{xt} u)]$$

$$P(x, t) = -\, F\, \dot{u}\, (\partial_x u) \quad \text{[taut string]}. \quad [5.48]$$

In the case of a massive spring, a similar analysis gives $\Delta E_{(C)} = \tfrac{1}{2}\, m_l\, \Delta x\, \dot{u}^2$ and $\Delta E_{(P)} = \tfrac{1}{2} K_l\, \Delta x (\partial_x u)^2$. We deduce the expressions of the energy density and the energy flux:

$$E_l = E_{(C)l} + E_{(P)l} = \tfrac{1}{2}\, m_l\, \dot{u}^2 + \tfrac{1}{2}\, K_l (\partial_x u)^2 \quad [5.49]$$

$$P(x, t) = -\, K_l\, \dot{u}\, (\partial_x u) \quad \text{[massive spring]}. \quad [5.50]$$

Similar relations may be written in the case of a wave in an elastic rod or in the case of sound in a pipe. For this, we only have to replace m_l with $m_v\mathcal{S}$ and K_l with $Y\mathcal{S}$ in the case of a rod or $B\mathcal{S}$ in the case of sound in a pipe. We obtain

$$E_l = \tfrac{1}{2}\mathcal{S}m_v\dot{u}^2 + \tfrac{1}{2}\mathcal{S}Y(\partial_x u)^2,\ P(x,t) = -\,Y\mathcal{S}\,\dot{u}\,(\partial_x u) \quad \text{[elastic rod]} \qquad [5.51]$$

$$E_l = \tfrac{1}{2}\mathcal{S}m_v\dot{u}^2 + \tfrac{1}{2}\mathcal{S}B(\partial_x u)^2,\ P(x,t) = -\,B\mathcal{S}\,\dot{u}\,(\partial_x u) \quad \text{[sound in a pipe].} \qquad [5.52]$$

Dividing by the section $\mathcal{S}$, we get the energy density per unit volume $E_v = E_l/\mathcal{S}$ and the energy flux density $S = P/\mathcal{S}$.

b) Three-dimensional waves

In the case of a wave propagating in three dimensions, the general expressions [4.82] for vector waves enable us to write the energy density per unit volume, the components of the vector energy flux density **S** and the momentum density. For instance, in the case of sound waves, we take $\mu = m_v$, $B_o = B$ (the gas bulk modulus), $C_o = 0$ and $\omega_o = 0$. We obtain:

$$E_v = E_{(C)\,v} + E_{(P)v} = \tfrac{1}{2}m_v\dot{\mathbf{u}}^2 + \tfrac{1}{2}\,B\,\Sigma_{ij}\,(\partial_i u_j)(\partial_i u_j),$$
$$S^i = -\,B\,\Sigma_j\ \dot{u}_j(\partial_i u_j), \qquad \mathbf{P}_v = -\,m_v\,\Sigma_j\,(\nabla u_j)\,\dot{u}_j \qquad \text{[sound waves].} \qquad [5.53]$$

In particular, in the case of a one-dimensional wave propagating in the direction Ox, **u** has one longitudinal component, which depends on x and t and we obtain the expressions:

$$E_v = \tfrac{1}{2}m_v\dot{u}^2 + \tfrac{1}{2}\,B\,(\partial_x u)^2, \qquad S^i = -\,B\,\dot{u}\,(\partial_x u), \qquad \mathbf{P}_v = -\,m_v\,\dot{u}\,(\partial_x u). \quad [5.54]$$

In the case of elastic waves, taking $B_o = -\,C_o = m_v v_T^2$, $D_o = m_v(2v_T^2 - v_L^2)$, $\omega_o = 0$ and $F_o = 0$, we obtain:

$$E_v = E_{(C)v} + E_{(P)v} = \tfrac{1}{2}m_v\dot{\mathbf{u}}^2 + \tfrac{1}{2}m_v[v_T^2\Sigma_{ij}\,(\partial_i\,u_j)(\partial_i u_j + \partial_j u_i) + (v_L^2 - 2v_T^2)(\nabla\mathbf{u})^2]$$
$$S^i = m_v\,[(2v_T^2 - v_L^2)\,\dot{u}_i\,(\nabla.\mathbf{u}) - v_T^2\Sigma_j\ \dot{u}_j\,(\partial_i u_j + \partial_j u_i)]$$
$$\mathbf{P}_v^{\,i} = -\,m_v\,\Sigma_j\ \dot{u}_j\,(\partial_i u_j) \qquad \text{[elastic waves].} \qquad [5.55]$$

Using the wave equation and the expression for the speed of propagation, we may easily verify the conservation of energy equation

$$\partial_t E_v + \nabla.\mathbf{S} = 0. \qquad [5.56]$$

If the elastic wave propagates in the direction Ox with a longitudinal component $u_L(x, t)$ and a transverse component $u_T(x, t)$ in the direction Oy, the displacement is $\mathbf{u} = u_L(x, t)\,\mathbf{e}_x + u_T(x, t)\,\mathbf{e}_y$ and we find:

$$\begin{aligned} E_v &= E_{(C)v} + E_{(P)v} = \tfrac{1}{2}m_v[\,\dot{u}_L^{\,2} + \dot{u}_T^{\,2} + v_L^{\,2}\,(\partial_x u_L)^2 + v_T^{\,2}(\partial_x u_T)^2\,] \\ \mathbf{S} &= -m_v\ \{[v_L^{\,2}\,\dot{u}_L\,(\partial_x u_L) + v_T^{\,2}\,\dot{u}_T\,(\partial_x u_T)]\,\mathbf{e}_x \\ &\qquad + [(v_L^{\,2} - 2v_T^{\,2})\,\dot{u}_T\,(\partial_x u_L) + v_T^{\,2}\,\dot{u}_L\,(\partial_x u_T)]\,\mathbf{e}_y\,\} \\ \mathbf{P}_v &= -\,m_v\,[\,\dot{u}_L\,(\partial_x u_L) + \dot{u}_T\,(\partial_x u_T)]\,\mathbf{e}_x \end{aligned} \qquad \text{[elastic waves]}. \quad [5.57]$$

We note that the energy propagation direction (which is the direction of $\mathbf{S}$) is not the same as the wave propagation direction $\mathbf{e}_x$, except in the case of purely longitudinal or transverse waves. If the solid body is anisotropic, the analysis is much more involved.

5.8. Progressive waves, impedance and intensity

To sustain a wave, the emitter must constantly exert a force at the entry and supply energy. As in the case of an oscillator, the emitted wave depends on the *impedance* of the medium, which is defined as the ratio of the excitation force to the generalized velocity $\partial_t u$ at the entry.

a) Waves on a string and a massive spring

In the case of a one-dimensional progressive wave on a string $u(x, t) = f(\xi)$ where $\xi = t - x/v$, the energy density [5.45] and the energy flux [5.48] are respectively $E_l = m_l\, f'(\xi)^2$ and $P(x, t) = (F/v)\, f'(\xi)^2$. The force $\mathbf{F}_{(ex)}$, that the emitter exerts, must be equal in magnitude to the string tension and tangent to the string at any time (Figure 5.10b). Indeed, a small element Δx of the string at the entry is subject to the total force $\mathbf{F} + \mathbf{F}_{(ex)}$. Its equation of motion is $m_l\,\Delta x\,\ddot{\mathbf{u}} = \mathbf{F} + \mathbf{F}_{(ex)}$. As Δx may be arbitrarily small, we must have $\mathbf{F} + \mathbf{F}_{(ex)} = 0$. Thus, the component of $\mathbf{F}_{(ex)}$ in the direction of vibration Oy is

$$F_{(ex)y} = -\,F \sin\theta \approx -\,F \tan\theta = -\,F\,(\partial_x u)\big|_{x=0} = (F/v)\, f'(\xi)\big|_{x=0} = (F/v)\, f'(t).$$

We define the *impedance* as the ratio of the excitation force $F_{(ex)y}$ to the velocity $\dot{u}(0, t) = f'(t)$ at the entry

$$Z = \frac{F_{(ex)y}}{\dot{u}(0,t)} = \frac{F}{v} = \sqrt{m_l F}\ . \qquad [5.58]$$

Z only depends on the characteristics of the string. At each point x, the part of the string which precedes x may be considered as a generator of the wave, which propagates on the part of the string that follows x. Thus, we have

$$Z = \frac{F_y(x,t)}{\dot{u}(x,t)} = -\frac{F\,\partial_x u(x,t)}{\dot{u}(x,t)}. \qquad [5.59]$$

The instantaneous power that is supplied by the emitter may be written in the form

$$P_{(ex)} = F_{(ex)\,y}\,\dot{u}\,|_{x=0} = -F\,\dot{u}\,(\partial u/\partial x)|_{x=0} = \frac{F}{v}f'(t)^2 = Z\dot{u}^2\,|_{x=0} = \frac{F_{(ex)y}^{\;2}}{Z}. \qquad [5.60]$$

Particularly, in the case of a simple harmonic wave $u = u_m \cos(\omega t - kx)$, we obtain:

$$E_l(x,t) = m_l\omega^2 u_m^{\;2}\sin^2(\omega t - kx),\quad P(x,t) = v\,E_l(x,t),\quad F_{(ex)y} = -\omega Z\,u_m \sin(\omega t),$$
$$P_{(ex)} = Z\omega^2 u_m^{\;2}\sin^2(\omega t - kx) \quad \text{and} \quad \langle P_{(ex)}\rangle = \tfrac{1}{2}Z\omega^2 u_o^{\;2}. \qquad [5.61]$$

For instance, a string of mass 4 g/m and tension 160 N has an impedance $Z = \sqrt{Fm_l} = 0.8$ kg/s. To sustain a wave of amplitude 0.5 cm and a frequency of 10 Hz, the emitter must have an average power $\langle P\rangle = \tfrac{1}{2}Z\omega^2 u_o^{\;2} = 0.039$ W.

In the case of a massive spring, we only have to replace F with K_l; we obtain

$$E_l = m_l f'^2 = (K_l/v^2)\,f'^2, \qquad P(x,t) = vE_l \quad \text{and } Z = \sqrt{m_l K_l}\,. \qquad [5.62]$$

b) Elastic wave in a rod and sound wave in a pipe

In the case of an elastic wave in a rod, the force exerted by the emitter is distributed on the area $\mathcal{S}$ of the entry, producing a stress $\Pi(0,t) = -F_{(ex)}/\mathcal{S} = Y(\partial_x u)_{x=0} = -(Y/v)f'(t)$. We define the *impedance per unit area*

$$Z_s = F_{(ex)}/\mathcal{S}\,\dot{u}(0,t) = -\Pi(x,t)/\,\dot{u}(x,t) = \sqrt{m_v Y}\,. \qquad [5.63]$$

If the wave is uniform on the sections of the rod, the expressions of [5.51] are equivalent to an energy density per unit volume and a vector energy flux density

$$E_v = m_v f'^2, \qquad \mathbf{S} = v\,m_v f'^2\,\mathbf{e}_x \qquad \text{[elastic rod]}. \qquad [5.64]$$

The instantaneous power that the emitter supplies is

$$P_{(ex)} = F_{(ex)}\ \dot{u}(0,t) = -\mathcal{S}Y\,\dot{u}\,(\partial_x u)|_{x=0} = \mathcal{S}Z_s\ \dot{u}^2\,|_{x=0} = F_{(ex)}^{\ 2}/\mathcal{S}Z_s. \qquad [5.65]$$

In particular, in the case of the simple harmonic wave $u = u_m \cos(\omega t - kx)$, we obtain

$$E_v(x,t) = m_v\omega^2 u_m^{\ 2}\sin^2(\omega t - kx), \qquad \mathbf{S}(x,t) = vE_v(x,t)\,\mathbf{e}_x, \qquad F_{(ex)} = -\omega Z u_m \sin(\omega t),$$
$$P_{(ex)} = Z\omega^2 u_m^{\ 2}\sin^2(\omega t - kx) \qquad \text{and} \qquad \langle P_{(ex)}\rangle = \tfrac{1}{2} Z\omega^2 u_o^{\ 2}. \qquad [5.66]$$

In the case of a sound wave in a pipe, we only have to replace Y with B and the stress Π with $-p_a = B(\partial_x u) = -(B/v)f'(\xi)$. The acoustic impedance of air per unit area is $Z_s(\text{air}) = p_a(x, t)/\dot{u}(x,t) = \sqrt{m_v B}$, i.e. $Z_s(\text{air}) = 428$ N.s/m^3 at STP. In particular, in the case of the wave $u = u_m\cos(\omega t - kx)$, we obtain $p_a(x, t) = p_m\cos(\omega t - kx + \pi/2)$ with $p_m = (\omega m_v v_s)u_m = (\omega Z_s)\,u_m$ and

$$S = v\,E_v = v_s\,m_v\omega^2 u_m^{\ 2}\sin^2(\omega t - kx) = (\,p_m^{\ 2}/v_s\,m_v)\,\sin^2(\omega t - kx) = v_s\,E_v. \qquad [5.67]$$

The sound intensity is the time-average of the energy flux density over a period

$$\mathcal{I}_s = (1/T)\int_t^{t+T} dt\; S(x, t) = \tfrac{1}{2}\,Z_s\,\omega^2 u_m^{\ 2} = p_m^{\ 2}/2Z_s. \qquad [5.68]$$

This is the time-averaged power that the unit area receives if it is normal to **S** (i.e. normal to the direction of propagation of the wave). In the case of a periodic wave in time, the intensity does not depend on t. It is also independent of x, as there is no production or loss of energy along the pipe (except at the entry).

c) three-dimensional waves

In the case of a progressive elastic wave $\mathbf{u} = f_L(t - x/v_L)\,\mathbf{e}_x + f_T(t - x/v_T)\,\mathbf{e}_y$, we find that $E_{(C)\,v} = E_{(P)\,v}$ and

$$E_v = m_v(f_L'^{\,2} + f_T'^{\,2}), \qquad \mathbf{S} = m_v(v_L f_L'^{\,2} + v_T f_T'^{\,2})\,\mathbf{e}_x + m_v f_L' f_T'(v_T + v_L - 2v_T^{\ 2}/v_L)\,\mathbf{e}_y$$
$$\mathbf{P}_v = m_v\,(f_L'^{\,2}/v_L + f_T'^{\,2}/v_T)\,\mathbf{e}_x \qquad \text{[elastic waves]}. \qquad [5.69]$$

In particular, in the case of a simple harmonic progressive wave of the form

$$\mathbf{u} = A_L\cos[\omega(t - x/v_L)]\,\mathbf{e}_x + A_T\cos[\omega(t - x/v_T) + \phi]\,\mathbf{e}_y, \qquad [5.70]$$

we find

$$E_v = m_v\,\omega^2\,\{A_L{}^2\sin^2[\omega(t - x/v_L)] + A_T{}^2\sin^2[\omega(t - x/v_T) + \phi],$$
$$\mathbf{S} = m_v\,\omega^2\,\{v_L\,A_L{}^2\sin^2[\omega(t - x/v_L)] + v_T\,A_T{}^2\sin^2[\omega(t - x/v_T) + \phi]\}\mathbf{e}_x$$
$$+\ m_v\,\omega^2\,A_L\,A_T\,(v_T + v_L - 2v_T{}^2/v_L)\sin[\omega(t - x/v_L)]\sin[\omega(t - x/v_T) + \phi]\,\mathbf{e}_y$$
$$\mathbf{P}_v = m_v\omega^2\{(A_L{}^2/v_L)\sin^2[\omega(t - x/v_L)] + (A_T{}^2/v_T)\sin^2[\omega(t - x/v_T)+\phi]\}\mathbf{e}_x. \quad [5.71]$$

In the case of sound, all that we have to do is take $f_T = 0$ and $v_L = v_s$, we obtain

$$E_v = m_v f'^2 = m_v\omega^2 A^2\sin^2[\omega(t - x/v_s)], \qquad \mathbf{S} = E_v\mathbf{e}_x, \qquad \mathbf{P}_v = \mathbf{S}/v_s{}^2 = (E_v/v_s)\mathbf{e}_x. \quad [5.72]$$

The fact that waves transport momentum implies that they exert a *radiation pressure* on a surface $\mathcal{S}$ which is exposed to the wave. For instance, if $\mathcal{S}$ is normal to the sound wave's direction of propagation, the energy that $\mathcal{S}$ receives during dt is $dE = \mathcal{S}\ (\mathbf{S}.\mathbf{n})\ dt = \mathcal{S}\ v_s\ E_v\ dt$. If $\mathcal{S}$ completely absorbs the wave, it receives a momentum $dP = dt\ \mathcal{S}\ v_s\ P_v = dt\ \mathcal{S}\ E_v$. This momentum transfer corresponds to an average force $f = dP/dt = \mathcal{S}E_v$ and an average *radiation pressure*

$$P_r = \langle E_v \rangle = \tfrac{1}{2}\,m_v\,\omega^2\,u_m{}^2 = p_m{}^2/2B. \quad [5.73]$$

If the surface $\mathcal{S}$ totally reflects the wave instead of absorbing it, $\mathcal{S}$ receives no energy but it is under a radiation pressure $P_r = p_m{}^2/B$, because the momentum density changes from $\mathbf{P}_v$ to $-\mathbf{P}_v$; thus, the surface receives a momentum $2\mathbf{P}_v$. This radiation pressure should not be confused with the acoustic pressure, which is an oscillatory harmonic quantity; thus its average value is equal to zero, while the radiation pressure is exerted in the propagation direction of the wave (repulsing the obstacle). For instance, in the case of an audible sound of frequency 1 kHz and acoustic pressure amplitude $p_m = 0.27$ Pa (a displacement amplitude $u_m = 10^{-7}$ m), the radiation pressure is only 2.5×10^{-7} Pa, but it may be much more important in the case of ultrasounds.

EXAMPLE 5.2. Calculate the energy density, the energy flux density and the momentum density: **a)** for a progressive plane sound wave, whose velocity potential is $\Psi = f(t - \mathbf{e}.\mathbf{r}/v_s)$; and **b)** for a progressive spherical diverging sound wave whose velocity potential is $\Psi = (R/r)\Psi_m\cos(\omega t - kr)$. Show that, at a large distance in this last case, E_v and the intensity decrease like $1/r^2$. Interpret this result.

SOLUTION

a) Writing $\Psi = f(\xi)$, where $\xi \equiv t - \mathbf{e}.\mathbf{r}/v_s$, the velocity is $\dot{\mathbf{u}} = -\nabla.\Psi = (\mathbf{e}/v_s)f'(\xi)$. Integrating with respect to time, we obtain the displacement $\mathbf{u} = (1/v_s)f(\xi)\mathbf{e}$. The acoustic pressure is $p_a = -B(\nabla.\mathbf{u}) = m_v f'(\xi)$. Using expressions [5.53], we obtain

$$E_{(C)\,v} = \tfrac{1}{2}\,(m_v/v_s^2)\,f'(\xi)^2$$
$$E_{(P)\,v} = \tfrac{1}{2}\,B\,\Sigma_{ij}\,[(e_i e_j/v_s^2)f'(\xi)]\,[(e_i e_j/v_s^2)f'(\xi)] = \tfrac{1}{2}(m_v/v_s^2)\,f'(\xi)^2$$
$$E_v = E_{(C)\,v} + E_{(P)\,v} = (m_v/v_s^2)\,f'(\xi)^2\,.$$
$$S_i(\mathbf{r},t) = -B\,\Sigma_j\,[-(e_j/v_s)f'(\xi)]\,[(e_i e_j/v_s^2)f'(\xi)] = (m_v/v_s)\,f'(\xi)^2\,e_i = E_v v_s\;e_i.$$
$$P_v^{\,i} = -m_v\,\Sigma_j\,[(-e_i/v_s^2)f'(\xi)\,e_j]\,(e_j/v_s)f'(\xi) = (m_v/v_s^3)f'(\xi)^2\,e_i = (E_v/v_s)\,e_i. \qquad [5.74]$$

b) If the velocity potential is $\Psi = (R/r)\Psi_m\cos(\omega t - kr)$ where $k = \omega/v_s$, the velocity, displacement and acoustic pressure at large distance are the $1/r$ terms in [5.42], i.e.

$$\partial_t\mathbf{u}(\mathbf{r},t) = -\,(\omega R/v_s r)\,\Psi_m\sin(\omega t - kr)]\,\mathbf{e}_r\,,\ \mathbf{u} = (R/v_s r)\,\Psi_m\cos(\omega t - kr)]\,\mathbf{e}_r$$
$$p_a(\mathbf{r},t) = (B\omega R/v_s^2 r)\,\Psi_m\cos(\omega t \pm kr + \pi/2). \qquad [5.75]$$

Using the expressions of [5.53], we obtain the dominant terms

$$E_{(C)\,v} = \tfrac{1}{2}\,m_v(\omega R\Psi_m/v_s r)^2\sin^2(\omega t - kr)$$
$$E_{(P)\,v} = \tfrac{1}{2}m_v v_s^2(R\Psi_m/v_s)^2\sin^2(\omega t - kr)\,\Sigma_{ij}(kr^i r^j/r^3)(kr^i r^j/r^3) = E_{(C)\,v}$$
$$E_v = E_{(C)\,v} + E_{(P)\,v} = m_v(\omega R\Psi_m/v_s r)^2\sin^2(\omega t - kr)$$
$$S_i = -B\,\Sigma_j\,(-\omega kR^2\Psi_m^2/v_s^2 r^3)\sin^2(\omega t - kr)\,\Sigma_j\,(r^i/r)\,r^j\,(r^j/r) = v_s E_v\,e_r^{\,i}$$
$$P_v^{\,i} = m_v\,\Sigma_j\,(k\omega R^2/v_s^2 r^2)\,\Psi_m^2\sin^2(\omega t - kr)](r^j/r)\,(r^i/r)(r^j/r) = (E_v/v_s)\,e_r^{\,i}\quad.[5.76]$$

The spherical wave, as specified by the velocity potential or the acoustic pressure p_a, propagates without a change of profile but its amplitude decreases like $1/r$. This decrease of the amplitude can be explained by the distribution of the emitted sound energy on a spherical surface of radius r. If P is the mean emitted power, the sound intensity, which is received at a point of this surface, is $P/4\pi r^2$. Thus, it decreases like $1/r^2$ and the wave amplitude decreases like $1/r$. Alternatively, if the wave is specified by $\mathbf{u}$ or $\dot{\mathbf{u}}$ in the equations of [5.42], it is a superposition of a term, which decreases like $1/r$ and a term, which decreases like $1/r^2$. The wave decreases as it propagates and its profile varies.

5.9. Elements of physiological acoustics

Hearing sounds is a complex (and not well-understood) physiological and psychological phenomenon, which occurs in the ear and the brain. Very briefly, sound waves are channeled to the eardrum – like a sound cavity – which amplifies them especially in the 3-4 kHz frequency band. The vibrations of the eardrum are transferred (via three pivoted bones) to the middle ear cavity and, through the oval window, into the inner ear. This is essentially a coiled cavity known as the *cochlea*, which is filled with a liquid and contains the basilar membrane formed by a large number of receptor cells, a kind of transducers transforming the sound vibrations into electric pulses, which are transmitted to the brain.

A normal human ear can hear sounds whose frequencies vary between about 16 Hz and 20 kHz (audible sounds). Usual sounds are superpositions of sinusoidal sounds. We may distinguish between *musical sounds*, which correspond to almost periodic vibrations produced by periodic musical instruments (organ pipes, strings, etc.), and the *noise* associated with non-periodic disturbances. A musical sound is a superposition of waves of discrete spectrum $p\tilde{\nu}$ (in a Fourier series), while a noise is a superposition of waves of continuous spectrum in a band $\Delta\tilde{\nu}$ (Fourier integral).

Each sinusoidal acoustic wave is called a *simple tone*, and the perception of this frequency of sound is called the *pitch*. The higher the frequency is, the higher the pitch is. A soprano voice is high-pitched and a bass voice is low-pitched. However, hearing is not so simple (the ear is not a simple spectrometer), the perception of the pitch depends on the sound intensity.

In the case of a periodic sound, the fundamental of its Fourier series corresponds to the *fundamental tone* and the other harmonics correspond to the *overtones*. Sounds of the same period have Fourier series with harmonics of the same frequencies but they may differ by the amplitudes and phases. This difference in the harmonics determines the characteristic *timber* of the sound produced by each musical instrument and by each person speaking.

The *loudness* of a sound is related to its intensity, i.e. to the amplitude of the acoustic pressure (or its effective value p_{eff}) but it also depends on the frequency, because the ear sensibility depends on the frequency. In particular, the *threshold of audibility* is the minimum intensity that can be perceived, this is about $\mathcal{I}_o = 10^{-12}$ W/m^2 ($p_m \approx 3 \times 10^{-5}$ Pa) at 0.7 – 6 kHz. This value of $\mathcal{I}_o$ is taken as the *standard threshold of audibility*. The maximum intensity (or acoustic pressure), at which sound is perceived without the sensation of pain is the *pain threshold.*

The ear can distinguish the ratio of intensities, not their difference. The sensation of loudness increases as this ratio increases but the relationship is not linear. A multiplication of the intensity by 10 is perceived as the doubling of the loudness. Thus, it is convenient to define the *sound intensity level* as $\mathcal{I}_{\text{ph}} = \log(\mathcal{I}_s/\mathcal{I}_o)$, where $\mathcal{I}_o = 10^{-12}$ W/m^2 is the standard audibility threshold. $\mathcal{I}_{\text{ph}}$ is expressed in *bel.* A multiplication of the physical intensity by a factor of 10 corresponds to an increase in the sound intensity level of 1 bel (the *decibel* (dB) is more commonly used). Thus, the intensity level in dB is

$$\mathcal{I}_{\text{ph}} = 10 \log (\mathcal{I}_s/\mathcal{I}_o) \qquad \text{(in dB).} \qquad [5.77]$$

The human ear can perceive a variation of about 1 dB in the intensity level. This corresponds to a multiplication of the physical intensity by a factor 1.26. At a frequency of 1 kHz, the audible sound intensity level varies approximately between 0 (for the threshold of audibility) and 120 dB (for the pain threshold). The pain threshold corresponds to $\mathcal{I}_s \approx 10^{12}\,\mathcal{I}_o = 1$ W/m^2, that is, an average sound energy density $< E_v > = \mathcal{I}_s/v_s = 3\times10^{-3}$ W/m^3, a displacement amplitude $u_m = (2\mathcal{I}_s/m_v\ v_s\omega^2)^{1/2} = 11$ µm and an acoustic pressure amplitude $p_m = (2m_v v_s \mathcal{I}_s)^{1/2} = 29$ Pa.

5.10. Infrasounds and ultrasounds

Infrasounds are inaudible sounds with frequencies of less than about 20 Hz. They are emitted by very low-frequency oscillating mechanical systems (a long metallic strip, for instance). They may have a high intensity and, thus, cause significant damage, particularly in the case of volcanic eruptions and earthquakes. They are part of our noisy environment, especially near airplanes, fast-moving cars, over amplified modern music bands, etc. Infrasounds may be harmful if the internal organs of the body go into resonance (at 8 Hz for the human chest, for instance) and cause dizziness, nausea, and even internal bleeding. Some animals (like elephants) use infrasounds to communicate over long distances.

Ultrasounds have a frequency higher than 20 kHz. Mechanical oscillators, such as sirens, may emit low-frequency ultrasounds (between 20 kHz and 200 kHz). They may also have a high intensity (of the order of watt.cm^{-2}), that is, a radiated power of several kilowatts. Ultrasounds with frequencies up to about 50 MHz may be emitted by piezoelectric oscillators using the strain produced is some crystals (quartz, for instance) if they are submitted to a strong electric field. Ultrasounds of several GHz may be emitted by a mechanical oscillator excited by an AC current. *Magnetostriction*, for instance, is the change in the physical dimensions of a magnetic material when its magnetization is changed in an intense magnetic field. Ultrasounds with frequencies higher than 1 GHz are called *hypersounds.* They correspond to a wavelength $\lambda < 0.1$ µm, that is, of the order of the inter-molecular distances in a gas at *STP*. Their frequency may be as high as 10^{13} Hz, which corresponds to a wavelength of the order of the inter-atomic distance in solids.

Ultrasounds with a very short wavelength, may be focused into very thin beams, exactly like light rays. They undergo refractions and reflections on the interface of two different mediums according to the same laws as geometrical optics (see Chapter 7). Thus, it is possible to conceive ultrasonic mirrors and lenses.

Ultrasounds are very attenuated in air and in certain media (especially those containing calcium, such as bones for instance). Because of their high frequency, the radiation pressure of ultrasounds may be high. For instance, an ultrasound of frequency 1 GHz and amplitude 0.1 μm (that is, p_m = 0.27 MPa) has a radiation pressure of 0.25 MPa (i.e. 2.5 atm). The interface of two non-miscible liquids and the liquid-gas interface are a discontinuity surface for the radiation pressure. This explains certain ultrasound properties.

Ultrasounds have many useful applications in physics (the study of certain properties of matter, acoustic microscopes, etc.), chemistry (activation of certain chemical reactions by provoking the formation of free-ions), technology (by producing cavitations, crushing, local heating, etc.), navigation (sonar) and medicine in both diagnosis (sonography) and treatment (destruction of living organisms, tumors and stones), etc.

5.11. Surface waves*

In this section we consider waves propagating on the surface of a liquid in contact with a low-density gas, under the effect of gravity and surface tension. If we immerse a visible powder (of a similar density to the liquid) in the liquid, we find that the liquid only moves up to a certain depth under the surface (Figure 5.11a). We take the *Oxy* plane coinciding with the surface of the liquid at equilibrium and the *Oz* axis oriented upward. We designate by $s(x, y, t)$ the height of the free surface of the liquid above the equilibrium plane *Oxy* (Figure 5.11b).

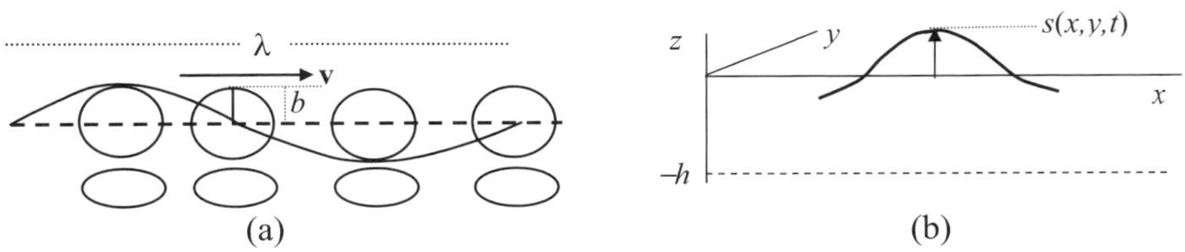

Figure 5.11. *Surface waves*

The propagations of surface waves are governed by two fundamental equations of fluid mechanics: the *equation of conservation of mass* and *Euler equation*:

$$\partial m_v/\partial t + \nabla.(m_v \dot{\mathbf{u}}) = 0, \qquad [5.78]$$

$$(\partial \dot{\mathbf{u}}/\partial t) + (\dot{\mathbf{u}}.\nabla)\dot{\mathbf{u}} = \mathbf{F}_{(ex)m} - \nabla p/m_v, \qquad [5.79]$$

where $\dot{\mathbf{u}}(\mathbf{r},t)$ is the mean velocity and $p(\mathbf{r}, t)$ is the pressure at point $\mathbf{r}$ and time t. $\mathbf{F}_{(ex)m}$ is the field of the external force per unit mass of the liquid, very often this is the acceleration of gravity $\mathbf{g} \equiv -\nabla(gz)$. The left-hand side of [5.79] is the *material derivative* (or *total derivative*) $D\dot{\mathbf{u}}/Dt$ (i.e. the rate of change of $\dot{\mathbf{u}}$ in the frame of the locally moving liquid). The total material derivative of a physical quantity $Q(\mathbf{r}, t)$ is defined by $DQ/Dt = \partial Q/\partial t + \dot{\mathbf{u}}\,.\nabla Q$.

To simplify, we neglect the compressibility of the liquid; thus, $\partial m_v/\partial t \cong 0$ in [5.78]. We also assume that the motion is irrotational; thus, we may introduce the velocity potential Ψ such that $\dot{\mathbf{u}} = -\nabla\Psi(\mathbf{r}, t)$. Finally, we assume that the motion is small, so that the term $(\dot{\mathbf{u}}\,.\nabla)\,\dot{\mathbf{u}}$ in [5.79] and all the second order terms in $\mathbf{u}$, $\dot{\mathbf{u}}$ and s are negligible. Equation [5.79] becomes $\nabla\dot{\Psi} = \nabla(gz) + \nabla p/m_v$, that we integrate. Fundamental equations [5.78] and [5.79] become

$$\Delta\Psi(\mathbf{r}, t) = 0, \qquad [5.80]$$

$$p(\mathbf{r}, t) - \overline{p} - m_v\,\dot{\Psi}(\mathbf{r}.t) + m_v\,g\,z = 0, \qquad [5.81]$$

where we take advantage of the determination of Ψ up to an arbitrary function of t to subtract the pressure $\overline{p}$ exerted by the gas on the free surface of the liquid.

In the case of a simple harmonic wave, we set $\Psi = F(\mathbf{r})\,e^{i\omega t}$. We deduce that $\dot{\mathbf{u}} = -\nabla F(\mathbf{r})\,e^{i\omega t}$ and $\mathbf{u} = (i/\omega)\nabla F(\mathbf{r})\,e^{i\omega t}$. We also set $p(\mathbf{r}, t) - \overline{p} + m_v\,g\,z = p(\mathbf{r})\,e^{i\omega t}$. This is the pressure in excess of the equilibrium value $\overline{p} - m_v\,gz$ within the liquid at depth z. Equations [5.80] and [5.81] become

$$\Delta F(\mathbf{r}) = 0, \qquad [5.82]$$

$$p(\mathbf{r}) = i\omega m_v\,F(\mathbf{r}). \qquad [5.83]$$

On the other hand, the wave must verify certain boundary conditions:

– On the surface $\mathcal{S}$ of any rigid obstacle, designating by x_n the normal coordinate to $\mathcal{S}$, the condition that the normal component of $\dot{\mathbf{u}}$ vanishes implies that

$$\mathbf{n}.\nabla F = 0, \qquad \text{i.e.} \qquad \partial F/\partial x_n = 0 \qquad \text{[on a rigid obstacle } \mathcal{S}\text{]}, \quad [5.84]$$

– The surface equation of the liquid is by definition $z - s(x, y, t) = 0$; hence, by taking the material (total) time derivative and neglecting the term $\dot{\mathbf{u}}\,.\nabla s$, which is of the second order, we obtain

$$\partial s/\partial t \approx \dot{u}_z = -\,\partial_z\Psi = -(\partial_z F)\;e^{i\omega t} \qquad \text{[on the free surface of the liquid]}. \quad [5.85]$$

However, on the free surface of the liquid, $p = \overline{p}$ and equation [5.83] takes the form $\mathrm{i}\omega F\, e^{\mathrm{i}\omega t} = gs$; thus, equation [5.85] can be written as

$$g\,\partial_z F - \omega^2 F = 0 \qquad \text{[on the free surface of the liquid].} \qquad [5.86]$$

We conclude that the function F must verify equations [5.82] in the liquid, [5.84] on rigid obstacles and [5.86] on the free surface of the liquid.

Let us consider the one-dimensional wave in a rectangular canal, which is very long in the direction Ox and has a width L in the direction Oy. We successively analyze the consequences of the conditions on the function F.

a) The Poisson equation [5.82] has solutions of the form $F(\mathbf{r}) = f(z)\,\zeta(x, y)$. In the case of a progressive wave propagating in an horizontal direction, whose wave vector has the components k_x and k_y, the function $\zeta(x, y)$ may be written in the complex form $\zeta(x, y) = e^{-\mathrm{i}(xk_x + yk_y - \alpha)}$. Substituting the expression $f(z)\,\zeta(x, y)$ for F in the equation $\Delta F = 0$, we find that $f(z)$ must be a solution of the differential equation

$$\partial^2{}_{zz} f - k^2 f = 0, \qquad [5.87]$$

where we have set

$$k^2 = k_x{}^2 + k_y{}^2. \qquad [5.88]$$

The general solution of [5.87] is of the form $f(z) = a\,e^{kz} + b\,e^{-kz}$ and the function $F(\mathbf{r})$ can be written as

$$F(\mathbf{r}) = [a\,e^{kz} + b\,e^{-kz}\,]\,e^{-\mathrm{i}(xk_x + yk_y - \alpha)}. \qquad [5.89]$$

In particular, the solution, which is progressive in the direction Ox of the canal, may be written in the form

$$F(\mathbf{r}) = [a\,e^{kz} + b\,e^{-kz}\,]\cos(k_y y - \alpha)\,e^{-\mathrm{i}xk_x}. \qquad [5.90]$$

b) The bottom ($z = -h$) being a rigid obstacle, condition [5.84] implies that

$$\partial_z F\,|_{z=-h} = 0, \quad \text{i.e.} \quad a\,e^{-kh} - b\,e^{kh} = 0.$$

Thus, we may express one of the amplitudes a and b in terms of the other. Setting $C = 2a\,e^{-kh}$, the function F may be written in the form

$$F(\mathbf{r}) = C\,\mathrm{ch}[k(z+h)]\cos(k_y y - \alpha)\,e^{-\mathrm{i}xk_x}\,. \qquad [5.91]$$

c) On the lateral walls of the canal $y = 0$ and $y = L$, we must have $\partial_y F\,|_{y=0} = 0$ and $\partial_y F|_{y=L} = 0$, that is, $k_y \sin\alpha = k_y \sin(k_y L - \alpha) = 0$. These conditions are satisfied if $\alpha = 0$ and $k_y = n\pi/L$ where $n = 0, 1, 2, \ldots$ Thus, the *mode n* solution may be written as

$$F_n(\mathbf{r}) = C_n\,\mathrm{ch}[k(z+h)]\cos(n\pi y/L)\,e^{-\mathrm{i}xk_x}\,. \qquad [5.92]$$

These modes correspond to waves, which are "stationary" in the direction Oy and progressive in the direction Ox. The wavelength in the direction of propagation Ox is $\lambda = 2\pi/k_x$. Relation [5.88] can then be written as

$$k^2 = k_x^{\,2} + n^2\pi^2/L^2 = 4\pi^2/\lambda^2 + n^2\pi^2/L^2. \qquad [5.93]$$

In particular, the mode $n = 0$ corresponds to a wave which is independent on y, that is, uniform in the transverse direction of the canal. Its wave number is $k = 2\pi/\lambda$.

d) On the free surface of the liquid ($z = s$), condition [5.86] is satisfied if

$$\tanh[k(s+h)] = \omega^2/gk. \qquad [5.94]$$

As s is a function of x and y, this condition may be only approximately verified if s is small, compared to h, that is, in the case of a deep canal. In this case, [5.94] takes the approximate form of a dispersion relation

$$\omega^2 = gk\tanh(kh). \qquad [5.95]$$

The phase velocity is

$$v_{(\mathrm{p})} = \omega/k = \sqrt{(g\lambda/2\pi)\tanh(kh)}\,. \qquad [5.96]$$

Setting $A_n = C_n(k/\omega)\,\mathrm{sh}(kh)$ and taking the real parts, we may write for a wave in a liquid and on its surface:

$$\begin{aligned}
\Psi_n &= A_n\,[\omega/k\,\mathrm{sh}(kh)]\,\mathrm{ch}[k(z+h)]\cos(n\pi y/L)\cos(\omega t - xk_x)\\
s_n &= -A_n\cos(n\pi y/L)\sin(\omega t - xk_x)\\
\dot{u}_{n.x} &= -\partial_x\Psi_n = -A_n\,[\omega k_x/k\,\mathrm{sh}(kh)]\,\mathrm{ch}[k(z+h)]\cos(n\pi y/L)\sin(\omega t - xk_x)\\
\dot{u}_{n.y} &= -\partial_y\Psi_n = A_n\,[n\pi\omega/kL\,\mathrm{sh}(kh)]\,\mathrm{ch}[k(z+h)]\sin(n\pi y/L)\cos(\omega t - xk_x)\\
\dot{u}_{n.z} &= -\partial_z\Psi_n = -A_n\,[\omega/\mathrm{sh}(kh)]\,\mathrm{sh}[k(z+h)]\cos(n\pi y/L)\cos(\omega t - xk_x).
\end{aligned} \qquad [5.97]$$

Integrating $\dot{\mathbf{u}}_n$ with respect to t, we obtain the components of the liquid displacement:

$$u_{n,x} = [k_x\, s_m / k\ \mathrm{sh}(kh)]\ \mathrm{ch}[k(z+h)]\ \cos(n\pi y/L)\ \cos(\omega t - xk_x)$$
$$u_{n,y} = [n\pi\, s_m/kL\ \mathrm{sh}(kh)]\ \mathrm{ch}[k(z+h)]\ \sin(n\pi y/L)\ \sin(\omega t - xk_x)$$
$$u_{n,z} = -\,[s_m/\ \mathrm{sh}(kh)]\ \mathrm{sh}[k(z+h)]\ \cos(n\pi y/L)\ \sin(\omega t - xk_x). \qquad [5.98]$$

Eliminating t between $u_{n,x}$ and $u_{n,z}$ in the case of the mode $n = 0$ (thus, $k_x = k$), we obtain the trajectory equation for the particles of the liquid in the Oxz plane:

$$\frac{\mathrm{sh}^2(kh)}{A_o^2\ \mathrm{ch}^2[k(z+h)]}\,u_x^2 + \frac{\mathrm{sh}^2(kh)}{A_o^2\ \mathrm{sh}^2[k(z+h)]}\,u_z^2 = 1. \qquad [5.99]$$

These are ellipses whose semi-minor axis in the vertical direction becomes smaller and smaller with the depth z (Figure 5.11a).

In a more general case of a very long but not very deep canal, we must take into account the surface tension γ. It may be shown that the phase velocity is then given by the expression

$$v_{(P)}^2 = \frac{\lambda g}{2\pi}\,[1 + \frac{\lambda_m^2}{\lambda^2}]\ \mathrm{th}(\frac{2\pi h}{\lambda}) \qquad \text{where } \lambda_m = 2\pi\sqrt{\gamma/g m_v}\ . \qquad [5.100]$$

λ_m is a characteristic wavelength of the gas-liquid interface. For instance, $\lambda_m = 1.71$ cm for air-water and 40.3 cm for air-mercury. In the case of *capillary waves* (that is of wavelength $\lambda << \lambda_m$), the surface tension plays a dominant part.

– If $\lambda << \lambda_m << h$, we get

$$v_{(P)} \approx \sqrt{2\pi\gamma/\lambda m_v} \qquad \text{(capillary waves, } \lambda << \lambda_m << h). \qquad [5.101]$$

– If the wavelength is large, compared to h ($kh << 1$), $\tanh(kh) \approx kh$ and

$$v_{(P)} = \sqrt{gh} \qquad (\text{if } kh << 1). \qquad [5.102]$$

– If the wavelength is much larger than λ_m, the free surface of the liquid has a small curvature and the surface tension has practically no effect. This is the case of *gravity waves* (like the usual water waves) whose speed of propagation is

$$v_{(P)} = \sqrt{\lambda g/2\pi} \qquad \text{(gravity waves, } \lambda_m << \lambda << h). \qquad [5.103]$$

In the case of the air-water interface, the speed [5.103] only differs from [5.100] by $1/10^6$ for $h = \lambda$. For instance, on the open sea ($h >> \lambda$), $v_{(p)}$ is equal to 12.5 m/s for $\lambda = 100$ m, this corresponds to waves of a frequency 0.125 Hz. In the case of a tsunami, $v_{(p)}$ may be as high as 200 m/s.

5.12. Summary

The propagation equation of elastic waves on a taut string of mass m_l per unit length and tension F is $\partial^2_{tt}u - v^2(\partial^2_{xx}u) = 0$, where $v = \sqrt{F/m_l}$ is the speed of propagation. The *polarization* of the wave is determined by the direction of $\mathbf{u}$.

If a solid body is subject to external forces, its particles are displaced from their equilibrium positions. The deformation determines the *strain* (defined as a dimensionless quantity). The internal forces are determined by the *stress* $\Pi = \delta F/\mathcal{S}$ (homogeneous to a pressure). *Hooke's law* states that if the deformation is small, the ratio of the stress to the strain is constant and equal to an *elasticity modulus* of the medium. In the case of compression or traction $\delta l/l = \Pi/Y = \delta F/Y\mathcal{S}$ and $\delta d/d = -\sigma\,\delta l/l = -(\sigma/Y)\,\Pi$. Y is *Young's modulus* and σ is the *Poisson ratio* ($0 < \sigma < 0.5$). If Π is uniform on all the faces of the solid, its volume varies according to the relationship $\delta\mathcal{V}/\mathcal{V} = \Pi/B$, where $B \equiv 1/\kappa = Y/3(1 - 2\sigma)$ is the *bulk modulus* and κ is the *compressibility*. The *shear modulus* (or *modulus of rigidity*) is $\mu = Y/2(1+\sigma)$.

The speed of propagation of elastic waves in a massive spring is $v = \sqrt{K_l/m_l}$, where K_l is the force constant if the spring has a unit length. The speed of propagation of elastic waves in a rod is $v = \sqrt{Y/m_V}$.

An acoustic wave is a displacement and pressure wave of small amplitude in fluids. The acoustic displacement is always longitudinal. The variation of pressure δp is related to the variation $\delta\mathcal{V}$ of volume by the relation $\delta p = -B(\delta\mathcal{V}/\mathcal{V})$. In the case of a one-dimensional wave (in a pipe, for instance), the *acoustic pressure* is $p_a(\mathbf{r}, t) = p(\mathbf{r}, t) - \overline{p} = -B(\partial_x u)$ where $\overline{p}$ is the equilibrium pressure. The *wave equation* is

$$\partial^2_{tt}u - v_s^2(\partial^2_{xx}u) = 0 \qquad \text{where } v_s = \sqrt{B/m_V} = \sqrt{\gamma p/m_V} = \sqrt{\gamma RT/M} \text{ (ideal gases)}$$

In air at STP, $v_s = 332$ m/s.

The wave equation on a membrane is $m_s\,\partial^2_{tt}u - F_x\,(\partial^2_{xx}u) - F_y\,(\partial^2_{yy}u) = 0$. The dispersion relation is $\omega^2 = (F_x/m_s)k_x^2 + (F_y/m_s)k_y^2$. If the direction of propagation makes a angle θ with *Ox,* the dispersion relation may be written in the form $\omega^2 = (k^2/m_s)[F_x \cos^2\theta + F_y \sin^2\theta]$. If $F_x = F_y \equiv F$, the membrane is isotropic and the propagation speed is $v = \sqrt{F/m_s}$.

In the case of a three-dimensional acoustic wave, $p_a = -B(\nabla.\mathbf{u})$ and the wave equations are $\ddot{\mathbf{u}} - v_s^2\,\Delta\mathbf{u} = 0$ and $\ddot{p}_a - v_s^2\,\Delta p_a = 0$. It is convenient to use the *velocity potential* Ψ defined by the relation $\dot{\mathbf{u}}(\mathbf{r},t) = -\nabla\Psi(\mathrm{r}, t)$. It obeys the d'Alembert wave equation $\ddot{\Psi} - v_s^2\,\Delta\Psi = 0$.

A plane simple harmonic acoustic wave, propagating in the direction $\mathbf{k} = \mathbf{k}/k = (\omega/v_s)\,\mathbf{e}$, may be written in the form:

$$\Psi = \Psi_m e^{i(\omega t - \mathbf{k.r})}, \qquad \dot{\mathbf{u}} = -\nabla\Psi = i\mathbf{k}\Psi = \mathbf{e}\,k\Psi_m e^{i(\omega t - \mathbf{k.r} + \pi/2)}$$

$$\mathbf{u} = \dot{\mathbf{u}}/i\omega = \mathbf{e}\,(\Psi_m/v_s)\,e^{i(\omega t - \mathbf{k.r})}, \qquad p_a = -B\,(\nabla.\mathbf{u}) = m_v\omega\Psi_m\,e^{i(\omega t - \mathbf{k.r} + \pi/2)}$$

with the relations of the amplitudes: $\dot{\mathbf{u}}_m = k\Psi_m\mathbf{e}$ and $\mathbf{u}_m = (\Psi_m/v_s)\mathbf{e}$, $p_m = m_v\,\omega\Psi_m$.

The three-dimensional wave equation for elastic waves is

$$\partial^2_{tt}u - v_T^2\,\Delta\mathbf{u} + (v_T^2 - v_L^2)\,\nabla\,(\nabla.\mathbf{u}) = 0,$$

where $v_L^2 = Y(1-\sigma)/m_v(1+\sigma)(1-2\sigma)$ and $v_T^2 = Y(1-\sigma)/2m_v(1+\sigma)$. The wave equations for longitudinal waves and for transverse waves are $\partial^2_{tt}u_L - v_L^2\,\partial^2_{xx}\mathbf{u}_L = 0$ and $\partial^2_{tt}u_T - v_T^2\,\partial^2_{xx}\mathbf{u}_T = 0$.

The energy density per unit length and the power or energy flux are:

$$E_l = \tfrac{1}{2}m_l\dot{u}^2 + \tfrac{1}{2}F(\partial_x u)^2 \quad \text{and} \quad P(x,t) = -F\,\dot{u}\,(\partial_x u) \quad \text{[taut string]}$$

$$E_l = \tfrac{1}{2}m_l\dot{u}^2 + \tfrac{1}{2}K_l\,(\partial_x u)^2 \quad \text{and} \quad P(x,t) = -K_l\,\dot{u}\,(\partial_x u) \quad \text{[massive spring]}$$

$$E_l = \tfrac{1}{2}\mathcal{S}m_v\dot{u}^2 + \tfrac{1}{2}\mathcal{S}Y(\partial_x u)^2 \quad \text{and} \quad P(x,t) = -Y\mathcal{S}\,\dot{u}\,(\partial_x u) \quad \text{[elastic rod]}$$

$$E_l = \tfrac{1}{2}\mathcal{S}m_v\dot{u}^2 + \tfrac{1}{2}\mathcal{S}B(\partial_x u)^2 \quad \text{and} \quad P(x,t) = -B\mathcal{S}\,\dot{u}\,(\partial_x u) \quad \text{[acoustic pipe]}$$

In the case of an elastic wave propagating in the direction Ox with a longitudinal component $u_L(x, t)$ and a transverse component $u_T(x, t)$ in the direction Oy, we have $\mathbf{u} = u_L(x, t)\,\mathbf{e}_x + u_T(x, t)\,\mathbf{e}_y$ and

$$E_v = \tfrac{1}{2}m_v[\dot{u}_L^{\,2} + \dot{u}_T^{\,2} + v_L^2(\partial_x u_L)^2 + v_T^2(\partial_x u_T)^2]$$

$$\mathbf{S} = -m_v\,\{[v_L^2\,\dot{u}_L\,(\partial_x u_L) + v_T^2\,\dot{u}_T\,(\partial_x u_T)]\mathbf{e}_x + [(v_L^2 - 2v_T^2)\,\dot{u}_T\,(\partial_x u_L) + v_T^2\,\dot{u}_L\,(\partial_x u_T)]\,\mathbf{e}_y\}$$

$$\mathbf{P}_v = -\,m_v\,[\,\dot{u}_L\,(\partial_x u_L) + \dot{u}_T\,(\partial_x u_T)]\mathbf{e}_x.$$

The law for the conservation of energy may be written in the form $\partial_t E_l + \partial_x P = 0$ or $\partial_t E_v + \nabla.\mathbf{S} = 0$.

We define the impedance Z of a medium as the ratio of the excitation force to the velocity at the entry, and the impedance per unit area Z_s as the ratio of the stress (or the acoustic pressure) to the velocity. In the case of a progressive wave $u = f(\xi)$ where $\xi = t - x/v$, we obtain

$$F_{(ex)y} = -F\,\partial_x u|_{x=0},\quad Z = \sqrt{m_l F}\quad \text{and } P_{(ex)} = F_{(ex)y}^2/Z = Zf'(t)^2 \qquad \text{[string]}$$

$$F_{(ex)} = -K_l\,\partial_x u|_{x=0},\quad Z = \sqrt{m_l K_l}\quad \text{and } P_{(ex)} = F_{(ex)y}/Z = Zf'(t)^2 \quad \text{[massive spring]}$$

$$F_{(ex)} = -\mathcal{S}Y\,\partial_x u|_{x=0},\quad Z = \sqrt{m_v Y}\quad \text{and } P_{(ex)} = F_{(ex)}^2/\mathcal{S}Z_s = \mathcal{S}Z_s f'(t)^2 \quad \text{[rod]}$$

$$F_{(ex)} = -\mathcal{S}B\,\partial_x u|_{x=0},\quad Z = \sqrt{m_v B}\quad \text{and } P_{(ex)} = F_{(ex)y}/Z = Zf'(t)^2 \qquad \text{[pipe]}$$

The sound intensity is the time-averaged value of $S(x, t)$. In the case of a simple harmonic wave, we find

$$\mathcal{I}_s = (1/T)\int_t^{t+T} dt\; S(x, t) = \tfrac{1}{2}\,Z_s\,\omega^2 u_m^2 = (p_m^2/2Z_s) = (m_v\omega^2/2v_s)\,\Psi_m^2$$

and the average *radiation pressure* is $P_r = \langle E_v \rangle = \tfrac{1}{2}\,m_v\omega^2 u_m^2 = p_m^2/2B$.

Infrasounds have frequencies lower than 20 Hz. *Ultrasounds* have frequencies higher than 20 kHz, up to about 10^{13} Hz. Ultrasounds may be produced and analyzed by mechanical, piezoelectric or magnetostrictive oscillators. As they have very short wavelengths, ultrasounds may be focused exactly like light rays.

In a deep liquid, the dispersion relation for surface waves is $\omega^2 = gk\tanh(kh)$ and the phase velocity is $v_{(p)} = \sqrt{(g\lambda/2\pi)\tanh(kh)}$. The particles of the liquid describe ellipses, which become increasingly smaller with the depth. Taking into account the surface tension γ, we find

$v_{(P)}^2 = (\lambda g/2\pi)\,[1 + \lambda_m^2/\lambda^2]\ \mathrm{th}(2\pi h/\lambda)$ where $\lambda_m = 2\pi\sqrt{\gamma/gm_v}$.

$v_{(P)} \approx \sqrt{2\pi\gamma/\lambda m_v}$ (if $\lambda << \lambda_m << h$, capillary waves).

$v_{(P)} = \sqrt{gh}$ (if $\lambda >> h$)

$v_{(P)} = \sqrt{\lambda g/2\pi}$ (if $\lambda_m << \lambda << h$, gravity waves).

5.13. Problem solving suggestions

1) Do not confuse the velocity of the molecules in a fluid (due to thermal agitation), the velocity $\dot{\mathbf{u}}$ of the fluid (due to the wave), the speed of propagation v (which appears in the wave equation), the phase velocity $\mathbf{v}_{(P)}$ and the group velocity $\mathbf{v}_{(g)}$. The velocity of the molecules is randomly oriented in all directions; thus, its average value is equal to zero. The velocity $\dot{\mathbf{u}}$ of the fluid is oscillatory. It is oriented in the direction of propagation for longitudinal waves, vertically for waves on the surface of a liquid and longitudinally or transversely for elastic waves in solids.

2) It is only in the case of progressive waves, that the kinetic energy density is equal to the potential energy density and the energy flux density is $\mathbf{S} = v_s\, E_v\, \mathbf{e}$. Do not confuse the acoustic pressure, which is oscillatory (thus, its average value is equal to zero) and the radiation pressure, which is always positive and acts in the direction of propagation.

5.14. Conceptual questions

Q1. Define Young's modulus Y, Poisson ratio σ and the shear modulus μ. Explain why σ always lies between 0 and 0.5. Very often mountains are cone shaped. Explain why such a configuration is more likely than a vertical structure to reach a great height. What should the maximum height of a vertical column be so that it does not crush itself, assuming that the rocks have a mass density of $2{,}5 \times 10^3$ kg.m^{-3} and can support a compression of 60 MPa.

Q2. A cubic block of aluminum with 2 cm sides is fixed on a table. What is its distortion angle β if a tangential force $F_1 = 10^3$ N is uniformly exerted on the upper face, and a force F_2 of equal magnitude is uniformly exerted on a vertical face?

Q3. A steel wire is enrolled as a helical spring with a given radius and step. Explain why its force constant is inversely proportional to its length.

Q4. Assume that a sound wave is emitted by a point-like source (or a pulsating sphere). Show that the sound intensity decreases like $1/r^2$ where r is the distance to the center of the source. How does the amplitude vary? The intensity of an underwater explosion is 0.5 GW.m^{-2} at a distance of 100 m. What is its intensity at 1 km?

Q5. If you can distinguish a sound intensity level of 1 dB, what is the ratio of the amplitudes of these sounds? If the intensity level of a person singing a note is 60 dB at 2 m what would be the intensity level if 4 people sing this note in unison?

5.15. Problems

Transverse waves on strings, springs and rods

P5.1*. A uniform string of length L and mass m_l per unit length hangs vertically under the influence of its own weight. A shake is produced at its lower end. How long does it take for this disturbance to reach the upper end? Calculate this time if $L = 2$ m, $g = 9.81$ m.s^{-2} and $m_l = 0.1$ kg/m.

P5.2. A line of masses equal to m are separated by a distances d and coupled by springs of constant $K \equiv m\omega_o^2$. **a)** Calculate the phase velocity $v_{(P)}$ and the group velocity $v_{(g)}$ as functions of the excitation angular frequency ω. **b)** Show that, at low frequency ($\omega << 2\omega_o$), the wavelength is very large ($\lambda >> d$) and the dispersion relation may be written in the form $\omega = \omega_o\, kd\, [1 - (kd)^2/24 + ...]$. What then is the expression of the group velocity? **c)** Assume that a force $F_m e^{i\omega t}$ acts on the first mass ($n = 1$). Show that the vibrations are of the form $u_n = A\, e^{i(\omega t - nkd - \phi)}$, where $A = F/m\omega\omega_o$ and $\tan\phi = \sqrt{4\omega_o^2/\omega^2 - 1}$ ($\pi/2 < \phi < \pi$). **d)** Calculate the impedance of this system and the average power supplied by a force $F_m \cos(\omega t)$ at the entry.

P5.3*. a) Consider a massive spring as the limit of the coupled masses by springs of the preceding problem. Show that the phase velocity of the longitudinal wave, which propagates on a spring of force constant K and mass M, is proportional to the length of the spring. Calculate the phase velocity in a spring of length 50 cm and mass 0.5 kg if it lengthens 4 cm under the weight of 2 kg. **b)** In order to directly analyze the propagation of waves on this spring, consider a small element of length Δx. Determine the force acting on this element and write its equation of motion. Deduce the wave equation. **c)** A line of masses coupled by springs may be used as a model for crystalline solids. For instance, consider a rod of length L whose cross-section is a square of side l. Assume that it has the cubic lattice structure of sides a, and treat the rod as a set of lines of masses m coupled by springs of constant K. Show that the Young's modulus of this material is $Y = K/a$.

P5.4*. Show that a torsional wave in an elastic rod propagates with the velocity $v = \sqrt{\mu/m_v}$, where μ is the shear modulus.

Sound Propagation in a pipe

P5.5. a) Show that the relative variation of the mass density and the volume of a gas are related by the equation $dm_v/m_v = -dV/V$. **b)** Show that the bulk modulus of an ideal gas is equal to p in the case of an isothermal process (constant temperature), and equal to γp in the case of an adiabatic process. **c)** What is the variation of the speed of sound in air if its temperature increases from 0° to 30°C? Compare the wavelength of the note 440 Hz at these temperatures. Assume that air is an ideal gas.

P5.6. Consider the sound wave $p_a(x, t) = p_m \cos(\omega t - kx)$ in a pipe. Write the corresponding expressions for the waves of displacement, temperature and mass density. Consider the propagation of sound in air at 0°C and for normal pressure. Determine the amplitude of the variation of temperature and mass density in the case of an audible sound of amplitude $p_m = 10^{-4}$ Pa and an ultrasound of amplitude 0.5 atm.

Waves on a taut membrane

P5.7. Two waves $u_1 = A \cos(\omega t - kx)$ and $u_2 = A \cos(\omega t - ky)$ propagate on a taut rectangular membrane with a uniform tension F per unit length on its boundaries. Show that the resulting vibration is a modulated wave, propagating in the direction of the bisector of the axes Ox and Oy, with a phase velocity $v\sqrt{2}$. What is its wavelength? Show that the amplitude vanishes on the straight lines of equations $x - y = (2n + 1)\pi/k$.

Energy of progressive mechanical waves, impedance and intensity

P5.8. Show that the energy density of an elastic wave $u = A \cos(\omega t - kx)$ on a string is $E_l = ½\, m_l A^2\omega^2[1 - \cos(2\omega t - 2kx)]$. What is its average value? Plot u and E_l as functions of x for a given value of t. What is the speed of propagation of the energy?

P5.9. An oscillator of amplitude 3 mm and period 2.5 ms acts at the origin $x = 0$ of a string of mass 50 g/m under a tension of 80 N. **a)** Write down the expression of the displacement as a function of x and t, knowing that the oscillator starts from rest in the negative y direction at $t = 0$. What is the wavelength? **b)** Calculate the impedance of this string and determine the energy density and the average power of the generator, which sustains this wave.

P5.10*. A longitudinal wave $u = u_o \sin(\omega t - kx)$ propagates in a rod of mass density m_v, sectional area S and Young's modulus Y. **a)** Write down the expressions of the kinetic energy density and the potential energy density as functions of x and t. **b)** Deduce that the potential energy and the kinetic energy per wavelength are both equal to $\frac{1}{4}m_v\mathcal{S}\lambda\omega^2u_o^2$. **c)** What is the average energy flux per unit time across a section of the rod and what should be the average power of the emitter sustaining this wave? **d)** Write down the expression of the force $F(x, t)$ acting at x on the part of the rod, which precedes x. Show that u and F are in quadrature. **e)** Plot u and F as functions of x.

P5.11. A sound wave, in an air-filled pipe, is represented by the displacement wave $u(x, t) = u_m \cos(\omega t - kx)$. Knowing that the frequency is 1 kHz, the amplitude is $u_m = 10^{-7}$ m/s and the propagation velocity is 332 m/s, write the expressions of the pressure and the velocity potential. Calculate the amplitude p_m of the acoustic pressure. Calculate the amplitude of the velocity and compare it to the propagation speed of the wave. Calculate the average energy density and the intensity of this wave. What is the intensity level in decibels?

P5.12. A piston vibrates with a frequency of 500 Hz and an amplitude of 10^{-2} mm at the entry of an air-filled pipe of cross-section 5 cm^2. **a)** Write the expressions of the displacement and the pressure waves. **b)** Calculate the energy density, the energy flux density and the intensity of this wave. **c)** What should the power of the generator emitting this wave be? The speed of sound is 332 m/s and $m_v = 1.293$ $kg.m^{-3}$.

P5.13. Compare the radiation pressure to the amplitude of the acoustic pressure in the case of a plane sound wave of frequency of 10 kHz and intensity level of 30 dB. This wave is incident normally on a surface of 20 cm^2. Calculate the force exerted on this surface if it is totally absorbent and if it is a perfect mirror.

P5.14.** A sound wave is emitted in a pipe of cross-section $\mathcal{S}$ by moving a piston at the entry of the pipe. The piston has a mass M, it is attracted to its equilibrium position by a force $-K\xi$, and is subject to a friction force $-b\dot{\xi}$ where ξ is the piston displacement. The motion of the piston is sustained with the force of a generator $F = F_m \cos(\Omega t)$. **a)** Determine the supplementary force, which acts on this piston due to the emission of the sound wave. Deduce its equation of motion and the expression of its displacement. Write the expression of the sound wave for the displacement u and the acoustic pressure. What is the impedance of this system? **b)** Determine the average power of the emitted wave, and the power which is required to sustain this sound. What is the acoustic efficiency of this emission? **c)** What should the answers to (b) be if the pipe is water-filled?

P5.15.** If a body oscillates, it emits a sound of the same frequency $\tilde{\nu}$. **a)** Assume that a pulsating sphere of radius R emits a spherical sound wave whose velocity potential is $\underline{\Psi} = \Psi_m(R/r)e^{i(\omega t - kr)}$, where $k = \omega/v_s$. Calculate the corresponding vector displacement and acoustic pressure. Deduce the vector energy flux density at large distance and the average emitted acoustic power. Verify that this power is due to the pressure force exerted on the gas by the sphere. **b)** A membrane, of radius R and mass m, vibrates with an amplitude u_m and a frequency $\tilde{\nu}$. Calculate the emitted power if $\lambda << R$. Deduce that the membrane is subject to a resistance $f = -b\,\dot{u}$ due to the radiation, where $b = \pi R^2 Z_a$. What is the impedance of the membrane?

P5.16*. Consider wave equation [5.38] of the velocity potential Ψ in the case of a spherical symmetric wave. **a)** Show that $\partial_x\Psi = (x/r)\,\partial_r\Psi$; and similarly for $\partial_y\Psi$ and $\partial_z\Psi$ deduce that $\Delta\Psi = \partial^2_{rr}\Psi + (2/r)\partial_r\Psi$. **b)** Setting $g(r, t) = r\Psi(r, t)$, show that $g(r, t)$ verifies the wave equation $\partial^2_{tt}g - v_s^2\,\partial^2_{rr}g = 0$. Deduce that the velocity potential may be written in the form $\Psi = (1/r)[F(t - r/v_s) + G(t + r/v_s)]$ where F and G are arbitrary functions. **c)** Consider the case $G = 0$ and $F(t) = A\cos(\omega t)$. Write the expression of the displacement $\mathbf{u}(\mathbf{r}, t)$ and the acoustic pressure $p_a(\mathbf{r}, t)$. **d)** Calculate the energy flux density and the average power, which is emitted by the source.

P5.17. A sound plane wave of angular frequency ω propagates in a fluid of bulk modulus B and mass density m_v. Let u_m be the amplitude of the displacement. **a)** Write down the expressions of the displacement wave, velocity wave and pressure wave and express the amplitudes p_m of the acoustic pressure and u'_m of the velocity in terms of u_m. **b)** Consider a slice of the medium of thickness Δl and area $\mathcal{S}$ perpendicular to the direction of propagation $\mathbf{e}$. Calculate the kinetic energy in this slice. Calculate the work of the pressure forces on the two faces of the slice between t and $t + dt$. Verify that the energy, which leaves this slice, is ($\nabla.\mathbf{S}\ \mathcal{S}\ \Delta l\ dt$) where $\mathbf{S}$ is the vector energy flux density. If $\mathbf{S}$ is a function of $(t - \mathbf{r}.\mathbf{e}/v_s)$, show that this energy may also be written in the form $(1/v_s)(\partial/\partial t)(\mathbf{e}.\mathbf{S})\ \mathcal{S}\ \Delta l\ dt$. **c)** Assume that the wave has an amplitude of 20 μm, calculate the amplitude of the acoustic pressure and the intensity of the wave in the case of air (v_s= 332 m/s) and in the case of water (v_s= 1,500 m/s) if the frequency is 500 Hz and 50 kHz respectively.

P5.18. The sound intensity level at 50 m from an aircraft is 130 dB. **a)** Neglecting the absorption of sound by air and assuming that the wave is spherical, for which distance is the intensity level 100 dB? **b)** Now assume that the air absorbs sounds at a rate of 7 dB/km, what is the sound intensity level at 5 km?

P5.19. The audibility threshold and the pain threshold correspond respectively to the amplitudes of acoustic pressure of 2×10^{-5} Pa and 30 Pa for a sound of frequency 500 Hz. Calculate the corresponding displacement amplitude, sound intensity and intensity level in dB. The air mass density is 1.29 kg/m^3 and the speed of sound in air is 332 m/s.

P5.20. The sound intensity level for a conversation is 60 dB at 1 m. Calculate the corresponding amplitudes p_m and u_m for a sound of 1 kHz. Assuming that the sound is emitted isotropically in all directions, calculate the power required to sustain this conversation. Compare it to the power required for a person of 75 kg to go upstairs, one-step of 20 cm per second.

P5.21.** While propagating, a sound wave is attenuated, essentially because of the *viscosity* which is a friction force setup by shear stresses within the flowing fluid. It may be shown that, in a viscous medium, the wave equations for the displacement **u** and for the acoustic pressure $p_a = -B(\nabla.\mathbf{u})$ are

$$\ddot{\mathbf{u}} - (B/m_v)\,\nabla(\nabla.\mathbf{u}) - (1/m_v)\,\left[(\xi + \tfrac{1}{3}\eta)\nabla(\nabla.\dot{\mathbf{u}}) + \eta\,\Delta\dot{\mathbf{u}}\right] = 0$$
$$\ddot{p}_a - (B/m_v)\,\Delta p_a - (1/m_v)\,(\xi + 4\eta/3)\,\Delta\dot{p}_a = 0$$

η is the *first coefficient of viscosity* of the fluid ($\eta > 0$) and ξ is the *second coefficient if viscosity* ($\xi > 0$). Verify that the plane wave $p_a(\mathbf{r}, t) = p_m\, e^{i(\omega t - q\mathbf{e}.\mathbf{r})}$ is a solution of the equation of propagation if ω and q are related by the equation $q^2[B + i\omega(\xi + 4\eta/3)] = m_v\omega^2$. Deduce that q is complex, of the form $q = k - i\delta$, where to the first order in ξ and η, $k = \omega\sqrt{m_v/B}$ and $\delta = \tfrac{1}{2}(\omega^2/B)\sqrt{m_v/B}\,(\xi + 4\eta/3)$. What is the phase velocity?

Surface waves

P5.22*. Using the dispersion relation $\omega^2 = gk\,\mathrm{th}(kh)$ of the surface waves in a deep canal, calculate the group velocity. Show that, if $\lambda \ll h$, the group velocity is $v_{(g)} = \sqrt{g\lambda/8\pi}$ and, if $\lambda \gg h$, the group velocity is $v_{(g)} = \sqrt{gh}$.

P5.23*. a) Plot the amplitude of the velocity potential [5.97] as a function of the depth z in the case $h = 10$ *m* and $\lambda = 31.4$ m. What is the limit of this expression if the liquid is very deep ($kh \gg 1$, as in the case of the sea, for instance)? Consider a surface wave, which has an amplitude of 4 m and a wavelength of 10 m. At which depth is the amplitude of the vertical motion of water reduced to 20 cm? **b)** Calculate the velocity at a point of coordinates (x, z). Integrating with respect to time, determine the displacement from equilibrium and deduce that the molecules describe ellipses [5.99]. Verify that these ellipses become smaller and smaller with depth and that the motion of water becomes negligible at a depth equal to several times the wavelength.

P5.24*. Calculate the phase velocity of surface waves in a channel of depth 5 m for 1 cm, 1 m, 10 m and 100 m wavelengths. Compare expressions [5.96], [5.100], [5.102] and [5.103].

P5.25*. a) What is the frequency of the capillary wave at the water surface if its wavelength is λ_m? **b)** Calculate the effect of the surface tension for wavelengths of 1 mm, 1 cm and 1 m.

Chapter 6

Electromagnetic Waves

A remarkable property of Maxwell's equations (that was not possible to foresee before the formulation of the electromagnetic theory) is that they have solutions representing electromagnetic waves propagating at the speed of light in matter and in vacuum. The existence of these waves was experimentally confirmed, in 1890 by Hertz, who succeeded in producing them and verifying that they have the same propagation, interference, diffraction and polarization properties as light waves. Thus, Maxwell theory has enabled us to understand the nature of light as electromagnetic waves with very short wavelengths. Actually, we can produce and detect electromagnetic waves of frequencies varying between 10^{-2} Hz and 10^{32} Hz, that is, of wavelengths varying between 10^{10} m and 10^{-24} m. They play a very important part in nearly all branches of physics and have a considerable impact on economic, social, political and intellectual life. In this chapter, we recall the principal results of the electromagnetic theory, namely Maxwell's equations. We discuss their wave solution and some applications of electromagnetic waves.

6.1. Principal results of the electromagnetic theory

The basic concepts of the electromagnetic theory are the *electric field* **E** and the *magnetic induction field* **B** (also called *magnetic field* for short) defined by their action on a charge q of velocity **v**

$$\mathbf{F} = q(\mathbf{E} + \mathbf{v} \times \mathbf{B}). \qquad [6.1]$$

In the presence of dielectrics, we also have to introduce the *electric displacement* (also called *electric induction*) $\mathbf{D} = \varepsilon_o \mathbf{E} + \mathbf{P}$ where $\varepsilon_o = 8.8541878 \times 10^{-12}\ \mathrm{A^2 s^4/kg.m^3}$

is the *electric permittivity* of vacuum and **P** is the *density of polarization* (or simply *polarization*), i.e. the electric dipole moment of the dielectric per unit volume. In the presence of magnetic materials, we also have to introduce the *magnetic field strength* (also called *magnetic intensity*) $\mathbf{H} = \mathbf{B}/\mu_o - \mathbf{M}$ where $\mu_o = \mu_o = 4\pi \times 10^{-7}$ kg.m/A^{2}s^2 is the *magnetic permeability* of vacuum and **M** is the *intensity of magnetization* (or simply *magnetization*), i.e. the magnetic dipole moment per unit volume. If the medium is linear and isotropic, we have the proportionality relations

$$\mathbf{D} = \varepsilon\mathbf{E}, \qquad \mathbf{B} = \mu\mathbf{H}. \tag{6.2}$$

ε is the electric permittivity of the medium and μ is its *magnetic permeability*.

If an electric field is set up in a conductor, the *conduction electrons* of charge density $q_v(\mathbf{r}, t)$ move with a drift velocity $\mathbf{v}(\mathbf{r}, t)$ and produce an electric current density

$$\mathbf{j}(\mathbf{r}, t) = q_v(\mathbf{r}, t)\, \mathbf{v}(\mathbf{r}, t). \tag{6.3}$$

A conductor is *Ohmic* if **E** is proportional to **j** (*Ohm's law*)

$$\mathbf{E} = \rho\, \mathbf{j}, \qquad \text{i.e.} \qquad \mathbf{j} = \sigma\mathbf{E}. \tag{6.4}$$

ρ is the *resistivity* of the conductor and $\sigma = 1/\rho$ is its *conductivity*. The conservation of the electric charge may be written in the form

$$\nabla . \mathbf{j} + \partial_t q_v = 0. \tag{6.5}$$

In the case of time-dependent phenomena, the electromagnetic fields, q_v and **j** are related by the four Maxwell equations, which are the fundamental laws of electromagnetism:

$$\nabla . \mathbf{D} = q_v \qquad \text{(Gauss's law),} \tag{6.6}$$
$$\nabla \times \mathbf{H} - \partial_t\mathbf{D} = \mathbf{j} \qquad \text{(Ampere's law),} \tag{6.7}$$
$$\nabla \times \mathbf{E} + \partial_t\mathbf{B} = 0 \qquad \text{(Faraday's law),} \tag{6.8}$$
$$\nabla . \mathbf{B} = 0 \qquad \text{(absence of magnetic charges).} \tag{6.9}$$

These are linear equations. Equations [6.8] and [6.9] are identically verified if we introduce the electric potential V and the vector potential **A**, such that

$$\mathbf{E} = -\nabla V - \partial_t\mathbf{A}, \qquad \mathbf{B} = \nabla \times \mathbf{A}. \tag{6.10}$$

The interface $\mathcal{S}$ of two media (1) and (2) may carry a charge density q_s and a surface current density $\mathbf{j}_s$ if one of the media, at least, is a conductor (Figure 6.1). $\mathbf{j}_s$

is related to the intensity dI across an infinitesimal segment $d\mathbf{L}$ by the relation $dI = d\mathbf{L}.\mathbf{j}_s$. It may be shown that, on the interface, the fields verify the boundary conditions:

$$\mathbf{E}_{//1} = \mathbf{E}_{//2}, \qquad (\mathbf{H}_1 - \mathbf{H}_2) \times \mathbf{n}_{12} = \mathbf{j}_s, \qquad [6.11]$$

$$\mathbf{D}_2.\mathbf{n}_{12} - \mathbf{D}_1.\mathbf{n}_{12} = q_s, \qquad \mathbf{B}_1.\mathbf{n}_{12} = \mathbf{B}_2.\mathbf{n}_1 \qquad [6.12]$$

where $\mathbf{n}_{12}$ is the unit vector that is normal to the interface and oriented from medium 1 toward medium 2.

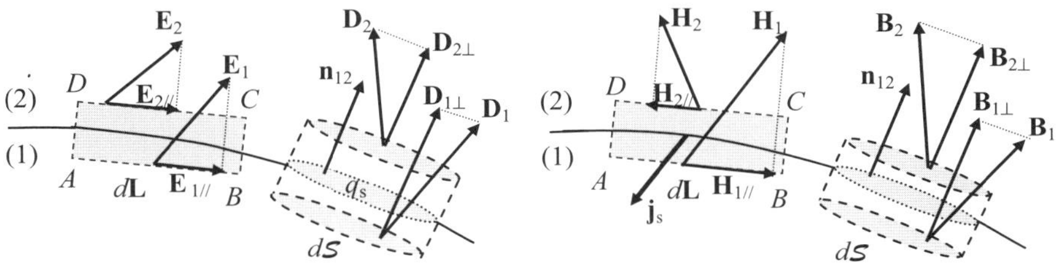

(a) (b)

Figure 6.1. *Electromagnetic fields at the interface of two media*

Now consider the electromagnetic energy. We may easily verify that the electric energy ½CV^2 of a capacitor and the magnetic energy ½LI^2 in a solenoid are respectively equivalent to an *electric energy density* ½**E.D** and a *magnetic energy density* ½**B.H**. These expressions hold in any space region where there are fields in any type of media. Thus the total electromagnetic energy density is

$$U_{(\mathrm{EM}),\,\mathrm{v}} = \tfrac{1}{2}\,\mathbf{E}.\mathbf{D} + \tfrac{1}{2}\,\mathbf{B}.\mathbf{H} \qquad [6.13]$$

This energy is supplied by the generators of the fields. If the fields are static, the energy densities ½**E.D** and ½**B.H** remain constant. Once the fields are setup, the generators no longer need to supply energy if the energy loss is negligible. On the other hand, if the fields vary in time, the energy densities vary. In this case, the energy is necessarily transmitted from one region to another or dissipated. To analyze the propagation of energy, we define the Poynting vector

$$\mathbf{S} = \mathbf{E} \times \mathbf{H}. \qquad [6.14]$$

It may be shown that **S** is the *electromagnetic energy flux density* (see problem P6.5).

6.2. The propagation equations of the fields in vacuum and infinite dielectrics

Maxwell's equations couple the electric fields **E** and **D** and the magnetic fields **B** and **H**. This coupling is responsible for the propagation of electromagnetic waves. In this chapter we consider the propagation in a medium (or vacuum) which is infinite, homogeneous, linear and isotropic. Let ε be its electric permittivity and μ its magnetic permeability. As $\mathbf{D} = \varepsilon\mathbf{E}$ and $\mathbf{H} = \mathbf{B}/\mu$, the electromagnetic fields may be specified solely by the fields **E** and **B**. In the case of a dielectric with no charge or current densities, the fields obey homogeneous Maxwell's equations:

$$\nabla . \mathbf{E} = 0, \quad [6.15]$$

$$\nabla \times \mathbf{B} - \mu\varepsilon\, \partial_t \mathbf{E} = 0, \quad [6.16]$$

$$\nabla \times \mathbf{E} + \partial_t \mathbf{B} = 0, \quad [6.17]$$

$$\nabla . \mathbf{B} = 0. \quad [6.18]$$

To establish the propagation equations, we evaluate the curl of both sides of [6.16] and use the vector identity $\nabla\times(\nabla\times\mathbf{B}) = \nabla(\nabla.\mathbf{B}) - \Delta\mathbf{B}$. Taking into account [6.18], we obtain $\nabla\times(\nabla\times\mathbf{B}) = -\Delta\mathbf{B}$. Similarly, we evaluate the curl of both sides of [6.17] and use [6.15]. We obtain the d'Alembert wave equation for **E** and **B**

$$\partial^2_{tt}\mathbf{E} - (1/\mu\varepsilon)\, \Delta\mathbf{E} = 0, \qquad \partial^2_{tt}\mathbf{B} - (1/\mu\varepsilon)\, \Delta\mathbf{B} = 0. \quad [6.19]$$

These equations indicate that the fields **E** and **B** propagate with the speed

$$v = 1/\sqrt{\mu\varepsilon}\,. \quad [6.20]$$

In the special case of vacuum, the speed of propagation is $1/\sqrt{\mu_o\varepsilon_o}$, which is numerically equal to the speed of light c in vacuum. This result cannot be fortuitous; it has incited Maxwell to consider light as electromagnetic waves. Designating by $\varepsilon_r = \varepsilon/\varepsilon_o$ the *relative electric permittivity* and $\mu_r = \mu/\mu_o$ the *relative magnetic permeability*, the speed of the electromagnetic wave in the medium may be written as $v = c/\sqrt{\mu_r\,\varepsilon_r}$. This corresponds to a *refraction index* $n = \sqrt{\mu_r\,\varepsilon_r}$. Most of the non-conducting media are non-magnetic; thus, $\mu_r \approx 1$ and $n \approx \sqrt{\varepsilon_r}$.

It is evident that the equations of [6.19] do not contain all the information of the four Maxwell equations. Indeed, the equations of [6.19] express no relation between **E** and **B**. Thus, to [6.19] we must add one of either equation [6.15] or [6.18] and one of either equation [6.16] or [6.17]. On the other hand, the wave equations [6.19] are partial differential equations, with an infinite number of solutions. To determine the solution corresponding to a given physical situation, we need the initial conditions and eventually the boundary conditions on the surfaces of the medium.

6.3. Electromagnetic simple harmonic plane waves

Consider a simple harmonic plane wave (Figure 6.2) propagating in the direction of $\mathbf{e} = \mathbf{k}/k$, and which is specified by the complex electric field

$$\underline{\mathbf{E}} = \underline{\mathbf{E}}_{\rm m}\, e^{\mathrm{i}(\omega t - \mathbf{k}.\mathbf{r})}. \qquad [6.21]$$

Substituting this expression in [6.15] and [6.17], we obtain the two equations

$$\mathbf{k}.\,\underline{\mathbf{E}}_{\rm m} = 0, \qquad \underline{\mathbf{B}} = (1/\omega)\,(\mathbf{k} \times \underline{\mathbf{E}}_{\rm m})\; e^{\mathrm{i}(\omega t - \mathbf{k}.\mathbf{r})}. \qquad [6.22]$$

Substituting the expressions for **E** and **B** in [6.16] and [6.18]; we find that they are verified if the phase velocity $v_{(\mathrm{p})} = \omega/k$ is equal to v given by [6.20]; thus, there is no dispersion in an infinite dielectric if the constants μ and ε do not depend on ω.

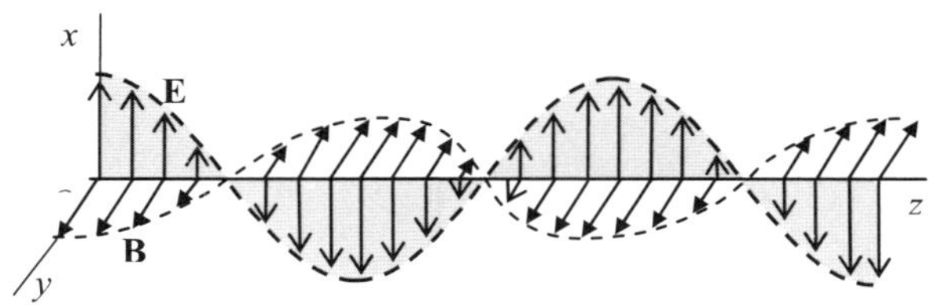

Figure 6.2. *Electromagnetic plane wave which is linearly polarized in the direction Ox and which propagates in the direction Oz*

Similarly, if the magnetic field of the wave is given by

$$\underline{\mathbf{B}} = \underline{\mathbf{B}}_{\rm m}\, e^{\mathrm{i}(\omega t - \mathbf{k}.\mathbf{r})}, \qquad [6.23]$$

Maxwell's equations [6.18] and [6.16] give respectively

$$\mathbf{k}.\,\underline{\mathbf{B}}_{\rm m} = 0, \qquad \underline{\mathbf{E}} = -\,(1/\mu\varepsilon\omega)(\mathbf{k} \times \underline{\mathbf{B}}_{\rm m})\; e^{\mathrm{i}(\omega t - \mathbf{k}.\mathbf{r})}. \qquad [6.24]$$

We may consider the real part of the fields

$$\mathbf{E} = \mathbf{E}_{\rm m} \cos(\omega t - \mathbf{k}.\mathbf{r} + \phi), \qquad \mathbf{B} = \mathbf{B}_{\rm m} \cos(\omega t - \mathbf{k}.\mathbf{r} + \phi),$$
$$\mathbf{B}_{\rm m} = (1/v)\,(\mathbf{e} \times \mathbf{E}_{\rm m}), \qquad \mathbf{E}_{\rm m} = -v\,(\mathbf{e} \times \mathbf{B}_{\rm m}). \qquad [6.25]$$

Some properties of the fields of an electromagnetic plane wave in dielectrics and vacuum may be deduced from relations [6.22] and [6.24] or [6.25]:

– **E** and **B** are orthogonal to each other and to the direction of propagation **e**. The trihedron (**E**, **B**, **e**) is right-handed.

– **E** and **B** are in phase.

– The amplitudes of **E** and **B** are related by the equation

$$E_{\mathrm{m}} = v\, B_{\mathrm{m}}. \qquad [6.26]$$

It should be noted that these properties concern the waves which propagate in vacuum or dielectrics, which are isotropic and infinite. If the medium is bounded by surfaces, the wave must obey some boundary conditions on these surfaces. For instance, if the wave propagates in a metallic tube or in a fibre, it undergoes successive reflections on the boundary; we say that the wave is *guided*. The fields of this wave are not always transverse and the phase velocity $v_{(\mathrm{p})} = \omega/k$ and $v_{(\mathrm{g})} = \partial\omega/\partial k$ are different from $v = 1/\sqrt{\mu\varepsilon}$. They depend on the dimensions of the medium and on the wave frequency, even if the infinite medium is non-dispersive (see sections 9.5, 9.6. and 9.9).

6.4. Energy density and the Poynting vector

In the case of an electromagnetic progressive plane wave whose real fields are [6.25], the electric energy density $U_{(\mathrm{E}),\mathrm{v}}$ and the magnetic energy density $U_{(\mathrm{M}),\mathrm{v}}$ are

$$U_{(\mathrm{E}),\mathrm{v}} = \tfrac{1}{2}\,\varepsilon E_{\mathrm{m}}^2 \cos^2(\omega t - \mathbf{k}.\mathbf{r} + \phi), \qquad U_{(\mathrm{M}),\mathrm{v}} = \tfrac{1}{2}(B_{\mathrm{m}}^2/\mu)\cos^2(\omega t - \mathbf{k}.\mathbf{r} + \phi). \quad [6.27]$$

As $E_{\mathrm{m}} = vB_{\mathrm{m}}$, we find that these densities are equal. Thus, the total electromagnetic energy density is

$$U_{(\mathrm{EM}),\mathrm{v}} = U_{(\mathrm{E}),\mathrm{v}} + U_{(\mathrm{M}),\mathrm{v}} = \varepsilon\, E_{\mathrm{m}}^2 \cos^2(\omega t - \mathbf{k}.\mathbf{r} + \phi) = (B_{\mathrm{m}}^2/\mu)\cos^2(\omega t - \mathbf{k}.\mathbf{r} + \phi). \quad [6.28]$$

The average value of $U_{(\mathrm{EM}),\mathrm{v}}$ in time over a period or in space over a wavelength is

$$< U_{(\mathrm{EM}),\mathrm{v}} > = \tfrac{1}{2}\,\varepsilon\, E_{\mathrm{m}}^2 = \tfrac{1}{2}\,(B_{\mathrm{m}}^2/\mu). \qquad [6.29]$$

The Poynting vector of this plane wave is

$$\mathbf{S} = \mathbf{E} \times \mathbf{H} = (\mathbf{E} \times \mathbf{B})/\mu = (E_{\mathrm{m}}^2/\mu v)\cos^2(\omega t - \mathbf{k}.\mathbf{r} + \phi)\, \mathbf{e} = v\, U_{(\mathrm{EM}),\mathrm{v}}\, \mathbf{e}, \qquad [6.30]$$

where $\mathbf{e} = \mathbf{k}/k$ is the direction of propagation of the wave. Thus, in a homogeneous and isotropic medium, $\mathbf{S}$ points in the direction of $\mathbf{e}$. This means that the energy propagates in this direction. The *intensity of the wave* is the average value of $|\mathbf{S}|$, i.e.

$$\mathcal{I} = < S > = E_m^2/\mu v = v < U_{(EM),v} >. \qquad [6.31]$$

For instance, if a wave of amplitude $E_m = 100$ V.m^{-1} propagates in glass of index $n = 1.50$, the average electromagnetic energy density is $<U_{(EM),v}> = \frac{1}{2}\varepsilon E_m^2 = \frac{1}{2}n^2\varepsilon_o E_m^2 \approx 10^{-7}$ J.m^{-3} and the intensity is $\mathcal{I} = (c/n) < U_{(EM),v} > \approx 20$ W.m^{-2}. It should be specified that relations [6.29], [6.30] and [6.31] are only valid in the case of a progressive wave.

We define the impedance per unit area of the medium as

$$Z = E/H = \mu E/B = \mu v = \sqrt{\mu/\varepsilon}\,. \qquad [6.32]$$

This enables us to write

$$\mathbf{S} = (E_m^2/Z)\cos^2(\omega t - \mathbf{k}.\mathbf{r} + \phi)\,\mathbf{e} \qquad \text{and} \qquad \mathcal{I} = < S > = (E_m^2/2Z). \qquad [6.33]$$

It should be noted that the energy density and the Poynting vector $\mathbf{S}(t)$ must be evaluated directly by using the real fields. The intensity of the wave may be evaluated either as the average value of $|\mathbf{S}(t)|$ in time by using the real fields or using the complex representation [6.21] for the electric field and the expression

$$\mathcal{I} = \frac{1}{2\mu v}\mathbf{E}.\mathbf{E}^* = \frac{1}{2Z}\mathbf{E}.\mathbf{E}^*. \qquad [6.34]$$

6.5. Polarization of electromagnetic waves

Most of the actions of electromagnetic waves (such as the photoelectric and photochemical effects) are exerted by the electric field. The reason is that B is of the order of E/c, where c is the speed of light and the magnitude of the velocity $\mathbf{v}$ of electric charges of the medium are often very small compared to c. Thus the electric force $q\mathbf{E}$ is much more important than the magnetic force $q\mathbf{v}_q \times \mathbf{B}$. For this reason, the electromagnetic wave is often specified by the electric field $\mathbf{E}$. Maxwell's equation [6.17] determines $\partial_t\mathbf{B}$ and an integration with respect to time gives $\mathbf{B}$. On the other hand, in the case of a nonlinear or a non-isotropic medium or a finite medium, the analysis is much more involved.

In an infinite, homogeneous, linear and isotropic medium, $\mathbf{E}$ and $\mathbf{B}$ are perpendicular to the direction of propagation $\mathbf{e}$. We say that the electromagnetic wave is *transverse*. The direction of **E** defines the *polarization* of the wave. Figure 6.2 represents an electromagnetic plane wave which propagates in the direction of *Oz*, such that **E** points everywhere in the direction of *Ox* and **B** points in the direction of *Oy*. We say that the wave is polarized linearly in the direction of *Ox*. In general, a wave of angular frequency ω, propagating in the direction of $\mathbf{e} = \mathbf{k}/k$ may be written as the superposition of two linearly polarized waves in two transverse directions $\mathbf{e}_1$ and $\mathbf{e}_2$ (i.e. perpendicular to **e**) and such that the trihedron ($\mathbf{e}_1$, $\mathbf{e}_2$, **e**) is right-handed:

$$\mathbf{E} = \mathbf{E}_1 + \mathbf{E}_2 \equiv E_1\, e^{i(\omega t - \mathbf{k.r})}\, \mathbf{e}_1 + E_2\, e^{i(\omega t - \mathbf{k.r} + \phi)}\, \mathbf{e}_2. \qquad [6.35]$$

Depending on the value of the phase shift ϕ of $\mathbf{E}_2$ with respect to $\mathbf{E}_1$, the resulting wave is polarized linearly if $\phi = 0$ or π, right-circularly if $\phi = \pi/2$, left-circularly if $\phi = -\pi/2$ and $E_1 = E_2$, and elliptically if ϕ is arbitrary or $E_1 \neq E_2$ (see section 4.4). The intensity of the resulting wave [6.35] is the sum of the intensities $\mathcal{I}_1$ and $\mathcal{I}_2$ of its components $\mathbf{E}_1$ and $\mathbf{E}_2$ if they are orthogonal. Indeed, field **B** may be written in the form $\mathbf{B} = (1/v)\ \mathbf{e}_z \times \mathbf{E} = (1/v)(E_1\, \mathbf{e}_y - E_2\, \mathbf{e}_x) \equiv \mathbf{B}_1 + \mathbf{B}_2$. The energy density and the Poynting vector of the resulting wave are respectively

$$U_{(EM),v} = \varepsilon \mathbf{E}^2 = \varepsilon\, (E_1^{\,2} + E_2^{\,2}) = U_{(EM),v\,1} + U_{(EM),v\,2},$$
$$\mathbf{S} = (1/\mu)\, \mathbf{E} \times \mathbf{B} = (1/\mu)\, (\mathbf{E}_1 \times \mathbf{B}_1 + \mathbf{E}_2 \times \mathbf{B}_2) = \mathbf{S}_1 + \mathbf{S}_2.$$

As both $\mathbf{S}_1$ and $\mathbf{S}_2$ point in the direction of *Oz*, the total intensity is

$$\mathcal{I} = |\,\mathbf{S}_1 + \mathbf{S}_2\,| = S_1 + S_2 = \mathcal{I}_1 + \mathcal{I}_2.$$

We define the *degree of polarization* in the direction of $\mathbf{e}_1$ as the ratio

$$\mathcal{P}_1 = \frac{\mathcal{I}_1 - \mathcal{I}_2}{\mathcal{I}_1 + \mathcal{I}_2}. \qquad [6.36]$$

If the wave is totally polarized in the direction of $\mathbf{e}_1$, we have $E_2 = 0$ and $\mathcal{P}_1 = 1$. On the contrary, if the wave is totally polarized in the direction of $\mathbf{e}_2$ perpendicular to $\mathbf{e}_1$, we find $E_1 = 0$ and $\mathcal{P}_1 = -1$. If the wave is circularly polarized, $\underline{E}_1$ and $\underline{E}_2$ have the same amplitude but a phase shift of $\pm\, \pi/2$, thus $\mathcal{P}_1 = 0$. A wave is *non-polarized* if it is a superposition of waves $\underline{\mathbf{E}}_1$ and $\underline{\mathbf{E}}_2$ whose amplitudes or phase shift ϕ vary in time, or if it is a superposition of right-circularly polarized and left-circularly polarized waves with a variable phase shift ϕ. The degree of polarization of non-polarized waves in any direction $\mathbf{e}_1$ is equal to zero.

6.6. Momentum density and angular momentum density, radiation pressure*

Consider a plane electromagnetic wave, which propagates in the direction of Oz and which is linearly polarized in the direction of Ox. It is incident normally on a metallic plate, which lies in the Oxy plane (Figure 6.3a). We assume that the plate absorbs all the radiation, so that all the physical quantities of the wave that the plate intercepts (energy, momentum, angular momentum, etc.) are transferred to the plate via the interaction of the wave with the conduction electrons of the plate. If the plate is thin, a conduction electron may only have significant motion in the plane Oxy. The electron is subject to a Lorentz force $\mathbf{F} = -e(\mathbf{E} + \mathbf{v}_e \times \mathbf{B})$ where $\mathbf{v}_e$ is the velocity of the electron. $\mathbf{E}$ points in the direction of Ox and $\mathbf{B}$ points in the direction of Oy. In the interval of time dt, the energy that the electron receives is

$$dW = \mathbf{F}.\mathbf{v}_e\, dt = -\, e(\mathbf{E}.\mathbf{v}_e)\, dt = -\, e\, v_e E \cos\alpha\, dt, \qquad [6.37]$$

where α is the angle of $\mathbf{v}_e$ with Ox. This energy is dissipated as Joule heat in the conductor. During the same time, the momentum of the electron varies by $d\mathbf{P} = \mathbf{F}\, dt = -\, e\, dt\, (\mathbf{E} + \mathbf{v}_e \times \mathbf{B}) = -\, e\, dt\, E\, \mathbf{e}_x - e v_e B \cos\alpha\, dt\, \mathbf{e}_z$. The first term lies in the plane of the plate and it is proportional to $\mathbf{E}$. Thus, it is a simple harmonic function of time and its average value over a period equals zero and we may write

$$d\mathbf{P}_z = -\, e\, (\mathbf{v}_e \times \mathbf{B})\, dt = -\, e\, v_e B \cos\alpha\, dt\, \mathbf{e}_z. \qquad [6.38]$$

This momentum is transferred to the plate and it is normal to the plate. Expressions [6.37] and [6.38] show that $dW/dP = E/B = v$. We deduce that the electromagnetic wave carries a momentum which points in the direction of propagation. The energy density $U_{(EM),v}$ and the momentum density $P_{(EM),v}$ of the wave are related by the equation (similar to that of sound waves)

$$\mathbf{P}_{(EM),v} = (U_{(EM),v}/v)\, \mathbf{e} = \mathbf{S}/v^2. \qquad [6.39]$$

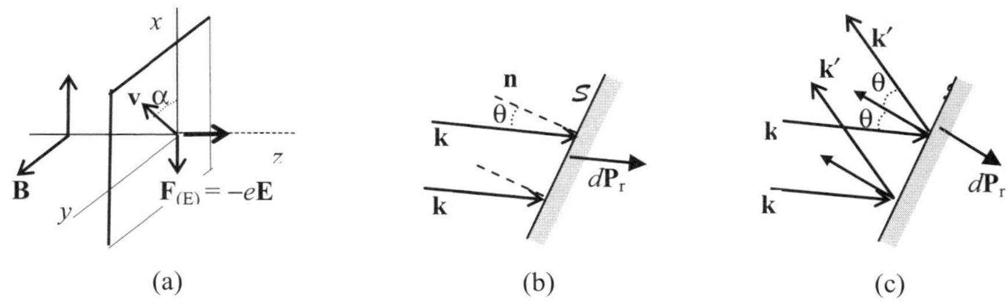

Figure 6.3. *Radiation pressure of an electromagnetic wave: a) in the case of normal incidence on a totally absorbing surface, b) in the case of oblique incidence on a totally absorbing surface and c) in the case of oblique incidence on a totally reflecting surface*

If a wave is incident at an angle θ on a body, an element of area $\mathcal{S}$ of this body receives during dt the energy $dU_{(EM)} = U_{(EM),v}\ \mathcal{S}v\cos\theta\ dt$ and the momentum $dP_r = P_{(EM),v}\ \mathcal{S}\ v\cos\theta\ dt$, which are contained in the cylinder of base $\mathcal{S}$ and length $v\ dt$ in the direction of propagation. If the surface is totally absorbing, this momentum transfer is equivalent to a force per unit area $F_{r,s} = vP_{(EM),v}\cos\theta$ pointing in the direction of propagation **e**. In particular, if the wave is incident normally ($\theta = 0$), the absorbing surface is subject to a *radiation pressure*

$$p_r = vP_{(EM),v} = U_{(EM),v} \qquad \text{(totally absorbing surface).} \qquad [6.40]$$

On the other hand, if the surface is totally reflecting (Figure 6.3c), there is no energy transfer to the body but a momentum transfer $2\mathcal{S}v\ P_{(EM),v}\cos\theta$ per unit time in the normal direction to the surface. This momentum transfer is equivalent to a radiation pressure

$$p_r = 2vP_{(EM),v}\cos\theta = 2U_{(EM),v}\cos\theta \qquad \text{(totally reflecting surface).} \qquad [6.41]$$

Although the radiation pressure is usually very small, it may have observable effects if it acts on very small particles (such as electrons) or for a very long interval of time. It may become important in the case of intense sources (such as lasers). In astrophysics, a very high temperature medium (at the core of stars, for instance) produces very intense radiation, and thus a radiation pressure which may counterbalance or even exceed the gravitational pressure.

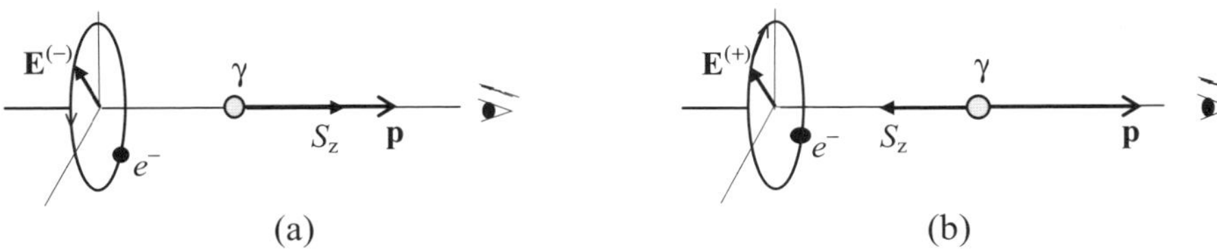

Figure 6.4. *a) A left-circularly polarized wave is equivalent to photons of helicity +1 (forward spin), b) a right-circularly polarized wave is equivalent to photons of helicity –1 (backward spin)*

In addition to the energy and momentum, a body may have an *orbital angular momentum* $\mathbf{L} = \mathbf{r} \times \mathbf{P}$ where **r** is the position of its center of mass and an *intrinsic angular momentum* **s** (associated with the rotation of the body about its centre of mass). It is known that the *total angular momentum* $\mathbf{J} = \Sigma_i(\mathbf{L}_i + \mathbf{s}_i)$ for all the bodies of an isolated system is conserved. In the case of a continuous system (a fluid for instance), we define a *density of orbital angular momentum* $\mathbf{L}_v = \mathbf{r} \times \mathbf{P}_v$ and a density of *intrinsic angular momentum* $\mathbf{s}_v$. We expect that the electromagnetic fields also have similar angular momentum densities. Let us again consider a wave, which is incident on a metallic plate (Figure 6.3a). Taking the position of the electron as the origin, the orbital angular momentum received by the electron equals zero but it may receive the intrinsic angular momentum of the wave. As **E** and **B** are orthogonal to Oz, the Lorentz force $-e(\mathbf{E} + \mathbf{v} \times \mathbf{B})$ acting on the electron, has a component $-e\mathbf{E}$ in the plane of the plate. If the wave is linearly polarized in a direction $\mathbf{e}_P$, the electron will move in this direction and there will be no variation of its angular momentum. Thus, a linearly polarized electromagnetic wave carries no intrinsic angular momentum. On the other hand, a circularly polarized wave may be written in the form (see Figure 6.4)

$$\mathbf{E}^{(\pm)} = E\cos(\omega t - kz)\,\mathbf{e}_x + E\cos(\omega t - kz \pm \pi/2)\,\mathbf{e}_y,$$
$$\mathbf{B}^{(\pm)} = -(E/c)\cos(\omega t - kz \pm \pi/2)\,\mathbf{e}_x + (E/c)\cos(\omega t - kz)\,\mathbf{e}_y. \qquad [6.42]$$

Neglecting the magnetic force, every conduction electron is subject to a force which lies in the plane of the plate

$$\mathbf{F}^{(\pm)}_{//} = -e\mathbf{E}^{(\pm)} = -eE\,[\cos(\omega t - kz)\,\mathbf{e}_x + \cos(\omega t - kz \pm \pi/2)\,\mathbf{e}_y]. \qquad [6.43]$$

The equation of motion of the electron in this plane is $m\ddot{\mathbf{r}}^{(\pm)} = \mathbf{F}^{(\pm)}_{//}$. Thus, we obtain by integration

$$\mathbf{r}^{(\pm)}_{//} = (eE/m\omega^2)\,[\cos(\omega t - kz)\,\mathbf{e}_x + \cos(\omega t - kz \pm \pi/2)\,\mathbf{e}_y]. \qquad [6.44]$$

This means that the electron describes a circle of radius $eE/m\omega^2$ in the direction of rotation of $\mathbf{E}^{(\pm)}$. The angular momentum, that the electron has received, is

$$\mathbf{s}_{(EM)}^{(+)} = m\mathbf{r}^{(\pm)} \times \dot{\mathbf{r}}^{(\pm)} = \mp\,(e^2E^2/m\omega^3)\,\mathbf{e}_z, \qquad [6.45]$$

while the received energy is $U_{(EM)} = e^2E^2/m\omega^2$. Thus, a circularly polarized wave carries an intrinsic angular momentum of density

$$\mathbf{s}_{(EM),v}^{(\pm)} = \mp\,\frac{U_{(EM),v}}{\omega}\,\mathbf{e}_z = \mp\,\frac{\mathbf{S}}{v\omega}. \qquad [6.46]$$

In the case of a left-circularly polarized wave $\mathbf{E}^{(-)}$, the received angular momentum points in the direction of propagation (*positive helicity*), while, in the case of a left-circularly polarized wave $\mathbf{E}^{(+)}$, the received angular momentum points in the opposite direction to the direction of propagation (*negative helicity*).

In addition to the intrinsic angular momentum, the electromagnetic wave also has an orbital angular momentum density $\mathbf{L}_{(EM),v} = \mathbf{r} \times \mathbf{P}_{(EM),v} = \mu\varepsilon\,\mathbf{r} \times \mathbf{S}$, which increases the effect. If a body intercepts an electromagnetic wave, the transferred angular momentum may cause its rotation. This was verified experimentally by R. Berth in 1936, by observing the rotation of a quartz plate receiving a circularly polarized wave. At present, this effect may be more easily observed using an intense laser beam.

6.7. Electromagnetic waves in plasmas*

A plasma is a gas that is totally or partially ionized. This ionization may be due to thermal agitation at very high temperatures (several thousand degrees). At low temperatures, it may be due to the bombardment by energetic particles. This is what happens effectively in the *ionosphere*, i.e. the upper Earth atmosphere at an altitude varying between 200 and 400 km: a small ionization (of the order of 1% of air molecules) is provoked mainly by the ultraviolet radiation from the Sun and by cosmic rays.

At equilibrium, the total charge density and the electric field in the plasma are equal to zero. If, for some reason, the electrons are displaced away from the positive ions, an electric field is produced with the effect of bringing the electrons back to their equilibrium positions. As they are accelerated, they gain some kinetic energy and continue their motion in the opposite direction and so on, producing the so-called *plasma oscillations*. Being very heavy, compared to electrons, the positive ions do not contribute to this oscillation and to the corresponding electric current and emission of secondary radiation, i.e. to the propagation of the wave in the plasma.

Consider an element of volume $\overline{\mathcal{V}}$ at equilibrium at the point **r**. If **u**(**r**, t) is the mean displacement, the volume $\overline{\mathcal{V}}$ becomes $\mathcal{V} = \overline{\mathcal{V}}\,(1+\nabla.\mathbf{u})$. If $\overline{N}_{\mathrm{V}}$ is the number of electrons per unit volume at equilibrium, the electronic charge in $\overline{\mathcal{V}}$ is $-e\overline{N}_{\mathrm{V}}\overline{\mathcal{V}}$ where $(-e)$ is the charge of the electron. It is counterbalanced at equilibrium by the positive ions charge $e\overline{N}_{\mathrm{V}}\overline{\mathcal{V}}$. After the displacement of the electrons, the total electrons charge being unchanged, the electronic charge density becomes $-e\overline{N}_{\mathrm{V}}\overline{\mathcal{V}}/\mathcal{V} \cong -e\overline{N}_{\mathrm{V}}(1-\nabla.\mathbf{u})$, while the charge density of the positive ions remains $e\overline{N}_{\mathrm{V}}$. Thus, the total charge density is $q_{\mathrm{V}} = e\overline{N}_{\mathrm{V}}(\nabla.\mathbf{u})$ and the current density due to the displacement of electrons is $\mathbf{j} = -\overline{N}_{\mathrm{V}}e\dot{\mathbf{u}}$. Thus we may write Maxwell's equations in the plasma as

$$\nabla.\mathbf{E} = (e/\varepsilon)\overline{N}_{\mathrm{V}}\nabla.\mathbf{u}, \qquad \nabla.\mathbf{B} = 0, \qquad [6.47]$$

$$\nabla\times\mathbf{B} = -\mu e\overline{N}_{\mathrm{V}}\dot{\mathbf{u}} + \mu\varepsilon\dot{\mathbf{E}}, \qquad [6.48]$$

$$\nabla\times\mathbf{E} + \dot{\mathbf{B}} = 0. \qquad [6.49]$$

These equations must be joined with the equation of motion for the electrons

$$m\,\ddot{\mathbf{u}} = -\,e\mathbf{E}, \qquad [6.50]$$

where we neglect the force exerted by the magnetic field.

We try a solution in the form of a plane wave

$$\underline{\mathbf{E}} = \underline{\mathbf{E}}_{\mathrm{m}}\, e^{\mathrm{i}(\omega t - \mathbf{k}.\mathbf{r})}, \qquad \underline{\mathbf{B}} = \underline{\mathbf{B}}_{\mathrm{m}}\, e^{\mathrm{i}(\omega t - \mathbf{k}.\mathbf{r})}, \qquad \underline{\mathbf{u}} = \underline{\mathbf{u}}_{\mathrm{m}}\, e^{\mathrm{i}(\omega t - \mathbf{k}.\mathbf{r})}. \qquad [6.51]$$

Maxwell's equations [6.47] to [6.49] and equation [6.50] are verified if

$$\underline{\mathbf{B}}_{\mathrm{m}} = (1/\omega)\, \mathbf{k} \times \underline{\mathbf{E}}_{\mathrm{m}}, \qquad \underline{\mathbf{E}}_{\mathrm{m}} = -\,(\omega/k^2)\, \mathbf{k} \times \underline{\mathbf{B}}_{\mathrm{m}},$$
$$\mathbf{k}.\underline{\mathbf{E}}_{\mathrm{m}} = 0,\ \mathbf{k}.\underline{\mathbf{B}}_{\mathrm{m}} = 0, \qquad \underline{\mathbf{u}}_{\mathrm{m}} = (e/m\omega^2)\, \underline{\mathbf{E}}_{\mathrm{m}}, \qquad [6.52]$$

where ω and k are related by the dispersion relation

$$\omega^2 = \omega_{\mathrm{p}}^{\,2} + v^2 k^2 \qquad \text{where} \qquad \omega_{\mathrm{p}}^{\,2} = (e^2/\varepsilon m)\, \overline{N}_{\mathrm{v}}, \qquad [6.53]$$

$v = 1/\sqrt{\mu\varepsilon}$ is the speed of propagation of the wave in the non-ionized gas and ω_{p} is called the *plasma angular frequency*.

It is also possible to write a wave equation for the fields in the plasma. For this, we differentiate [6.48] with respect to time and use [6.50]; we obtain

$$\nabla \times \dot{\mathbf{B}} = e^2\, (\mu/m)\, \overline{N}_{\mathrm{v}}\, \mathbf{E} + \mu\varepsilon\, \ddot{\mathbf{E}}\,. \qquad [6.54]$$

Taking the curl of [6.49] and eliminating field $\mathbf{B}$, using [6.54], we obtain the equation

$$\ddot{\mathbf{E}} - v^2\, \Delta\mathbf{E} + \omega_{\mathrm{p}}^{\,2}\, \mathbf{E} + v^2\, \nabla(\nabla.\mathbf{E}) = 0. \qquad [6.55]$$

To get rid of the last term in this equation, we evaluate the divergence of this equation and get $(\omega_{\mathrm{p}}^{\,2} + \partial^2/\partial t^2)(\nabla.\mathbf{E}) = 0$. If $\mathbf{E}$ is a simple harmonic function of time with an angular frequency $\omega \neq \omega_{\mathrm{p}}$, this equation can only be verified if $(\nabla.\mathbf{E}) = 0$ and equation [6.55] reduces to the wave equation of field $\mathbf{E}$

$$\ddot{\mathbf{E}} - v^2\, \Delta\mathbf{E} + \omega_{\mathrm{p}}^{\,2}\, \mathbf{E} = 0. \qquad [6.56]$$

The field **B** and the displacement **u** obey a similar equation, called the *Klein-Gordon* equation, with dispersion relation [6.53].

In the case of the ionosphere, $\overline{N}_v$ is of the order to 10^{12} to 10^{13} electrons per m^3, depending on the time of day, the seasons and the solar activity. This corresponds to a plasma frequency $\tilde{\nu}_p$ = 10 to 30 MHz. If the wave frequency $\tilde{\nu}$ is higher than $\tilde{\nu}_p$, the ionosphere is dispersive and the wave propagates without attenuation with a wave number $k = (1/v)\sqrt{\omega^2 - \omega_o^2}$, a phase velocity and a group velocity given by [4.90]. On the contrary, if $\tilde{\nu} < \tilde{\nu}_p$, the wave is attenuated with an attenuation coefficient $\delta = (1/v)\sqrt{\omega_o^2 - \omega^2}$. For instance, if $\tilde{\nu}_p = 20$ MHz ($\omega = 1.3 \times 10^8$ rad/s), a radiowave of wavelength 1 m ($\omega = 3 \times 10^8$ rad/s) is progressive with a phase velocity $v_{(p)} = 3.20 \times 10^8$ m/s and a group velocity $v_{(g)} = 2.72 \times 10^8$ m/s, while a radiowave of wavelength 200 m ($\omega = 9.4 \times 10^6$ rad.s^{-1}) is attenuated with an attenuation coefficient $\delta = 0.42$ m^{-1}. The penetration depth of this wave is $l = 1/\delta = 2.3$ m.

6.8. Electromagnetic waves in Ohmic conductors*

In an Ohmic conducting medium, the current density is related to the electric field by Ohm's law $\mathbf{j} = \sigma\mathbf{E}$. Thus Maxwell's equations may be written as:

$$\nabla.\mathbf{E} = q_v/\varepsilon, \qquad [6.57]$$

$$\nabla \times \mathbf{B} = \mu\sigma\mathbf{E} + \mu\varepsilon\dot{\mathbf{E}}, \qquad [6.58]$$

$$\nabla \times \mathbf{E} + \dot{\mathbf{B}} = 0, \qquad [6.59]$$

$$\nabla.\mathbf{B} = 0. \qquad [6.60]$$

The continuity equation of the electric charge $\nabla.\mathbf{j} + \partial_t q_v = 0$, the relation $\mathbf{j} = \sigma\mathbf{E}$ and equation [6.57] give the differential equation for q_v

$$\dot{q}_v + \beta q_v = 0, \qquad \text{where} \quad \beta = \sigma/\varepsilon, \qquad [6.61]$$

whose solution is $q_v = q_{vo}\, e^{-\beta t}$. Thus, an eventual charge density in the conductor will decrease exponentially over time. In a typical metallic conductor of conductivity $\sigma \approx 10^7\ \Omega^{-1}.\text{m}^{-1}$, the relaxation time is $\tau = 1/\beta \approx 10^{-18}$ s. Thus, the charge density in metallic conductors is practically equal to zero (this is the so-called *quasi-permanent regime*, which holds for phenomena with frequencies much lower than 10^{18} Hz).

Setting $q_v = 0$ in [6.57], calculating the curl of both sides of [6.58] and [6.59] and taking into account the other equations, we obtain the wave equations of the fields (called *equation of telegraphy*, with $\omega_o = 0$)

$$\ddot{\mathbf{E}} + \beta\,\dot{\mathbf{E}} - v^2\,\Delta\mathbf{E} = 0, \qquad \ddot{\mathbf{B}} + \beta\,\dot{\mathbf{B}} - v^2\,\Delta\mathbf{B} = 0. \qquad [6.62]$$

Assume that an electromagnetic plane wave of angular frequency ω is incident normally on the surface Oxy of an Ohmic conductor (Figure 6.5a). We shall see in Chapter 7 that a reflected plane wave and a transmitted plane wave inside the conductor are produced with the same angular frequency ω. The plane wave $\underline{\mathbf{E}} = \underline{\mathbf{E}}_{\text{m}}\, e^{\mathrm{i}(\omega t - q\mathbf{e}.\mathbf{r})}$ is a solution of propagation equation [6.62] for $\mathbf{E}$ if q verifies the equation $v^2q^2 = \omega^2 - \mathrm{i}\beta\omega$. This equation shows that q is always complex, of the form $q = k - \mathrm{i}\delta$. Thus,

$$\underline{\mathbf{E}} = \underline{\mathbf{E}}_{\text{m}}\, e^{-\delta x}\, e^{\mathrm{i}(\omega t - kx)}$$

$$k = \frac{\omega}{v\sqrt{2}}\left\{\sqrt{1+\beta^2/\omega^2} + 1\right\}^{1/2}, \qquad \delta = \frac{\omega}{v\sqrt{2}}\left\{\sqrt{1+\beta^2/\omega^2} - 1\right\}^{1/2}. \qquad [6.63]$$

k is the wave number. The phase velocity is $v_{(\text{p})} = \omega/k$. The wave is attenuated with an attenuation coefficient δ, i.e. a penetration depth (or *skin depth*) $l = 1/\delta$.

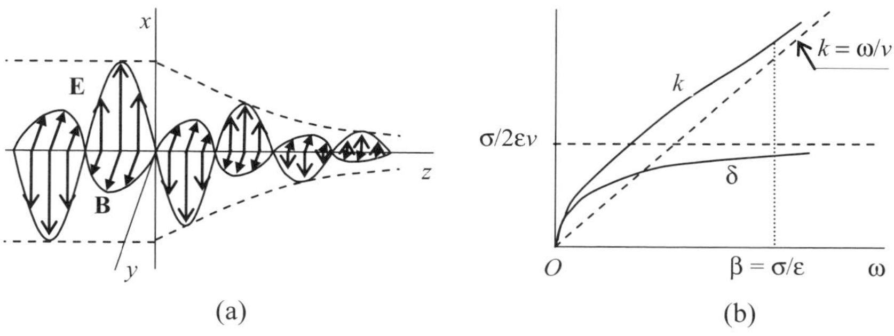

Figure 6.5. *a) Electromagnetic wave incident normally on a conductor. The reflected wave is not represented. b) k and δ in the conductor versus the angular frequency ω*

The variations of k and δ as functions of ω are illustrated in Figure 6.5b. $\beta = \sigma/\varepsilon$ is a characteristic angular frequency of the conductor. At low frequency ($\omega \ll \beta$), k^2 and δ^2 are approximately $\frac{1}{2}\mu\sigma\omega$, and at high frequency ($\omega \gg \beta$), k has the asymptotic expression ω/v and δ has the asymptotic value $\sigma/2\varepsilon v$. For $\omega = \beta$, the expressions [6.63] give $k \approx 1.1\ \sigma/\varepsilon v$ and $\delta \approx 0.45\ \sigma/\varepsilon v$. For instance, in the case of copper ($\sigma \cong 6{,}0 \times 10^7\ \Omega^{-1}.\text{m}^{-1}$), the characteristic angular frequency is $\beta = \sigma/\varepsilon \approx \sigma/\varepsilon_o \cong 6.8 \times 10^{18}$ rad.s^{-1}. Using the equations of [6.63], we obtain $\delta = 1.5 \times 10^4$ m^{-1} and $l = 65$ μm for a 1 MHz radiowave, $\delta = 3.8 \times 10^8$ m^{-1} and $l = 2.7$ nm for 500 nm visible light and $\delta = 1.2 \times 10^9$ m^{-1} and $l = 0.84$ nm for 50 nm ultraviolet. After travelling a distance $x = 10$ μm in copper, the intensity is reduced by a factor of $e^{-2x/l}$. Thus it is reduced by a factor of 1.4 in the case of a radiowave and it practically becomes zero in the case of visible and ultraviolet light.

The fields in the conductors have the following properties:

– $\underline{\mathbf{E}}$ and $\underline{\mathbf{B}}$ are orthogonal to the direction of propagation **e**, as indicated by the equations of [6.57] (with $q_v \approx 0$) and [6.60].

– If $\underline{\mathbf{E}} = \underline{\mathbf{E}}_m\, e^{-\delta\mathbf{e.r}}\, e^{i(\omega t - k\mathbf{e.r})}$, equation [6.59] gives the magnetic field

$$\underline{\mathbf{B}} = (k/\omega - i\delta/\omega)\,(\mathbf{e} \times \underline{\mathbf{E}}_m)\, e^{-\delta\mathbf{e.r}}\, e^{i(\omega t - k\mathbf{e.r})}$$
$$= (1/v)\,(1 + \beta^2/\omega^2)^{1/4}\,(\mathbf{e} \times \mathbf{E}_m)\, e^{-\delta\mathbf{e.r}}\, e^{i(\omega t - k\mathbf{e.r} - \alpha)}. \qquad [6.64]$$

The phase lag of the magnetic field over the electric field is

$$\alpha = \text{Arctan}(\delta/k) = \text{Arctan}[\sqrt{\omega^2/\beta^2 + 1} + \omega/\beta] \qquad (0 < \alpha < \pi/2). \qquad [6.65]$$

The amplitudes of **B** and **E** are related by the equation

$$\mathbf{B}_m = (1/v)\,(1 + \beta^2/\omega^2)^{1/4}\,(\mathbf{e} \times \mathbf{E}_m). \qquad [6.66]$$

Thus, **B** is orthogonal to **E** and to the direction of propagation **e**. The trihedron (**e**, **E**, **B**) is right-handed.

We define the *complex impedance per unit area* of the conducting medium as the ratio of the complex amplitudes of the fields $\underline{\mathbf{E}}$ and $\underline{\mathbf{H}}$

$$\underline{Z}_c = \frac{\underline{E}}{\underline{H}} = \mu\frac{\underline{E}}{\underline{B}} = \frac{\mu\omega(k + i\delta)}{k^2 + \delta^2} \equiv Z_c\, e^{i\alpha} \qquad \text{where } Z_c = |\underline{Z}_c| = \frac{\mu v}{(1 + \beta^2/\omega^2)^{1/4}}. \qquad [6.67]$$

The attenuation of the wave in conductors is due to the dissipation of energy as Joule heat. Indeed, let us consider a real wave whose fields are

$$\mathbf{E} = \mathbf{E}_m\, e^{-\delta\mathbf{e.r}} \cos(\omega t - k\mathbf{e.r}),$$
$$\mathbf{B} = (1/\omega)\sqrt{k^2 + \delta^2}\,(\mathbf{e} \times \mathbf{E}_m)\, e^{-\delta\mathbf{e.r}} \cos(\omega t - k\,\mathbf{e.r} - \alpha). \qquad [6.68]$$

The energy density in the conductor is $U_{(EM),v} = \frac{1}{2}\varepsilon E^2 + \frac{1}{2}B^2/\mu$, i.e.:

$$U_{(EM),v} = \tfrac{1}{2}\varepsilon E_m^2\, e^{-2\delta\mathbf{e}.\mathbf{r}}\, \{\cos^2(\omega t - k\mathbf{e}.\mathbf{r}) + (v/\omega)^2\, [k\cos(\omega t - k\,\mathbf{e}.\mathbf{r}) + \delta\sin(\omega t - k\,\mathbf{e}.\mathbf{r})]^2\}, \qquad [6.69]$$

where $v = 1/\sqrt{\mu\varepsilon}$. The variation of this energy density in the interval of time dt is

$$dU_{(EM),v} = \tfrac{1}{2} E_m^2\, e^{-2\delta\mathbf{e}.\mathbf{r}}\, \{\sigma\cos(2\omega t - 2k\,\mathbf{e}.\mathbf{r}) - 2\omega\varepsilon\sin(2\omega t - 2k\,\mathbf{e}.\mathbf{r})]\}\,dt. \quad [6.70]$$

It is easy to verify that the power entering a volume element is equal to the sum of the variation rate of the stored electromagnetic energy and the power $dP_J = \mathbf{E}.\mathbf{j}\, d\mathcal{V}$, which is dissipated as Joule heat in this volume.

The Poynting vector and the intensity of the wave may be written as

$$\mathbf{S} = \mathbf{e}\,(E_m^2/Z_c)\, e^{-2\delta\mathbf{e}.\mathbf{r}}\, \cos(\omega t - k\mathbf{e}.\mathbf{r})\cos(\omega t - k\mathbf{e}.\mathbf{r} - \alpha), \qquad [6.71]$$

$$\mathcal{I} = \langle S\rangle = (E_m^2/2Z_c)\, e^{-2\delta\mathbf{e}.\mathbf{r}} \cos\alpha = (k/2\mu\omega)\, E_m^2\, e^{-2\delta\mathbf{e}.\mathbf{r}} \equiv \mathcal{I}_o\, e^{-2\delta\mathbf{e}.\mathbf{r}}, \qquad [6.72]$$

where $\mathcal{I}_o$ is the intensity at the entry of the conductor ($\mathbf{r} = 0$). The intensity $\mathcal{I}$ decreases exponentially with the distance travelled.

A medium is considered to be a *good conductor* for a wave of angular frequency ω if the conduction current density $j = \sigma|E|$ is much larger than the displacement current $|\partial_t\mathbf{D}| = \varepsilon\omega|\mathbf{E}|$, thus, $\omega \ll \beta$. For instance, in the case of copper, the characteristic frequency $\beta \approx 7\times 10^{18}$ rad.s^{-1} corresponds to the frequency of X-rays. Thus, copper is a good conductor up to ultraviolet.

Neglecting the displacement current, Maxwell's equations in a good conductor may be written as:

$$\nabla.\mathbf{E} = 0, \qquad \nabla\times\mathbf{B} = \mu\sigma\mathbf{E}, \qquad \nabla\times\mathbf{E} + \dot{\mathbf{B}} = 0, \qquad \nabla.\mathbf{B} = 0 \qquad [6.73]$$

and the wave equations become the so-called *diffusion equations*

$$\Delta\mathbf{E} - \mu\sigma\dot{\mathbf{E}} = 0, \qquad \Delta\mathbf{B} - \mu\sigma\dot{\mathbf{B}} = 0. \qquad [6.74]$$

In the case of a simple harmonic wave of angular frequency ω, these equations are the limits of [6.62] if $\omega \ll \beta$. Thus, $k \approx \delta \approx \sqrt{\mu\sigma\omega/2}$ and the fields in the good conductor may be written as

$$\underline{\mathbf{E}} = \underline{\mathbf{E}}_m\, e^{-\delta\mathbf{e}.\mathbf{r}}\, e^{i(\omega t - \delta\mathbf{e}.\mathbf{r})}, \quad \underline{\mathbf{B}} = (1/v)\sqrt{\beta/\omega}\,(\mathbf{e}\times\mathbf{E}_m)\, e^{-\delta\mathbf{e}.\mathbf{r}}\, e^{i(\omega t - k\mathbf{e}.\mathbf{r} - \pi/4)}. \quad [6.75]$$

A *superconductor* has zero resistivity; thus, its conductivity σ, its characteristic angular frequency β and the attenuation coefficient δ are infinite. The penetration depth is equal to zero. Thus, there are no fields in the superconductor.

6.9. Quantization of electromagnetic radiation

According to the classical concepts, a wave is extended in space and time with a certain distribution of physical quantities (energy, momentum, etc.) and a continuous flux of these quantities. However, some effects of emission and absorption of electromagnetic waves, such as the *photoelectric effect* (discovered by Hertz in 1887 and interpreted by Einstein in 1905), may be understood only if the wave is made up of discrete energy quantities called *photons*. We say that electromagnetic radiation is *quantized*. The photons, constituting an electromagnetic wave of frequency $\tilde{\nu}$, have an energy

$$E_\gamma = h\tilde{\nu} \qquad \text{where } h = 6.626\,176 \times 10^{-34}\,\text{J.s} = 4.135\,669 \times 10^{-15}\,\text{eV.s.} \quad [6.76]$$

h is the *Planck constant*. The radiation effects are due to the interaction of the individual photons with matter. Thus, they depend on the energy of the photons, that is, its frequency (or its wavelength). The shorter the wavelength is, the higher the photons energy is and the more important the effects of the radiation become. In particular, the physiological effects of the photons on the cells of the retina depends on their frequencies and this produces the phenomena of colors.

If N_v is the number of photons per unit volume, the energy density is $U_{(EM),v} = N_v h\tilde{\nu}$. According to [6.39], the wave also has a momentum density $P_{(EM),v} = U_{(EM),v}/c = N_v h\tilde{\nu}/c$. Thus, the momentum of a single photon is

$$p_\gamma = h/\lambda. \quad [6.77]$$

According to the special theory of relativity, the energy and the momentum of a particle of mass m are related by the equation $E = \sqrt{p^2c^2 + m^2c^4}$. The relation $E_\gamma = cp_\gamma$ means that the mass of the photon equals zero. The expressions [6.76] and [6.77] of the energy and the momentum of photons were confirmed by the *Compton effect* in 1923.

On the other hand, each photon has an angular momentum equal to the sum of its orbital angular momentum $\mathbf{r} \times \mathbf{p}_\gamma$ and its intrinsic angular momentum (or *spin*). The angular momentum density of the wave is that of the N_v photons per unit volume. Thus, we are lead to admit that each photon has a *spin*

$$s^{(\pm)} = \mathbf{L}_{(EM)}{}^{(\pm)}/N_v = \mp \hbar\, \mathbf{e}_z \qquad \text{where} \quad \hbar = h/2\pi. \qquad [6.78]$$

The photon spin in any direction z may have one of the two values: $s = +\hbar$ if the wave is left-circularly polarized (positive helicity) and $s = -\hbar$ if the wave is right-circularly polarized (negative helicity).

6.10. Electromagnetic spectrum

An electromagnetic wave carries energy; thus, it cannot be emitted by charges at rest or in uniform motion: only accelerated charges (or variable currents) may emit them. If the current is a simple harmonic function of time, the emitted wave is a simple harmonic of equal frequency. The emitter of a wavelength λ is always a system whose dimensions are comparable to λ. For instance, waves with wavelengths of roughly more than 1 mm are emitted by macroscopic systems (electronic or electric).

While propagating, the electromagnetic waves experiences the various effects due to the propagation medium and eventual obstacles. The most important effects are the decrease of intensity with the increasing distance like $1/r^2$ (in the case of isotropic emission), reflection on surfaces with large dimensions compared to the wavelength, diffraction by apertures and obstacles with dimensions comparable to λ, and absorption by the medium, etc.

The most important applications of electromagnetic waves are in the domain of telecommunications. Information (sound, image, etc.) may be transmitted by *modulated waves*, that is, of amplitude, frequency or phase that vary according to the information. The modulation is realized using an *input transducer*, which transforms the information into an electric signal. Microphones and photoelectric cells are examples of input transducers. In the receiver, an *output transducer* transforms the electrical signal of the modulated wave into a non-electrical signal. A loudspeaker and a liquid crystal screen are examples of output transducers. The extracted signal reproduces the original signal with more or less deformations depending on the quality of the system.

Here is a classification of electromagnetic waves according to their frequency $\tilde{\nu}$ (or their wavelength $\lambda = c/\tilde{\nu}$) and some of their principal uses.

1) Waves with *industrial frequencies* (30 Hz $< \tilde{\nu} <$ 3 kHz and 10^5 m $< \lambda < 10^7$ m): they are emitted by *ac* generators and electrical setups. They are used in traditional telephony and for the transport of electrical energy.

2) *Radiowaves* (or Hertzian waves). They include:

– very low frequency (*VLF*) waves (3 kHz $< \tilde{\nu} <$ 30 kHz and 10^4 m $< \lambda < 10^5$ m),

– low frequency (*LF*) waves (30 kHz $< \tilde{\nu} <$ 300 kHz and 10^3 m $< \lambda < 10^4$ m),

– medium frequency (*MF*) waves (0.3 MHz $< \tilde{\nu} <$ 3 MHz and 10^2 m $< \lambda < 10^3$ m),

– high frequency waves (*HF*) (3 MHz $< \tilde{\nu} <$ 30 MHz and 10 m $< \lambda < 10^2$ m),

– very high frequency waves (*VHF*) (30 MHz $< \tilde{\nu} <$ 300 MHz and 1 m $< \lambda <$ 10 m).

Hertzian waves are emitted by macroscopic antennas, and are particularly used with amplitude modulation in telephony and *AM* radio emissions with a band width of the order of 10 kHz (which is the frequency domain of the usual audible sounds). Frequency modulation is used for *FM* radio emissions and for television emissions.

3) *Microwaves* or ultra high frequency (*UHF*) waves (0.3 GHz $< \tilde{\nu} <$ 300 GHz and 1 mm $< \lambda <$ 1 m) are emitted by some atoms and molecules (vibration spectrum and rotation spectrum). The ground state of the cesium atom consists of two very close energy levels separated by 4.14×10^{-15} eV. In the transition from the higher to the lower level, the atom emits a precise frequency of $9.192\,631\,77 \times 10^9$ Hz which is the basis of atomic clocks. The polar water molecule absorbs microwaves easily; hence, their use in microwave ovens ($\lambda = 12.2$ cm) to cook food which contains a large amount of water. They are also used by diathermy machines (to warm muscles and joints in order to relieve soreness), and telecommunications with aircrafts and satellites, radioastronomy and *radar* systems.

4) *Infrared* radiation ($3 \times 10^{11} < \tilde{\nu} < 3 \times 10^{14}$ Hz and 1 μm $< \lambda <$ 1 mm) is emitted and absorbed by molecules. Most hot materials emit heat in the form of infrared radiation. It has many applications in medicine (diagnostics, thermotherapy and thermography), industry and for infrared teledetection.

5) *Light* ($4 \times 10^{14} < \tilde{\nu} < 7.5 \times 10^{14}$ Hz and 400 nm $< \lambda <$ 750 nm) is a major agent for the transport of solar energy to Earth, photosynthesis and life itself. Its color is associated with its frequency, and it is emitted by atoms and molecules.

6) *Ultraviolet* radiation ($3 \times 10^{15} < \tilde{\nu} < 3 \times 10^{16}$ Hz and 10 nm $< \lambda <$ 100 nm) is emitted by atoms. It is used to study atoms and efficiently initiate some photochemical reactions (such as the combination of chloride and hydrogen, the breaking of the carbon-carbon bond, etc.). It depolymerizes nucleic acids, destroys proteins and inhibits the body's immune system, it has bad effects on the skin and it may cause skin cancer. It constitutes a component of solar radiation, the large part of which is absorbed by the atmosphere, particularly the ozone (O_3) layer.

7) *X-rays* (20 pm $< \lambda <$ 10 nm and $3 \times 10^{16} < \tilde{\nu} < 15 \times 10^{20}$ Hz) are emitted by atoms and charged particles (especially electrons) as they collide with other atoms.

They are used in the study of crystalline structures and in medicine (radioscopy and radiotherapy). *Gamma rays* ($\lambda < 20$ pm and $\tilde{\nu} > 15 \times 10^{20}$) are emitted by atomic nuclei and by decelerating charged particles. They are used in scientific research and in medicine (gammascopy and gammatherapy).

6.11. Emission of electromagnetic radiations

One of the basic ideas of quantum theory is that atoms and molecules can only be in discrete states of well-defined energy. Normally, they are in the lowest energy level called *ground state* and they emit no radiation. If they are in an excited energy state E_j, they may undergo a transition to a lower energy level E_i by emitting a photon of energy

$$E_j - E_i = h\tilde{\nu}_{ji}, \qquad [6.79]$$

where $\tilde{\nu}_{ji}$ is the frequency of the emitted radiation. For instance, the ground state energy of the hydrogen atom is $E_1 = -13.6$ eV and that of the first excited level is $E_2 = -3.40$ eV. Thus, the wavelength of the emitted radiation in the transition $2 \rightarrow 1$ is $\lambda_{21} = hc/(E_2 - E_1) = 0.122$ µm. The emission spectrum of atoms and molecules is discrete and it depends only on their energy levels. It is almost independent of the physical or chemical conditions. This is true if the atoms or the molecules are well separated (in the gaseous state, for instance). In the case of dense media (liquids or solids), the emission is rather collective and the spectrum is almost continuous.

In reality, the waves, which are emitted in atomic transitions, are not monochromatic. The reason is that the exact transition time cannot be predicted. Quantum theory asserts that only, if the atom is excited at $t = 0$, the probability of staying in the excited state decreases exponentially according to $e^{-t/\tau}$, where τ is the *mean life* of the excited state. τ may be considered to be the duration of the emission. It is usually of the order of 10^{-9} to 10^{-8} s. According to Fourier's theory (section 4.8), this wave packet may be considered as a superposition of monochromatic waves of frequencies $\tilde{\nu}$ and wave numbers $k = \omega/v$. A wave packet, which propagates in the direction of Oz, may be written as

$$\underline{E}(z, t) = \int_0^\infty d\tilde{\nu}\ \underline{\mathcal{E}}(\tilde{\nu})\ e^{i(\omega t - kz)} = (1/2\pi)\int_0^\infty d\omega\ \underline{\mathcal{E}}(\omega)\ e^{i(\omega t - kz)}. \qquad [6.80]$$

In the case of a monochromatic wave $a_o\, e^{i(\omega_o t - k_o z)}$, the spectral amplitude may be written as $\underline{\mathcal{E}}(\omega) = 2\pi a_o\ \delta(\omega - \omega_o) = a_o\ \delta(\tilde{\nu} - \tilde{\nu}_o)$ and the spectral intensity is $\mathcal{I}(\omega) = \mathcal{I}_o\ \delta(\omega - \omega_o)$, where δ is the Dirac delta-function (section 2.8). A three-dimensional wave may be written as

$$\underline{\mathbf{E}}(\mathbf{r}, t) = (2\pi)^{-3/2} \iiint d^3\mathbf{k}\ \hat{\mathbf{E}}(\mathbf{k})\ e^{i(\omega t - \mathbf{k}.\mathbf{r})} \qquad \text{where } \omega = ck\ . \qquad [6.81]$$

We may easily show the Parseval relation

$$\iiint d^3\mathbf{r}\ |\underline{\mathbf{E}}(\mathbf{r}, t)|^2 = \iiint d^3\mathbf{k}\ |\hat{\mathbf{E}}(\mathbf{k})|^2 \qquad [6.82]$$

which expresses the intensity of the packet as the integral of its spectral energy.

The spectral function $\hat{\mathbf{E}}(\mathbf{k})$ is important in a band of width Δk_x, Δk_y and Δk_z related to the extensions Δx, Δy and Δz of the wave packet in the direction of Ox, Oy and Oz by the uncertainty relations

$$\Delta x.\ \Delta k_x \approx 2\pi, \qquad \Delta y.\ \Delta k_y \approx 2\pi, \qquad \Delta z.\ \Delta k_z \approx 2\pi. \qquad [6.83]$$

The band width $\Delta\omega$ is related to the duration Δt of the packet by the uncertainty relation $\Delta\omega.\Delta t \approx 2\pi$. If we identify Δt with τ, we find a band width $\Delta\tilde{\nu}_o \approx 1/\tau \approx 10^8$ to 10^9 Hz. This *quantum width* is very small, compared to the frequency of the radiation ($\approx 10^{15}$ Hz for visible light). In this case, the radiation is said to be *quasi-monochromatic*. In fact, $\Delta\tilde{\nu}$ is wider than $\Delta\tilde{\nu}_o$ because of the thermal agitation of atoms (producing collisions and a widening of the band by the Doppler effect).

The radiation, which is emitted by a macroscopic body, is constituted of many wave packets, which are emitted by the atoms of the body. There are two types of sources:

– *Incoherent sources* in which the atoms emit independently, thus the emitted packets have no relation of polarization, amplitude, frequency and phase. In this case, the total wave is $\mathbf{E} = \Sigma_j\ \underline{\mathbf{E}}_j(\mathbf{r}, t)$ where $\underline{\mathbf{E}}_j$ are packets of the form of [6.81] whose random phases ϕ_j are related to the time of emission by the atoms. Thus, the total wave $\mathbf{E}(\mathbf{r}, t)$ is neither polarized, nor *coherent* (Figure 6.6a). The intensity of the total wave is the sum of intensities emitted by the atoms

$$\mathcal{I}(\mathbf{r}, t) = \Sigma_j\ \mathcal{I}_j(\mathbf{r}, t). \qquad [6.84]$$

This is the case of traditional (non-laser) sources, such as thermal sources (flames and incandescent lamp) and electric discharge sources (electric arcs, neon discharge tubes and spectral lamps).

– *Coherent sources*, such as lasers, in which the wave packets $\mathbf{E}_j(\mathbf{r}, t)$ that are emitted by the atoms have well-defined polarization and phase relations. The atoms in these sources cannot be considered to be completely independent, because the wave, which is emitted by one atom, acts on the others. The total emitted wave $\underline{\mathbf{E}} = \Sigma_j \underline{\mathbf{E}}_j(\mathbf{r}, t)$ is polarized and coherent (Figure 6.6b). Its intensity is given by

$$\mathcal{I}(\mathbf{r}, t) = (1/2\mu v)\,|\underline{\mathbf{E}}(\mathbf{r}, t)|^2 = (1/2\mu v)\,|\Sigma_j\,\underline{\mathbf{E}}_j(\mathbf{r}, t)|^2. \qquad [6.85]$$

It may be much larger (or smaller) than the sum of the intensities $\Sigma_j\ \mathcal{I}_j$ of the individual atoms.

(a) (b)

Figure 6.6. *a) Non-coherent wave, and b) coherent wave*

6.12. Spontaneous emission and stimulated emission

The radiation emission by an atom as it undergoes the transition from an energy level E_j to a lower level E_i may be spontaneous, if it occurs without the influence of an incident wave. This is the type of emission by the common natural or artificial sources. The atoms emit independently according to a probability law without any amplitude, polarization and phase correlation. The emitted radiation is *incoherent*. Conversely, if an atom is in an energy level E_i and it receives a photon of energy $E_j - E_i = h\tilde{\nu}_{ji}$, it may undergo an excitation to a higher level E_j followed by the transition back to the level E_i with the emission of a photon of frequency $\tilde{\nu}_{ji}$ but generally without any relation to the initial photon. Thus a good radiation emitter is also a good absorber of this radiation.

In 1917, Einstein discovered that, if the atom is in an excited level $E_j > E_i$ and it receives a photon of energy $E_j - E_i$, instead of getting excited to a higher level E_j, there is a large probability that the atom undergoes a transition to the lower level E_i with the emission of a photon, which is identical to the incident photon. This *stimulated emission* amplifies the incident wave and it is the basic principle of the operation of *lasers*. The emitted photon may excite another atom and so on, producing an avalanche of emissions. Noting that, if N identical waves $\mathbf{E}_j$ of the same intensity $\mathcal{I}_1$ superpose, the resulting intensity is, according to [6.85], $\mathcal{I}(\mathbf{r}, t) = (1/2\mu v)\,|N\underline{E}_1(\mathbf{r}, t)|^2 = N^2\,\mathcal{I}_1$; thus, it may be very large. This large energy is evidently supplied by the external system, which excites the atoms coherently. The word "laser" is an acronym for "*light amplification by stimulated emission of radiation*". The emitted wave is extremely phase-coherent and directional with a very narrow spectral band.

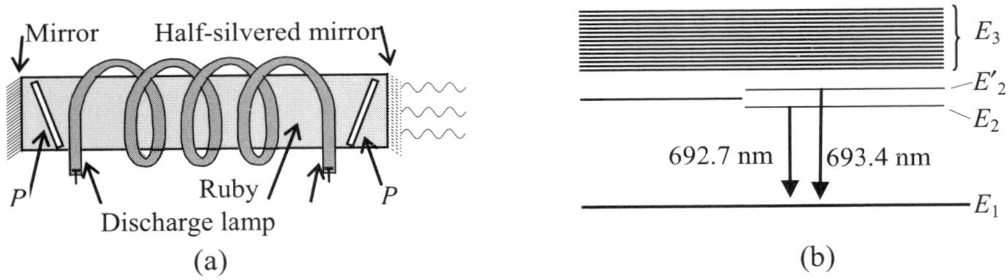

Figure 6.7. *a) Schematic representation of a ruby laser. The Brewster polarizers P polarize the beam. b) The energy levels*

In conventional light emitters, the stimulated emission is not significant because of the extremely small probability that an atom be in an excited state. The situation is different in the case of *population inversion*, i.e. many atoms in an excited state caused by a process of *optical pumping*. For this, the excited state must have a sufficiently long lifetime (it is then said to be a *metastable state* with a lifetime τ of the order of 10^{-3} s or more). To reinforce the process of light amplification, the lasing medium is placed between two plane or spherical mirrors (Figure 6.7a). The wave propagates back and forth with a period of $2en/c$, where n is the index of refraction of the medium and e is the distance between the mirrors. The system constitutes an *optical cavity* with standing waves of frequency $\tilde{\nu}_p = pc/2ne$ similar to the waves on a stretched string. Here p is an integer, which labels the mode of the standing wave. The waves, which are emitted in the oblique directions with respect to the mirrors axis, or whose frequencies are different from $\tilde{\nu}_p$, are taken out of the beam by reflection on the mirrors or are rapidly attenuated. Only the waves propagating perpendicularly to the mirrors remain in the beam after many reflections. The beam intensity increases by a resonance phenomenon. Two parallel plates placed at the ends of the cavity receive the wave at the Brewster incidence and reflect only the wave that is polarized perpendicularly to the incidence plane (see the section 7.5.4). One of the mirrors (the *output coupler*) is partially transparent; it transmits a laser beam which is highly directional (if the cavity is long enough), polarized and quasi-monochromatic. The high energy of the beam is provided by the optical pumping system, which maintains the population inversion.

A lasing medium with only two levels E_i and E_j is not possible. A typical three level laser (and the first laser to be realized) is the *ruby laser*. This is a crystal of aluminum oxide Al_2O_3 where some aluminum atoms are replaced with chromium atoms in a ratio of $1/10^4$ (Figure 6.7a). The energy levels are illustrated in Figure 6.7b. The intermediary level is in fact two very close levels E_2 and E'_2. The levels E_3 form a wide band covering the entire visible spectrum. The atoms of the lasing medium absorb any visible light and get excited to one level of the continuous band. Then they undergo a rapid transition to one of the levels E_2 and E'_2, which have a relatively long lifetime ($\approx$ 2 ms); this favors the inversion. Afterwards, they undergo transitions to the ground state E_1 with the emission of 692.7 nm and 693.4 nm radiations respectively. The difference between the energy absorbed in the excitation, and the emitted radiation energy is dissipated as heat in the medium and must be evacuated.

The output of a laser is a very narrow and intense beam. For instance, a beam of power 10 mW on a section of 1 mm^2 corresponds to an intensity of 10^4 W.m^{-2}, compared to about 150 W.m^{-2} for sunlight. Being very coherent, this beam can travel a very long distance without significantly spreading, or it can be focused to a very tiny spot of very high irradiance. The laser may be operating continuously or in a *pulsed mode*, emitting flashes of very short duration (of the order of 1 μs). In the so-called Q-switched laser, the emission is held-up while the population inversion is allowed to build up to the maximum level; then, a rapid lasing is allowed, producing a pulse of very short duration (of the order of a nonosecond) but with a very high power (up to 10^9 W).

Actually, we have many types of lasers (*gas lasers*, *solid-state lasers, semiconductor lasers, optical fiber lasers* etc.), practically for all electromagnetic radiations: visible light, infrared laser, ultraviolet laser, X-ray laser, and so on. Similar devices operating at microwave and radio frequencies are called *masers* rather than lasers. The power of lasers varies between several milliwatts for the common type to considerable values for pulsed lasers.

Due to the almost monochromatism of their radiation, coherence, polarization and high intensity, laser beams have many applications: in research they include spectroscopy, laser scattering and interferometry, holography, etc. Due to their high intensity, they may be used, for instance, to study the interaction of the electric field with nonlinear materials, produce very hot plasma and realize nuclear fusion. Laser beams enable us to realize high precision measurements of small angles, small lengths (for machine tools, for instance) and also large distance measurements (for instance the distance of Earth-Moon), to realize the precise alignment in mechanical constructions, etc. In industry, well-focused laser beams are used for cutting, welding, marking parts, micro piercing of metals and very hard materials, realization of semi-conductors, non-contact measurement of parts etc. Military applications for lasers includes guiding munitions, missile defense, as an alternative to radar, etc. In medicine, they are used in bloodless surgery, laser healing, kidney stone treatment, eye treatment etc. Their daily use includes compact disc players and engravers, laser printers, barcode scanners in supermarkets. In telecommunications, we are in the first stages of a new era of optical communications combining lasers and fiberoptics with an incredible increase in data-handling capacity for *TV* transmission, phone conversations, etc.

6.13. Summary

In vacuum, the electromagnetic phenomena depend on the electric field **E** and the magnetic induction field **B**, defined by their action on a particle of charge q and velocity **v** (*Lorentz force*) $\mathbf{F} = q\,\mathbf{E} + q\,\mathbf{v} \times \mathbf{B}$.

The charge density per unit volume q_v and the current density **j** verify the continuity equation (or conservation of charge equation) $\partial_t q_v + \nabla.\mathbf{j} = 0$.

In Ohmic conductors, the motion of the free charges (i.e. the conduction charges) produces a *current density* $\mathbf{j}(\mathbf{r}, t) = q_v(\mathbf{r}, t)\ \mathbf{v}(\mathbf{r}, t)$ given by Ohm's law $\mathbf{j} = \sigma\mathbf{E}$ or $\mathbf{E} = \rho\mathbf{j}$, where σ is the *conductivity* and $\rho = 1/\sigma$ is the *resistivity*.

A dielectric medium gets polarized in an external **E** field and a magnetic material gets magnetized in an external **B** field. We introduce the *electric displacement field* **D** and the *magnetic field strength* **H**. If the medium is linear and isotropic, $\mathbf{D} = \varepsilon\mathbf{E}$ and $\mathbf{B} = \mu\mathbf{H}$ where ε is the *electric permittivity* and μ is the *magnetic permeability*. In vacuum, $\varepsilon_o = 8.8541878 \times 10^{-12}$ C^2/N.m and $\mu_o \equiv 4\pi \times 10^{-7}$ Henry.m^{-1}.

If one of the fields **E** and **B** depends on time, the other field must exist. **D**, **H**, **E** and **B** are related to the charge and current densities by *Maxwell's equations*:

$\nabla.\mathbf{D} = q_v$	(Gauss' law),
$\nabla \times \mathbf{H} = \mathbf{j} + \dot{\mathbf{D}}$	(Ampere's law),
$\nabla \times \mathbf{E} + \dot{\mathbf{B}} = 0$	(Faraday's induction law),
$\nabla.\mathbf{B} = 0$	(conservation of the magnetic flux).

At the interface of two dielectrics (1) and (2), the fields verify the boundary conditions $\mathbf{D}_{\perp 1} = \mathbf{D}_{\perp 2}$, $\mathbf{B}_{\perp 1} = \mathbf{B}_{\perp 2}$, $\mathbf{E}_{//1} = \mathbf{E}_{//2}$ and $\mathbf{H}_{//1} = \mathbf{H}_{//2}$. At the interface $\mathcal{S}$ of a conductor (1) and a dielectric (or the vacuum) (2) the fields verify the conditions: $\mathbf{D}_{\perp 2} = \mathbf{D}_{\perp 1} + q_s$, $\mathbf{B}_{\perp 1} = \mathbf{B}_{\perp 2}$, $\mathbf{E}_{//1} = \mathbf{E}_{//2}$ and $\mathbf{e}.\mathbf{H}_1 = \mathbf{e}.\mathbf{H}_2 + \mathbf{j}_s(\mathbf{e} \times \mathbf{n}_{12})$.

The electromagnetic energy density is $U_{(EM),v} = ½\mathbf{E}.\mathbf{D} + ½\mathbf{B}.\mathbf{H}$. The vector energy flux density (*Poynting vector*) is $\mathbf{S} = \mathbf{E} \times \mathbf{H}$. The conservation of energy equation may be written in the form $\nabla.\mathbf{S} + \partial_t U_{(EM),v} + \mathbf{E}.\mathbf{j} = 0$. The dissipated power as Joule heat per unit volume is $\mathbf{E}.\mathbf{j} = \sigma\ \mathbf{E}^2 = \rho\ \mathbf{j}^2$.

In vacuum or in a linear and isotropic dielectric in the absence of charges and currents, **E** and **B** verify the wave equations

$$\ddot{\mathbf{B}} - v^2\, \Delta\mathbf{B} = 0, \qquad \ddot{\mathbf{E}} - v^2\, \Delta\mathbf{E} = 0.$$

The propagation speed of electromagnetic waves is $v = 1/\sqrt{\mu\varepsilon} = c/\sqrt{\mu_r\varepsilon_r} = c/n$.

The fields of a plane harmonic wave may be written as

$$\underline{\mathbf{E}} = \underline{\mathbf{E}}_m\, e^{i(\omega t - \mathbf{k}.\mathbf{r})}, \qquad \underline{\mathbf{B}} = \underline{\mathbf{B}}_m\, e^{i(\omega t - \mathbf{k}.\mathbf{r})},\ \omega/k = v.$$

They can be seen to verify the relations $\mathbf{k}.\underline{\mathbf{E}}_m = \mathbf{k}.\underline{\mathbf{B}}_m = 0$, $\underline{\mathbf{B}}_m = (\mathbf{k}\times\underline{\mathbf{E}}_m)/\omega$, $\underline{\mathbf{E}}_m = -(v^2/\omega)(\mathbf{k}\times\mathbf{B}_m)$ and $E_m = vB_m$. The fields **E** and **B** are in phase and orthogonal to each other and to the direction of propagation $\mathbf{e} = \mathbf{k}/k$. The trihedron (**e**, **E**, **B**) is right-handed.

In the case of a progressive electromagnetic wave, the electric energy density and the magnetic energy density are equal. Thus the total energy density is $U_{(EM),v} = \varepsilon E_m^2 \cos^2(\omega t - \mathbf{k}.\mathbf{r}) = (B_m^2/\mu)\cos^2(\omega t - \mathbf{k}.\mathbf{r})$ and its average value is $< U_{(EM),v} > = \frac{1}{2}\,\varepsilon\, E_m^2 = \frac{1}{2}\, B_m^2/\mu$.

The Poynting vector is $\mathbf{S} = (E_m^2/\mu v)\cos^2(\omega t - \mathbf{k}.\mathbf{r})\,\mathbf{e} = (E_m^2/Z)\cos^2(\omega t - \mathbf{k}.\mathbf{r})\,\mathbf{e}$. The intensity of the wave is $\mathcal{I} = < S > = E_m^2/2\mu v = E_m^2/2Z$ or $\mathcal{I} = \mathbf{E}.\mathbf{E}^*/2\mu v = \mathbf{E}.\mathbf{E}^*/2Z$. The impedance of the medium is $Z = \mu E/B = \mu v = \sqrt{\mu/\varepsilon}$. The radiation momentum density is $\mathbf{P}_{(EM),\,v} = \varepsilon\mu\mathbf{S} = \mathbf{S}/v^2 = \mathbf{e}U_{(EM),v}/v$. It exerts a radiation pressure. If it is circularly polarized, it has an intrinsic angular momentum density related to the energy density by the relationship $\mathbf{s}_{(EM),v}^{(\pm)} = \mp (1/\omega)\, U_{(EM),v}^{(\pm)}\,\mathbf{e}$.

The *polarization* of an electromagnetic wave is determined by the direction of **E**. Any wave propagating in the direction of **e**, may be considered as the superposition of two waves of amplitudes E_1 and E_2, that are polarized in the directions of $\mathbf{e}_1$ and $\mathbf{e}_2$, which are perpendicular to each other and to **e**. The *degree of polarization* in the direction of $\mathbf{e}_1$ is $\mathcal{P}_1 = (\mathcal{I}_1 - \mathcal{I}_2)/(\mathcal{I}_1 + \mathcal{I}_2) = (E_1^2 - E_2^2)/(E_1^2 + E_2^2)$.

Plasma is a partially or totally ionized gas. In plasma, the electromagnetic field displaces the charges (especially the electrons). The wave equation of **E** is:

$$\ddot{\mathbf{E}} - v^2\, \Delta\mathbf{E} + \omega_p^2\,\mathbf{E} = 0 \qquad \text{where} \qquad v = 1/\sqrt{\mu\varepsilon} \quad \text{and} \quad \omega_p^2 = N_v\, e^2/\varepsilon m.$$

ω_p is the oscillation angular frequency of the plasma and N_v is the number of electrons per unit volume. The wave may be written in the form

$$\underline{\mathbf{E}} = \underline{\mathbf{E}}_m\, e^{i(\omega t - \mathbf{k.r})}, \qquad \underline{\mathbf{B}} = \underline{\mathbf{B}}_m\, e^{i(\omega t - \mathbf{k.r})} \qquad \text{where } \mathbf{k.}\underline{\mathbf{E}}_m = 0 \text{ and } \mathbf{k.}\underline{\mathbf{B}}_m = 0.$$

The dispersion relation is $\omega^2 = \omega_o^{\,2} + v^2 k^2$. The plasma is dispersive if $\omega > \omega_p$, and reactive if $\omega < \omega_p$, with an attenuation coefficient $\delta = (1/v)\sqrt{\omega_p^{\,2} - \omega^2}$.

The wave equations in a conductor are

$$\ddot{\mathbf{E}} + \beta\,\dot{\mathbf{E}} - v^2\Delta\mathbf{E} = 0 \qquad \text{and} \qquad \ddot{\mathbf{B}} + \beta\,\dot{\mathbf{B}} - v^2\,\Delta\mathbf{B} = 0$$

where $\beta = \sigma/\varepsilon$ and $v = 1/\sqrt{\mu\varepsilon}$. The wave is attenuated in the form

$$\underline{\mathbf{E}} = \underline{\mathbf{E}}_m\, e^{-\delta \mathbf{e.r})}\; e^{i(\omega t - k\mathbf{e.r})}, \qquad \underline{\mathbf{B}} = (1/v)(1 + \beta^2/\omega^2)^{1/4}\,(\mathbf{e} \times \mathbf{E}_m)\, e^{-\delta\mathbf{e.r}}\; e^{i(\omega t - k\mathbf{e.r} - \alpha)}$$

$$k = \frac{\omega}{v\sqrt{2}}\,\{\sqrt{1 + \beta^2/\omega^2} + 1\}^{1/2}, \qquad \delta = \frac{\omega}{v\sqrt{2}}\,\{\sqrt{1 + \beta^2/\omega^2} - 1\}^{1/2} \text{ and } \alpha = \text{Arctan}(\delta/k).$$

The wave penetrates the conductor to a very small distance in the order of the penetration depth $l = 1/\delta$, also called the *skin depth*. The complex impedance of the conductor is $\underline{Z}_c = \mu\underline{E}/\underline{B} = \mu\omega/(k - i\delta)$. The intensity of the wave decreases according to the exponential law $\mathcal{I} = (\delta/2\mu\omega)\,E_m^{\,2}\,e^{-2\delta\mathbf{e.r}} \equiv \mathcal{I}_o\,e^{-2\delta\mathbf{e.r}}$.

The energy of an electromagnetic wave is *quantized*. The wave is made up of photons of energy $E_\gamma = h\tilde{\nu}$ and momentum $p_\gamma = h/\lambda$. In addition to the orbital angular momentum, the photon has an intrinsic angular momentum, which may have only two values $\pm\,\hbar$ (where $\hbar = h/2\pi$) depending on whether the photon is left or right polarized.

Electromagnetic waves are emitted by accelerated charges or variable currents. Hertzian waves and microwaves are emitted by macroscopic systems. Infrared, visible light, ultraviolet and X-rays are emitted by atoms and molecules, while γ-rays are emitted by atomic nuclei.

In their transition from a level E_j to a level E_i, the atoms and molecules emit or absorb a photon of energy $E_j - E_i = h\,\tilde{\nu}_{ji}$, where $\tilde{\nu}_{ji}$ is the frequency of the radiation. The emission is *spontaneous* if it is not under the influence on an external wave. In this case, the transition occurs according to a statistical quantum theory law, and the emitted electromagnetic photons have no phase or polarization correlation; they form a *non-coherent* radiation.

A *stimulated emission* occurs under the influence of an external primary wave. The emitted wave is in phase with the primary wave. Thus, the wave is amplified and it is coherent. A stimulated emission is possible, if the number of atoms in the excited state is significant (*population inversion* by energy pumping). There are many types of lasers, depending on the nature of the lasing medium. A laser beam is almost monochromatic, polarized, coherent and it has a large intensity. Lasers have many applications in research, holography, metrology, industry, medicine and telecommunications.

6.14. Problem solving suggestions

1) In an infinite, homogeneous and isotropic medium, the electromagnetic wave is *transverse* (**E** and **B** are perpendicular to the direction of propagation **e**). This property does not hold in the case of guided waves or in the case of an anisotropic medium. The *polarization* of the wave is determined by the direction of **E** (not **B**).

2) In the case of a progressive electromagnetic wave, the electric energy density and the magnetic energy density are equal. The vector *energy flux density* is the Poynting vector $\mathbf{S} = vU_{(EM)v}\ \mathbf{e}$. The radiation energy, which is received in unit time by an element of area $d\mathcal{S}$, is $\mathbf{S.n}\ d\mathcal{S}$, where $\mathbf{n}$ is the unit vector normal to $d\mathcal{S}$. The *intensity* $\mathcal{I}$ of the wave is the time-average of the magnitude S of the Poynting vector. In the case of a progressive wave, the intensity is $\mathcal{I} = |S| = v\langle U_{(EM),v}\rangle$ where $\langle U_{(EM),v}\rangle$ is the average electromagnetic energy density. The power that is dissipated as Joule heat per unit volume is $\mathbf{j.E} = \sigma\mathbf{E}^2 = \rho\mathbf{j}^2$. In the case of a plane wave in a dielectric, the important energy relations are $\langle U_{(EM)v}\rangle = \frac{1}{2}\varepsilon E_m^2 = \frac{1}{2}B_m^2/\mu$ and $\mathcal{I} = E_m^2/2\mu v = E_m^2/2Z$, where Z is the impedance of the medium. The mean values over a period may be directly evaluated using the complex expressions of the fields via the relations

$$\langle U_{(EM),v}\rangle = \tfrac{1}{2}\varepsilon\mathbf{E.E}^{*2} = \tfrac{1}{2}\mathbf{B.B}^*/\mu \quad \text{and} \quad \mathcal{I} = \tfrac{1}{2}\mathbf{E.E}^*/\mu v = \tfrac{1}{2}\,\mathbf{E.E}^*/Z \text{ with } Z = \sqrt{\mu/\varepsilon}$$

6.15. Conceptual questions

Q1. The resistivity analysis of common metallic conductors shows that the drift velocity of the free electrons carrying the electric current is of the order of a fraction of mm/s. However, if the current is switched on, a lamp lights up almost immediately. Explain why.

Q2. Using Maxwell's equations, establish the equation of propagation in a dielectric and the properties of plane wave solutions. Verify that $1/\sqrt{\mu\varepsilon}$ has the dimensions of a velocity. Water is non-magnetic and its refraction index is $n = 1.33$. What is its electric permittivity?

Q3. What is the physical interpretation of the Poynting vector and the intensity of the wave? Verify that the magnitude of the Poynting vector is related to the electromagnetic energy density by the relation $S = vU_{(EM),v}$ if the wave is progressive.

Q4. To transfer the electric energy from a battery to a resistor R, we use an electric line constituted by two parallel conducting wires of negligible resistance. Qualitatively show that the energy is transferred not within the wires themselves but in the dielectric that separates them. What happens if the wires have a small resistance per unit length? Make the same analysis for the case of a transmission line used to transmit electric signals and formed by a cylindrical wire of radius a surrounded by a cylindrical shell of internal radius b and external radius c.

Q5. Explain why a small sphere (whose diameter is of the order of a μm) may be suspended in an upward laser beam (optical levitation) and how the solar radiation pressure may explain the queue of comets. It was suggested that it might be possible to travel around in space not far from the Sun using a sail craft. Explain.

Q6. Explain what is meant by *plasma oscillation*. What is its effect on the propagation of waves? Explain why light and radiowaves of wavelengths less than about 10 m propagate easily in the ionosphere.

Q7. Is a conducting medium dispersive, reactive or resistive for electromagnetic waves? To which characteristics of the medium is the constant β of the telegraphy equation related? What happens to a radiowave that is incident on a conductor? At high frequency, a hollow conductor is no more resistant than a full conductor. A copper wire whose surface is oxidized is much more resistant than a non-oxidized wire or a silvered copper wire. Explain.

Q8. What is meant by quantization of electromagnetic waves? What are the fundamental relations of this quantization?

Q9. Does an isolated electric charge emit an electromagnetic wave? Do the household appliances emit electromagnetic waves? If yes, what is their wavelength? If you want to detect these waves, what should the size of the antenna be?

Q10. A car radio's reception is poor in a metal-encased tunnel or on a steel bridge. *AM* radio reception is inadequate behind mountains while *FM* and television have a range of several tens of km. It is possible to cook a piece of meat on a paper plate in a microwave oven but not in a metal saucepan. Explain these effects.

Q11. Explain what is meant by amplitude modulation and frequency modulation? What are the band widths for radio emissions and for TV emissions? How many TV channels can be transmitted in a fiber using the entire spectrum of visible light (400 nm $< \lambda <$ 700 nm), if the band width is 25 MHz for each emission? How many telephone calls can be transmitted simultaneously?

Q12. Atoms emit light in wave packets with durations of about 10^{-8} s. What is the length of these packets in the direction of propagation? How many wavelengths does a wave packet contain?

Q13. A laser emits light in impulses with durations of 2 μs, forming a beam of diameter 2 mm. Each impulse has an energy of 1 W. What is the length of an impulse in the direction of propagation? What is the average energy density during the impulse?

6.16. Problems

Electromagnetic waves in vacuum and in a dielectric

P6.1. Determine the polarization state of the following waves:

a) $\mathbf{E} = A\cos(\omega t - kz)\,\mathbf{e}_x + B\cos(\omega t - kz)\,\mathbf{e}_y$,
b) $\mathbf{E} = A\cos(\omega t - kz)\,\mathbf{e}_x + A\sin(\omega t - kz)\,\mathbf{e}_y$,
c) $\mathbf{E} = A\cos(\omega t - kx)\,\mathbf{e}_y + A\sin(-\omega t + kx)\,\mathbf{e}_z$,
d) $\mathbf{B} = A\cos(\omega t - kz + \pi/3)\,\mathbf{e}_x + A\sin(\omega t - kz)\,\mathbf{e}_y$,
e) $\mathbf{B} = -A\cos(\omega t - ky)\,\mathbf{e}_x + A\sin(\omega t - ky)\,\mathbf{e}_z$,
f) $\mathbf{B} = A\cos(\omega t + kx)\,\mathbf{e}_z - A\sin(\omega t + kx)\,\mathbf{e}_y$.

P6.2. Write down the expression for a plane wave propagating in a direction, which lies in the plane Oxz making an angle θ with Oz if: **a)** it is polarized in the plane Oxz, **b)** polarized perpendicularly to this plane, **c)** polarized right-circularly, and **d)** polarized left-circularly.

Energy density and the Poynting vector

P6.3. The electric field of a wave in a dielectric is $\mathbf{E} = \mathbf{E}_m\cos(\omega t - \mathbf{k}.\mathbf{r})$. **a)** Show that $\mathbf{k}.\mathbf{E}_m = 0$, which means that $\mathbf{E}$ is orthogonal to the direction of propagation. Assume that the magnetic field is $\mathbf{B} = \mathbf{B}_m\cos(\omega t - \mathbf{k}.\mathbf{r} + \alpha)$. Show that $\mathbf{B}_m = (\mathbf{k}\times\mathbf{E}_m)/\omega$ and $\alpha = 0$. Deduce that $\mathbf{B}$ is orthogonal to $\mathbf{k}$ and to $\mathbf{E}$. **b)** Calculate the electric energy density and the magnetic energy density of this wave.

P6.4*. A parallel plate capacitor is formed by two discs of area $\mathcal{S}$ separated by a distance d. The stored energy in this capacitor is $\frac{1}{2}CV^2$, where V is the voltage. **a)** Verify that this energy is equal to the integral of the volume energy density $\frac{1}{2}\mathbf{E}.\mathbf{D}$. **b)** Show that, while charging the capacitor, the energy flows across the cylindrical lateral surface rather than in the connection wires. Calculate the power crossing this surface.

P6.5*. **a)** Using Maxwell's equations in the case of a linear, homogeneous and isotropic medium, show that $-\nabla.\mathbf{S} = \partial_t U_{(EM),v} + \mathbf{j}.\mathbf{E}$. Integrate this equation over a volume $\mathcal{V}$ bounded by a surface $\mathcal{S}$ and transform the volume integral of $\nabla.\mathbf{S}$ to the flux of $\mathbf{S}$ across $\mathcal{S}$. Interpret the two terms of the right-hand side. Deduce that the Poynting vector $\mathbf{S}$ is the *radiation energy flux density*. **b)** Determine the Poynting vector and analyze the energy flow inside a metallic cylinder carrying a current of uniform density $\mathbf{j}$.

P6.6. An electromagnetic wave of angular frequency ω propagates in a non-magnetic medium of index n. Assume that the direction of propagation lies in the plane Oyz, making an angle θ with Oz and that the wave is polarized in this plane. **a)** Write down the expressions for the corresponding fields $\mathbf{E}$ and $\mathbf{B}$. **b)** Calculate the energy density and the Poynting vector of this wave. What is its intensity? What is the average power received by the unit area of the plane Oxy?

Radiation pressure

P6.7. A He-Ne laser beam has a wavelength $\lambda = 632.8$ nm, a section $\mathcal{S} = 0.10$ cm^2 and a power $P = 1$ W. Calculate the amplitude of the corresponding electric field and that of the magnetic field. What should be the power of an incandescence lamp of efficiency 10% to produce the same light intensity at 1 m from the lamp? What should the radius of a particle of mass density 10^3 kg.m^{-3} be in order for it to be suspended by this upward laser beam? Assume that the particle totally absorbs light.

P6.8*. The electric field of a wave is $\mathbf{E} = \mathbf{E}_m \cos(\omega t - kz)$ with $\mathbf{E}_m$ pointing in the direction of Ox. **a)** Calculate the corresponding energy density and the Poynting vector. **b)** Determine the energy of the photons of this wave, their momentum and their average number per unit volume. A totally absorbing surface $\mathcal{S}$ which lies in the plane Oxy is exposed to this wave. Determine the average number of photons that are absorbed in unit time per unit area of this plate, the momentum that it receives and the pressure force exerted on it. **c)** Assume that this plate is metallic with the conduction electrons subject to a friction force $-b\mathbf{v}$. Show that, at low frequency ($\omega << b/m_e$), the electrons reach a terminal speed $\mathbf{v} = -e\mathbf{E}/b$. Show that the electrons are under a magnetic force $F_{(M)} = (e^2/2b)\, EB$ pointing in the direction of Oz. **d)** This force is the origin of the radiation pressure. Show that it is equivalent to a momentum transfer $\Delta p = (e^2/b)\, EB$ per electron and per unit time. Show that the work exerted by the electric field on the moving electron is $\Delta W = (e^2/b)E^2$. Verify that, in the case of an electromagnetic wave in vacuum, $\Delta W = c\, \Delta p$. This is the relationship between the energy density of the wave and its momentum density or the energy of the momentum of the photons making the wave.

Electromagnetic waves in plasmas

P6.9. The frequency of plasma oscillation in the ionosphere is about $\tilde{\nu}_p \approx 30$ MHz.

a) Verify that the ionosphere is reactive for the wavelengths 10 m $< \lambda <$ 1 km used in *AM* radio emissions. This explains why such waves may be received at points that are not within direct sight of the emitter due to Earth's curvature. **b)** Verify that the ionosphere is dispersive for the waves that are used in *FM* and *TV* emissions.

P6.10*. If we neglect the friction force exerted on the conduction electrons in a metal and, consequently, the dissipation of energy as Joule heat, the metal may be treated as plasma. Assume that the number of conduction electrons is 1 electron per atom in silver. **a)** Calculate the number of conduction electrons per unit volume and the cut-off frequency $\tilde{\nu}_p$. Can visible light propagate in solid silver? **b)** What is the attenuation coefficient in silver for light of wavelength $\lambda = 580$ nm in vacuum? Verify that visible light penetrates silver only by a fraction of a μm. **c)** What should the thickness of silver on a glass plate be in order to have a mirror that reflects half the light intensity at 99%? At what wavelength does silver become transparent?

Electromagnetic waves in Ohmic conductors

P6.11. a) Using Maxwell's equations in conductors, show the propagation equations $\ddot{\mathbf{E}} + \beta\dot{\mathbf{E}} - v^2\Delta\mathbf{E} = 0$ and $\ddot{\mathbf{E}} + \beta\dot{\mathbf{B}} - v^2\Delta\mathbf{B} = 0$, where $v^2 = 1/\mu\varepsilon$ and $\beta = \sigma/\varepsilon$. Verify that these equations have solutions of the form

$$\underline{\mathbf{E}} = \underline{\mathbf{E}}_m\, e^{-\delta\mathbf{e.r}}\, e^{i(\omega t - k\mathbf{e.r})} \text{ and } \underline{\mathbf{B}} = (1/\omega)\sqrt{k^2+\delta^2}\,(\mathbf{e}\times\mathbf{E}_m)\, e^{-\delta\mathbf{e.r}}\, e^{i(\omega t - k\mathbf{e.r}-\alpha)},$$

where k, δ and α are given by equations [6.63] and [6.65]. Verify that $\mathbf{E}_m$ is orthogonal to the direction of propagation $\mathbf{e}$ and that $\mathbf{B}_m$ is orthogonal to $\mathbf{e}$ and $\mathbf{E}_m$. **b)** Study the variation of the attenuation constant as a function of the angular frequency of the wave. Show that, at low frequency ($\omega << \sigma/\varepsilon$), $\delta^2 \approx \frac{1}{2}\mu\sigma\omega$; while, at high frequency ($\omega >> \sigma/\varepsilon$), $\delta \approx \sigma/2\varepsilon v$. The resistivity of copper is 1.7×10^{-8} Ω.m. Calculate the characteristic angular frequency σ/ε. Calculate the attenuation depth for visible light of wavelength 0.6 μm.

P6.12*. a) Using the equation of propagation of electromagnetic waves in conductors, show that an electric field set up at $t = 0$ decreases exponentially in time according to the expression $\mathbf{E}(\mathbf{r}, t) = \mathbf{E}_m(\mathbf{r})\, e^{-t/\tau}$. Express the relaxation time τ as a function of the dielectric constant ε and the conductivity σ of the medium. **b)** Deduce that a charge density $q_v(0)$, which is set up in the conductor at $t = 0$, decreases with a relaxation time τ, to become distributed on the conductor surface. Show this result using the charge continuity equation. **c)** Give an estimate of τ in the case of copper ($\varepsilon_r \approx 1$ and $\sigma = 5.9 \times 10^5\ \Omega^{-1}\,\text{m}^{-1}$).

P6.13*. A simple model for interpreting Ohm's law assumes that the conduction electrons are subject to a friction force $-bv$. It can be shown that they reach a terminal speed (drift velocity) $\mathbf{v} = -e\mathbf{E}/b$. In this model, the conductivity is given by the expression $\sigma = Ne^2/b$, where N is the number of conduction electrons per unit volume. This result holds for a time-independent field. We expect that this result remains valid in the case of slowly time-varying fields but not for rapidly varying fields. In this problem we consider the case of a simple harmonic field of angular frequency ω. **a)** Neglecting the magnetic force, write the equation of motion of the electron and its solution. Deduce that Ohm's law is valid in this conductor but with a complex conductivity $\underline{\sigma} = Ne^2/(b + im\omega)$. **b)** Calculate the dissipated power per unit volume in this conductor. **c)** Give an estimate of b in the case of copper ($\rho = 1.7 \times 10^{-8}\ \Omega$.m), assuming one free electron per atom. **d)** A medium may be considered to be "metallic" if $b >> m\omega$, and a "plasma" if $b << m\omega$. For which frequencies can copper be treated as a plasma?

P6.14*. Let the plane Oxy be the surface of a conductor and Oz be pointing toward the interior of the conductor. Assume that this conductor is placed in an external electric field which is simple harmonic of angular frequency ω and oriented in the direction Ox. **a)** Determine the fields inside the conductor. **b)** In the case of common metallic conductors, it is assumed that the charge density is negligible. For which frequencies is this a good approximation? What then are the expressions of the fields and the current density?

Emission and reception of electromagnetic waves

P6.15. An electromagnetic plane wave of angular frequency ω and amplitude E_m for the electric field is polarized in the direction Ox and it propagates in the direction Oz. **a)** Write down the expressions of **E**, **B** and the Poynting vector **S**. **b)** An antenna of length L lies in the plane Oxz at an angle α with Ox. Calculate the induced voltage between its ends. How should this antenna be oriented in order to have the maximum voltage amplitude. What is the value of this maximum?

P6.16*. The emitter of a radar station has a power P and it consists of a small electric dipole located at the focus of a parabolic antenna whose aperture radius is R. As we shall see in section 8.7, the emitted wave of wavelength λ is diffracted in a cone of half-angle θ, such that $\sin\theta = 0.6\ \lambda/R$. **a)** Assuming that the wave is emitted isotropically in this cone, calculate the intensity of the wave $\mathcal{I}(r)$ at a distance r from the antenna. **b)** The wave encounters an obstacle of area $\mathcal{S}$ at a distance D. The reflected wave is assumed to be isotropic in all directions on one side of this surface, it is intercepted by the same parabolic antenna. Calculate the intercepted power P_i.

P6.17. The energy of the hydrogen atom in the ground state is $E_1 = -13.6057$ eV and its first excited level is at $E_2 = -3.4014$ eV. What is the wavelength emitted in the transition from E_2 to E_1? If an electron is in collision with an atom exciting it from the fundamental level to the E_2 level what should its kinetic energy be? What should the kinetic energy be to ionize the atom if it is initially in the ground state?

Chapter 7

Reflection and Refraction of Waves

If a wave propagating in a medium encounters the face of another medium, it may be partially or totally reflected and a part may be transmitted across the interface. If $\underline{u}$ is the complex incident wave, the amplitudes and the phase shifts of the reflected wave $\underline{u}'$ and those of the transmitted wave $\underline{u}''$ are determined by the boundary conditions on the interface, which depend on the nature of the wave and the media. We define the *reflection coefficient* $\underline{\mathcal{R}}$ and the *transmission coefficient* $\underline{\mathcal{T}}$ respectively as the ratios of the complex amplitudes of the reflected wave and the transmitted wave to that of the incident wave:

$$\underline{\mathcal{R}} = \underline{u}'_{\mathrm{m}}/\underline{u}_{\mathrm{m}}, \qquad \underline{\mathcal{T}} = \underline{u}''_{\mathrm{m}}/\underline{u}_{\mathrm{m}}. \qquad [7.1]$$

Our purpose in this chapter is to study the laws of reflection and transmission of waves. We start by studying the simple case of a one-dimensional elastic wave on two joined strings and sound in two pipes connected via a membrane; then, we study the case of three-dimensional sound waves. Afterwards, we consider the reflection and transmission of electromagnetic waves on the interface of two dielectrics: a dielectric with a plasma and a dielectric with a conductor.

7.1. Reflection of an elastic wave on two joined strings

Consider two strings of masses per unit length m_{l1} and m_{l2} respectively. They are joined at point O which we take as the origin of the x coordinate along the strings (Figure 7.1a). The strings are assumed to be under a tension F and infinite toward the negative x and positive x. The propagation speeds on these strings are $v_1 = \sqrt{F/m_{l1}}$ and $v_2 = \sqrt{F/m_{l2}}$ respectively. A simple harmonic wave

$$u = u_m \cos(\omega t - x/v_1), \quad [7.2]$$

propagating toward O on the first string, excites the origin O and produces a wave

$$u'' = u''_m \cos[\omega''(t - x/v_2) + \phi''] \quad [7.3]$$

on the second string. ω'' is the angular frequency on the second string and ϕ'' is a possible phase shift.

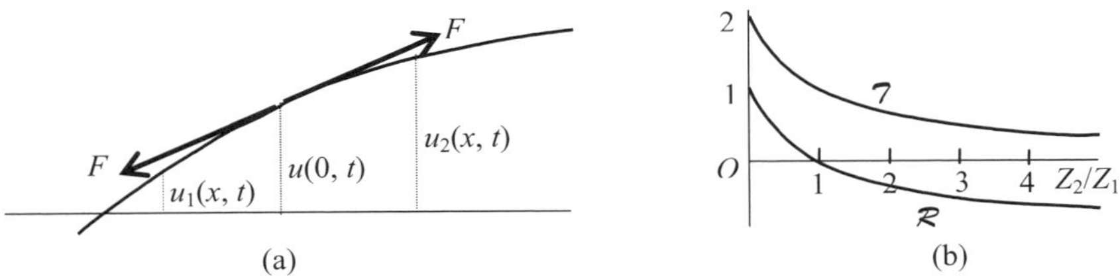

Figure 7.1. *a) Boundary conditions at the junction point of two strings, and b) variations of $\mathcal{R}$ and $\mathcal{T}$ versus Z_2/Z_1*

At the junction point O, we must have the *boundary conditions*

$$u_1(0, t) = u_2(0, t), \qquad \partial_x u_1 = \partial_x u_2 \,|_{x=0}, \quad [7.4]$$

where u_1 and u_2 are the displacements on the strings. The first equation expresses the evident equality of the displacements on both sides of O and the second expresses that the tangent to the strings at O is the same. The physical reason for the second condition is that, if the strings form a certain angle at O, a small element of the string centered on O will be subject to two forces of magnitude F and tangent to the strings. Their resultant will be different from zero. As this element may be as small as we want, it would have an infinite acceleration and this is impossible.

Using expressions [7.2] and [7.3], conditions [7.4] may be written as

$$u_m \cos(\omega t) = u''_m \cos(\omega'' t + \phi''), \qquad (\omega/v_1)\, u_m \sin(\omega t) = (\omega''/v_2)\, u''_m \sin(\omega'' t + \phi'').$$

These equations must be verified at any moment t. Thus, we must have $\omega = \omega''$, $\phi'' = 0$, $u_m = u''_m$ and $u_m/v_1 = u''_m/v_2$. The last two relations are only compatible if $v_1 = v_2$ and this is only possible if $m_{l1} = m_{l2}$. In this special case, the strings cannot be distinguished and the wave continues to propagate on the second string as it did on the first string. If $m_{l1} \neq m_{l2}$, there must be a reflected wave on the first string, propagating toward the negative x, of the form

$$u' = u'_{\mathrm{m}} \cos[\omega'(t + x/v_1) + \phi']. \qquad [7.5]$$

Thus, on string (1) we have a total wave $u_1 = u + u'$ and, on string (2) the wave is $u_2 = u''$. The boundary conditions [7.4] become

$$u(0, t) + u'(0, t) = u''(0, t), \qquad \partial_x u + \partial_x u' = \partial_x u''|_{x=0}. \qquad [7.6]$$

Using expressions [7.2], [7.3] and [7.5], we obtain

$$u_{\mathrm{m}} \cos(\omega t) + u'_{\mathrm{m}} \cos(\omega' t + \phi') = u''_{\mathrm{m}} \cos(\omega'' t + \phi'')$$
$$(\omega/v_1)\, u_{\mathrm{m}} \sin(\omega t) - (\omega'/v_1)\, u'_{\mathrm{m}} \sin(\omega' t + \phi') = (\omega''/v_2)\, u''_{\mathrm{m}} \sin(\omega'' t + \phi'').$$

These relations must be verified at any moment t. This gives the conditions

$$\omega = \omega' = \omega'', \qquad \phi' = \phi'' = 0, \qquad u_{\mathrm{m}} + u'_{\mathrm{m}} = u''_{\mathrm{m}}, \quad u_{\mathrm{m}}/v_1 - u'_{\mathrm{m}}/v_1 = u''_{\mathrm{m}}/v_2. \qquad [7.7]$$

If we take the solutions such that ϕ' or ϕ'' is equal to π, the amplitudes u'_{m} or u''_{m} would have the opposite sign and the result is the same. In this chapter, we take the amplitudes positive or negative. The reflection and transmission coefficients are

$$\mathcal{R} = \frac{u'_{\mathrm{m}}}{u_{\mathrm{m}}} = \frac{v_2 - v_1}{v_1 + v_2} = \frac{Z_1 - Z_2}{Z_1 + Z_2}, \qquad \mathcal{T} = \frac{u''_{\mathrm{m}}}{u_{\mathrm{m}}} = \frac{2v_2}{v_1 + v_2} = \frac{2Z_1}{Z_1 + Z_2}, \qquad [7.8]$$

where we used the strings impedances $Z_1 = \sqrt{Fm_{l1}} = F/v_1$ and $Z_2 = \sqrt{Fm_{l2}} = F/v_2$.

Figure 7.1b illustrates the variations of $\mathcal{R}$ and $\mathcal{T}$ versus Z_2/Z_1. We note that the amplitude of the transmitted wave may be larger than the amplitude of the incident wave (this is the case if $Z_1 > Z_2$) but the amplitude of the reflected wave is always less than the amplitude of the incident wave. This may be understood if we consider the energy flux $P = -F\dot{u}\,(\partial_x u)$. For the waves u, u' and u'' we obtain:

$$P = Z_1\omega^2 u_{\mathrm{m}}^2 \sin^2(\omega t - \omega x/v_1), \qquad P' = -Z_1\omega^2 u'^2_{\mathrm{m}} \sin^2(\omega t + \omega x/v_1),$$
$$P'' = Z_2\omega^2 u''^2_{\mathrm{m}} \sin^2(\omega t - \omega x/v_2), \qquad [7.9]$$

where P and P'' are positive as they correspond to the energy propagation toward the positive x, while P' is negative as it corresponds to an energy propagation toward the negative x. As $u'_{\mathrm{m}} = \mathcal{R} u_{\mathrm{m}}$ and $u''_{\mathrm{m}} = \mathcal{T} u_{\mathrm{m}}$, the conservation of energy at O may be written in the form

$$P = |P'| + P'', \qquad \text{i.e.} \qquad 1 = \mathcal{R}^2 + (Z_2/Z_1)\mathcal{T}^2, \qquad [7.10]$$

which is verified by the expressions [7.9]. The relationship $P = |P'| + P''$ requires that $|P'< P$ and this leads to $|u'_m| < |u_m|$, that is, $|\mathcal{R}| < 1$.

The ratios of the reflected power and the transmitted power to the incident power are, respectively, the *energy reflection factor* f_R and the *energy transmission factor* f_T. The conservation of energy requires that the sum of these factors be equal to 1

$$f_R = |P'|/P\,, \qquad f_T = P''/P\,; \qquad f_R + f_T = 1. \qquad [7.11]$$

We may consider the following special cases:

– If $Z_1 = Z_2$, $\mathcal{R}$ is equal to zero and the wave is totally transmitted. We say that the impedances are *adapted*. This case corresponds to $m_{l1} = m_{l2}$ and the strings are indistinguishable from the wave point of view.

– If $Z_1 > Z_2$ (that is, $m_{l1} > m_{l2}$ and $v_1 < v_2$), $\mathcal{R}$ is positive and u'_m has the same sign as u_m. In this case, the wave is reflected without a change of phase.

– If $Z_1 < Z_2$ (that is, $m_{l1} < m_{l2}$ and $v_1 > v_2$), $\mathcal{R}$ is negative and u'_m has the opposite sign to u_m. In this case, the wave is reflected with a phase change of π.

– If $Z_2 \to \infty$ (that is, $m_{l2} \to \infty$), $\mathcal{T}$ vanishes and there is no transmitted wave. Thus, the wave is totally reflected with a change of sign. This is the case for a string fixed to a rigid obstacle, in which $u''_m = 0$; thus, $u'_m = - u_m$. In this case, only the first boundary condition $u_1 = u + u' = 0$ must be kept. The second condition $\partial_x u_1 = \partial_x u_2|_{x=0}$ is meaningless and it is incompatible with the first condition.

– If $Z_2 = 0$, $\mathcal{R}$ is equal to 1. In this case, the wave is totally reflected without a change of sign. This is the case for a string with a "free" end. To realize this situation, the end of the string may be attached to a very light ring, which can slide without friction on a rigid rod, which is fixed in the direction of oscillation.

7.2. Reflection and transmission of a one-dimensional acoustic wave

Assume that an acoustic wave propagating in fluid (1) is incident on the surface of fluid (2). We assume in this section that the wave is one-dimensional, in a long and uniform pipe, for instance. If the fluids are miscible, we can imagine that they are separated by a thin membrane of negligible mass, which exerts no force on the fluids (Figure 7.2). Assume that the wave (represented by the displacement u) propagates in the positive x direction in fluid (1) toward the membrane. The membrane reflects a wave u' propagating toward the negative x in fluid (1) and transmits a wave u'' propagating in the positive x direction in fluid (2). The complex displacements of these waves are

$$\underline{u}(x,t)=\underline{u}_m\,e^{i(\omega t-kx)},\qquad \underline{u}'(x,t)=\underline{u}'_m\,e^{i(\omega' t+k'x)},\qquad \underline{u}''(x,t)=\underline{u}''_m\,e^{i(\omega'' t-k''x)},\quad [7.12]$$

where $k=\omega/v_1$, $k'=\omega'/v_1$ and $k''=\omega''/v_1$. Using the relationship $\underline{p}_a(x,t)=-B\,\partial_x u$ and the acoustic impedances $Z_i=B_i/v_i$, the corresponding pressure waves are

$$\underline{p}_a(x,t)=i\omega Z_1\underline{u}_m\,e^{i(\omega t-kx)},\qquad \underline{p}_a'(x,t)=-i\omega Z_1\,\underline{u}'_m\,e^{i(\omega' t+k'x)}$$

$$\underline{p}_a''(x,t)=i\omega Z_2\,\underline{u}''_m\,e^{i(\omega'' t-k''x)}.\qquad [7.13]$$

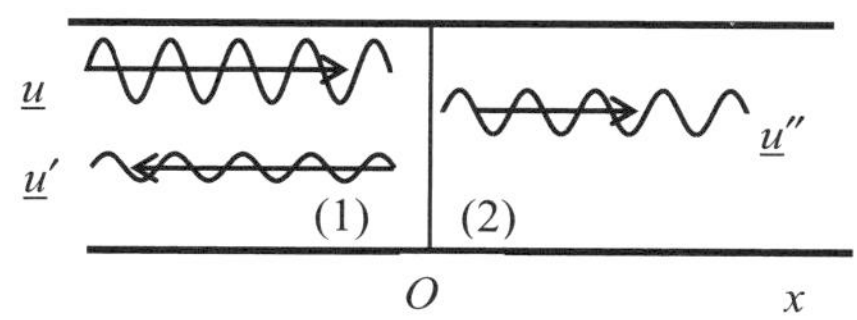

Figure 7.2. *Reflection and transmission of an acoustic wave in a pipe*

The boundary conditions at the membrane ($x=0$) are

$$\underline{u}(0,t)+\underline{u}'(0,t)=\underline{u}''(0,t),\qquad \underline{p}_a(0,t)+\underline{p}_a'(0,t)=\underline{p}_a''(0,t).\qquad [7.14]$$

The first condition expresses that the displacements of the fluids are equal (they are both equal to the displacement of the membrane). The second condition expresses that the fluids exert equal forces of pressure on the membrane. This last condition results from the equation of motion of the membrane $F_1-F_2=ma=0$ (as it has a negligible mass m).

Substituting expressions [7.12] and [7.13] for the wave functions $\underline{u}$, $\underline{u}'$, $\underline{u}''$, $\underline{p}_a$, $\underline{p}_a'$ and $\underline{p}_a''$ in [7.14], we find that these equations are verified at any time if $\omega=\omega'=\omega''$. Thus, the angular frequency of the wave does not change if it is reflected or transmitted at the interface of two fluids. The boundary conditions are then reduced to the conditions on the amplitudes

$$\underline{u}_m+\underline{u}'_m=\underline{u}''_m,\qquad Z_1\,\underline{u}_m-Z_1\,\underline{u}'_m=Z_2\,\underline{u}''_m.\qquad [7.15]$$

These equations determine the reflection and transmission coefficients:

$$\mathcal{R}_u\equiv\frac{\underline{u}'_m}{\underline{u}_m}=\frac{Z_1-Z_2}{Z_1+Z_2},\qquad \mathcal{T}_u\equiv\frac{\underline{u}''_m}{\underline{u}_m}=\frac{2Z_1}{Z_1+Z_2}.\qquad [7.16]$$

The corresponding coefficients for the pressure waves are

$$\mathcal{R}_p \equiv \frac{\underline{p}'_m}{\underline{p}_m} = -\frac{\underline{u}'_m}{\underline{u}_m} = \frac{Z_2 - Z_1}{Z_1 + Z_2}, \qquad \mathcal{T}_p \equiv \frac{\underline{p}''_m}{\underline{p}_m} = \frac{Z_2}{Z_1}\frac{\underline{u}''_m}{\underline{u}_m} = \frac{2Z_2}{Z_1 + Z_2}. \quad [7.17]$$

The energy flux density of an acoustic wave is given by the expression [5.72]. The power that it carries along the pipe is $P = \mathcal{S}S = \mathcal{S}Z\dot{u}^2 = \mathcal{S}p_a^2/Z$. Using the real expressions of the displacements, we obtain that the powers carried by the incident wave, the reflected wave and the transmitted wave are respectively:

$$P(0, t) = \mathcal{S}Z_1\omega^2 u_m^2 \sin^2(\omega t), \qquad P'(0, t) = \mathcal{S}Z_1\omega^2 u'^2_m \sin^2(\omega t)$$
$$P''(0, t) = \mathcal{S}Z_2\omega^2 u''^2_m \sin^2(\omega t). \quad [7.18]$$

The energy reflection and transmission factors are respectively:

$$f_R = \frac{P'}{P} = \mathcal{R}_u^2 = \mathcal{R}_p^2, \qquad f_T = \frac{P''}{P} = \frac{Z_2}{Z_1}\mathcal{T}_u^2 = \frac{Z_1}{Z_2}\mathcal{T}_p^2 \qquad \text{with } f_R + f_T = 1. [7.19]$$

We may consider the following cases:

– If $Z_1 = Z_2$, $\mathcal{R}_u$ and $\mathcal{R}_p$ are equal to zero and $\mathcal{T}_u = \mathcal{T}_p = 1$, this result is expected in the case of the same gas on both sides of the massless membrane.

– If $Z_1 > Z_2$, $\mathcal{R}_u$ is positive and $\mathcal{R}_p$ is negative. The acoustic wave is reflected without a change of sign (i.e. without any phase shift) for the displacement wave and a change of sign (i.e. a phase shift of π) for the pressure wave.

– If $Z_1 < Z_2$, $\mathcal{R}_u$ is negative and $\mathcal{R}_p$ is positive. The acoustic wave is reflected with a change of sign (i.e. a phase shift of π) for the displacement and without a change of sign (i.e. without any phase shift) for the pressure wave.

– If the second medium is rigid, its face plays the part of the membrane but there is no displacement and no transmitted wave in the second medium. Thus, we must replace u'' by 0 in the first boundary condition [7.14]. As an ideal rigid body can support any pressure without deformation, the pressure continuity does not necessarily hold. The rigid body has an infinite bulk modulus B, thus, infinite impedance. We get in this case $\mathcal{R}_u = -1$ and $\mathcal{R}_p = 1$. The acoustic wave is totally reflected with a change of sign (i.e. a phase shift of π) for the displacement wave and without a change of sign (i.e. without any phase shift) for the pressure wave.

– If the pressure in the second fluid is constant and the membrane exerts ideally no surface tension, the displacement of the second gas is arbitrary, thus the first boundary condition [7.14] is meaningless. In the second boundary condition, we must set $p''_a(0, t) = 0$. This is equivalent to set $Z_2 = 0$ in equations [7.16] and [7.17];

thus, $\mathcal{R}_u = 1$ and $\mathcal{R}_p = -1$. The acoustic wave is totally reflected without a change of sign (i.e. without any phase shift) for the displacement and with a change of sign (i.e. a phase shift of π) for the pressure.

EXAMPLE 7.1.– A membrane of mass per unit area m_s separates two gases but it exerts no force of pressure on them (Figure 7.2). Write the boundary conditions on the membrane. Deduce the reflection coefficient and the transmission coefficient. Write down the expressions of the real waves of displacement in the gases and the expression of displacement of the membrane.

SOLUTION – In this case, the displacement $\xi(t)$ of the membrane is the same as that of the gases on both sides of it and the gases exert two forces of pressure on it. Thus, we have the boundary conditions

$$\underline{\xi}(t) = \underline{u}(t, 0) + \underline{u}'(t, 0) = \underline{u}''(t, 0), \qquad m_s \ddot{\underline{\xi}} = \underline{p}_a(t, 0) + \underline{p}_a'(t, 0) - \underline{p}_a''(t, 0) .$$

Using expressions [7.12] and [7.13], these equations may be written as

$$\underline{\xi}(t) = (\underline{u}_m + \underline{u}'_m)\, e^{i\omega t} = \underline{u}_m''\, e^{i\omega t}, \qquad m_s \ddot{\underline{\xi}} = i\omega Z_1(\underline{u}_m - \underline{u}'_m)\, e^{i\omega t} - i\omega Z_2\, \underline{u}''_m\, e^{i\omega t} .$$

We deduce that

$$\mathcal{R}_u \equiv \frac{\underline{u}'_m}{\underline{u}_m} = \frac{Z_1 - Z_2 - i\omega m_s}{Z_1 + Z_2 + i\omega m_s}, \quad \mathcal{T}_u \equiv \frac{\underline{u}''_m}{\underline{u}_m} = \frac{2Z_1}{Z_1 + Z_2 + i\omega m_s}, \quad \underline{\xi}(t) = \frac{2Z_1 \underline{u}_m}{Z_1 + Z_2 + i\omega m_s}\, e^{i\omega t}$$

Assuming that the incident wave is $u = u_m\, e^{i(\omega t - k_1 x)}$ and taking the real parts, the reflected wave, the transmitted wave and the displacement $\xi(t)$ may be written as

$$u' = u'_m \cos(\omega t - k_1 x - \phi^- - \phi^+), \;\; u'' = u''_m \cos(\omega t - k_2 x - \phi^+) \text{ and } \xi(t) = \xi_m \cos(\omega t - \phi^+).$$

The amplitudes are $u'_m = u_m Z^- / Z^+$ and $u''_m = \xi_m = 2u_m\, Z_1 / Z^+$, where we have set $Z^{\pm} = \sqrt{(Z_1 \pm Z_2)^2 + \omega^2 m_s{}^2}$ and $\phi^{\pm} = \text{Arctan}[\omega m_s / (Z_1 \pm Z_2)]$ $(0 < \phi^{\pm} < \pi)$.

7.3. General laws of reflection and transmission of three-dimensional waves

In 1678, Huygens formulated a geometrical interpretation of the laws of reflection and refraction of light (see Problem P7.5) and qualitatively explained its diffraction (see section 8.7). This interpretation is actually known as *Huygens' principle*. It assumes that each point P of a wavefront Σ_o at time t_o behaves like a point source, emitting a secondary wavelet S_P. The envelope Σ of these wavelet is the wavefront at a later time. It is not necessary for the sources to be material emitters; the principle holds even in the case of the propagation of light in vacuum. From this general principle, it is possible to deduce the following laws of reflection and refraction, which are very well verified experimentally:

– The direction of propagation of the incident wave, that of the reflected wave and that of the refracted wave lie in the same plane containing the normal to the interface $\mathcal{S}$ at each point of incidence.

– The angle of reflection θ' (between the direction of propagation of the reflected wave and the normal Oz' to $\mathcal{S}$) is equal to the angle of incidence θ (between the direction of propagation of the incident wave and the normal Oz to $\mathcal{S}$). The angle of refraction θ'' (between the direction of propagation of the refracted wave and the normal Oz to $\mathcal{S}$) is related to θ by Snell's law

$$n_1 \sin\theta = n_2 \sin\theta'', \qquad [7.20]$$

where n_i is the *index of refraction* of medium (i) defined as the ratio of the phase velocity c of the wave in a reference medium (the vacuum in the case of light) to its phase velocity v_i in medium (i):

$$n_i = c/v_i. \qquad [7.21]$$

Equation [7.20] determines the angle of refraction θ'' if $(n_1/n_2)\sin\theta < 1$. This condition is always satisfied if $n_1 < n_2$. In the case $n_1 > n_2$, we must have $\sin\theta < n_1/n_2$. Thus, θ must be less than a *critical angle* (or *limiting angle*) i_L, given by

$$\sin i_L = n_2/n_1. \qquad [7.22]$$

At the angle of incidence equal to i_L the angle of refraction is $\theta'' = 90°$. If the angle of incidence θ is larger than i_L, the wave undergoes *total reflection*.

The wave theory enables us to establish the laws of reflection and refraction using the *boundary conditions* (or *continuity equations*) on the interface $\mathcal{S}$. We have only to assume that these conditions are expressed as *linear relations* between the incident wave $\underline{u}$, the reflected wave $\underline{u}'$ and the transmitted wave $\underline{u}''$ and their partial derivatives with respect to time or space coordinates. In the case of simple harmonic waves $u = u_m\, e^{i(\omega t - \mathbf{k.r})}$, the time derivative is simply $i\omega u$ and the derivative with respect to x, for instance, is $-ik_x u$. Thus, the boundary conditions are linear, of the general form $\underline{au} + \underline{bu'} + \underline{cu''} = 0$, that is,

$$\underline{a}\,\underline{u}_m\, e^{i(\omega t - \mathbf{k.r})} + \underline{b}\,\underline{u}'_m\, e^{i(\omega' t - \mathbf{k}'.\mathbf{r})} + \underline{c}\,\underline{u}''_m\, e^{i(\omega'' t - \mathbf{k}''.\mathbf{r})} = 0. \qquad [7.23]$$

A relation of this form can be verified at any t and at any point $\mathbf{r}$ of the interface $\mathcal{S}$ only if the angular frequency and the scalar product $(\mathbf{k}.\mathbf{r})$ on $\mathcal{S}$ are the same for the three waves u, u' and u''. We deduce that the angular frequency undergoes no change in the reflection and transmission ($\omega = \omega' = \omega''$) and the tangential component of $\mathbf{k}$, $\mathbf{k}'$ and $\mathbf{k}''$ (i.e. their components which are parallel to $\mathcal{S}$) are the same. As for the normal component of the wave vectors, they are obtained using the dispersion relation $\omega = \omega(\mathbf{k})$ for each medium.

Consider a wave that is incident on the interface $\mathcal{S}$ of two isotropic media, in which the speeds of propagation are v_1 and v_2 respectively (Figure 7.3). Let $\mathbf{k}$, $\mathbf{k}'$ and $\mathbf{k}''$ be the wave vectors of the incident wave, the reflected wave and the transmitted wave respectively. We choose the axes of coordinates in such a way that Oxy is tangent to the interface $\mathcal{S}$ with Oz oriented from medium (1) toward medium (2) and the x-axis in the plane of incidence (formed by $\mathbf{k}$ and the normal axis Oz). Due to the symmetry with respect to the plane of incidence Oxz, the vectors $\mathbf{k}'$ and $\mathbf{k}''$ are in this plane (there is no physical reason for them to point to one side and not the other). This same result may also be obtained by analysis. Indeed, let θ be the angle of $\mathbf{k}$ with Oz, θ' the angle of $\mathbf{k}'$ with Oz' and θ'' the angle of $\mathbf{k}''$ with Oz (so they lie between 0 and 90° as usual). Let us assume that the azimuthal angles of the reflection plane and the refraction plane are ϕ' and ϕ'' respectively. Thus, we have

$$\mathbf{k} = k\,(\sin\theta\,\mathbf{e}_x + \cos\theta\,\mathbf{e}_z),\quad \mathbf{k}' = k'\,(\sin\theta'\cos\phi'\,\mathbf{e}_x + \sin\theta'\sin\phi'\,\mathbf{e}_y - \cos\theta'\,\mathbf{e}_z),$$
$$\mathbf{k}'' = k''\,(\sin\theta''\cos\phi''\,\mathbf{e}_x + \sin\theta''\sin\phi''\,\mathbf{e}_y + \cos\theta''\,\mathbf{e}_z).$$

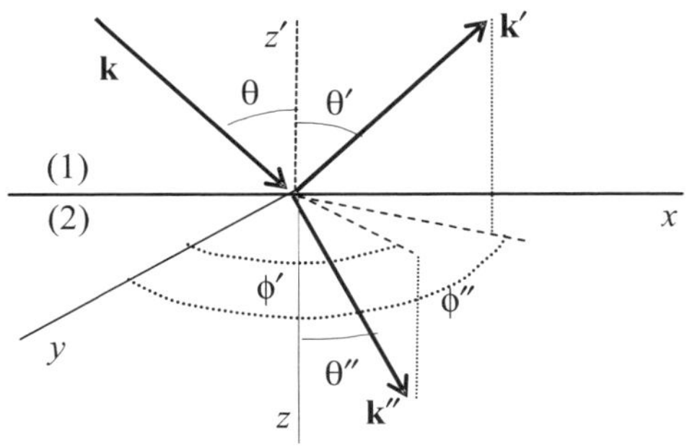

Figure 7.3. *Boundary conditions and laws of reflection and refraction*

The equality of the scalar products $(\mathbf{k}'.\mathbf{r})$, $(\mathbf{k}''.\mathbf{r})$ and $(\mathbf{k}.\mathbf{r})$ on $\mathcal{S}$ ($z = 0$) requires that

$$k'\,(x\sin\theta'\cos\phi' + y\sin\theta'\sin\phi') = k''(x\sin\theta''\cos\phi'' + y\sin\theta''\sin\phi'') = k(x\sin\theta).$$

These relations are satisfied for all values of the coordinates x and y if

$$k' \sin\theta' \cos\phi' = k'' \sin\theta'' \cos\phi'' = k \sin\theta, \qquad k' \sin\theta' \sin\phi' = k'' \sin\theta'' \sin\phi'' = 0.$$

As $k = k' = \omega/v_1$ and $k'' = \omega/v_2$, we obtain the relations

$$\phi' = \phi'' = 0, \qquad \theta = \theta', \qquad (1/v_1)\sin\theta = (1/v_2)\sin\theta''. \qquad [7.24]$$

The equations $\phi' = 0$ and $\phi'' = 0$ are the expressions that the vectors **k**, **k**′ and **k**" lie in the plane of incidence Oxz. The equality $\theta = \theta'$ is the law of reflection and the last relationship [7.24] is the expression of Snell's law [7.20]. Consequently, the wave vectors may be written as:

$$\mathbf{k} = k(\sin\theta\,\mathbf{e}_x + \cos\theta\,\mathbf{e}_z), \qquad \mathbf{k}' = k(\sin\theta\,\mathbf{e}_x - \cos\theta\,\mathbf{e}_z)$$

$$\mathbf{k}'' = k''(\sin\theta''\,\mathbf{e}_x + \cos\theta''\,\mathbf{e}_z). \qquad [7.25]$$

Taking into account the equality of the phases $(\omega t - \mathbf{k.r})$ on $\mathcal{S}$, the boundary conditions [7.23] reduce to linear relationships between the complex amplitudes

$$\underline{a}\,\underline{u}_m + \underline{b}\,\underline{u}'_m + \underline{c}\,\underline{u}''_m = 0. \qquad [7.26]$$

We must have two relations of this type to determine the amplitude and the phase of the transmitted wave and those of the reflected wave.

Instead of the boundary conditions, it is possible to use the fundamental physical law of the *conservation of energy*. This law requires that the normal components of the vector energy flux verify the boundary condition

$$S_z + S'_z = S''_z \qquad (\text{for } z = 0). \qquad [7.27]$$

The tangential components of the vectors **S**, **S**′ and **S**" play no part in the transfer of energy across the interface, as they correspond to a propagation of energy in a direction, which is parallel to the interface in each medium. **S** being quadratic in **u**, relationship [7.27] can be verified at each point of the interface $\mathcal{S}$ only if the phase $\omega t - \mathbf{k.r}$) is the same for the three waves on the interface $\mathcal{S}$. Thus, we must have $\omega = \omega' = \omega''$ and $\mathbf{k.r} = \mathbf{k'.r} = \mathbf{k''.r}$ for $z = 0$. Thus, the laws of reflection and refraction are closely related to the principle of conservation of energy.

7.4. Reflection and refraction of a three-dimensional acoustic wave

Consider an acoustic wave that is incident from a medium (1) on the surface $\mathcal{S}$ of a medium (2). Let ω be the angular frequency, **k**, **k**′ and **k**" the wave vectors of the incident wave, the reflected wave and the transmitted wave respectively, **e**, **e**′ and **e**″ the unit vectors in their direction of propagation (Figure 7.4). The acoustic waves being longitudinal in fluids, the displacements $\underline{\mathbf{u}}$, $\underline{\mathbf{u}}'$ and $\underline{\mathbf{u}}''$ point in their respective propagation directions. Thus, the displacement waves are

$$\underline{\mathbf{u}} = \mathbf{e}\, \underline{u}_m\, e^{i(\omega t - k\mathbf{e}.\mathbf{r})}, \qquad \underline{\mathbf{u}}' = \mathbf{e}'\, \underline{u}_m'\, e^{i(\omega t - k'\mathbf{e}'.\mathbf{r})}, \qquad \underline{\mathbf{u}}'' = \mathbf{e}''\underline{u}_m''\, e^{i(\omega t - k''\mathbf{e}''.\mathbf{r})}. \quad [7.28]$$

On the interface $\mathcal{S}$, the normal component of the total displacement **u** and the total acoustic pressure $p_a = - B(\nabla.\mathbf{u})$ must be continuous. Using the acoustic impedances $Z_i = B_i/v_i$ and dropping the common phase factor $e^{i(\omega t - \mathbf{k}.\mathbf{r})}$, the boundary conditions may be written as

$$\underline{u}_m \cos\theta - \underline{u}_m' \cos\theta' = \underline{u}_m'' \cos\theta'', \qquad Z_1\underline{u}_m + Z_1\underline{u}_m' = Z_2\, \underline{u}_m'' \quad [7.29]$$

We deduce the reflection and transmission coefficients for the displacement wave

$$\mathcal{R}_u \equiv \frac{\underline{u}'_m}{\underline{u}_m} = \frac{Z_2\cos\theta - Z_1\cos\theta''}{Z_2\cos\theta + Z_1\cos\theta''}, \qquad \mathcal{T}_u \equiv \frac{\underline{u}''_m}{\underline{u}_m} = \frac{2Z_1\cos\theta}{Z_2\cos\theta + Z_1\cos\theta''}. \quad [7.30]$$

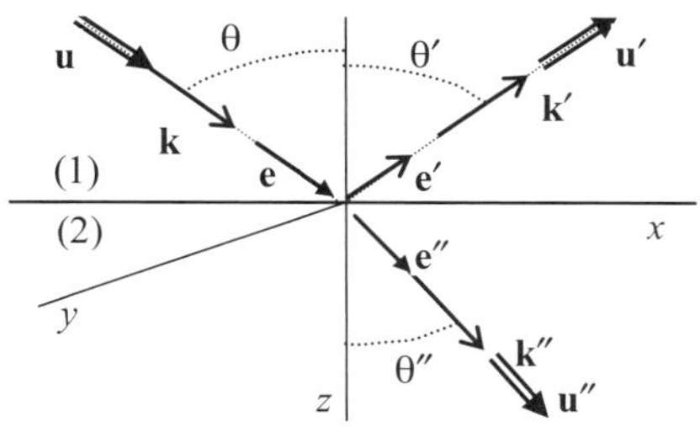

Figure 7.4. *Reflection and refraction of sound*

The amplitude of the acoustic pressure is related to the displacement amplitude by the equation

$$\underline{p}_m = i(B\omega/v)\,(\mathbf{e}.\underline{\mathbf{u}}_m) = i\omega Z\, \underline{u}_m. \quad [7.31]$$

Thus the reflection and transmission coefficients for the pressure wave are

$$\mathcal{R}_p \equiv \frac{\underline{p}'_m}{\underline{p}_m} = \frac{Z_2\cos\theta - Z_1\cos\theta''}{Z_2\cos\theta + Z_1\cos\theta''}, \qquad \mathcal{T}_p \equiv \frac{\underline{p}''_m}{\underline{p}_m} = \frac{2Z_2\cos\theta}{Z_2\cos\theta + Z_1\cos\theta''}. \quad [7.32]$$

Expressions [7.30] and [7.32] determine the amplitudes of the reflected wave and those of the transmitted wave. In particular, at normal incidence ($\theta = \theta' = \theta'' = 0$), we again find expressions [7.16] and [7.17]. The impedances being real and positive in the absence of absorption, the reflection and transmission coefficients are real. The transmission coefficient is always positive, while the reflection coefficient may be positive, negative or zero but always its absolute value is less than 1.

EXAMPLE 7.2.– Determine the energy flux densities of the incident, reflected and transmitted waves. Verify the conservation of energy at the interface.

SOLUTION – The vector energy flux density of an acoustic wave has the components $S_i = -B\Sigma_j \dot{u}_j(\partial u_j/\partial x_i)$. Taking the real parts of the waves [7.28] and using the expressions of the reflection and transmission coefficients, we obtain:

$$S_i = e_i kB_1\omega\, u_m^2 \sin^2(\omega t - k\,\mathbf{e}.\mathbf{r}) \quad\Rightarrow\quad \mathbf{S} = \mathbf{e}\, Z_1\omega^2 u_m^2 \sin^2(\omega t - k\,\mathbf{e}.\mathbf{r})$$
$$S'_i = e'_i kB_1\omega\, u'^2_m \sin^2(\omega t - k\mathbf{e}'.\mathbf{r}) \quad\Rightarrow\quad \mathbf{S}' = \mathbf{e}' Z_1\omega^2 \mathcal{R}_u^2 u_m^2 \sin^2(\omega t - k\,\mathbf{e}'.\mathbf{r})$$
$$S''_i = e''_i k'' B_2\omega\, u''^2_m \sin^2(\omega t - k''\mathbf{e}''.\mathbf{r}) \Rightarrow \mathbf{S}'' = \mathbf{e}'' Z_2\,\omega^2 \mathcal{T}_u^2 u_m^2 \sin^2(\omega t - k''\mathbf{e}''.\mathbf{r}).$$

To verify the conservation of energy, let us consider an element of area $d\mathcal{S}$ of the interface. The power that $d\mathcal{S}$ receives from medium (1) toward the positive z is

$$d\mathcal{S}\,(\mathbf{S} + \mathbf{S}').\mathbf{e}_z = d\mathcal{S}\,[(\mathbf{e}.\mathbf{e}_z) + (\mathbf{e}'.\mathbf{e}_z)\,\mathcal{R}_u^2]\, Z_1\omega^2 u_m^2 \sin^2(\omega t - k\,\mathbf{e}.\mathbf{r})$$
$$= d\mathcal{S}\,(\cos\theta - \cos\theta'\,\mathcal{R}_u^2)\, Z_1\omega^2 u_m^2 \sin^2(\omega t - k\,\mathbf{e}.\mathbf{r}).$$

The transmitted power across $d\mathcal{S}$ toward the positive z in medium (2) is

$$d\mathcal{S}\,(\mathbf{S}''.\mathbf{e}_z) = d\mathcal{S}(\mathbf{e}''.\mathbf{e}_z)\mathcal{T}_u^2\, Z_2\omega^2 u_m^2 \sin^2(\omega t - k''\mathbf{e}''.\mathbf{r})$$
$$= d\mathcal{S}\cos\theta''\,\mathcal{T}_u^2\, Z_2\omega^2 u_m^2 \sin^2(\omega t - k''\,\mathbf{e}''.\mathbf{r}).$$

As $\omega t - k''\mathbf{e}''.\mathbf{r} = \omega t - k\mathbf{e}.\mathbf{r}$ on the interface, the conservation of energy may be written in the form $Z_1(1 - \mathcal{R}_u^2)\cos\theta = Z_2\mathcal{T}_u^2 \cos\theta''$, which is verified by the expressions of [7.30] for $\mathcal{R}_u$ and $\mathcal{T}_u$.

7.5. Reflection and refraction of an electromagnetic wave at the interface of dielectrics

Consider an electromagnetic plane wave that is incident from a dielectric (1) at an angle θ on the face $\mathcal{S}$ of a dielectric (2). The primary electromagnetic fields act on the electric charges in medium (2); they oscillate and emit secondary waves. The superposition of the secondary waves constitutes the reflected wave toward medium (1) and their superposition with the primary wave constitutes the refracted wave in medium (2). Globally, this process produces the boundary conditions on the interface $\mathcal{S}$

$$\mathbf{D}_{\perp}(1) = \mathbf{D}_{\perp}(2), \qquad \mathbf{B}_{\perp}(1) = \mathbf{B}_{\perp}(2), \qquad \mathbf{E}_{//}(1) = \mathbf{E}_{//}(2), \qquad \mathbf{H}_{//}(1) = \mathbf{H}_{//}(2). \quad [7.33]$$

An electromagnetic wave has two independent states of linear polarization determined by the direction of the electric field **E**. Any wave is a superposition of these two states. Thus, it is sufficient to study the reflection and refraction of plane waves in these two states of polarization. As the media are assumed to be isotropic, if we choose the first direction of polarization in the plane of incidence *Oxz* and the second perpendicular to this plane, the incident wave is symmetric with respect to the plane of incidence (for the direction of propagation and the direction of polarization). The reflected wave and the transmitted wave must have the same symmetry. Thus their wave vectors are in the plane of incidence (and given by the [7.25]) and they must have the same linear polarization as the incident wave. We write the fields in the form $\mathbf{E} = \underline{\mathbf{E}}_m e^{i(\omega t - \mathbf{k.r})}$, $\mathbf{B} = \underline{\mathbf{B}}_m e^{i(\omega t - \mathbf{k.r})}$, etc., and, for simplicity, we omit the exponential which has the same value on the interface.

7.5.1. *Case of linear polarization in the plane of incidence*

In this case, the fields **E**, **E′** and **E"** of the incident, reflected and refracted waves are in the plane of incidence *Oxz* (Figure 7.5a), while the fields **B**, **B′** and **B"** are parallel to *Oy*. Thus, it is convenient to determine the waves by the magnetic fields. The electric fields are then $\mathbf{E} = \mathbf{B} \times \mathbf{k}/\omega\mu_1\varepsilon_1$, $\mathbf{E}' = \mathbf{B}' \times \mathbf{k}'/\omega\mu_1\varepsilon_1$ and $\mathbf{E}'' = \mathbf{B}'' \times \mathbf{k}''/\omega\mu_2\varepsilon_2$. The complex amplitudes of the fields may be written as

$$\begin{aligned}
&\underline{\mathbf{B}}_m = \underline{B}_m\, \mathbf{e}_y, && \underline{\mathbf{E}}_m = v_1 \underline{B}_m (\cos\theta\, \mathbf{e}_x - \sin\theta\, \mathbf{e}_z),\\
&\underline{\mathbf{B}}'_m = -\underline{B}'_m\, \mathbf{e}_y, && \underline{\mathbf{E}}'_m = v_1 \underline{B}'_m (\cos\theta'\, \mathbf{e}_x + \sin\theta'\, \mathbf{e}_z),\\
&\underline{\mathbf{B}}''_m = \underline{B}''_m\, \mathbf{e}_y, && \underline{\mathbf{E}}''_m = v_2 \underline{B}''_m (\cos\theta''\, \mathbf{e}_x - \sin\theta''\, \mathbf{e}_z),
\end{aligned} \qquad [7.34]$$

where we introduced a (–) sign into the expression of $\mathbf{B}'_m$ in order for $\mathbf{E}'_m$ and $\mathbf{E}_m$ to have the same sign at the limit $\theta = 0$. If the media are not magnetic ($\mu_1 = \mu_2 = \mu_o$), the boundary conditions [7.33] on the interface $\mathcal{S}$ ($z = 0$) may be written as

$$\begin{aligned}
&\varepsilon_1 v_1(\underline{B}_m \sin\theta - \underline{B}'_m \sin\theta') = \varepsilon_2 v_2 \underline{B}''_m \sin\theta'', && \underline{B}_m - \underline{B}'_m = \underline{B}''_m,\\
&v_1(\underline{B}_m \cos\theta + \underline{B}'_m \cos\theta') = v_2\, \underline{B}''_m \cos\theta''.
\end{aligned} \qquad [7.35]$$

These equations determine the amplitudes B'_m and B''_m of the reflected wave and the transmitted wave and, consequently, the electric fields as functions of B_m. We obtain the reflection and transmission coefficients

$$\mathcal{R}_{//} = \frac{\underline{E}'_{\rm m}}{\underline{E}_{\rm m}} = \frac{\underline{B}'_{\rm m}}{\underline{B}_{\rm m}} = \frac{n_1 \cos\theta'' - n_2\cos\theta}{n_1\cos\theta'' + n_2\cos\theta} = \frac{\tan(\theta''-\theta)}{\tan(\theta+\theta'')},$$

$$\mathcal{T}_{//} = \frac{\underline{E}''_{\rm m}}{\underline{E}_{\rm m}} = \frac{v_2}{v_1}\frac{\underline{B}''_{\rm m}}{\underline{B}_{\rm m}} = \frac{2n_1\cos\theta}{n_1\cos\theta'' + n_2\cos\theta} = \frac{2\cos\theta\sin\theta''}{\sin(\theta+\theta'')\cos(\theta-\theta'')}, \quad [7.36]$$

where we have used Snell's law. We find that $\underline{B}_{\rm m}$, $\underline{B}'_{\rm m}$ and $\underline{B}''_{\rm m}$ are in phase or in opposite phase; thus, we may take them to be real.

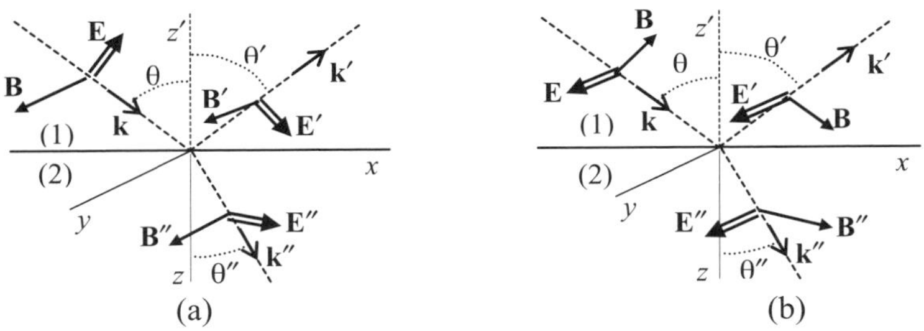

Figure 7.5. *Reflection and refraction of an electromagnetic wave that is linearly polarized: a) in the plane of incidence, and b) in the direction perpendicular to the plane of incidence*

7.5.2. *Case of linear polarization perpendicular to the plane of incidence*

In this case, **E** is parallel to Oy and $\mathbf{B} = (\mathbf{k}\times\mathbf{E})/(vk)$ is in the plane of incidence (Figure 7.5b). Thus, it is convenient to determine the waves directly by the electric fields and write the amplitudes

$$\underline{\mathbf{E}}_{\rm m} = \underline{E}_{\rm m}\,\mathbf{e}_{\rm y}, \qquad \underline{\mathbf{B}}_{\rm m} = (\underline{E}_{\rm m}/v_1)(-\cos\theta\,\mathbf{e}_{\rm x} + \sin\theta\,\mathbf{e}_{\rm z}),$$
$$\underline{\mathbf{E}}'_{\rm m} = \underline{E}'_{\rm m}\,\mathbf{e}_{\rm y}, \qquad \underline{\mathbf{B}}'_{\rm m} = (\underline{E}'_{\rm m}/v_1)(\cos\theta'\,\mathbf{e}_{\rm x} + \sin\theta'\,\mathbf{e}_{\rm z})$$
$$\underline{\mathbf{E}}''_{\rm m} = \underline{E}''_{\rm m}\,\mathbf{e}_{\rm y}, \qquad \underline{\mathbf{B}}''_{\rm m} = (\underline{E}''_{\rm m}/v_1)(-\cos\theta''\,\mathbf{e}_{\rm x} + \sin\theta''\,\mathbf{e}_{\rm z}). \quad [7.37]$$

If the media are not magnetic, the boundary conditions of [7.33] give

$$E_{\rm m} + E'_{\rm m} = E'', \qquad n_1(E_{\rm m}\cos\theta - E'_{\rm m}\cos\theta') = n_2 E''_{\rm m}\cos\theta''. \quad [7.38]$$

These equations determine the reflection and transmission coefficients:

$$\mathcal{R}_{\perp} = \frac{\underline{E}'_{\rm m}}{\underline{E}_{\rm m}} = \frac{\underline{B}'_{\rm m}}{\underline{B}_{\rm m}} = \frac{\tan\theta'' - \tan\theta}{\tan\theta'' + \tan\theta} = \frac{n_1\cos\theta - n_2\cos\theta''}{n_1\cos\theta + n_2\cos\theta''} = \frac{\sin(\theta''-\theta)}{\sin(\theta''+\theta)},$$

$$\mathcal{T}_{\perp} = \frac{\underline{E}''_{\rm m}}{\underline{E}_{\rm m}} = \frac{v_2}{v_1}\frac{\underline{B}''_{\rm m}}{\underline{B}_{\rm m}} = \frac{2n_1\cos\theta}{n_1\cos\theta + n_2\cos\theta''} = \frac{2\cos\theta\sin\theta''}{\sin(\theta''+\theta)}. \quad [7.39]$$

Note that the reflection and transmission coefficients in the case of incidence from medium (1) on medium (2) and from medium (2) on medium (1), for any state of polarization, verify the relations

$$\mathcal{R}_{12} = -\mathcal{R}_{21}, \qquad \mathcal{T}_{12}\,\mathcal{T}_{21} = 1 - \mathcal{R}_{12}\,\mathcal{R}_{21}. \qquad [7.40]$$

If the angle of incidence is small ($\theta \approx \theta' \approx \theta'' \approx 0$), $\mathcal{R}_{//}$ and $\mathcal{R}_{\perp}$ approach the same limit $\mathcal{R}_o$, while $\mathcal{T}_{//}$ and $\mathcal{T}_{\perp}$ approach the same limit $\mathcal{T}_o$ given by

$$\mathcal{R}_o \approx \frac{\theta'' - \theta}{\theta'' + \theta} \approx \frac{n_1 - n_2}{n_1 + n_2}, \qquad \mathcal{T}_o \approx \frac{2\theta''}{\theta'' + \theta} \approx \frac{2n_1}{n_1 + n_2}. \qquad [7.41]$$

The variations of the reflection and transmission coefficients as functions of the angle of incidence θ are illustrated in Figure 7.6. $\mathcal{R}_{\perp}$ decreases from $\mathcal{R}_o$ to -1 and $\mathcal{R}_{//}$ increases from $\mathcal{R}_o$ to $+1$. If $n_1 < n_2$, $\mathcal{T}_{\perp}$ and $\mathcal{T}_{//}$ are positive and decrease from $\mathcal{T}_o$ for $\theta = 0$ to 0 for $\theta = \pi/2$. If $n_1 > n_2$, we have similar variations, but θ cannot be larger than the critical angle i_L given by [7.22]. If θ is larger than i_L, we shall see in section 7.6 that the wave undergoes a *total internal reflection*.

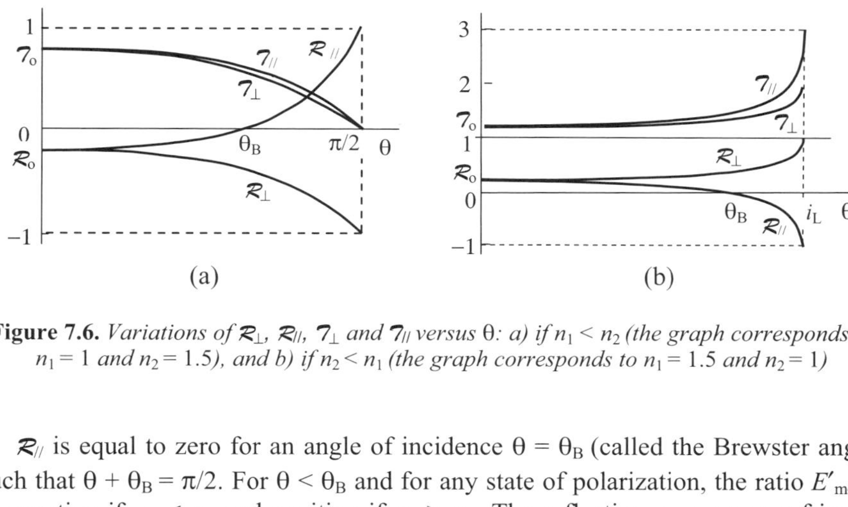

Figure 7.6. *Variations of $\mathcal{R}_{\perp}$, $\mathcal{R}_{//}$, $\mathcal{T}_{\perp}$ and $\mathcal{T}_{//}$ versus θ: a) if $n_1 < n_2$ (the graph corresponds to $n_1 = 1$ and $n_2 = 1.5$), and b) if $n_2 < n_1$ (the graph corresponds to $n_1 = 1.5$ and $n_2 = 1$)*

$\mathcal{R}_{//}$ is equal to zero for an angle of incidence $\theta = \theta_B$ (called the Brewster angle) such that $\theta + \theta_B = \pi/2$. For $\theta < \theta_B$ and for any state of polarization, the ratio E'_m/E_m is negative if $n_1 < n_2$ and positive if $n_1 > n_2$. The reflection on a more refringent medium ($n_2 > n_1$) occurs with a change of sign (i.e. a phase shift of π). The reflection on a less refringent medium ($n_2 < n_1$) occurs without a change of sign (i.e. without phase shift). As the transmission coefficient is always positive, the transmitted wave has no phase shift.

7.5.3. *Conservation of energy*

The Poynting vector of an electromagnetic wave is

$$\mathbf{S} = (1/\mu)\, \mathbf{E} \times \mathbf{B} = (1/\mu v)\, E^2\, \mathbf{e} = (n/\mu c)\, E^2\, \mathbf{e}. \qquad [7.42]$$

The conservation of energy on the interface $\mathcal{S}$ of two non-conducting media requires that the power that any element of area $d\mathcal{S}$ receives from medium (1) to be equal to the transmitted power toward medium (2). If the axis Oz is normal to $\mathcal{S}$ and oriented from medium (1) toward medium (2), this condition can be expressed by the equation $S_z + S'_z = S''_z$, that is,

$$n_1\, E^2 \cos\theta - n_1\, E'^2 \cos\theta = n_2\, E''^2 \cos\theta''. \qquad [7.43]$$

Using the reflection and transmission coefficients, this relation may be written as

$$n_1\,(1 - \mathcal{R}^2) \cos\theta = n_2\, \mathcal{T}^2 \cos\theta''. \qquad [7.44]$$

Any wave may be considered as the superposition of two waves, which are polarized in the plane of incidence and perpendicularly to this plane respectively. The Poynting vector is the sum of the corresponding Poynting vectors (see section 6.2.4). The equation of conservation of energy [7.44] holds for each one of them, as it may be easily verified by using expressions [7.36] and [7.39]. Thus, [7.44] holds for any state of polarization and for non-polarized waves.

We define the *energy reflection factor* as the ratio of the reflected power to the incident power and the *energy transmission factor* as the ratio of the transmitted power to the incident power per unit area of the interface $\mathcal{S}$

$$f_R \equiv \frac{S'_z}{S_z} = \frac{E'^2}{E^2} = \mathcal{R}^2, \qquad f_T \equiv \frac{S''_z}{S_z} = \frac{E''^2}{E^2}\,\frac{n_2}{n_1}\,\frac{\cos\theta''}{\cos\theta} = \mathcal{T}^2\,\frac{n_2}{n_1}\,\frac{\cos\theta''}{\cos\theta}. \qquad [7.45]$$

The conservation of energy on the interface $\mathcal{S}$ requires that

$$f_R + f_T = 1. \qquad [7.46]$$

The variations of f_R and f_T as functions of the angle of incidence are illustrated in Figure 7.7a. f_R and f_T vary very little for small angles of incidence and, in the case of non-magnetic media, their values are given by

$$f_{Ro} \cong \left(\frac{n_1 - n_2}{n_1 + n_2}\right)^2, \qquad f_{To} \cong \frac{4n_1 n_2}{(n_1 + n_2)^2}. \qquad [7.47]$$

f_{Ro} is usually small (4% at the air-glass interface and 2% at the air-water interface).

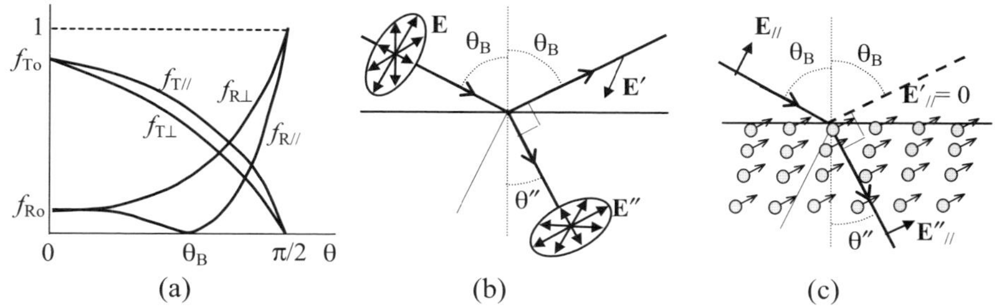

Figure 7.7. *a) Variations of the energy reflection factors $f_{R\perp}$ and $f_{R//}$ and the transmission factors $f_{T\perp}$ and $f_{T//}$ versus θ, b) Brewster polarization, and c) its interpretation*

7.5.4. *Brewster's law*

For any values of the indices n_1 and n_2, the coefficient $\mathcal{R}_{//}$ is equal to zero if $\tan(\theta + \theta'') = \infty$, that is, $\theta + \theta'' = \pi/2$. Thus, the intensity of the reflected wave vanishes if it is polarized in the plane of incidence. Using Snell's law, we deduce that the Brewster angle verifies the relationship

$$\tan \theta_B = n_2/n_1. \qquad [7.48]$$

This result may be used to obtain a linearly polarized wave from waves of any polarization or from non-polarized waves. Writing such waves as a superposition of waves polarized in the plane of incidence and waves polarized perpendicularly to this plane, at Brewster incidence only the component that is polarized perpendicularly to the plane of incidence produces a reflected wave with the same polarization (Figure 7.7b). This phenomenon was first observed by Malus in 1807 and was analyzed by Brewster in 1815. It may be explained by the polarization of the molecules of the second medium in the direction of **E** and the vanishing of the dipole radiation in the direction of the dipole. As the reflected wave is the superposition of these dipole radiations, it vanishes in this direction (Figure 7.7c).

EXAMPLE 7.3.– Any incident wave on the interface of two dielectrics may be written in the form $\mathbf{E} = \underline{E}_{//}\,\mathbf{e}_{//} + \underline{E}_{\perp}\mathbf{e}_{\perp}$, where $\mathbf{e}_{//}$ and $\mathbf{e}_{\perp}$ are unit vectors that are perpendicular to the direction of propagation. $\mathbf{e}_{//}$ is in the plane of incidence and $\mathbf{e}_{\perp}$ is perpendicular to this plane. **a)** Write in this form the expression of a plane wave that is incident at an angle θ and polarized linearly, polarized circularly and non-polarized. **b)** Consider the case of a non-polarized incident wave. Calculate the energy reflection and transmission factors. It will be convenient to set $n = n_2/n_1$. Determine the degree of polarization perpendicularly to the incidence plane for the reflected wave and for the transmitted wave. What are their values in the case of the incidence at the Brewster angle and at $\theta = 30°$ on the surface of the water.

SOLUTION –

a) The wave may be written in the form $\underline{\mathbf{E}} = \underline{E}_{//}\,\mathbf{e}_{//} + \underline{E}_{\perp}\mathbf{e}_{\perp}$ where the components are complex, of the form $\underline{E}_{//} = A_{//}\,e^{\mathrm{i}(\omega t - \mathbf{k.r})}$ and $\underline{E}_{\perp} = A_{\perp}\,e^{\mathrm{i}(\omega t - \mathbf{k.r}+\phi)}$. In the case of a linearly polarized wave, $\phi = 0$ or π. In the case of a circularly polarized wave, we have $A_{//} = A_{\perp}$ and $\phi = \pm\,\pi/2$. In the case of a non-polarized wave, we may assume that $A_{//} = A_{\perp}$ and ϕ vary at random. The wave vector **k** lies in the plane of incidence and it makes with the normal to the interface an angle equal to the angle of incidence θ.

b) In the case of a non-polarized wave that is incident at an angle θ, the amplitudes of the incident, reflected and transmitted waves may be written as

$$\underline{\mathbf{E}} = \underline{E}_{//}\,\mathbf{e}_{//} + \underline{E}_{\perp}\mathbf{e}_{\perp}, \qquad \mathbf{E}' = \mathcal{R}_{//}\,\underline{E}_{//}\,\mathbf{e}_{//} + \mathcal{R}_{\perp}\,\underline{E}_{\perp}\mathbf{e}_{\perp}, \qquad \mathbf{E}'' = \mathcal{T}_{//}\,\underline{E}_{//}\,\mathbf{e}_{//} + \mathcal{T}_{\perp}\,\underline{E}_{\perp}\mathbf{e}_{\perp}$$

with $\underline{E}_{//} = A\,e^{\mathrm{i}(\omega t - \mathbf{k.r})}$ and $\underline{E}_{\perp} = A\,e^{\mathrm{i}(\omega t - \mathbf{k.r}+\phi)}$. The powers that are received, reflected and transmitted by the unit area of the interface are respectively

$$<S_z> = \mathcal{I}\cos\theta = (1/\mu_o v_1)\,<(\mathcal{R}e\,\mathbf{E})^2>\cos\theta = (n_1/\mu_o c)\,E_m^{\,2}\cos\theta,$$
$$<S'_z> = \mathcal{I}'\cos\theta' = (1/\mu_o v_1)\,<(\mathcal{R}e\,\mathbf{E}')^2>\cos\theta' = (n_1/2\mu_o c)\,E_m^{\,2}(\mathcal{R}_{//}^{\,2} + \mathcal{R}_{\perp}^{\,2})\cos\theta,$$
$$<S''_z> = \mathcal{I}''\cos\theta'' = (1/\mu_o v_2)\,<(\mathcal{R}e\,\mathbf{E}'')^2>\cos\theta'' = (n_2/2\mu_o c)\,E_m^{\,2}\,(\mathcal{T}_{//}^{\,2} + \mathcal{T}_{\perp}^{\,2})\cos\theta''.$$

Thus, the energy reflection and transmission factors are

$$f_R = \frac{\mathcal{I}'\cos\theta'}{\mathcal{I}\cos\theta} = \frac{1}{2}\,(\mathcal{R}_{//}^{\,2} + \mathcal{R}_{\perp}^{\,2}) = \tan^2(\theta - \theta'')\,\frac{1 + \tan^2\theta\,\tan^2\theta''}{(\tan\theta + \tan\theta'')^2},$$

$$f_T = \frac{\mathcal{I}''\cos\theta''}{\mathcal{I}\cos\theta} = \frac{n_2\cos\theta''}{2n_1\cos\theta}(\mathcal{T}_{//}^{\,2} + \mathcal{T}_{\perp}^{\,2}) = \frac{2\tan\theta\,\tan\theta''}{(\tan\theta + \tan\theta'')^2}\,[2 + \tan^2(\theta - \theta'')].$$

They verify the conservation of energy equation $f_T + f_R = 1$.

Using equations [7.36] and [7.39], we obtain the degree of polarization perpendicularly to the plane of incidence for the reflected and transmitted waves:

$$\mathcal{P}' = \frac{\mathcal{R}_{\perp}^{\,2} - \mathcal{R}_{//}^{\,2}}{\mathcal{R}_{\perp}^{\,2} + \mathcal{R}_{//}^{\,2}} = \frac{2n(1-n^2)\cos\theta''\cos\theta\,(\cos^2\theta - \cos^2\theta'')}{(1-n^2)^2\cos^2\theta''\cos^2\theta + n^2(\cos^2\theta'' - \cos^2\theta)^2},$$

$$\mathcal{P}'' = \frac{\mathcal{T}_\perp^2 - \mathcal{T}_{//}^2}{\mathcal{T}_\perp^2 + \mathcal{T}_{//}^2} = \frac{(n^2-1)(\cos^2\theta - \cos^2\theta'')}{(n^2+1)(\cos^2\theta + \cos^2\theta'') + 4n\cos\theta\cos\theta''}.$$

In the case of the air-water interface ($n_1 = 1$ and $n_2 = 1.33$), the Brewster angle is such that $\tan\theta_B = n$. In the case of Brewster incidence we obtain

$$\mathcal{P}' = 1 \qquad \text{and} \qquad \mathcal{P}'' = \frac{(n^2-1)\cos(2\theta_B)}{n^2+1+2n\,\sin(2\theta_B)} = -\frac{(n^2-1)^2}{n^4+1+6n^2} = -0.04.$$

In the case of a non-polarized wave and an incidence angle $\theta = 30°$, we find $\theta'' = 22°.08$; then, $\mathcal{P}' = 0.44$ and $\mathcal{P}'' = -0.0097$. Thus, at the Brewster incidence, the reflected wave is totally polarized perpendicularly to the plane of incidence, while the transmitted wave is slightly polarized in the plane of incidence. At an angle of incidence of 30°, the reflected wave is partially polarized perpendicularly to the plane of incidence, while the transmitted wave is very slightly polarized in the plane of incidence. It is the polarization of the reflected wave, which has an important variation as a function of the angle of incidence. The polarization of the transmitted wave varies very little because of the small energy reflection factor.

7.6. Case of attenuated waves in the second medium*

In this section we consider the case of damped waves in the second medium. This includes the case of total internal reflection, reflection on a reactive medium and reflection on a resistive medium. The incident wave and the reflected wave in the first medium are of the form $\underline{u} = \underline{u}_m e^{i(\omega t - \mathbf{k}.\mathbf{r})}$ and $\underline{u}' = \underline{u}'_m e^{i(\omega t - \mathbf{k}'.\mathbf{r})}$ where $\mathbf{k}$ and $\mathbf{k}'$ lie in the plane of incidence Oxz. They have the form [7.25] with a magnitude ω/v_1. We may always write the wave in the second medium in the general form

$$\underline{u}'' = \underline{u}''_m e^{i(\omega t - p_1 x - p_2 y - p_3 z)}. \qquad [7.49]$$

The equality of phases for $z = 0$ gives the relationships

$$\theta' = \theta, \qquad k' = k, \qquad p_1 = k\sin\theta = (\omega/v_1)\sin\theta \qquad \text{and} \qquad p_2 = 0. \qquad [7.50]$$

As for the parameter p_3, it is determined by the wave equation in the second medium. The reflection and refraction coefficients are determined by the boundary conditions on the interface.

1) Total internal reflection on a dispersive medium

This is the case of the reflection of electromagnetic waves on a less refringent medium, if the angle of incidence exceeds the critical angle i_L given by [7.22]. Wave [7.49] verifies the wave equation of the second medium $\ddot{\underline{u}} - (c^2/n_2^2)\,\Delta\underline{u} = 0$ if

$$p_1^2 + p_2^2 + p_3^2 = n_2^2\omega^2/c^2; \qquad [7.51]$$

that is,

$$p_3^2 = n_2^2\,\frac{\omega^2}{c^2}\,[1 - \frac{n_1^2}{n_2^2}\sin^2\theta\,] = \frac{4\pi^2}{\lambda_2^2}\,[1 - \frac{\sin^2\theta}{\sin^2 i_L}\,], \qquad [7.52]$$

where $\lambda_2 \equiv 2\pi c/n_2\omega$ is the wavelength in the second medium. As long as $\theta < i_L$, we have $\sin\theta < \sin i_L$ and p_3 is real. In this case, the three parameters (p_1, p_2, p_3) are the components of a wave vector in the second medium and the wave is progressive. On the other hand, if $\theta > i_L$, p_3 is imaginary of the form

$$p_3 = \pm\,i\delta \qquad \text{where} \qquad \delta = (2\pi/\lambda_2)\,\sqrt{\sin^2\theta/\sin^2 i_L - 1}\,. \qquad [7.53]$$

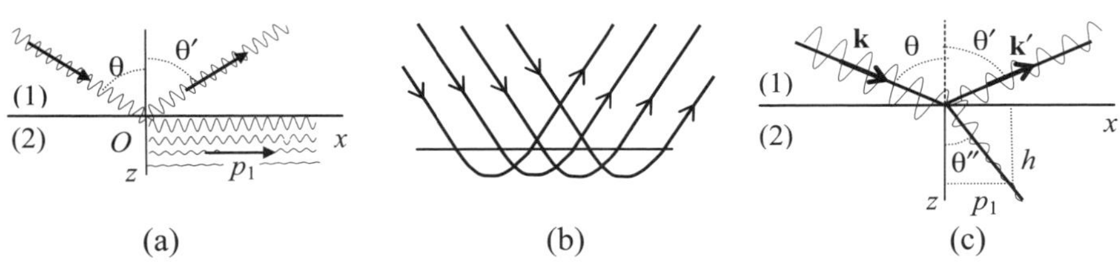

Figure 7.8. *a) Total internal reflection, b) propagation of the energy in the case of total reflection, and c) reflection on a resistive medium*

In this case, p_3 cannot be interpreted as the component of a wave vector. In general, the wave in the second medium is $\underline{u}'' = [\underline{B}''_m e^{-\delta z} + \underline{C}''_m e^{\delta z}\,]\,e^{i(\omega t - p_1 x)}$. The oscillatory factor of the wave is $e^{i(\omega t - p_1 x)}$. This indicates that $\underline{u}''$ propagates with a wave vector p_1 in the Ox direction (i.e. the intersection of the interface with the plane of incidence, see Figure 7.8a). Its phase velocity is $v_{(p)} = \omega/p_1 = v_1/\sin\theta$. The wave u'' is the sum of a term of amplitude $\underline{B}''_m e^{-\delta z}$, which decreases exponentially with z and a term of amplitude $\underline{C}''_m e^{\delta z}$ which increases exponentially with z. If the second medium is finite in the direction Oz, the two terms are possible, but if the second medium is infinite in the direction Oz, the second term is physically impossible and the wave exponentially decreases as $\underline{u}'' = u''_m e^{-\delta z} e^{i(\omega t - p_1 x)}$. According to [7.53], the attenuation coefficient δ of the wave in the direction of Oz

is proportional to the inverse of the wavelength λ_2 in the second medium. For instance, in the case of total reflection on the interface water-air, $\delta = 3.6\ \lambda_2^{-1}$ for $\theta = 60°$. The wave only penetrates a distance of the order of λ_2 in the second medium.

It may be shown that the reflected wave has the same amplitude as the incident wave but with a certain phase shift. Thus, the energy is totally reflected. In the second medium, we have a propagation of energy in the direction Ox but only at a depth of a few wavelengths (*evanescent wave*). The energy propagates as if it slightly penetrates the second medium where it propagates near the surface and returns to the first medium without any loss (Figure 7.8b). Thus, the wave undergoes *total reflection*, as if from a perfect mirror.

*2) Reflection on a reactive medium**

We consider, for instance, the case of an electromagnetic wave incident on a plasma where the wave equation is of the form $\ddot{\underline{u}} - v_2^2\,\Delta\underline{u} + \omega_p^2\underline{u}_2 = 0$. The solution in the plasma is in the form of [7.49] with the same relations [7.50] but, instead of [7.51], we have the relation $p_1^2 + p_2^2 + p_3^2 = (\omega^2 - \omega_p^2)/v_2^2$; hence,

$$p_3 = (\omega/v_2)\,\sqrt{1-(\omega_p/\omega)^2 - (v_2/v_1)^2\sin^2\theta}\ . \qquad [7.54]$$

The quantity inside the square root may be negative. This is the case if $\omega < \omega_p$ for any angle θ or if $\omega > \omega_p$ and θ is larger than a critical angle θ_L such that

$$\sin\theta_L = (v_1/v_2)\,\sqrt{1-(\omega_p/\omega)^2}\ . \qquad [7.55]$$

The situation is similar to internal reflection. The wave is damped in the plasma with a damping coefficient $\delta = (\omega/v_2)\sqrt{(v_2/v_1)^2\sin^2\theta + (\omega_p/\omega)^2 - 1}$. If the plasma is infinite toward the positive z, the energy is totally reflected with an amplitude equal to that of the incident wave. The wave only penetrates the plasma a distance of the order of λ_2.

If the quantity inside the square root of [7.54] is positive, p_3 is real and the wave in the plasma is progressive with a wave vector $\mathbf{k}'' = p_1\,\mathbf{e}_x + p_3\,\mathbf{e}_z$. This is the case if $\omega > \omega_p$ and $\theta < \theta_L$. The direction of propagation in the second medium forms an angle θ'' with the normal to the interface, such that

$$\sin\theta'' = \frac{p_1}{\sqrt{p_1^2 + p_3^2}} = \frac{v_2}{v_1}\frac{\sin\theta}{\sqrt{1-(\omega_p/\omega)^2}} = \frac{\sin\theta}{\sin\theta_L}\ . \qquad [7.56]$$

*3) Reflection on a resistive medium**

This is the case of an electromagnetic wave that is incident on a conductor (Figure 7.8c). The wave equation in the second medium is $\ddot{\underline{u}}_2 - v_2^2\,\Delta\underline{u}_2 + 2\beta\,\dot{\underline{u}}_2 = 0$ where $\beta = \sigma/\varepsilon_2$. The solution of this equation is in the form of [7.49] with relations [7.50] and $p_1^2 + p_2^2 + p_3^2 = \omega^2/v_2^2 - i\omega\beta/v_2^2$ instead of [7.51]; hence,

$$p_3^2 = (\omega^2/v_2^2)\,(\eta - i\beta/\omega) \qquad \text{with} \qquad \eta = 1 - (v_2^2/v_1^2)\sin^2\theta. \qquad [7.57]$$

This equation shows that p_3 is always complex of the form

$$\underline{p}_3 = h - i\delta \qquad [7.58]$$

$$h = \frac{\omega}{v_2\sqrt{2}}\,[\sqrt{\eta^2 + \beta^2/\omega^2} + \eta]^{1/2} \quad \text{and} \quad \delta = \frac{\omega}{v_2\sqrt{2}}\,[\sqrt{\eta^2 + \beta^2/\omega^2} - \eta]^{1/2}. \qquad [7.59]$$

The wave in the conductor is of the form $\underline{\mathbf{E}}'' = \underline{E}''_m\, e^{-\delta z}\, e^{i(\omega t - p_1 x - hz)}\,\mathbf{e}_y$. It propagates with an attenuation coefficient δ in the direction of Oz and a wave vector $p_1\,\mathbf{e}_x + h\,\mathbf{e}_z$, which makes with Oz an angle θ'' given by

$$\tan\theta'' = p_1/h = \sqrt{2}\,(\omega v_2/\beta v_1)\sin\theta\,[\sqrt{\eta^2 + \beta^2/\omega^2} - \eta]^{1/2}. \qquad [7.60]$$

For instance, in the case of a thick conductor of conductivity $\sigma \approx 5 \times 10^7\,\text{m}^{-1}\,\Omega^{-1}$, the characteristic angular frequency β is about 10^{18} rad/s. In the case of light, the ratio β/ω is of the order of 10^4, thus much larger than η in [7.59] and we may write $h \approx \delta \approx \sqrt{\beta\omega/2v_2^2}$. It may be shown that the boundary conditions give the reflection and refraction coefficients

$$\underline{\mathcal{R}} = \frac{E'_m}{E_m} \approx -1 + \gamma, \quad \underline{\mathcal{T}} = \frac{E''_m}{E_m} \approx \gamma \qquad \text{where } \gamma = 2\frac{\mu_2 v_2}{\mu_1 v_1}\sqrt{\frac{\omega}{\beta}}\; e^{i\pi/4}\cos\theta. \qquad [7.61]$$

The wave is nearly totally reflected with a phase shift near π. In this case, there is a small transfer of energy across the surface of the conductor. This energy is totally dissipated in the conductor as Joule heat. In the case of a superconductor, σ and β are infinite, $\mathcal{T} = 0$ and $\mathcal{R} = -1$. Thus, there is no transmitted wave and the superconductor is a perfect obstacle.

7.7. Summary

If a wave of amplitude $\underline{u}_m$ is incident on the interface of two media, the amplitude $\underline{u}'_m$ of the reflected wave and the amplitude $\underline{u}''_m$ of the transmitted wave are determined by the *boundary conditions* on the interface. The *reflection coefficient* is $\mathcal{R} = \underline{u}'_m/\underline{u}_m$ and the *transmission coefficient* is $\mathcal{T} = \underline{u}''_m/\underline{u}_m$.

In the case of an elastic wave on two joined strings, the continuity conditions are $u + u' = u''|_{x=0}$ and $\partial_x u + \partial_x u' = \partial_x u''|_{x=0}$. If P, P' and P'' are, respectively, the incident power, reflected power and transmitted power, the energy reflection and transmission factors are $f_R = |P'|/P = \mathcal{R}^2$ and $f_T = |P''|/P = (v_1/v_2)\mathcal{T}^2$ with $f_R + f_T = 1$. If $Z_1 = Z_2$, we find $\mathcal{R} = 0$ and there is no reflected wave (*adapted impedances*).

In the case of an acoustic wave in a pipe, the continuity conditions are $\underline{u} + \underline{u}' = \underline{u}''|_{x=0}$ and $\underline{p}_a + \underline{p}_a' = \underline{p}_a''|_{x=0}$.

In the case of a three-dimensional wave, the Huygens principle allows us to establish the following laws of reflection and refraction:

– The directions of propagation of the incident, reflected and refracted waves are in the incidence plane, which is normal to the interface $\mathcal{S}$.

– The angle of reflection θ' is equal to the angle of incidence θ. The angle of refraction θ'' is given by Snell's law, $n_1 \sin\theta = n_2 \sin\theta''$.

There is always a refracted wave if $n_1 < n_2$ but, if $n_1 > n_2$, the angle θ must be less than a *critical angle* i_L given by $\sin i_L = n_2/n_1$, which corresponds to an angle of refraction $\theta'' = 90°$. If $\theta > i_L$, the wave undergoes *total internal reflection.*

Linear boundary conditions on the interface $\mathcal{S}$ are verified if the phases of the three waves are the same on $\mathcal{S}$: $\omega t - \mathbf{k}.\mathbf{r} = \omega' t - \mathbf{k}'.\mathbf{r} = \omega'' t - \mathbf{k}''.\mathbf{r}$. The same boundary conditions determine the coefficients $\mathcal{R}$ and $\mathcal{T}$ and the energy factors f_R and f_T.

In the case of a three-dimensional acoustic wave, the boundary conditions are: $u_z + u'_z = u''_z$ and $p_a + p_a' = p_a''$.

In the case of an electromagnetic wave that is polarized in the plane of incidence or perpendicularly to this plane, the reflected wave and the refracted wave have the same polarization as the incident wave. On the interface of two dielectrics, the normal components of $\mathbf{B}$ and $\mathbf{D} = \varepsilon\mathbf{E}$ and the tangential components of $\mathbf{H} = \mathbf{B}/\mu$ and $\mathbf{E}$ are continuous. We deduce the values of $\mathcal{R}$ and $\mathcal{T}$ and the energy factors f_R and f_T in each case. If the media are non-magnetic, we may write for small angles

$$\mathcal{R}_{//} \approx \mathcal{R}_{\perp} \approx \frac{\theta''-\theta}{\theta''+\theta} \approx \frac{n_1 - n_2}{n_1 + n_2}, \qquad \mathcal{T}_{//} \approx \mathcal{T}_{\perp} \approx \frac{2\theta''}{\theta''+\theta} \approx \frac{2n_1}{n_1 + n_2}.$$

If $n_1 > n_2$, the reflection occurs without any phase shift for **E**; but if $n_1 < n_2$, there is a phase shift of π. The transmitted wave is always in phase with the incident wave. The energy reflection and transmission factors are

$$f_R \equiv S'_z/S_z = \mathcal{R}^2, \qquad f_T \equiv S''_z/S_z = \mathcal{T}^2\,(n_2/n_1)\,(\cos\theta''/\cos\theta).$$

The conservation of energy on $\mathcal{S}$ may be written as

$$f_R + f_T = 1, \qquad \text{i.e.} \qquad n_1\,(1 - \mathcal{R}^2)\cos\theta = n_2\,\mathcal{T}^2\cos\theta''.$$

$\mathcal{R}_{//}$ is equal to zero at Brewster incidence θ_B such that $\theta_B + \theta'' = \pi/2$, that is, $\tan\theta_B = n_2/n_1$. A non-polarized wave, which is incident at the Brewster angle, is reflected polarized perpendicularly to the plane of incidence.

If θ is larger than the critical angle i_L, the boundary conditions show that the wave propagates in the second medium parallel to the surface with an amplitude which decreases exponentially as a function of the distance to the interface. The attenuation coefficient is $\delta = \eta\omega/v_2 = (2\pi/\lambda_2)\sqrt{\sin^2\theta/\sin^2 i_L - 1}$, where λ_2 is the wavelength in the second medium. The amplitude of the reflected wave is equal to that of the incident wave (total reflection).

If an electromagnetic wave is incident on a plasma, a part of it is transferred as a progressive wave in the plasma if $\omega > \omega_p$ and θ is less than a limiting angle θ_L, such that $\sin\theta_L = (v_1/v_2)\sqrt{1 - \omega_p^2/\omega^2}$; otherwise, the wave is totally reflected.

If an electromagnetic wave is incident on a conductor, a part of it is transferred as an attenuated wave in the conductor. In the case of the usual metallic conductors when $\omega << 10^{18}$ rad/s, the wave is nearly totally reflected with a phase shift close to π.

7.8. Problem solving suggestions

1) It is often more applicable to use the complex representation of waves rather than the trigonometric real representation. The boundary conditions require that the phase $(\omega t - \mathbf{k.r})$ be the same for the incident, reflected and transmitted waves at each point of the surface and at any time. Knowing **k** for the incident wave, the equality of the phases and the wave equations in the media (or the dispersion relation), we can determine **k**′ and **k**″, that is, the direction of the reflected wave and that of the transmitted wave. If the angle of incidence is larger than the critical angle i_L or if the second medium is reactive or resistive, the equality of phases and the wave

equations in the media require that k'' be complex. Thus, the wave is attenuated in the second medium. The boundary conditions also determine the coefficients $\mathcal{R}$ and $\mathcal{T}$.

2) Do not confuse the *reflection coefficients* $\mathcal{R}$ and the *transmission coefficients* $\mathcal{T}$ with the *energy reflection factor* f_R and *transmission factor* f_T. $\mathcal{R}$ and $\mathcal{T}$ are defined respectively as the ratios of the reflected amplitude and transmitted amplitude to the incident amplitude. In the case of a one-dimensional wave, f_R and f_T are respectively the ratios of the reflected power and the transmitted power to the incident power (with the relation $f_R + f_T = 1$). In the case of a three-dimensional wave, f_R and f_T are not defined as the ratios of the reflected intensity $\mathcal{I}'$ and transmitted intensity $\mathcal{I}''$ to the incident intensity $\mathcal{I}$; rather f_R is the ratio of the reflected power $\langle S'_z \rangle$ to the incident power $\langle S_z \rangle$ per unit area of the interface, and f_T is the ratio of the transmitted power $\langle S''_z \rangle$ to the incident power $\langle S_z \rangle$ per unit area. It is only with this definition that the conservation of energy law takes the form $f_R + f_T = 1$.

3) On the interface of two media of the same impedance, there is no reflection. A medium, in which the wave cannot propagate (that is, an obstacle), has an *infinite impedance*. A wave, which is incident on an obstacle, is totally reflected ($|\mathcal{R}| = 1$) or diffused back to the incidence medium. It should be noted that a conductor is not an obstacle for electromagnetic waves; a fraction of the energy penetrates the conductor and is dissipated as Joule heat. Thus, a conductor has a *complex impedance* and the energy reflection factor is less than 1. A superconductor is a perfect obstacle.

7.9. Conceptual questions

Q1. Linear boundary conditions always lead to the equality $\omega = \omega' = \omega''$. Do you expect this result considering the response of the medium? Do the reflected wave and the refracted wave have the same wavelength as the incident wave? Explain how a boundary condition, which is linear in the waves and their derivatives, allows us to establish the laws of reflection and refraction for the direction of propagation. Analyze the symmetries and show that the wave vectors $\mathbf{k}$, $\mathbf{k}'$ and $\mathbf{k}''$ are in the incidence plane. What happens if the boundary conditions are quadratic? How many independent boundary conditions are required on the interface of two media, on a rigid obstacle and on the surface of a medium with vacuum?

Q2. An electromagnetic wave is incident at an angle θ on the interface $\mathcal{S}$ of two dielectrics and it is polarized in the plane of incidence. Specifying the waves by the magnetic fields, write down the expressions of the wave vectors and the fields. Write the boundary conditions and verify that they reduce to two independent equations, which allow us to determine the coefficients $\mathcal{R}_{//}$ and $\mathcal{T}_{//}$. Make a similar analysis in the case of a wave that is polarized perpendicularly to the plane of incidence by specifying the waves according to the electric fields.

Q3. The Brewster angle is defined by tan $\theta_B = n_2/n_1$. Verify that the reflected ray and the refracted ray are perpendicular. Compare the angles θ_{B12} for the incidence from medium (1) on medium (2), and θ_{B21} from medium (2) on medium (1).

Q4. A prism of index n = 1.5 has a right isosceles triangular section with the angle A equal to 90°. It receives a light beam nearly normal to the face AB. Show that this beam emerges from the face AC. Discuss its polarization and its amplitude compared to the reflection on a mirror BC.

Q5. Figure 7.9 illustrates a setup of two identical prisms, which are separated by a thin air layer. A light beam is incident on the face AB. It propagates in the first prism and it falls on the face AC at an incidence angle larger than the critical angle. If the air layer is sufficiently thin, a fraction of the light beam may enter the second prism and finally emerge parallel to the incident beam. Explain this effect. What should the thickness of the air layer be? Can you think of an application of this setup?

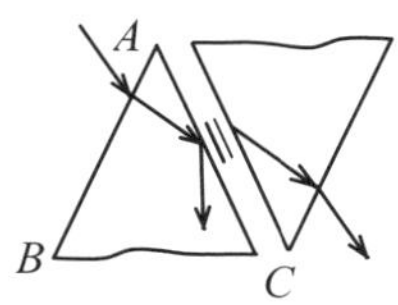

Figure 7.9. *Question Q5*

7.10. Problems

Elastic wave on two joined strings

P7.1. A copper string and an iron string of the same radius r = 2 mm are soldered together and taut at a tension of 500 N. A wave of amplitude 1 cm and frequency 20 Hz propagates on the copper string and arrives at the junction point O. **a)** Write down the expressions of the incident, reflected and transmitted waves and calculate their amplitudes. The mass density of copper is 8.9 g/cm^3 and that of iron is 7.8 g/cm^3. **b)** Calculate the energy flux for the three waves and verify the conservation of energy.

P7.2. A string lies at equilibrium on the x-axis and it is attached to a ring of mass M, which slides on a transverse rigid rod in the direction of Oy with a friction force $-b\dot{y}$ (Figure 7.10). We assume that the string is infinite toward the negative x. **a)** Write the boundary conditions on the ring. Deduce the reflection coefficient. **b)** Consider the limiting case $M = \infty$ and $M = 0$ with $b = 0$ and find the known results for the reflection on an obstacle. **c)** Is it possible to choose the ring in order to have no reflected wave? Analyze the energy conservation in this case.

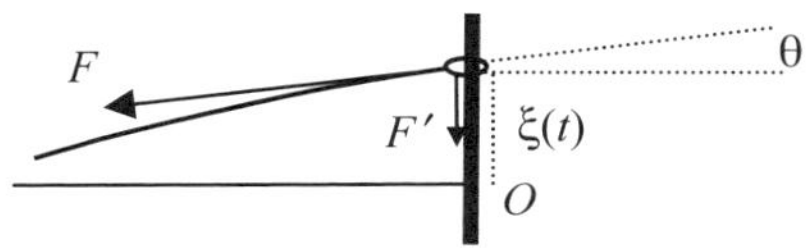

Figure 7.10. *Problem P7.2*

Reflection of an acoustic wave

P7.3. An acoustic wave is incident normally on the surface of water. The speed of sound is 332 m/s in air and 1,500 m/s in water. Determine the reflection and transmission coefficients. What is the fraction of energy entering the water?

P7.4. A pipe of section $\mathcal{S}$ is closed by a membrane of mass m, subject to a friction force of constant b and a restoring force of constant K. Assume the space behind this membrane is empty. **a)** What is the impedance of this membrane? **b)** An acoustic wave u propagates along this pipe and it is reflected on the membrane. Determine the reflection coefficient. **c)** Calculate the incident and the reflected energy fluxes. What happens to the difference between these energies? How should we choose the membrane in order to have the minimum reflected intensity and to have the maximum reflected intensity? **d)** Assume that $m = 5$ g, $b = 10$ N.s.m^{-1}, $K = 1{,}000$ N.m^{-1}, $\mathcal{S} = 50$ cm^2, the pipe is filled with air ($m_v = 1.293$ kg/m^2, $v_s = 332$ m/s) and the wave has a frequency $\tilde{\nu} = 500$ Hz and a displacement amplitude of 1.0×10^{-5} m. Calculate the impedance of the membrane, the reflection coefficient and the reflected intensity.

P7.5. A light plane wave is incident on the interface $\mathcal{S}$ of two media (1) and (2). Consider three rays, which reach $\mathcal{S}$ at times 0, τ and 2τ respectively, and the secondary waves that are emitted according to Huygen's principle in media (1) and (2). Establish the reflection and refraction laws.

P7.6. An acoustic wave is incident on the air-water interface from the air side at an angle θ. The speed of sound is 332 m/s in air and 1,500 m/s in water. **a)** Analyze the variation of the angle of refraction as a function of θ. **b)** Analyze the variation of the reflection coefficient as a function of θ. Make the same analysis if the wave is incident from the water side.

P7.7*. An acoustic wave is incident on a glass plate of thickness e and Young's modulus Y. **a)** Write the boundary conditions on the faces of the plate and show that we have to consider the elastic waves in both directions in the plate. **b)** Calculate the global transmission coefficient for this plate. What should the thickness e of the plate be in order to have the maximum energy transmission factor?

Reflection and refraction on the interface of two dielectrics

P7.8*. An electromagnetic plane wave is polarized in a direction that makes an angle α with the plane of incidence. **a)** Show that the reflected and transmitted electric fields form with the plane of incidence respectively the angles α' and α'' given by the relations $\tan\alpha' = (\mathcal{R}_\perp/\mathcal{R}_{//})\tan\alpha$ and $\tan\alpha'' = (\mathcal{T}_\perp/\mathcal{T}_{//})\tan\alpha$ where $\mathcal{R}_\perp$ and $\mathcal{R}_{//}$ are the reflection coefficients, while $\mathcal{T}_\perp$ and $\mathcal{T}_{//}$ are the transmission coefficients for parallel polarization and perpendicular polarization to the plane of incidence. **b)** Calculate the coefficients $\mathcal{R}_\perp$, $\mathcal{T}_\perp$, $\mathcal{R}_{//}$ and $\mathcal{T}_{//}$ for light that is incident on water ($n = 1.33$) at 60°. Calculate α' and α'' if $\alpha = 45°$.

P7.9*. An electromagnetic wave is incident at an angle θ on the interface of two dielectrics of indices n_1 and n_2. The Fresnel formulas show that the amplitude of the transmitted electric field may be larger than that of the incident field. **a)** Verify that the incident power is always equal to the sum of the reflected and transmitted powers. **b)** A light wave is incident normally on a plate of index 1.5. Neglecting multiple reflections, calculate the global energy transmission factor. What should this factor be for a setup of four parallel but separated plates?

P7.10. A non-polarized wave is incident at 45° on a medium of index 1.5. Calculate the energy reflection and transmission factors. Determine the degree of polarization in the plane of incidence for the reflected and the transmitted waves.

P7.11. A non-polarized wave is incident on a dielectric of index n at the Brewster angle. Calculate the energy reflection and transmission factors.

P7.12.** **a)** An electromagnetic wave that is polarized perpendicularly to the plane of incidence falls on a glass plate of index 1.5. Show that the intensity of the reflected wave varies from 4% to 100% as θ varies from 0 to 90°. Calculate the reflected intensity at $\theta = 45°$. **b)** Show that, in the case of a polarized wave in the plane of incidence, the reflected intensity decreases from 4% at $\theta = 0$ to zero at Brewster incidence, then it increases to 100% at $\theta = 90°$. What is the reflected intensity at $\theta = 45°$? **c)** A wave is incident on the glass-air interface from the glass side. Analyze the reflected intensity if the wave is polarized in the plane of incidence and if it is polarized perpendicularly to this plane. **d)** A loss of 4% is sometimes intolerable if light crosses the plate many times. Show that, at the Brewster incidence, after a large number of crossings, the light component that is polarized perpendicularly to the plane of incidence is eliminated and only the component that is polarized in this plane remains with an intensity nearly equal to the incident intensity.

Total reflection

P7.13.** An electromagnetic wave is incident from a medium of index n_1 onto a medium of index n_2 ($n_1 > n_2$) at an angle $\theta > i_L$. The wave is polarized in the plane of incidence. **a)** Write down the expressions of the incident wave and the reflected wave and specify their wave vectors. **b)** Write down the expression of the transmitted wave in the case of an infinite second medium and an incidence angle $\theta > i_L$. Interpret this expression. Calculate the attenuation coefficient of the wave in the case of the reflection on the water-air interface at $\theta = 60°$. **c)** Determine the amplitude of the reflected wave and that of the transmitted wave and their phase shifts. **d)** Calculate the mean values of the Poynting vector over a period and discuss the propagation of energy.

P7.14.** **a)** What is the critical angle on the glass-air interface if the glass index is 1.5? A wave is incident at 50° on the glass-air interface from the glass side. Calculate the attenuation coefficient in air for light of wavelength 0.6 μm in vacuum. How deeply does it penetrate the air? **b)** Consider a thin layer of air between the planes $z = 0$ and $z = L$ separating two glass plates. L is of the order of the wavelength of light. A wave propagates in the first plate and falls on the surface $z = 0$ at an angle of incidence $\theta > i_L$. Assume that the wave is polarized in the plane of incidence. Write the boundary conditions on the faces of the air layer. Show that the wave may cross this layer, produce a wave in the second plate and propagate parallel to the direction of propagation in the first plate. Calculate the amplitude of the reflected wave and that of the transmitted wave for $L = 1$ μm. Note that, in the air layer, the wave has the form $f\,e^{-\delta z} + g\,e^{\delta z}$.

Reflection on a plasma and a conductor

P7.15. An electromagnetic wave is incident at an angle θ from a medium of index n_1 on a plasma (the ionosphere, for instance) in which the propagation equation is $\ddot{\mathbf{E}}_2 - v_2^2\Delta\mathbf{E}_2 + \omega_p^2\mathbf{E}_2 = 0$. The wave is polarized in the plane of incidence. **a)** Write down the expressions of the incident wave and the reflected wave and specify the wave vectors. b) Show that the transmitted wave may be written in the form $\underline{\mathbf{B}}'' = \underline{B}''_m\,e^{i(\omega t - px - qy - sz)}\,\mathbf{e}_y$ where

$$p = (\omega/v_1)\sin\theta, \qquad q = 0 \qquad \text{and} \quad s = (\omega/v_2)\sqrt{1 - \omega_p^2/\omega^2 - (v_2^2/v_1^2)\sin^2\theta}\,.$$

Discuss the nature of the wave in the plasma depending on the values of ω and the angle θ. Show that there is total reflection if $\omega < \omega_p$, for an arbitrary angle θ or if $\omega > \omega_p$ and $\theta > \theta_L$ such that $\sin\theta_L = (v_1/v_2)\sqrt{1-\omega_p^2/\omega^2}$.

P7.16. In this problem we analyze the reflection and refraction of a one-dimensional scalar wave that is incident from a dispersive medium (1) on a reactive or resistive medium (2). We assume that wave equation takes the form of $\ddot{u}_1 - v^2(\partial^2_{xx}u_1) = 0$ in medium (1) and $\ddot{u}_2 + 2\beta\dot{u}_2 - v^2(\partial^2_{xx}u_2) + \omega_o^2 u_2 = 0$ in the medium (2). The wave obeys to the boundary conditions: $u + u' = u''$ and $\partial_x u + \partial_x u' = \partial_x u''$ at $x = 0$. **a)** Determine the reflection and transmission coefficients. **b)** Write the expressions of these coefficients for the case of a purely reactive medium ($\beta = 0$) according to the values of ω. **c)** Write these coefficients for the case of a purely resistive medium ($\beta \neq 0$ but $\omega_o = 0$) according to the values of ω.

P7.17*. An electromagnetic wave is incident on a conducting medium, in which the wave equation is $\ddot{\mathbf{E}}_2 - v_2^2\Delta\mathbf{E}_2 + \beta\mathbf{E}_2 = 0$ where $\beta = \sigma/\varepsilon_2$. The wave is polarized perpendicularly to the plane of incidence. **a)** Write down the expressions of the incident wave and the reflected wave and specify the wave vectors. **b)** Show that the transmitted wave may be written in the form of $\underline{\mathbf{B}}'' = \underline{B}''_m e^{i(\omega t - px - qy - sz)}\,\mathbf{e}_y$ where $p = (\omega/v_1)\sin\theta$, $q = 0$ and $\underline{s} = h - i\delta$, such that

$$h = \frac{\omega}{v_2\sqrt{2}}\left[\sqrt{\eta^2+\beta^2/\omega^2} + \eta\right]^{1/2} \quad \text{and} \quad \delta = \frac{\omega}{v_2\sqrt{2}}\left[\sqrt{\eta^2+\beta^2/\omega^2} - \eta\right]^{1/2}$$

with $\eta = 1 - (v_2^2/v_1^2)\sin^2\theta$. Interpret this result. Write down the expressions of $\underline{\mathbf{E}}''$ and $\underline{\mathbf{B}}''$. **c)** Write the boundary conditions, assuming that the surface of the conductor has no charge density or surface current density. Deduce the reflection and transmission coefficients. **d)** Show that, in the case of a good conductor ($\sigma \approx 5\times10^6$ m$^{-1}\,\Omega^{-1}$) and for a frequency that is not very high, we have $\beta/\omega \ll 1$, $h \approx \delta \approx \sqrt{\omega_o\omega/2}\,/v_2$ and:

$$\underline{\mathcal{R}} = \frac{\underline{E}'_m}{\underline{E}_m} \approx -1 + 2\frac{\mu_2 v_2}{\mu_1 v_1}\sqrt{\frac{\omega}{\omega_o}}\,e^{i\pi/4}\cos\theta, \qquad \underline{\mathcal{T}} = \frac{\underline{E}''_m}{\underline{E}_m} \approx 2\frac{\mu_2 v_2}{\mu_1 v_1}\sqrt{\frac{\omega}{\omega_o}}\,e^{i\pi/4}\cos\theta$$

Thus, the wave is nearly totally reflected with a phase shift π.

P7.18*. An electromagnetic wave is incident from a dielectric (medium 1) normally on the surface of a good conductor (medium 2). **a)** Show that Maxwell's equations in the conductor are the same as those in a dielectric with a complex dielectric constant $\underline{\varepsilon} = \varepsilon_2 - i\sigma/\omega$. Write the solution representing a plane wave, and deduce the usual properties of waves in conductors. **b)** Is it possible to define a complex index using $\underline{\varepsilon}$, and write Snell's law to determine the direction of propagation in the conductor? **c)** Consider a good conductor that is not magnetic (like silver with $\sigma = 6.29 \times 10^7\ \Omega.\mathrm{m}^{-1}$). Assume that $v_2 = v_1 = c$ and use the exact expression of $\mathcal{R}$. For which frequencies is the electromagnetic wave reflected with more than 95% in intensity? Estimate the modulus and the phase of the reflection coefficient on silver in the case of visible light of wavelength 0.5 µm. What should be the thickness of a silver layer to allow the transfer of no more than 10% of light intensity through the face of entry?

Chapter 8

Interference and Diffraction

If two beams of particles are intercepted on the same region of a screen, the energy, momentum and other physical quantities, which are received at each point, are the sums of the corresponding quantities for the beams if they are intercepted separately. This is not the case if two waves superpose in a region of space; the experiment shows that there is a redistribution of physical quantities. We say that the waves *interfere*. On the other hand, a free particle has rectilinear motion even if it passes near a body or through an aperture with which it does not interact. The experiment shows that a wave deviates from its straight-line propagation if it encounters an obstacle, or it passes through an aperture whose dimensions are comparable to the wavelength. This effect is known as *diffraction*. Interference and diffraction are two characteristic properties of waves; they are consequences of the *superposition principle* of waves and the quadratic dependence of the physical quantities on u.

In this chapter we study the interference and diffraction of mechanical and electromagnetic waves. In the case of vector waves, we assume that they are either longitudinal or transverse but polarized linearly in the same direction; this allows us to replace their vector sums by algebraic sums, just like in the case of scalar waves. We will not discuss the questions that are specific to optics, such as coherence, interference setups, effects of diffraction on the formation of images, polarization, etc.

8.1. Order and fringes of interference of two waves

Consider two waves with equal frequencies and equal amplitudes, that are emitted by two sources S_1 and S_2 at a sufficiently long distance from the region of observation, so that the amplitude of the waves varies little in this region (Figure 8.1a). By "sources", we mean either the emitters of the waves or two small apertures through which the waves pass before their superposition at a point M. If a wave travels a distance r in a homogeneous medium, it undergoes a phase lag $\omega r/v$, were v is the speed of propagation. If the waves at the sources are $u_1 = A\cos(\omega t - \phi_1)$ and $u_2 = A\cos(\omega t - \phi_2)$, their expressions at M are $u_1 = A\cos[\omega(t - r_1/v) - \phi_1]$ and $u_2 = A\cos[\omega(t - r_2/v) - \phi_2]$, where r_1 and r_2 are the distances of M to the sources, measured along the wave path. The phases ϕ_1 and ϕ_2 of the sources may always be chosen in such a way that the amplitudes are positive. The resultant wave at M is

$$u(M) = u_1 + u_2 = 2a\cos(½\Delta\phi)\cos(\omega t - \phi_1 - ½\Delta\phi), \qquad [8.1]$$

where $\Delta\phi$ is the phase lag of the wave (2) over the wave (1) at M. It is given by

$$\Delta\phi = \Delta_o\phi + \Delta_s\phi \qquad \text{with} \qquad \Delta_o\phi \equiv \phi_2 - \phi_1 \qquad \text{and} \qquad \Delta_s\phi \equiv (\omega/v)\,\Delta r = 2\pi\,(\Delta r/\lambda). \quad [8.2]$$

$\Delta_o\phi$ is the initial phase lag (at $t = 0$ and $r_1 = r_2 = 0$) and $\Delta_s\phi$ is the phase lag due to the *path difference* $\Delta r \equiv r_2 - r_1$ at M. Thus, each path difference Δr equal to a wavelength introduces a phase lag of 2π. If the wave propagates in different media, the phase shift due to the path difference is the sum of the phase shifts due to the path difference in each medium (evaluated with the corresponding wavelength).

Expression [8.1] shows that the total wave at M is a sinusoidal function of time, whose angular frequency is ω and amplitude is $A(M) = 2A\cos(½\,\Delta\phi)$. If $\Delta\phi$ is an integer multiple of 2π, the amplitude at M is maximum in absolute value and equal to $2a$; the waves at M are *in phase* and the interference is *constructive*. Alternatively, if $\Delta\phi$ is a half-integer multiple of 2π, the amplitude $A(M)$ vanishes; the waves at M are *in opposite phase* and the interference is *destructive*:

$$\Delta\phi = 2p\pi \qquad p = \text{integer} \qquad [\text{maximums}] \qquad [8.3]$$

$$\Delta\phi = 2p\pi \qquad p = \text{half-integer} \qquad [\text{minimums}]. \qquad [8.4]$$

Thus, the state of interference at M is determined by the phase shift between the waves at M, that is, the *order of interference* p defined by

$$p \equiv \Delta\phi/2\pi = \Delta r/\lambda + \Delta_o\phi/2\pi. \qquad [8.5]$$

The interference is constructive if the order p is an integer number (positive, negative or zero) and destructive if the order p is a half-integer number. The locus of the maxima of given order p is an *interference fringe*. In particular, the order 0 corresponds to the waves reaching M with a phase shift equal to 0. Each increase of the path difference by $\Delta r = \lambda$ increases the order of interference by 1. In optics, the constructive interference fringes are bright while destructive interference fringes are dark. They may be easily observed on a screen or on a photographic plate. They form the *interference pattern.*

In the case of a homogeneous medium, the wave paths are straight lines; the locus of the maxima of given order p is a hyperboloid of foci S_1 and S_2. Figure 8.1a shows their intersections with a plane containing S_1 and S_2. These are branches of hyperbola, each one corresponding to an integer value of the order of interference p. If the sources are in phase $\phi_1 = \phi_2$, the fringe of order $p = 0$ lies in the median plane of S_1S_2. For instance, Figure 8.1a is the interference pattern on the surface of a liquid if S_1 and S_2 are the end points of a U-shaped rod excited by a vibrator and slightly touching the liquid surface.

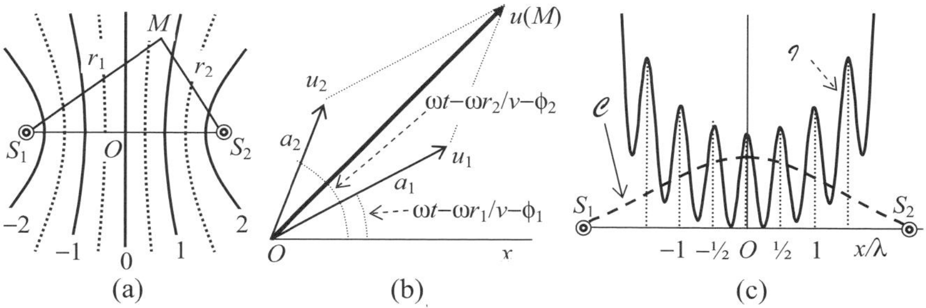

Figure 8.1. *a) Interference pattern of two waves in a plane containing two sources in phase. The fringes of the minima are drawn as dotted lines and the fringes of the maxima are drawn as solid lines. The number indicates the order of interference p. b) Phasor diagram in the general case $a_1 \neq a_2$. c) The intensity and the contrast as functions of the position x/λ on the axis S_1S_2. Near a source, the amplitude of the wave that it emits and the total intensity become very large and the contrast approaches zero.*

8.2. Intensity and contrast

The energy flux density S associated with a sound wave, represented by the acoustic pressure p_a, is proportional to p_a^2, and that of an electromagnetic wave is proportional to E^2. Thus, for a wave of any type $u = A\cos(\omega t - \mathbf{k.r} - \phi)$, we may write

$$S = \alpha\, u^2 = \alpha\, a^2 \cos^2(\omega t - \mathbf{k.r} - \phi), \qquad [8.6]$$

where α is a factor that depends on the nature of the wave. The intensity of the wave $\mathcal{I}$ is the average of the energy flux density over a period

$$\mathcal{I} = \tfrac{1}{2}\alpha\, a^2. \qquad [8.7]$$

It is often convenient to use the complex representation of the wave

$$u = A\,e^{i(\omega t - \omega r/v - \phi)}. \qquad [8.8]$$

Its intensity may be written as

$$\mathcal{I} = \tfrac{1}{2}\,\alpha u u^*. \qquad [8.9]$$

In the case of the superposition of two waves of equal amplitudes but with a phase shift $\Delta\phi$, the resultant wave is given by [8.1]. Its intensity is

$$\mathcal{I}(M) = \mathcal{I}_{max}\cos^2(\tfrac{1}{2}\Delta\phi) \qquad \text{where} \quad \mathcal{I}_{max} = 2\alpha a^2. \qquad [8.10]$$

As a function of $\Delta\phi$, $\mathcal{I}(M)$ oscillates between minimums equal to zero and maximums equal to $\mathcal{I}_{max}$. As the value of $\cos^2(\tfrac{1}{2}\Delta\phi)$ is ½, the average value of $\mathcal{I}(M)$ over space is $<\mathcal{I}(M)> = \alpha a^2$. Thus, the interference produces a redistribution of the intensity, the average intensity being the sum of the intensities emitted by the sources.

In general, the amplitudes of the waves may be different and they may depend on the observation point M. Using the complex representation, the waves at M are

$$\underline{u}_1 = a_1(M)\,e^{i(\omega t - \phi_1 - \omega r_1/v)} \equiv \underline{a}_1(M)\,e^{i\omega t}, \qquad \underline{u}_2 = a_2(M)\,e^{i(\omega t - \phi_2 - \omega r_2/v)} \equiv \underline{a}_2(M)\,e^{i\omega t}.$$

The resultant wave may be written as $\underline{u}(M) = u_1 + u_2 = [\underline{a}_1(M) + \underline{a}_2(M)]\,e^{i\omega t}$. The phase shift between $\underline{a}_1(M)$ and $\underline{a}_2(M)$ is just $\Delta\phi$, given by [8.2]. The intensity at M can then be written as

$$\begin{aligned}\mathcal{I}(M) &= \tfrac{1}{2}\alpha|\underline{a}_1(M) + \underline{a}_2(M)|^2 = \tfrac{1}{2}\alpha[a_1(M)^2 + a_2(M)^2 + 2a_1(M)a_2(M)\cos(\Delta\phi)\\ &= \mathcal{I}_1(M) + \mathcal{I}_2(M) + 2\sqrt{\mathcal{I}_1(M)\,\mathcal{I}_2(M)}\,\cos(\Delta\phi). \qquad [8.11]\end{aligned}$$

$\mathcal{I}_1(M)$ and $\mathcal{I}_2(M)$ are the intensities of the waves if they are observed separately. The third term is the *interference term*. It may be positive or negative. Thus, the resultant intensity varies between a minimum $\mathcal{I}_{min}$ and a maximum $\mathcal{I}_{max}$ given by

$$\mathcal{I}_{min} = [\sqrt{\mathcal{I}_1(M)} - \sqrt{\mathcal{I}_2(M)}\,]^2 \qquad \text{and} \quad \mathcal{I}_{max} = [\sqrt{\mathcal{I}_1(M)} + \sqrt{\mathcal{I}_2(M)}\,]^2. \qquad [8.12]$$

Figure 8.1b is the phasor diagram for the interference of the waves. The parallelogram rotates about O at the angular velocity ω without deformation. If the waves have different amplitudes but the same frequency and a fixed phase shift, the resultant amplitude varies between a minimum $|a_1 - a_2|$ if the order $p_M = \Delta\phi/2\pi$ is equal to a half-integer (then the vectors point in opposite directions), and a maximum ($a_1 + a_2$) if the order p_M is equal to an integer (then, the vectors point in the same direction). We define the *contrast* or *visibility factor* by

$$\mathcal{C} = \frac{\mathcal{I}_{max} - \mathcal{I}_{min}}{\mathcal{I}_{max} + \mathcal{I}_{min}} = \frac{2\sqrt{\mathcal{I}_1(M)\mathcal{I}_2(M)}}{\mathcal{I}_1(M) + \mathcal{I}_2(M)}. \qquad [8.13]$$

$\mathcal{C}$ always lies between 0 and 1. The maximum value $\mathcal{C} = 1$ corresponds to $\mathcal{I}_{min} = 0$, that is, the waves have equal amplitudes. The minimum value $\mathcal{C} = 0$ corresponds to $\mathcal{I}_{max} = \mathcal{I}_{min}$, that is, the total absence of interference fringes near the point M.

8.3. Interference of light waves, Young's experiment

Mechanical waves are emitted by macroscopic sources, whose frequency and phase are easy to control and their interference is very easy to observe. On the contrary, light sources are constituted by an extremely large number of atoms, which are not monochromatic and coherent (before the discovery of lasers). To observe light interference, Young had the brilliant idea of splitting the same light wave into two coherent parts and observing their interference after travelling different paths. This historic experiment in 1807 enabled us to verify the wave nature of light and even to measure its wavelength. Figure 8.2a shows Young's experimental setup: The same light wave (emitted by a source S at finite or infinite distance) illuminates two small apertures S_1 and S_2 (pinholes or parallel narrow slits), pierced in an opaque screen P a distance d apart and at equal distance from S. Light is diffracted by these apertures exactly like two waves that are emitted by two identical sources. If S_1 and S_2 are sufficiently narrow and close to each other, the diffracted beams interfere in a common region in front of P. The interference may be observed on a screen that is parallel to P at a long distance D ($D >> d$). In the corpuscular theory of light, the trajectories of the free light particles are straight lines and, according to the laws of geometrical optics, we should observe two bright spots identical to the sources S_1 and S_2. First using sunlight, Young observed a central white band bordered on both sides by dark bands followed by colored bands from violet to red slightly shifted from one another, with partial overlapping. Repeating the experiment with colored light, he observed nice equidistant fringes parallel to the mediator plane of the sources.

Initially, Young used two pinholes S_1 and S_2 pierced in the opaque screen P; the setup is similar to that of Figure 8.1a. The intersections of the hyperboloids of the maximums with the observation screen are bright hyperbolas corresponding to integer values $\pm p$ of the order of interference. In the case of the interference of light, we only observe an almost rectilinear part of the hyperbolas near their apexes. The fringe of order 0 is the straight line Oy that is equidistant from S_1 and S_2. Let SOz be the axis of symmetry, which is normal to the screens and Ox the axis that is parallel to S_2S_1 (or perpendicular to the slits S_1 and S_2). The setup is symmetric with respect to the planes Oxz and Oyz. By using slits instead of pinholes as sources, the fringes are parallel to the slits and the luminosity is considerably increased. If the source S, which illuminates the apertures S_1 and S_2, is at equal distances from the apertures, the waves u_1 and u_2 have the same initial phase (thus, $\Delta_o\phi \equiv \phi_2 - \phi_1 = 0$).

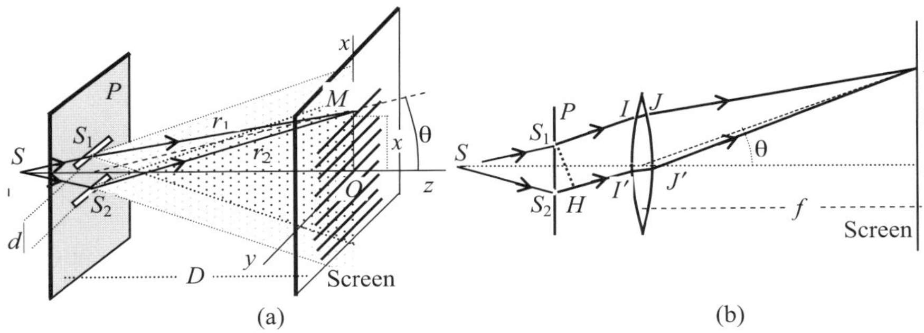

Figure 8.2. *a) Interference of light waves that are diffracted by Young's slits, b) interference at infinity realized using a converging lens*

Let M be a point of the x-axis. If θ is the angle that the direction of observation makes with the Oz axis, the coordinate of M is $x = D \tan \theta$. If $d << D$, the distances r_1 and r_2 of M to the pinholes S_1 and S_2 (or the mid-points of the slits S_1 and S_2) are

$$r_1 = \sqrt{D^2 + (x - d/2)^2} = \sqrt{D^2/\cos^2\theta + d^2/4 - dD\,\mathrm{tg}\theta} \approx D/\cos\theta - \tfrac{1}{2}d\sin\theta$$

$$r_2 = \sqrt{D^2 + (x + d/2)^2} = \sqrt{D^2/\cos^2\theta + d^2/4 + dD\,\mathrm{tg}\theta} \approx D/\cos\theta + \tfrac{1}{2}d\sin\theta.$$

Thus, the path difference at M is $\Delta r \equiv r_2 - r_1 \approx d \sin\theta$. The points of the screen where the path difference is equal to a multiple integer of the wavelength ($\Delta r = p\lambda$), receives the waves in phase. They form bright fringes in the directions θ_p such that

$$\sin\theta_p = p\lambda/d \qquad \text{where} \quad p = \text{integer}. \qquad [8.14]$$

The Oy axis verifies this condition with $p = 0$, this is the central bright fringe (*fringe of order* 0); then, come the fringes of order ±1, ±2, ±3, etc., on both sides of the central fringe. Using the relation $\sin\theta \approx \tan\theta = x/D$, we deduce that the bright fringe of order p is at the distance $x_p = pD\lambda/d$. The spacing between two adjacent fringes is the *interference fringe width* (or the *spacing between fringes*)

$$i = D\lambda/d. \qquad [8.15]$$

A measurement of the spacing between fringes allows us to determine the distance d if the wavelength λ is known or to determine λ if d is known. For instance, if $i = 1.4$ mm, $d = 0.5$ mm and $D = 1.5$ m, we obtain $\lambda = i\,d/D = 0.47$ µm.

As $|\sin\theta| < 1$, the order of interference is less than $p_{max} = d/\lambda$. A sufficient number of fringes may be observed if the distance d is large, compared to λ. If u_1 and u_2 have equal amplitudes, the minimum of the resultant amplitude $|a_1 - a_2|$ is zero; thus, $\mathcal{I}_{min} = 0$ and the contrast is $\mathcal{C} = 1$. This is the optimal situation to observe the interference. On the contrary, if the amplitudes of u_1 and u_2 are different, $\mathcal{I}_{min} \neq 0$ and $\mathcal{C} < 1$ (the fringes of the minimums are not completely dark). This is the case if the sources have different widths, or if the observation point M is at very different distances from the sources S_1 and S_2 (producing a high order of interference). On the other hand, for a given ratio λ/d, the fringes are well separated if D is very large. The limit of the long distance D is equivalent to having S_1M and S_2M nearly parallel. This can easily be realized by observing the interference in the focal plane of a converging lens with optical axis Oz (Figure 8.2b). The rays, that are emitted by the sources in a direction making an angle θ with the axis Oz, emerge from the lens and meet each other at the point M of the screen situated at a distance $x = f\tan\theta$, where f is the focal length. The phase shift between the rays S_1IJM and $S_2I'J'M$ at M is the same as at the points S_1 and H. A travelled distance r in a medium of index n corresponds to a phase lag $\omega rn/c = 2\pi(rn/\lambda)$. The quantity nr is called the *optical path.* Although the ray $HI'J'M$ is geometrically shorter than S_1IJM, it travels a distance $I'J'$ in the lens that is longer than the distance IJ travelled by the ray S_1IJM. The diminution of the wavelength in the glass of the lens compensates for the increase of the path outside the lens. In other words, the optical path S_1M is the same as HM and the phase shift is only due to $S_2H = d\sin\theta$. If the sources S_1 and S_2 are in phase, the order at M is $p = \Delta_s\phi/2\pi = (S_2H)/\lambda = (d/\lambda)\sin\theta$. Thus, the bright fringes are in the directions θ_p such that $\sin\theta_p = p\lambda/d$, where p is an integer ($|p| < d/\lambda$). The points of the screen, situated at short distances x, correspond to small angles θ_p. Thus, the bright fringes are at $x_p = f\tan\theta_p \approx pf\lambda/d$. The spacing between fringes is

$$i = f\lambda/d. \qquad [8.16]$$

In section 8.7 we shall see that the diffraction by a slit of width ε produces an angular spreading of light on both sides of the geometrical optics direction by an angle $\Theta \approx \lambda/\varepsilon$. If the angular spacing between adjacent fringes is $\delta\theta \approx \lambda/d$, we deduce that the number of observable fringes is about $2\Theta/\delta\theta = 2d/\varepsilon$.

If S_2 has the same intensity $\mathcal{I}_1$ as S_1 but it has a phase lag $\Delta_o\phi$, the total phase lag at M is $\Delta\phi = \Delta_o\phi + 2\pi(d/\lambda)\sin\theta$ and intensity [8.11] may be written as

$$\mathcal{I}(M) = \mathcal{I}_{max}\cos^2[\tfrac{1}{2}\Delta_o\phi + \pi(d/\lambda)\sin\theta] \qquad \text{where} \quad \mathcal{I}_{max} = 4\mathcal{I}_1. \qquad [8.17]$$

The variation of $\mathcal{I}$ versus $\sin\theta$ and x is illustrated in Figure 8.3 for the case of $\Delta_o\phi = 0$. It reaches a maximum value $\mathcal{I}_{max}$ on the bright fringes in the directions θ_p such that $\sin\theta_p = p\lambda/d$, that is, $x_p = pi$. The intensity $\mathcal{I}$ vanishes at the positions of the dark fringes, in the directions such that $\sin\theta = (p + ½)\lambda/d$, that is, $x_p = (p + ½)i$. If the sources have a phase shift $\Delta_o\phi$, the fringes are displaced a distance $\delta x = i|\Delta_o\phi|/2\pi$ toward the source which has a phase lag. The fringe of order 0 is displaced to a direction θ_o such that $\Delta\phi = 0$, that is, $\sin\theta_o = -(\lambda/2\pi d)\,\Delta_o\phi$. In particular, if the sources are in opposite phase ($\Delta_o\phi = \pi$), the bright fringes are displaced a distance $i/2$, to occupy the positions of the dark fringes of the setup with $\Delta_o\phi = 0$.

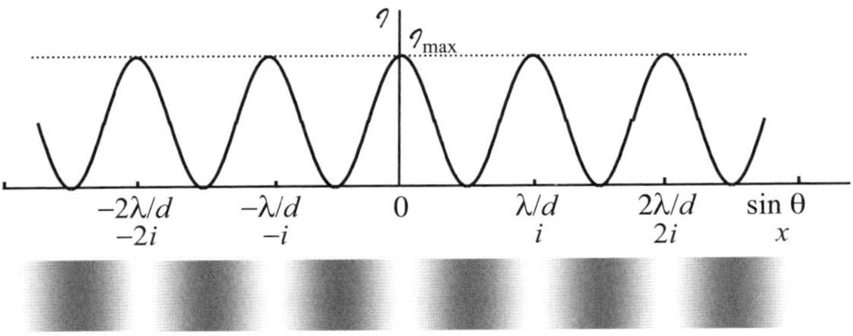

Figure 8.3. *The intensity of light versus* $\sin\theta$ *and* x*, where* θ *is the angle of the observation direction and* x *is the position on the screen measured from the central point* O

EXAMPLE 8.1.– In a Young's experiment the sources are two slits of width $\varepsilon = 1.0$ µm, at 0.2 mm apart. They are illuminated by a mercury lamp of wavelength $\lambda = 546$ nm. The interference pattern is observed on a screen in the focal plane of a converging lens of focal length $f = 100$ cm. **a)** Determine the spacing between the fringes on this screen. What is the maximum number of observable fringes? What is the position of the fringe of order 100? **b)** What is the displacement of the fringes if we cover one of the slits with a transparent sheet of thickness $e = 3$ µm and an index of refraction $n = 1.5$?

SOLUTION –

a) The direction θ_p of the fringe of order p is such that $\sin\theta_p = p\lambda/d$ (Figure 8.4a). The fringe of order 0 is in the direction $\theta_o = 0$ and the first fringe is in the direction θ_1 such that $\theta_1 = \lambda/d = 2.73 \times 10^{-3}$. The spacing between fringes on the screen near O is $i = f\tan\theta_1 = 2.73$ mm. The angular width of diffraction is $2\Theta = 2\lambda/\varepsilon$. This corresponds to a linear diffraction width $2\lambda f/\varepsilon$ on the screen. Thus, the maximum number of observable fringes is $N = 2\lambda f/\varepsilon i = 2d/\varepsilon = 400$.

The direction of the fringe of order 100 is such that $\sin \theta_{100} = 100\, \lambda/d = 0.273$; hence, $\theta_{100} = p\lambda/d = 15.84$. This corresponds to a linear position on the screen $x_{100} = f \tan \theta_{100} = 28.4$ cm; this is slightly different from $100\, i = 27.3$ cm.

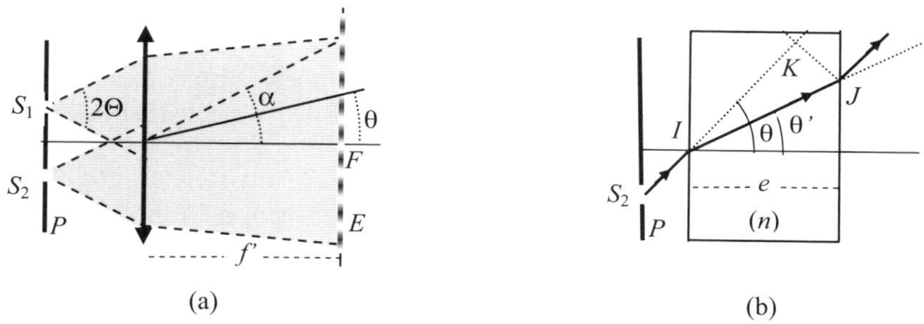

Figure 8.4. *Example 8.1*

b) Let us assume that the sheet covers the lower slit S_2 (Figure 8.4b). If light is observed in the direction θ, the angle of refraction in the sheet is θ′ such that $\sin \theta = n \sin \theta'$. Light travels in the sheet a distance $IJ = e/\cos \theta'$ instead of $IK = IJ \cos(\theta - \theta') = e(\cos \theta + \tan \theta' \sin \theta)$. Thus, we have to add an optical path difference to S_2H:

$$n\, IJ - IK = ne/\cos \theta' - e(\cos \theta + \tan \theta' \sin \theta) = e[\sqrt{n^2 - \sin^2\theta} - \cos \theta\,] \approx e(n-1),$$

where we have assumed that the angles are small. This optical path difference modifies the order of interference by $\delta p = (n-1)(e/\lambda)$. Thus, it displaces the fringes by $i(n-1)(e/\lambda) = 7.5$ mm or more.

8.4. Multiwave interference, conditions for interference

Consider the superposition of N waves $u = \Sigma_j\, a_j \cos(\omega_j t - \phi_j)$. Its square may be written in the form

$$\begin{aligned} u^2 &= [\Sigma_j\, a_j \cos(\omega_j t - \phi_j)]^2 = \Sigma_{kj}\, a_k a_j \cos(\omega_j t - \phi_j) \cos(\omega_k t - \phi_k) \\ &= \Sigma_j\, a_j^2 \cos^2(\omega_j t - \phi_j) + \tfrac{1}{2} \Sigma_{k\neq j}\, a_k a_j \cos[(\omega_k + \omega_j)t - \phi_k - \phi_j] \\ &\quad + \tfrac{1}{2} \Sigma_{k\neq j}\, a_k a_j \cos[(\omega_k - \omega_j)t - \phi_k + \phi_j], \end{aligned}$$

where we separated the terms $k = j$ from the terms $k \neq j$. To evaluate the resultant intensity, we take the average over the time of observation. The first term gives the sum of the intensities of each wave taken separately $\mathcal{I}_j = \frac{1}{2}\alpha a_j^2$. The second term has an average equal to zero. Thus, we obtain

$$\mathcal{I} = \Sigma_j \mathcal{I}_j + \Sigma_{k \neq j} \sqrt{\mathcal{I}_k \mathcal{I}_j} < \cos[(\omega_k - \omega_j)t + (\phi_j - \phi_k)] >, \quad [8.18]$$

where $< f(t) >$ designates the average of $f(t)$ over the time of observation. The second term is the interference term. As an application of [8.18] we consider the following cases:

a) The waves have different angular frequencies: Then, the time-average of the interference term is zero and the total intensity is equal to the sum of the intensities of the individual waves:

$$\mathcal{I} = \Sigma_j \mathcal{I}_j \qquad \text{(waves of different frequencies).} \quad [8.19]$$

b) The waves have equal frequency but different phases: in this case we obtain a generalization of equation [8.11]:

$$\mathcal{I} = \Sigma_j \mathcal{I}_j + \Sigma_{k \neq j} \sqrt{\mathcal{I}_k \mathcal{I}_j} \cos(\phi_k - \phi_j) \qquad \text{(waves of equal frequency).} \quad [8.20]$$

c) The number of waves N is large with different phases, or N is small but the phases ϕ_j change at random during the observation time (*non-coherent waves*). In this case, the average of the function $\cos[(\omega_k - \omega_j)t + (\phi_j - \phi_k)]$ for all the waves is equal to zero and the total intensity is equal to the sum of the intensities $\mathcal{I}_i$

$$\mathcal{I} = \Sigma_j \mathcal{I}_j \qquad \text{(large number of waves or non-coherent waves).} \quad [8.21]$$

Only in case (b) does the intensity $\mathcal{I}$ depend on the observation point via the intermediary of the phases ϕ_k. In the other cases, the intensity is uniform. Thus, the interference is observable if the following conditions are verified:

– in the case of vector waves (such as electromagnetic waves), they must have the same polarization, otherwise the minimums are not zero and the contrast is small;

– the waves must be synchronous (i.e. they have the same frequency);

– the waves must have constant phases (i.e. independent of time), we say that they are *coherent temporally*.

– in the case of light, the *spatial coherence* requires that the sources are not very large, in order for its different points to be approximately coherent.

Let us briefly discuss these conditions:

1) Consider the interference of two waves $\underline{\mathbf{E}}_{m1}\, e^{i\omega(t-r_1/v)}$ and $\underline{\mathbf{E}}_{m2}\, e^{i\omega(t-r_2/v)}$. The intensity of the resultant wave is

$$\mathcal{I}(M) = \tfrac{1}{2}\alpha[|\mathbf{E}_{m1}|^2 + |\mathbf{E}_{m2}|^2 + 2\mathcal{R}e(\underline{\mathbf{E}}_{m1}.\underline{\mathbf{E}}_{m2}{}^*)\cos(\Delta\phi_s) - 2\mathcal{I}m(\underline{\mathbf{E}}_{m1}.\underline{\mathbf{E}}_{m2}{}^*)\sin(\Delta\phi_s)]$$

If the waves are non-polarized, the angle between the vectors $\underline{\mathbf{E}}_{m1}$ and $\underline{\mathbf{E}}_{m2}$ varies rapidly. The time-average of their scalar product $\underline{\mathbf{E}}_{m1}.\underline{\mathbf{E}}_{m2}{}^*$ is equal to zero, $\mathcal{I}(M) = \mathcal{I}_1 + \mathcal{I}_2$ is uniform and no interference pattern is observable. Thus, the waves must be polarized, ideally in the same state of linear polarization. The resultant intensity is given by the same expression [8.11] as if the waves were scalar.

2) To discuss the condition of synchronism, we consider the double slit Young's experiment but with the sources illuminated by two close wavelengths λ_1 and λ_2 ($\lambda_1 < \lambda_2$). We find a common central bright fringe (of order $p = 0$) followed by the fringe of order $p = 1$ of λ_1, then of λ_2 and so on. A fringe of a certain order p of λ_2 may precede or coincide with the fringe of order $(p + 1)$ of λ_1. At a given point of the screen, the difference of order is $\delta p \equiv p_1 - p_2 = \Delta r\, \delta\tilde{\nu}\, /c$. The fringes at this point can be distinguished if δp is smaller than a certain small fraction ε (0.1, for instance). Thus, we must have $\Delta r < \varepsilon c/ \delta\tilde{\nu}$; that is, $p < \varepsilon\, \tilde{\nu} \,/\, \delta\tilde{\nu} \;\approx \varepsilon\, \lambda/\delta\lambda$. If sunlight is used, the zero order fringes for all colors coincide, producing a central white band. Then comes the first order bright spread fringe for colors from violet to red at distances $x_\lambda = \lambda f/d$. Next comes that of the second order at distances $2\lambda f/d$. The bright second order fringe for violet almost coincides with the bright first order fringe for red. At a relatively long distance x, bright fringes of various orders are found for almost all colors: this is the *white of high order*. If we examine this point using a microscope, we find white with dark lines corresponding to some wavelengths. This is a so-called *channeled spectrum*.

3) Now consider the condition of temporal coherence. On the contrary to mechanical waves, light from traditional sources (not lasers) is emitted at random by the individual atoms. Thus, it is formed by a large number of *wave packets*, each one having an average duration τ of about 10^{-8} s, depending on the atom and the excited state. According to Fourier's theory, its bandwidth is $\Delta\omega \approx 1/\tau$ and its *coherence time* τ_c is at least of the order of τ. The only way to have coherent light waves from traditional sources is to use an *aperture splitting* setup or an *amplitude splitting* setup. In the first setup-type (such as Young's double slit experiment), two or several secondary waves originate from different parts of the same primary wave front (Figure 8.5a). In the second setup-type, the secondary waves originate from the same part of the primary wave front (the half-silvered mirror of Figure 8.5b is an example of such setups). In all cases, we have twin wave packets which originate from the same atom and which have a short duration τ_c, called the *coherence time*.

During an observation time of the order of $t_{ob} \approx 1$ ms, some $t_{ob}/\tau_c \approx 10^6$ different wave packets are received at each point of the screen or by the receiver.

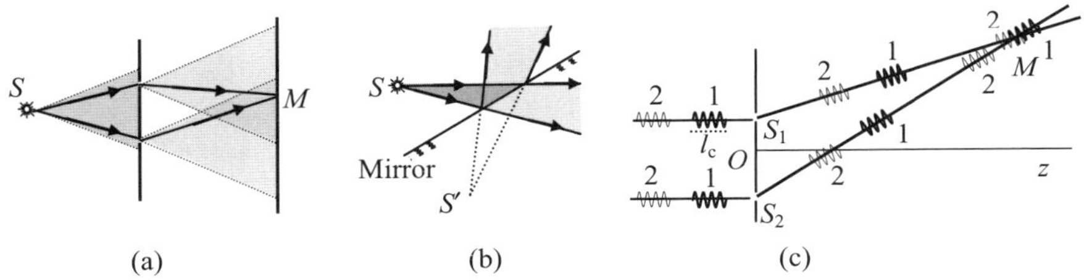

Figure 8.5. *Temporal coherence: a) aperture splitting in Young's experiment, b) amplitude splitting by a half-silvered mirror, c) if the direction of observation at M is very tilted away from the z-axis, wave packet (2) of the ray S_1M arrives at M at the same time as packet (1) of the more tilted ray S_2M. In this situation, the temporal coherence is lost*

A wave packet whose coherence time is τ_c has a space extension $l_c = c\tau_c$, called the *coherence length*. For instance, if $\tau_c \approx 10^{-9}$ s the coherence length is $l_c \approx 10$ cm. If the path difference Δr at the observation point M exceeds l_c, non-twin wave packets may interfere (Figure 8.5c). Thus, temporal coherence requires the order of interference to be much less than $p_{max} = l_c/\lambda$. If the screen is placed at a long distance D from the slits (compared to the distance d separating the slits), the interference is only observable in the central region of the screen (i.e. close to the z-axis).

4) Finally let us consider the condition of spatial coherence. In order to increase the luminosity of S_1 and S_2 and consequently the interference pattern, we use an extended source of light S. The intensity at a point M of the screen depends on the excitation of the slits S_1 and S_2 that is produced by each atom of the source (Figure 8.6a). The wave emitted by the atom A, situated at an equal distance from S_1 and S_2, produces an excitation of the slits in phase while the wave emitted by an atom at the extreme point B, situated at a distance x' from the z-axis, reaches S_1 and S_2 with a path difference

$$b_2 - b_1 = \sqrt{L^2 + (x'+d/2)^2} - \sqrt{L^2 + (x'-d/2)^2} \approx x'd/L \approx \beta d,$$

where L is the distance of the light source S from the slits, $\beta = x'/L$ is the angle subtended by AB as seen from the midpoint of S_1S_2, and we have assumed that $d << L$ and $y << L$. As the atoms A and B are completely independent, we must add the intensities that are received at M (not the amplitudes). Thus, on the screen we observe a superposition of two interference patterns produced by A and B and separated by a distance $\Delta x = i(b_2 - b_1)/\lambda = i\beta d/\lambda$. In particular, if $\beta = \lambda/2d$, that is, $x' = \lambda L/2d$, the bright fringes of A coincide with the dark fringes of B; we say that there is complete blurring. We note that the extension of the source in the direction O_1y' does not affect the spatial coherence (Figure 8.6b). Thus, to increase the luminosity without affecting the coherence, we may use a source of light in the form of a narrow slit along the O_1y' axis parallel to the slits. We may also use a source S at infinity by placing the source in the focal plane of a converging lens, then $\beta = 0$.

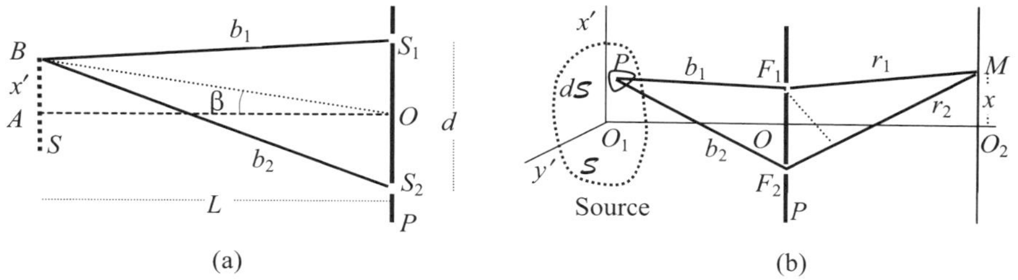

Figure 8.6. *a) Spatial coherence, and b) interference in the case of an extended source*

8.5. Holography

Traditional photography is a recording of the light intensity and color that is emitted or diffracted by each point M of an object and impinged on the corresponding point image M' on a flat surface. In fact, it records the average intensity, which is received during the exposure time, i.e. $\langle E^2 \rangle$, where $\mathbf{E}$ is the emitted electric field. It does not exactly reproduce the light being emitted by the object because any information concerning the phase of the light waves is lost. The phase depends on the distance to the point M of the object. For this reason, the image is the same as if the visible face of the object was flat. A process, called *holography*, was proposed by Gabor in 1947 and realized only 15 years later, after the discovery of the laser. It consists of capturing both the amplitude and the phase using interference and recording the interference pattern on a photographic plate. For this, a beam of coherent light is split into two parts: the first part is diffracted by each point A of the object; and the second part, called the *reference wave*, is reflected by a mirror or deviated by a prism (Figure 8.7a). These two waves interfere at points M of a photographic plate. After the development, we obtain a *hologram*, i.e. a record of the interference patterns for each point of the object. In contrast to the traditional photograph, the interference pattern of each point A is recorded on the whole hologram. Thus, any part of the hologram contains information on the intensity and the phase of the wave being diffracted by all points of the object. If the hologram is illuminated, the light that is diffracted by these interference patterns interferes constructively for the wavelength that was used to form the hologram. Thus, it produces a virtual image in the direction of the original point A and a three-dimensional perception of the whole object (Figure 8.7b). It is also possible to realize acoustic holograms (using ultrasounds), X-rays holograms, etc.

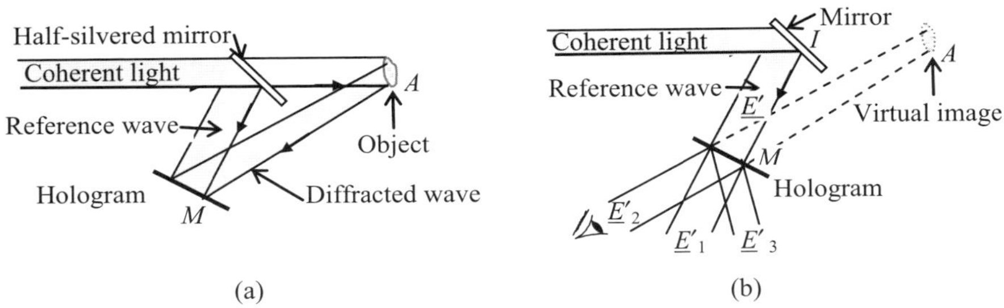

Figure 8.7. *a) Formation of a hologram, b) reconstitution of the diffracted wave*

8.6. Thin film interference

The interference of light on thin films produces the intense colors of a soap bubble or oil film spread on the surface of water, butterfly wings, etc. This interference occurs between the parts of the light wave that are reflected on both surfaces of the thin film or transmitted by this film after internal reflections on these surfaces. The fringes may be observed directly, photographed or projected on a screen parallel to the thin film (Figure 8.8a).

Consider a plane parallel plate of index n_2 and thickness e separating a medium of index n_1 on one side and a medium of index n_3 on the other. A wave, represented by the ray SJ_1, is incident from medium (1) (Figure 8.8b). A fraction R_1 of this wave is reflected off the front face $\mathcal{S}_1$ and the other R_2 emerges after its transmission as J_1P_1, this is followed by subsequent reflection as P_1J_2 on the back face $\mathcal{S}_2$ and transmission through $\mathcal{S}_1$.

Similarly, a fraction T_1 of the wave J_1P_1 is transmitted at P_1 of $\mathcal{S}_2$ and the other part P_1J_2 undergoes partial internal reflection as J_2P_2 and a transmission through $\mathcal{S}_2$ as the wave T_2. We neglect the fractions of the wave, which undergo more reflections on the surfaces. We designate the wavelength of light in vacuum by λ_o, the wavelength in a medium of index n_i by $\lambda_i = \lambda_o/n_i$ and the angles of the light rays with the normal axis to the plate by θ_i. These angles are related by Snell's law.

If the incident wave is emitted by a pinhole source, circular interference rings may be observed on a screen placed at a long distance from the plate. By increasing the dimensions of the source, the rings become blurred, but they become visible again if the screen is moved to a larger distance. The long distance limit is obtained by placing the screen at the focal length of a converging lens. The fringes may be observed with the naked eye, the crystalline lens and the retina playing the parts of the lens and the screen.

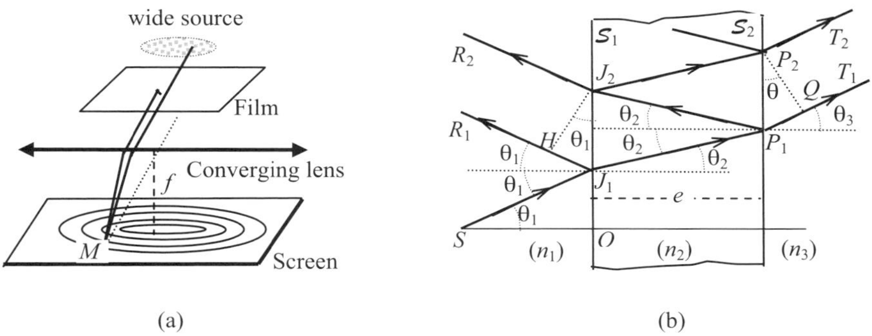

Figure 8.8. *Thin film interference: a) circular fringes localized at infinity and produced by a uniform thin film, and b) interference of the reflected waves and the transmitted waves*

The optical path difference between R_2 and R_1 is the same as at H and J_2, that is

$$n_2\, J_1P_1J_2 - n_1\, HJ_1 = 2n_2e/\cos\theta_2 - n_1J_1J_2\sin\theta_1 = 2n_2e\cos\theta_2.$$

This is also the optical path difference between T_2 and T_1. We note that a reflection on a more refringent medium introduces a phase shift equal to $\pm\pi$. It is easy to verify that the additional phase $\hat{\phi}$, between the reflected waves R_2 and R_1, is equal to zero if n_2 lies between n_1 and n_3, otherwise it is equal to $\pm\pi$. The additional phase shift between the transmitted waves T_2 and T_1 is $\hat{\phi}+\pi$. Thus, the total phase shift ϕ_R between the reflected waves R_2 and R_1 and the total phase shift ϕ_T between the transmitted waves T_2 and T_1 are

$$\phi_R = \phi + \hat{\phi} \qquad \text{and} \qquad \phi_T = \phi + \hat{\phi} + \pi$$

$$\text{where} \quad \phi = \phi_2 - \phi_1 = 4\pi n_2\,(e/\lambda_o)\cos\theta_2 = 4\pi\,(e/\lambda_2)\cos\theta_2. \qquad [8.22]$$

The interference of the reflected waves is constructive (and that of the transmitted waves is destructive) if $p_R = \phi_R/2\pi$ is equal to an integer and conversely if p_R is equal to a half-integer. For instance, in the case of the normal incidence ($\theta_i = 0$) on a film immersed in the same medium ($n_1 = n_3$, thus, $\hat{\phi} = -\pi$), we find maximums of the reflected intensity (thus, a minimum of the transmitted intensity) if $e = \frac{1}{4}\lambda_2(1+2q)$, where q is an integer. On the contrary, the interference is destructive for the reflected intensity and constructive for the transmitted intensity if $e = \frac{1}{2} q\lambda_2$. If the film is illuminated by sunlight, the wavelengths corresponding to $\phi_R/2\pi$ equal to an integer produce constructive interference by reflection and destructive interference by transmission.

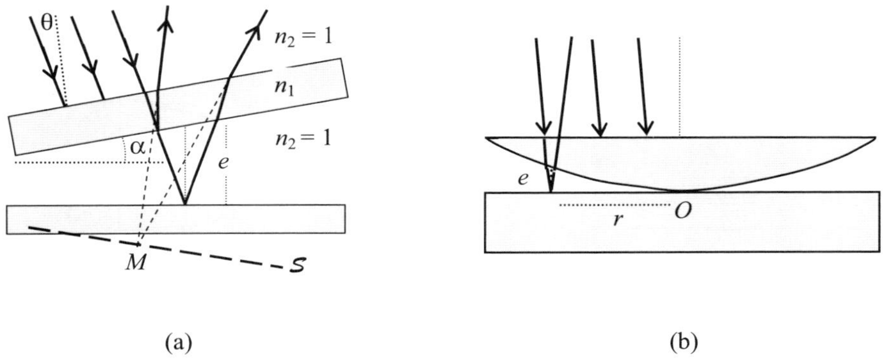

Figure 8.9. *Fringes of equal thickness: a) in a wedge-shaped air film, and b) Newton's rings*

If the thickness e is large compared to λ, the difference of path $2n_2e \cos \theta_2$ is large and the condition of coherence is lost. This is the reason why no interference of light is observed with a thick glass plate, for instance. A sound interference with a frequency of 1 kHz is possible for a plate with a thickness of the order of $\lambda = 0.3$ m.

If the thickness of the film is not uniform, we may show that ϕ_R and ϕ_T are the same as those of a uniform film of thickness e at each point. This explains the different colors of films of variable thickness such as soap bubbles. The air film between two plates of glass (Figure 8.9a) produces straight fringes that are parallel to the wedge. The thin air film, between a convex lens and the flat plate of glass it rests on (Figure 8.9b), produces circular colored fringes called *Newton's rings*. This property enables us to use thin films as very efficient *interference filters*.

Optical coating is one of the important applications of thin films. It consists of applying a transparent film to a glass surface in order to reduce the amount of light reflected at the surface or, on the contrary, to increase the reflected intensity. The thickness and the index of the film are chosen in such a way that the reflected waves interfere destructively or constructively. Taking into account the amplitudes of the reflected and transmitted waves, it may be shown that, if an intensity $\mathcal{I}_i$ is incident at a small angle, the transmitted intensity may be written as

$$\mathcal{I}_T \approx \frac{4n_1 n_3}{(n_1+n_3)^2} \frac{1}{1+M\sin^2(\phi/2)} \mathcal{I}_i \quad \text{where} \quad M = \frac{(n_2{}^2-n_1{}^2)(n_2{}^2-n_3{}^2)}{n_2{}^2(n_1+n_3)^2}, \quad [8.23]$$

while the reflected intensity is $\mathcal{I}_R = \mathcal{I}_i - \mathcal{I}_T$. To reduce $\mathcal{I}_R$ (*antireflection coating*), we must choose a film of index $n_2 = \sqrt{n_1 n_3}$ and thickness e such that $\sin^2(\phi/2) = 1$, that is, $\phi = \pi + 2q\pi$, where q is a positive integer or zero. This corresponds to a thickness $e = \frac{1}{4}\lambda_2(1 + 2q)$ (particularly a *quarter wave* for $q = 0$). On the other hand, to increase $\mathcal{I}_R$ (i.e. reduce $\mathcal{I}_T$) the coating must have $e = \frac{1}{4}\lambda_2(1 + 2q)$ (quarter wave for the minimum thickness) and the largest or the smallest possible index (not lying between n_1 and n_3). This is the principle of *dielectric mirrors.*

8.7. The Huygens-Fresnel principle and diffraction by an aperture

Diffraction refers to the bending of waves near obstacles, i.e. their deviation from rectilinear propagation in homogeneous media. The mathematical problem consists of determining the solution of the wave equation, which verifies some given boundary conditions, which depend on the nature of the obstacle. This proves to be a considerably difficult problem; hence, approximation methods must be used.

Huygens' principle (see section 7.3) qualitatively explains the diffraction by assuming that the points of the aperture behave like sources, emitting spherical wavelets whose envelope is the wave front of the diffracted wave (Figure 8.10a). However, this simple geometrical formulation does not allow us to determine the angular distribution of the diffracted intensity and, in particular, it does not explain the absence of a diffracted wave in the backward direction, i.e. the opposite direction to the incident wave.

In order to quantitatively analyze the diffraction, Fresnel assumed that the diffracted wave results from the interference of the spherical wavelets $d\underline{u}_d$ emitted by the elements of area $d\mathcal{S}$ of the aperture. According to the *Fresnel-Huygens principle*, these wavelets may be written as

$$d\underline{u}_d(M) = d\mathcal{S}\, \eta_P(\theta', \theta)\, \underline{u}(P) \frac{1}{PM} e^{-i\mathbf{k}.\overrightarrow{PM}}. \quad [8.24]$$

In this expression, we have omitted the factor $e^{i\omega t}$ that is common to all the wavelets. $\underline{u}(P)$ is the incident wave at P and $(1/PM)\, e^{-i\mathbf{k}.\overrightarrow{PM}}$ characterizes the spherical wavelet that is emitted by the element of area $d\mathcal{S}$. θ and θ' are respectively the angles of the direction of the diffracted wave and the incident wave with the normal to the aperture oriented in the direction of propagation. Intuitively, Fresnel has chosen the *inclination factor* $\eta_P(\theta', \theta)$ equal to $(1/2\lambda)(\cos\theta + \cos\theta')$ to favor the direction of geometrical optics. The factor $1/2\lambda$ is necessary in order for η_P to have the dimensions of the inverse of a length, which can only be λ. More recently, Kirchhoff explained the origin of this factor but with a phase of $\pi/2$; thus, it may be written as

$$\eta_P(\theta', \theta) = (i/2\lambda)(\cos\theta + \cos\theta'). \quad [8.25]$$

The diffracted wave is obtained by the integration of [8.24] over the aperture

$$\underline{u}_d(M) = \iint_S dS\, \eta_P(\theta', \theta)\, \underline{u}(P)\, \frac{1}{PM} e^{-i\mathbf{k}.\overrightarrow{PM}}. \quad [8.26]$$

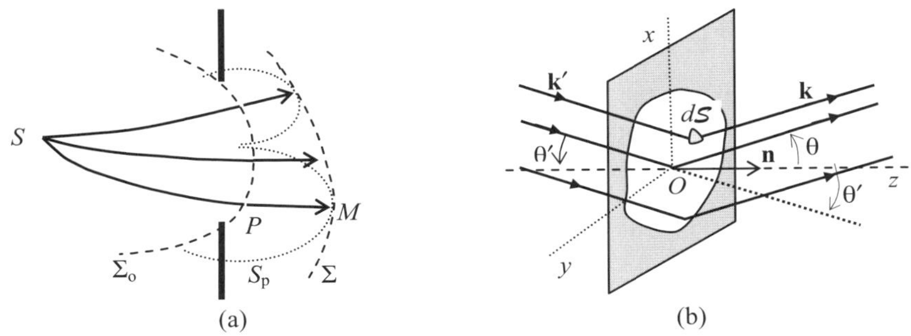

Figure 8.10. *a) Huygens' principle, and b) Fraunhofer diffraction by an aperture*

The mathematical analysis of the diffraction according to [8.26] is simplified if the source S and the point of observation are at long distances from the aperture and the angles θ and θ' are small. This is the so-called *Fraunhofer diffraction*. In this case, the incident wave and the total diffracted wave are plane waves (Figure 8.10b). In expression [8.26], the distance PM and the inclination factor η_P are approximately independent of P, the incident wave is a plane wave $\underline{u}(P) = Ae^{-i\mathbf{k}'.\overrightarrow{OP}}$, where $\mathbf{k}'$ is its wave vector. Thus, the total diffracted wave may be written as

$$\underline{u}_d(M) = \chi \iint_S dS\, e^{i(xK_x + yK_y)} \qquad \text{where} \quad \mathbf{K} = \mathbf{k} - \mathbf{k}'. \quad [8.27]$$

χ is a constant and the wave vectors $\mathbf{k}$ and $\mathbf{k}'$ have the same magnitude $k = k' = 2\pi/\lambda$.

In practice, this is realized by placing an intense pinhole source at the focal point of a converging lens and letting the outgoing parallel beam illuminate the diffraction aperture or the small diffracting obstacle. The diffracted wave is intercepted by a second converging lens and the diffraction pattern is observed on a screen, which coincides with the focal plane of the lens (Figure 8.11a).

In the following, we apply [8.27] in the case of a narrow slit, a rectangular aperture and a circular aperture.

a) Consider the case of a narrow slit of width d in the direction Ox with length L in the direction Oy (Figure 8.11a). If L is large and the wave vector $\mathbf{k}'$ of the incident wave is parallel to the plane Oxz, the system has translation symmetry in the direction Oy. Thus, we consider the observation points M in the $O'Xz$ plane in the direction making an angle θ with Oz (then, $k_x = k \sin\theta$ and $k_y = 0$). If the incident wave is normal to the aperture, $k'_x = k'_y = 0$, the expression [8.27] may be written as

$$\underline{u}_d(M) = \chi \int_{-d/2}^{d/2} dx \int_{-L/2}^{L/2} dy\ e^{ixk_x} = \chi L d\,\frac{\sin(\Phi/2)}{\Phi/2}, \qquad \Phi = k_x d = 2\pi\frac{d}{\lambda}\sin\theta. \qquad [8.28]$$

As $d \sin\theta$ is the path difference between the wavelets that are emitted from the extreme points of the slit in the width direction Ox, Φ is the phase shift between these wavelets. The diffracted intensity may be written as

$$\mathcal{I}(\theta) = \mathcal{I}_o\,\mathcal{F}_d(\Phi) \qquad \text{where} \qquad \mathcal{F}_d(\Phi) = \left[\frac{\sin\Phi/2}{\Phi/2}\right]^2. \qquad [8.29]$$

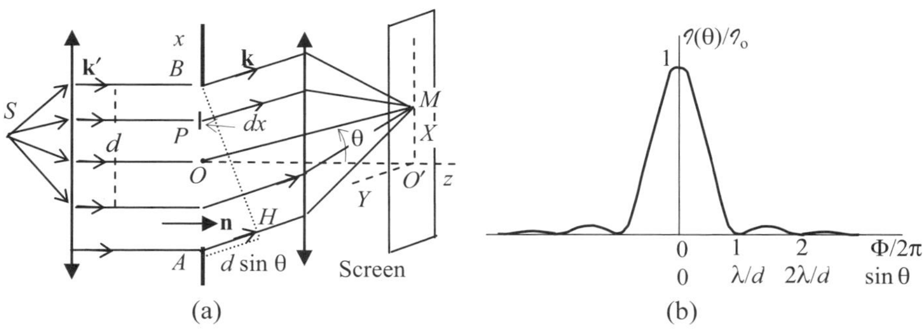

Figure 8.11. *a) Diffraction by a slit AB of width d in the case of a normal incident wave, and b) variation of the intensity as a function of* $\Phi/2$ *or* $\sin\theta$

The variation of the diffracted intensity $\mathcal{I}(\theta)$ is illustrated in Figure 8.11b. For $\Phi = 0$, $\mathcal{F}_d(\Phi) = 1$ and the intensity has a sharp principal maximum equal to $\mathcal{I}_o$. The intensity $\mathcal{I}(\theta)$ vanishes for $\Phi = 2p\pi$, i.e.

$$\sin\theta_p = p\lambda/d, \qquad p = \pm1, \pm2, \pm3... \qquad \text{[zeros of the intensity]}. \qquad [8.30]$$

Approximately, halfway between two zeros of $\mathcal{I}(\theta)$, i.e. for $\sin\theta \approx (p + ½)(\lambda/d)$, the intensity has a secondary maximum approximately equal to $\mathcal{I}_o^2/[(p + ½)\pi]^2$ which decreases if p increases. Nearly all the diffracted intensity is concentrated in the principal maximum. In the case of light, precise photometric measurements using photomultipliers confirm this distribution of the intensity. We may define the half-width as the distance between the center of the principal maximum ($\Phi = 0$) and the first zero of $\mathcal{I}$ ($\Phi = 2\pi$, i.e. $\sin\theta = \lambda/d$). Note that the width of the principal maximum is twice the width of the secondary maximums.

If the angle of incidence on the aperture is θ' (algebraic and measured from the normal **n** to the aperture pointing in the direction of the incident wave), the previous expressions remain valid but with a phase shift $\Phi = 2\pi\,(d/\lambda)(\sin\theta - \sin\theta')$. Then, the direction of the principal maximum is $\theta = \theta'$.

The first zero of the intensity corresponds to $\Phi = 2\pi$, i.e. the wavelets emitted by the end points A and B in phase. To interpret this result, imagine that the slit is divided into elements of infinitesimal width in the direction AB. If the direction θ_1 is such that the phase shift between the extreme wavelets emitted by A and B is 2π, the phase shift between the wavelets emitted by A and by the middle O of the slit is π. Their interference is destructive and it will be so for all the elements of the upper half of the slit and the corresponding elements of the lower half. The diffracted intensity is thus equal to zero in the direction θ_1.

This argument may be repeated for the other zeros of the intensity: it is sufficient that the slit be formed by an even number of zones, such that the wavelets emitted by the points of a zone and the corresponding points of the next zone are in opposite phase in the direction θ_p. Conversely, if the direction θ is such that the slit is formed by an odd number of zones, the wavelets emitted by the unpaired last zone do not interfere destructively with any other zone, and we have a secondary maximum. The emitted intensity is that of this unpaired zone and it decreases as the number of zones increases because the area of the unpaired zone decreases.

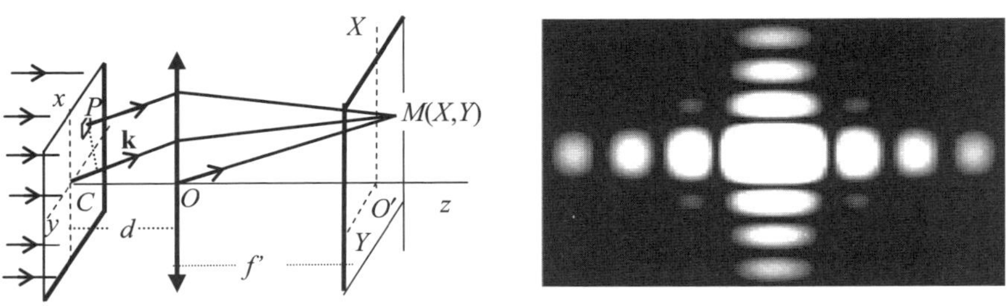

(a) (b)

Figure 8.12. *a) Diffraction by a rectangular aperture, and b) the diffraction pattern*

b) Consider the case of a rectangular aperture of sides a and b illuminated by a plane wave incident normally ($k'_x = k'_y = 0$) as in Figure 8.12a. In this case, expression [8.27] may be written in the form

$$\underline{u}_d(M) = \chi \int_{-a/2}^{a/2} dx \int_{-b/2}^{b/2} dy \; e^{i(xk_x + yk_y)} = \chi \int_{-a/2}^{a/2} dx \; e^{ixk_x} \int_{-b/2}^{b/2} dy \; e^{iyk_y}$$

$$= u_o \frac{\sin(\Phi_x/2)}{\Phi_x/2} \frac{\sin(\Phi_y/2)}{\Phi_y/2}, \qquad \Phi_x = ak_x \quad \text{and} \quad \Phi_y = bk_y. \qquad [8.31]$$

Φ_x and Φ_y, respectively, are the phase shifts of the wavelets emitted by the extreme points of the aperture in the directions Ox and Oy. The intensity in this direction is

$$\mathcal{I}(M) = \mathcal{I}_o \, \mathcal{F}_d(\Phi_x) \, \mathcal{F}_d(\Phi_y). \qquad [8.32]$$

The diffraction pattern is a double series of maximums aligned in the directions Ox and Oy with decreasing intensity from the centre (Figure 8.12b).

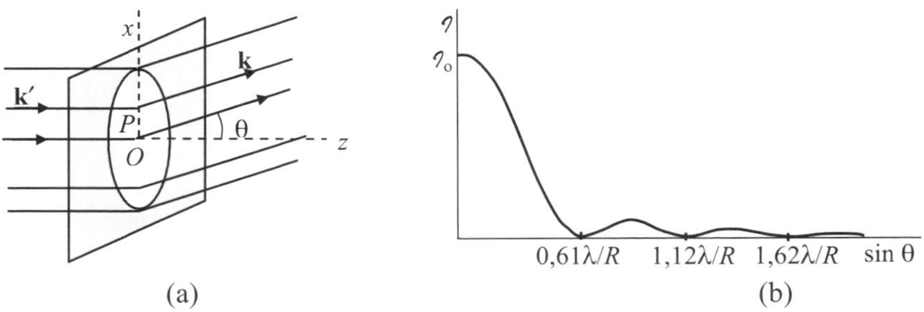

Figure 8.13. *a) Diffraction by a circular aperture, and b) variation of the intensity as a function of* sin θ, *where* θ *is the angle of observation*

c) Consider the case of a circular aperture of radius R illuminated by a plane wave incident normally (Figure 8.13a). This setup has rotation symmetry around the axis of the aperture Oz; thus, the intensity of the diffracted wave has the same symmetry. We divide the aperture into small elements of area and we use the polar coordinates (ρ, ϕ) for a running point P of the aperture. In this case, diffracted wave [8.27] may be written as

$$\underline{u}_d(M) = \chi \int_0^R d\rho \, \rho \int_0^{2\pi} d\phi \; e^{ik_x \rho \cos\phi} \qquad \text{where} \quad k_x = (2\pi/\lambda) \sin\theta. \qquad [8.33]$$

The double integral may be expressed in terms of the Bessel function J_1:

$$u_d(M) = 2\pi\chi R^2 J_1(s)/s \qquad \text{where} \qquad s = k_x R = 2\pi(R/\lambda)\sin\theta. \qquad [8.34]$$

The intensity at M can then be written as

$$\mathcal{I}(M) = 4\mathcal{I}_o\,[J_1(s)/s]^2. \qquad [8.35]$$

The function $J_1(s)$ oscillates like a sine function but its amplitude decreases and its zero s_{1p} are not equally spaced. The ratio $J_1(s)/s$ has a principal maximum equal to ½ for $s = 0$. Figure 8.13b illustrates the variation of $\mathcal{I}$ as a function of $\sin\theta$. It has a sharp maximum equal to some value $\mathcal{I}_o$ for $\sin\theta = 0$, vanishes for a series of values $\sin\theta_{min} = (s_{1p}/2\pi)(\lambda/R)$ and it has a secondary maximum at $\sin\theta_{max}$ between each two consecutive zeros:

$$\sin\theta_{max} = p_{max}\,(\lambda/R), \qquad \sin\theta_{min} = p_{min}\,(\lambda/R). \qquad [8.36]$$

The first values of p_{max}, p_{min} and the relative intensities of the first maximums, are

$p_{min} =$		0.61		1.12		1.62		etc.	
$p_{max} =$	0		0.82		1.32		1.84,	etc.	
$\mathcal{I}_{max}/\mathcal{I}_o =$	1		0.0175		0.0042		0.0016,	etc.	[8.37]

Diffraction by small apertures and obstacles is encountered in all types of waves: visible light, X rays, microwaves, radio waves, sound waves, surface waves, electronic waves, etc. If we increase the dimensions of the apertures or obstacles, the zeros and secondary maximums approach the direction $\theta = 0$ and the phenomenon of diffraction becomes less important. At the limit of large dimensions ($d >> \lambda$) the rectilinear propagation of geometrical optics is recovered.

8.8. Diffraction grating

We now consider the interference of N waves with equal phase shift ϕ between consecutive waves. These N waves may be emitted by sources or apertures regularly distributed in the direction x with a spacing d. This analysis can be generalized to two-dimensional and three-dimensional periodic series of sources.

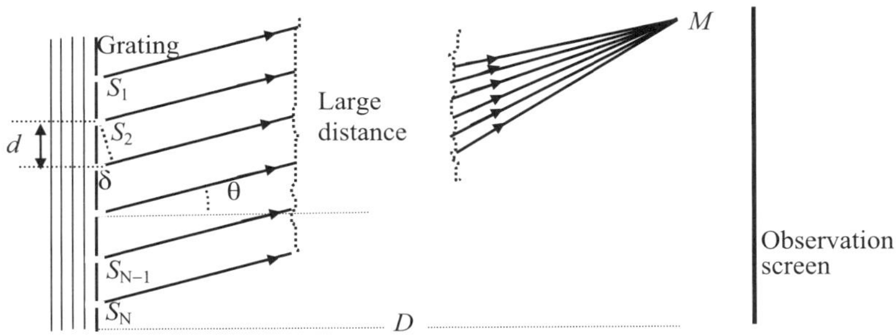

Figure 8.14. *Interference at infinity of several waves*

A *diffraction grating* is usually a series of N parallel and identical slits with a spacing d between adjacent slits or, more generally, an optical structure, which is repeated N times in the x direction with a period d of the order of the wavelength λ (Figure 8.14). They produce, by reflection or transmission, N coherent waves of equal amplitude a with the same phase-shift between consecutive waves. Their diffraction properties are essentially determined by the spacing d and the number N of the slits. We may realize diffraction gratings for light waves, infrared and ultraviolet waves, microwaves, acoustic waves, etc. The characteristic effects of diffraction gratings are easily observed on a compact disc, fish scale, bird's fine feather, etc.

The resultant wave at an observation point M (of the observation screen in the case of light) is the superposition of N waves emitted or diffracted by the slits. If the distance D from the diffraction grating to the observation screen is large, compared to Nd, the rays S_iM are nearly parallels. The setup is called *Fraunhofer diffraction* (or *diffraction at infinity*). If the observation point M is in the direction θ (with the normal Oz to the diffraction grating), the path difference between two consecutive waves is $\delta = d \sin\theta$ and their phase shift at M is

$$\phi = 2\pi \frac{d}{\lambda} \sin\theta. \qquad [8.38]$$

Using the complex representation for the N waves, the resultant wave is

$$\underline{u} = a\,e^{i(\omega t - \mathbf{k.r})} + a\,e^{i(\omega t - \mathbf{k.r} - \phi)}, \ldots, + a\,e^{i[\omega t - \mathbf{k.r} - (N-1)\phi]}$$

$$= a\,e^{i(\omega t - \mathbf{k.r})}\,[\,1 + \zeta + \zeta^2 + \ldots + \zeta^{N-1}] \qquad \text{where} \quad \zeta = e^{-i\phi}.$$

The expression in the brackets is a geometrical progression; it may be written as

$$[...] = \frac{1-\zeta^N}{1-\zeta} = \frac{1-e^{-\mathrm{i}N\phi}}{1-e^{-\mathrm{i}\phi}} = \frac{e^{-\mathrm{i}N\phi/2}(e^{\mathrm{i}N\phi/2}-e^{-\mathrm{i}N\phi/2})}{e^{-\mathrm{i}\phi/2}(e^{\mathrm{i}\phi/2}-e^{-\mathrm{i}\phi/2})} = ae^{-\mathrm{i}(N-1)\phi/2}\,\frac{\sin(N\phi/2)}{\sin\phi/2}\,.$$

Thus, the real part of $\underline{u}$ is

$$\mathcal{R}e\;\underline{u} = A(\phi)\cos[\omega t - \mathbf{k}.\mathbf{r} - \tfrac{1}{2}(N-1)\phi] \qquad \text{where} \quad A(\phi) = a\,\frac{\sin(N\phi/2)}{\sin(\phi/2)}$$

and the resultant intensity is

$$\mathcal{I} = \mathcal{I}_o\,\mathcal{F}_{dg}(\phi) \qquad \text{where} \quad \mathcal{I}_o = \alpha N^2 a^2 \quad \text{and} \quad \mathcal{F}_{dg}(\phi) \equiv \frac{\sin^2(N\phi/2)}{N^2\sin^2(\phi/2)}. \qquad [8.39]$$

$\mathcal{F}_{dg}(\phi)$ is the *diffraction grating function*. The amplitude $A(\phi)$ has principal maximums equal to Na and $\mathcal{F}_{dg}(\phi)$ has principal maximums equal to 1 for $\phi = 2p\pi$. Between two consecutive principal maximums, $A(\phi)$ and $\mathcal{F}_{dg}(\phi)$ have $(N-1)$ zeros (corresponding to directions such that $\sin(N\phi/2) = 0$, other than those of the principal maximums) and $(N-2)$ secondary maximums of relatively small amplitudes. The principal maximums correspond to all the slit waves in phase. The zeros correspond to phase shifts, such that the phasor representation is a closed regular polygon described one or several times. For instance, between the principal maximums of order $p = 0$ and $p = 1$, the zeros correspond to $\phi = 2\pi/N$, $4\pi/N$, ... to $2\pi(N-1)/N$. The width of a principal maximum is thus $\delta\phi = 4\pi/N$. The larger N is, the sharper the principal maximums are and the more numerous and smaller the secondary maximums are. This makes multiple slit interferometry (using an optical diffraction grating) very useful in spectroscopy. Figure 8.15 illustrates the function $\mathcal{F}_{dg}$ as a function of the phase shift $\phi/2\pi$ in the case $N = 6$.

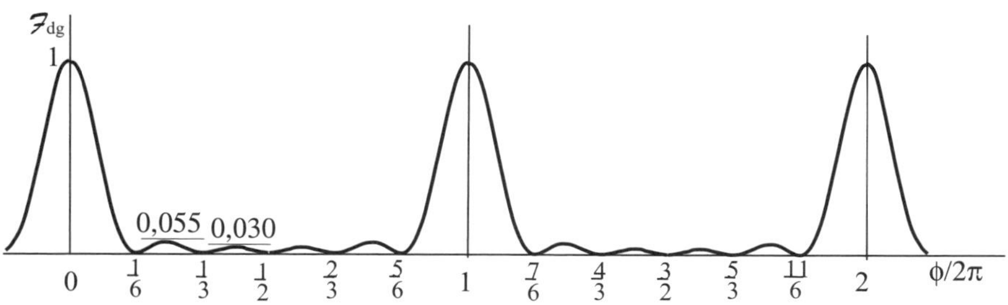

Figure 8.15. *Variation of A^2/a^2 versus $\phi/2\pi$ in the case of 6 slits*

As in the case of diffraction, if the angle of incidence on the aperture is θ' with the normal **n** to the diffraction grating (θ' is algebraic measured from **n** pointing in the direction of the incident wave), the previous expressions remain valid but with a phase shift $\phi = 2\pi\,(d/\lambda)(\sin\theta - \sin\theta')$. Then, the direction of the principal maximum of order $p = 0$ is $\theta_o = \theta'$.

Instead of having the superposition of waves that are diffracted by N slits, we may have the superposition of waves received by N receptors. For instance, a radio telescope is a set of highly orientable antennas (often parabolic), which may be directed toward the sky to receive radio waves emitted by celestial bodies and to measure their wavelength (going from 1 mm to more than 1 km) and their intensity. The parabolic antenna must have a large aperture area to pick up more energy and reduce the diffraction by the aperture of each antenna (see section 8.7). The largest interferometric radio telescope in operation is the Allen Telescope Array in California. It is composed of 42 antennas of diameter 6.1 m and distributed over an area of diameter 1 km. It was conceived to detect waves of frequencies from 0.5 to 11 GHz (i.e. a wavelength going from 2.7 to 60 cm).

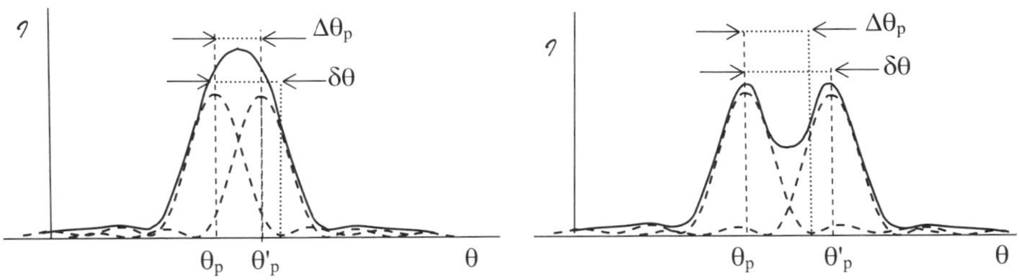

Figure 8.16. *Rayleigh criterion and resolving power*

One of the important uses of a diffraction grating is as a dispersive element for spectroscopic instruments, enabling very precise measurement of wavelengths. The direction θ_p of the order p is such that $\phi_p = 2\pi(d/\lambda)\sin\theta_p = 2p\pi$. Thus, a measurement of θ_p allows us to determine the wavelength $\lambda = (d/p)\sin\theta_p$. The first zeros of intensity correspond to $\phi_p \pm \delta\phi = 2p\pi \pm 2\pi/N$. Thus, the half-width of the principal maximum of order p is $\delta\theta_p = \lambda/Nd\cos\theta_p$. If the incident wave is a superposition of two wavelengths λ and $\lambda' = \lambda + \Delta\lambda$, the principal maximum of order 0 is in the same direction $\theta = 0$ for both waves; while the other principal maximums are in slightly different directions such that $\lambda = (d/p)\sin\theta_p$. By differentiating this relationship we obtain the angular separation $\Delta\theta_p = p\,\Delta\lambda/d\cos\theta_p$ for their principal maximums of order p. According to the *Rayleigh criterion*, the two radiations can be distinguished by observing the order p if their principal maximums are separated by an angle $\Delta\theta_p$ larger than the half-width of these principal maximums (Figure 8.16). Thus, we must have $\Delta\theta_p > \delta\theta$, that is, $\Delta\lambda > \lambda/pN$.

This is the smallest wavelength difference that may be measured with this instrument. The *resolving power* in the pth-order is

$$\mathcal{R}_m \equiv \lambda/\Delta\lambda = pN. \qquad [8.40]$$

The higher the order and the larger the number of lines N are, the higher the resolving power is.

EXAMPLE 8.2.– A radar station uses an emitter that consists of 30 rectilinear parallel wires separated by a distance $d = 20$ cm between consecutive wires. The station uses a wavelength of 10 cm. **a)** What are the directions of the principal maximums of order 0 and 1 and what is their angular width. **b)** Instead of rotating the array, it is possible to use a *phase command*, which consists of producing a phase shift $\phi_i(t)$ between adjacent antennas. What then is the direction of the principal maximum of order 0? How should we choose $\phi_i(t)$ in order to have sweeping at an angular speed ω?

SOLUTION –

a) If the wave is emitted in a direction making an angle θ with the normal to the array of wires (Figure 8.17), the phase shift between two consecutive waves is $\phi = \phi_o + 2\pi(d/\lambda) \sin\theta$, where ϕ_o is their initial phase shift. The direction of the pth principal maximum is such that $\phi = 2p\pi$; thus, $\sin\theta_p = (\lambda/d)(p - \phi_o/2\pi)$. If the initial phase shift ϕ_o is equal to zero, the principal maximum of order 0 is in the normal direction ($\theta = 0$), while the maximum of order 1 is in the direction such that $\sin\theta_1 = (\lambda/d) = 0.5$, thus $\theta_1 = 30°$. The angular half-width of the principal maximum is the direction of the first zero, given by $N\phi/2 = \pi$, that is, $\sin(\delta\theta) = (1/30)(\lambda/d)$ or $\delta\theta = 0.95°$.

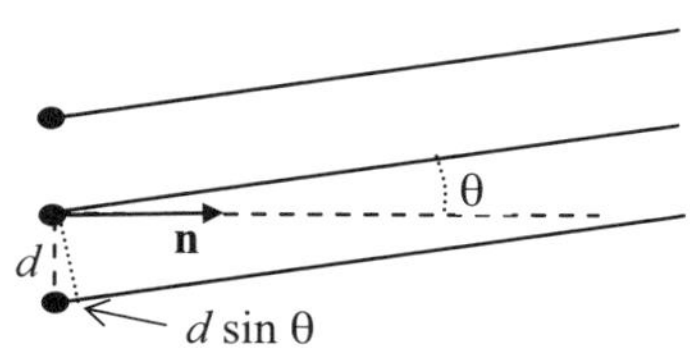

Figure 8.17. *Example 8.2*

b) If the consecutive antennas have an initial phase shift $\phi_i(t)$, the direction of the principal maximum of order 0 is given by $\sin\theta_o = -\lambda\phi_i(t)/2\pi d$. The sweeping has a constant angular speed ω if $\theta_o = \omega t$; hence, $\phi_i(t) = -2\pi(d/\lambda)\sin(\omega t)$.

8.9. Diffraction of X-rays

A diffraction grating may be two-dimensional or three-dimensional if the diffracting centers are periodically distributed over a plane or in a volume in all directions with a spacing of the order of λ. X-rays are electromagnetic waves of very short wavelength, of the order of 0.1 nm – that is, of the order of the inter-atomic distances in solids. It is impossible to have a one-dimensional diffraction grating with such small spacing. In 1912, Max von Laue observed the diffraction of X-rays by a crystal. He proposed that a crystal is a regular array of atoms. Thus, a crystal may act as a three-dimensional diffraction grating for X-rays. By observing the diffraction of X-rays by a crystal of known structure, it is possible to measure their wavelength; and conversely, by observing the diffraction of X-rays of known wavelength by a crystal, it is possible to explore its crystalline structure.

The fundamental structure of a crystal is the *unit cell*, which is an arrangement of a certain number of atoms (or ions) in a characteristic geometrical configuration. A macroscopic crystal is a periodical three-dimensional juxtaposition of unit cells in all directions. For instance, the unit cell of sodium chloride is a center-faced cubic of sides $d = 0.562737$ nm. It is formed by Cl^- and Na^+ ions. Each of these cells is a diffracting center for X-rays and the total diffracted wave results from the interference of these partial waves. The direction of the principal maximums of interference is determined by the geometry of the configuration of the unit cells, while the intensity of the diffracted wave in a given direction is determined by the distribution of the electric charge within the unit cell. Thus, by observing the direction of the diffraction maximums, we obtain information on the geometry of the diffraction centers, and by observing the intensity distribution, we obtain information on the distribution of charge within the unit cell.

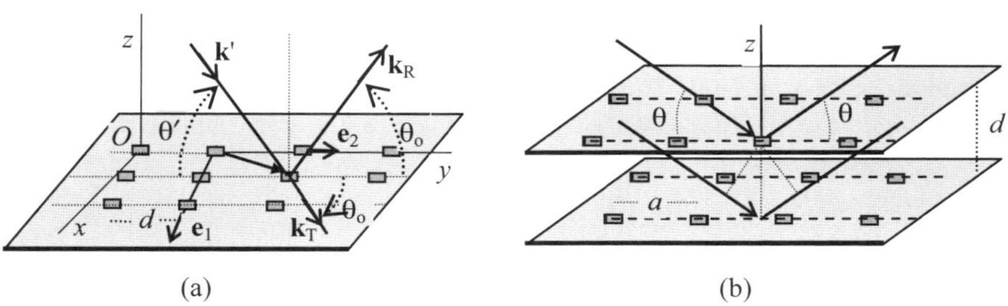

Figure 8.18. *a) Diffraction by the unit cells in a crystallographic plane, and b) diffraction by a three-dimensional crystal*

First we consider the diffraction by a chain of unit cells of spacing d along an axis parallel to Oy (Figure 8.18a). Let θ' be the angle of the incident wave vector $\mathbf{k}'$

with the chain. The diffracted intensity has a principal maximum if the diffracted wavelets by all the cells are in phase; this is the case if the direction of observation makes an angle θ_p with the chain such that

$$\cos\theta_p - \cos\theta' = p\lambda/d \qquad p = \text{integer}. \tag{8.41}$$

In particular, $p = 0$ corresponds to $\cos\theta_o = \cos\theta'$. If we make our observations in the plane of incidence, we may have either $\theta_o = \theta'$ (which corresponds to a "transmitted" wave without deviation) or $\theta_o = -\theta'$ (which corresponds to a "reflected" wave).

Now consider the wavelets diffracted by all the unit cells of the plane *Oxy* (or any plane parallel to *Oxy*). The total diffracted wave has a principal maximum if all the diffracted wavelets are in phase. It may be shown that this is the case if we make our observations in the direction of reflection $\theta_o = -\theta'$ (as on a plane mirror) or the direction of transmission $\theta = \theta'$ (Figure 8.18a). Finally, consider the superposition of the diffracted waves by all the planes of unit cells, which are parallel to *Oxy* with a spacing d (Figure 8.18b). The path difference of the waves diffracted by two consecutive planes is $2d \sin\theta$. Thus, the total diffracted wave has a principal maximum if this path difference is an integer multiple of λ; hence, the Bragg law:

$$2\, d \sin\theta_p = p\,\lambda \qquad p = \text{integer}. \tag{8.42}$$

This condition must be verified by the incidence and observation angles ($\theta' = \theta_p$) in order to have a principal maximum. In the other directions, there is some phase difference between the waves diffracted by the planes. As the number of these planes is very large, the reflected intensity in these directions is negligible. We conclude that, if a wave (X-rays, electronic wave, etc.) is incident on the crystal in a direction making an angle θ_p with the crystallographic plane, there is a diffracted wave only in the directions verifying Bragg's law [8.42]. For instance, in the case of X-rays of wavelength $\lambda = 0.200$ nm incident on a ClNa crystal, the Bragg condition gives $\sin\theta = p\lambda/2d = 0.178\, p$, thus, $\theta = 0 - 10.2° - 20.8°$, etc.

The diffraction of X-rays of known wavelength allows us to study the crystalline structure. If the spacing of a particular crystal (ClNa or $CaCO_3$, for instance) is determined by a method other than X-rays, this crystal may then be used as a spectrometer for X-rays. Observation of the diffraction of a beam of X-rays by this crystal allows us to determine the wavelength of the X-rays. Then, the diffraction of this beam of known λ by other crystals can be used to study their structures.

8.10. Summary

Interference is a redistribution of the physical quantities of waves, which results from their superposition. Diffraction is the deviation of a wave if it encounters an obstacle or it passes through an aperture whose dimensions are comparable to λ.

The superposition of two waves $a\cos[\omega(t - r_1/v) - \phi_1]$ and $a\cos[\omega(t - r_2/v) - \phi_2]$ may be written as $u(M) = 2a\cos(\frac{1}{2}\Delta\phi)\cos(\omega t - \phi_1 - \frac{1}{2}\Delta\phi)$, where $\Delta\phi$ is the phase lag of wave (2) over wave (1). It is given by $\Delta\phi = \Delta_o\phi + \Delta_m\phi$ with $\Delta_o\phi = \phi_2 - \phi_1$ and $\Delta_m\phi = 2\pi(r_2 - r_1)/\lambda$. *The order of interference* at M is $p \equiv \Delta\phi/2\pi = \Delta\phi_o/2\pi + \Delta r/\lambda$.

The superposition of two waves of amplitudes a_1 and a_2 has an amplitude that varies between a minimum $|a_1 - a_2|$ (*destructive interference*) if p is equal to a half-integer, and a maximum $|a_1 + a_2|$ (*constructive interference*) if p is equal to an integer. The resultant intensity, its minimum, its maximum and the *contrast* are

$$\mathcal{I}(M) = \mathcal{I}_1(M) + \mathcal{I}_2(M) + 2\sqrt{\mathcal{I}_1(M)\mathcal{I}_2(M)}\cos(\Delta\phi)$$

$$\mathcal{I}_{\min} = [\sqrt{\mathcal{I}_1(M)} - \sqrt{\mathcal{I}_2(M)}]^2 \quad \text{and} \quad \mathcal{I}_{\max} = [\sqrt{\mathcal{I}_1(M)} + \sqrt{\mathcal{I}_2(M)}]^2.$$

$$\mathcal{C} = \frac{\mathcal{I}_{\max} - \mathcal{I}_{\min}}{\mathcal{I}_{\max} + \mathcal{I}_{\min}} = \frac{2\sqrt{\mathcal{I}_1(M)\mathcal{I}_2(M)}}{\mathcal{I}_1(M) + \mathcal{I}_2(M)}, \quad 0 < \mathcal{C} < 1.$$

Young's experiment uses two slits (or pinholes) of spacing d and the interference is observed at points M situated at long distance in directions θ. The path difference from the slits to M is $\Delta r \equiv r_2 - r_1 = d\sin\theta$, the spacing between adjacent fringes is $i = D\lambda/d$. The long distance limit is realized by observing the interference on a screen placed at the focal length f of a converging lens, then $i \cong f\lambda/d$. The observed intensity is $\mathcal{I} = \mathcal{I}_{\max}\cos^2[\pi(d/\lambda)\sin\theta + \frac{1}{2}\Delta_o\phi]$. If white light is used, the fringe of order 0 is common (white) and it is bordered on both sides by dark bands, then colored bands of order 1 at distances $x_\lambda = \lambda f/d$. The fringes of higher orders overlap. For a large x, we again find white light with dark lines (a *channeled spectrum*).

The interference is observable, if the waves are synchronous, temporally coherent and have the same polarization (if they are transverse vector waves). In the case of light, the source must not be wide (spatial coherence).

The sharp colors of thin films are due to the interference of the reflected waves on both faces of the film. The phase shift of the first two reflected waves and two transmitted waves are $\phi_R = \phi + \hat{\phi}$ and $\phi_T = \phi + \hat{\phi} + \pi$, where $\phi = 4\pi n_2 (e/\lambda_o)\cos\theta_2$ and $\hat{\phi}$ is equal to zero if n_2 lies between n_1 and n_3, and is equal to $\pm\pi$ otherwise. The interference of the reflected waves is constructive (and that of the transmitted waves is destructive) if $p_R = \phi_R/2\pi$ is equal to an integer and vice versa if p_R is equal to a half-integer. If the film is illuminated with white light, the wavelengths that correspond to an integer p_R produce a constructive interference by reflection, and a destructive interference by transmission. This explains the different colors of a wedge-shaped air layer or Newton rings. This property enables us to use thin films as *interference filters*. Optical coating consists of applying a transparent film to a glass surface in order to reduce the reflected light or to increase it. The thickness of the film must be $\lambda/4$ and the index conveniently chosen for that purpose.

The diffraction is due to the interference of the wavelets that are emitted by the elements of area $d\mathcal{S}$ of the aperture. The amplitude of the wavelet $d\underline{u}_d$ that is emitted by $d\mathcal{S}$ near a point P is proportional to $d\mathcal{S}$, to the amplitude of the incident wave $\underline{u}(P)$ and to an inclination factor, which favors the direction of geometrical optics. If the light source S and the screen are at long distances from the aperture, and the angles of **k** and **k′** with the normal to the aperture are small (*Fraunhoffer diffraction*), the diffracted wave may be written in the form:

$$\underline{u}_d(M) = \chi \iint_S d\mathcal{S}\ e^{ix(k_x - k'_x) + iy(k_y - k'_y)} .$$

The diffracted intensity by a narrow slit of width d is:

$$\mathcal{I}(\theta) = \mathcal{I}_o\, \mathcal{F}_d(\Phi) \qquad \text{where } \mathcal{F}_d(\Phi) = \left[\frac{\sin \Phi/2}{\Phi/2}\right]^2 \qquad \text{and } \Phi = 2\pi(d/\lambda) \sin \theta.$$

Φ is the phase shift between the extreme wavelets. $\mathcal{I}$ has a principal maximum $\mathcal{I}_o$ for $\Phi = 0$ (i.e. $\theta = 0$), zeros for $\Phi/2 = p\pi$ (i.e. $\sin \theta = p\lambda/d$, where $p = \pm 1, \pm 2, \pm 3$ ...) and secondary maximums for $\Phi/2 \approx (p + ½)\pi$ [i.e. $\sin \theta \approx (p + ½)(\lambda/d)$].

The diffracted intensity by a *rectangular aperture* of sides a and b, in the case of normal incidence is $\mathcal{I}(M) = \mathcal{I}_o\, \mathcal{F}_d(\Phi_x)\, \mathcal{F}_d(\Phi_y)$, where $\Phi_x = ak_x$ and $\Phi_y = bk_y$.

In the case of a *circular aperture* of radius R illuminated normally, the diffracted intensity in the direction θ with the axis of the aperture is $\mathcal{I}(M) = \mathcal{I}_o[2J_1(s)/s]^2$, where J_1 is a Bessel function of order 1 and $s = k_x R = 2\pi(R/\lambda) \sin \theta$. The diffraction has a principal maximum $\mathcal{I}_o$ bordered successively by zeros and secondary maximums of decreasing intensity. The angle of the first minimum is given by $\sin \theta_{min} = 0.61\ \lambda/R$.

If a diffraction grating is formed by N lines, emitting waves of amplitude a and phase shift ϕ between adjacent lines, the amplitude and the resultant intensity are

$$A(\phi) = a\frac{\sin(N\phi/2)}{\sin(\phi/2)} \quad \text{and} \quad \mathcal{I}(\theta) = \mathcal{I}_o\, \mathcal{F}_{dg}(\phi) \quad \text{where} \quad \mathcal{F}_{dg}(\phi) \equiv \frac{\sin^2(N\phi/2)}{N^2 \sin^2(\phi/2)} .$$

If θ' and θ, respectively, are the algebraic angles that the incident wave vector $\mathbf{k}'$ and the diffracted wave vector $\mathbf{k}$ make with the normal $\mathbf{n}$ to the diffraction grating, oriented in the direction of light propagation, the phase shift is

$$\phi = 2\pi(d/\lambda)(\sin\theta - \sin\theta') = \mathbf{K.d} \qquad \text{where} \quad \mathbf{K} = \mathbf{k} - \mathbf{k}'.$$

For $\phi_p = 2p\pi$, $\mathcal{I}(\theta)$ has principal maximums $\mathcal{I}_o$ equal to N^2 times the intensity of a single slit. The principal maximums have a small half-width $\delta\phi = 2\pi/N$. If $\delta\lambda$ is the difference of two wavelengths that may just be distinguished in the pth order, the resolving power of the diffraction grating is $\mathcal{R}_p \equiv \lambda/\delta\lambda = pN$.

A crystal is a three-dimensional diffraction grating for X-rays. Each unit cell is a diffracting center. The interference of the diffracted X-rays by the crystallographic planes of spacing d is constructive if the angles θ' and θ of the incident and diffracted rays with the planes are equal and verify Bragg's law $2d \sin\theta_p = p\lambda$, where p is an integer.

8.11. Problem solving suggestions

1) To analyze the interference of waves, we first evaluate the *path difference* Δr at the point of observation. Each path difference equal to the wavelength λ corresponds to a *phase lag* of 2π. Thus, the phase lag is $2\pi(\Delta r/\lambda)$. Do not forget that the wave frequency does not depend on the medium. Usually, in the case of light, the wavelength λ_o is specified for the propagation in vacuum. In a medium of index n (i.e. propagation velocity $v = c/n$), the wavelength is $\lambda = v/\tilde{\nu} = \lambda_o/n$. The phase lag due to Δr is $2\pi(\Delta r/\lambda) = 2\pi(n\,\Delta r/\lambda_o)$. The quantity $n\,\Delta r$ is called the *optical path*.

2) Do not forget the initial phase shift (of the sources) and an eventual phase shift of π if the wave undergoes a reflection on a more refringent medium (see Chapter 7) or if a vector wave passes by a focal point (see problem P4.8). A phase shift of π is equivalent to a change of the sign of the amplitude, and thus, to a change of bright fringes to dark fringes. If you use the reflection coefficient, the reflection change of sign is already included in this coefficient, so you do not have to add a phase shift π for the reflection. As the phase of a wave is determined up to 2π, the phase shift due to the reflection may be taken as $+\pi$ or $-\pi$. By making one choice or the other, you modify the order of interference by 1.

3) By writing the oscillation of a source in the form $u = a\cos(\omega t - \phi)$, ϕ is the initial phase lag. While propagating, the wave undergoes an additional phase lag $\mathbf{k.r}$. To calculate the intensity resulting from the superposition of the waves that are emitted by two sources (1) and (2), we may take $\phi_1 - \phi_2$ or $\phi_2 - \phi_1$ because $\mathcal{I}(M)$ is symmetric in the exchange of the sources. If you add phase shifts, the order is

important. If source (2) has an initial phase lag $\Delta_o\phi$ over source (1), the total phase shift at M is $\Delta_o\phi + \Delta_s\phi$, where $\Delta_s = 2\pi(\Delta r/\lambda)$ with $\Delta r = r_2 - r_1$ (not $r_1 - r_2$).

4) In all optical interference systems illuminated by white light, the order 0 is the same for all wavelengths. The corresponding fringe is white. At observation points where the order of interference is high ($\Delta r >> \lambda$), the fringes are so dense that we find bright fringes for almost all wavelengths; this is the *high-order white* but the light spectrum is channeled (it has dark lines).

5) In the case of Fraunhofer diffraction, the diffracted intensity by a narrow slit is $\mathcal{I} = \mathcal{I}_o \mathcal{F}_d(\Phi)$, where Φ is the phase shift of the extreme wavelets (emitted by the edges of the slit), $\mathcal{F}_d(\Phi) = \sin^2(\Phi/2)/(\Phi/2)^2$ and $\mathcal{I}_o$ is the maximum intensity at the center of the diffraction pattern. In the case of a rectangular aperture, we find $\mathcal{I} = \mathcal{I}_o \mathcal{F}_d(\Phi_x) \mathcal{F}_d(\Phi_y)$ where Φ_x and Φ_y are the phase shifts of the extreme wavelets of the slit in the directions Ox and Oy respectively. In the case of a circular aperture of radius R, the angular half-width θ_{min} of the principal maximum is given by $\sin\theta_{min} = 0.61\,\lambda/R$.

6) In the case of X-rays diffracted by a crystal, the fundamental relation is Bragg's law, $2d\sin\theta = p\lambda$. Note that d is not the spacing of the atomic planes but the crystallographic planes (i.e. the planes of the unit cells).

8.12. Conceptual questions

Q1. In which case is the intensity that is emitted by two sources equal to the sum of the intensities emitted by each one separately; and in which case is the amplitude equal to the sum of the amplitudes?

Q2. Two waves of amplitudes A_1 and A_2, of equal angular frequency ω and in phase are emitted by two pinholes situated at a distance d from a light source. Taking into account the variation of the amplitudes with the travelled distance, determine the contrast of the interference of these waves at the distances r_1 and r_2 from the pinholes. What are the conditions required for the fringes to be observable?

Q3. Two identical loudspeakers are at the same height and separated by a distance d in an auditorium. They emit, in phase, two sounds of equal frequency $\tilde{\nu}$. Draw the crest lines in the horizontal plane at a given instant of time. Deduce the interference fringes. Discuss the sound intensity that is perceived, depending on the position of a person in the hall. Is it possible that he hears the sound with one ear and not the other? Does he perceive interference effects if the loudspeakers emit sounds of several frequencies?

Q4. How do we locate the fringe of order 0 in the case of light interference? In a Young's double-slit experiment, the slits are illuminated by sunlight. We can only distinguish colored fringes of the first order. Explain why.

Q5. In a Young's double slit experiment, the incident wave makes an angle α with the z-axis (which is normal to the opaque screen of the slits). What is the initial phase shift of the slits? How far are the fringes displaced from their positions at normal incidence? What is the displacement of the principal maximums in the case of diffraction by a single slit and in the case of a diffraction grating. Compare this with geometrical optics.

Q6. The reception of *AM* emissions behind a hill is good but that of *FM* and *TV* emissions is poor. Similarly, you may hear a sound even if the source is behind a wall, but you cannot see a light source if it is behind a disk of diameter 10 cm and you cannot see the shadow of a speck of dust. Explain why.

Q7. What should the spacing of a diffraction grating be if it is conceived to observe the first five orders in the case of infrared of wavelength $\lambda = 10^4$ nm ? What should we observe if an infrared wave is incident on a diffraction grating conceived for visible light? What should we observe if visible light is incident on a diffraction grating conceived for infrared?

Q8. Compare the diffraction patterns by a single slit of width L, two slits of spacing L and a diffraction grating of N lines and total width L.

Q9. Is it possible that the 0-order spectrum given by a diffraction grating overlap on the spectrum of order of 1 if it is illuminated by sunlight (390 nm $< \lambda <$ 760 nm)? Compare the width of the principal maximums and the resolving power of a diffraction grating in the different orders.

8.13. Problems

Interference of two waves

P8.1. Two identical sound emitters lying at $x = -d/2$ and $x = +d/2$ on the x axis have equal frequency $\tilde{\nu}$ and are in phase. **a)** Show that d must be larger than $\lambda/2$ in order to have a destructive interference somewhere in the plane Oxy. **b)** Assuming that $\tilde{\nu}$ = 200 Hz, d = 5 m and the speed of sound is equal to 332 m/s, determine the points where the intensity is at a minimum in the plane Oxy. **c)** The sources emit the same audible sounds (20 Hz $< \tilde{\nu} <$ 4 kHz) and a person is sitting at 6 m from the first source and 8 m from the second. What are the frequencies that he does not hear?

P8.2. In order to protect a metallic mirror, it is covered by a film of thickness 5 μm and index 1.5. Light of wavelength 0.6 μm (in vacuum) is incident normally on this mirror. Determine the positions of the bright points within the film. We recall that the metallic mirror reflects light with a phase shift of π.

Young's experiment

P8.3. In a Young's setup, the slits are separated by a distance $d = 0.150 \pm 0.002$ mm and the interference is observed on a screen situated at a distance $D = 1.000 \pm 0.003$ m (Figure 8.19). **a)** Using sunlight, we find a bright white fringe at O'. We make a pinhole in the screen at a distance of 1 cm from O'. Show that the light spectrum passing through this hole has dark lines (channeled spectrum). Determine the corresponding wavelengths which lie in the visible spectrum 390 nm < λ < 760 nm. **b)** Using quasi-monochromatic light, we find that the fourth bright fringe is at a distance $x_4 = 12.7 \pm 0.1$ mm from O'. What is the value of the wavelength? Estimate the precision of this measurement. **c)** Assume that the source is at a distance $b = 10$ cm from the slits. Determine the variation of the order at $x = 2$ cm if the source is displaced transversally by a distance $a = 0.05$ mm from S to S'.

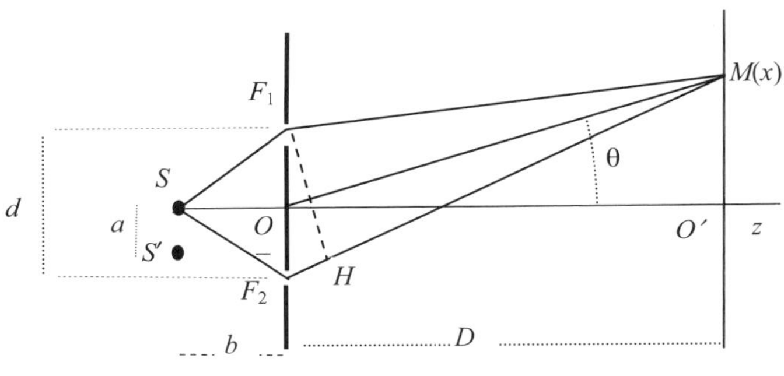

Figure 8.19. *Problem P8.3*

Multiwave interference, conditions for interference

P8.4*. A source is situated at an equal distance from two slits. It emits two coherent waves of equal amplitudes and close wavelengths λ and λ + δλ. The interference is observed on a screen at a long distance D from the slits. **a)** Determine the intensity at a point M situated at a distance x from the symmetry plane. **b)** Calculate the contrast $\mathcal{C}$ at x and study its variation as a function of x. Express $\mathcal{C}$ as a function of the path difference at the point x. Is it possible to use these results to determine λ and its precision δλ?

Spatial coherence

P8.5. a)** Two slits are separated by a distance $d = 0.5$ mm and illuminated by 0.5 μm light. What is the number of fringes that may be observed if the screen is placed at a distance $D = 1$ m? **b)** Assume that the slits are illuminated by a circular source of radius $R = 0.5$ mm situated at a distance a from the slits. Show that all the atoms A lying on the z-axis, although they emit independently, produce oscillations of the slits in phase. What is the phase shift of the slits if light is emitted by an atom B situated at the distance R from the axis? Show that, on the screen, this phase shift produces a displacement of the fringes by a distance $Rid/\lambda\sqrt{a^2+d^2}$, where i is the spacing between fringes. What should the distance a of the source be for the displacement due to its extreme points B and B' not to exceed $i/4$? **c)** Assume that the condition of question (b) is satisfied but light has a bandwidth $\Delta\lambda$ near $\lambda = 0.5$ μm. What is the relative variation of i due to $\Delta\lambda$? Is it possible to use this result to measure the bandwidth of radiations if we are capable of observing a displacement of $i/4$? **d)** Assume that a glass plate of thickness e and index $n = 1.5$, independent of the wavelength, is placed after one of the slits. Show that it introduces a phase shift given by $2\pi(n-1)(e/\lambda)$. What is the displacement of the fringes if $e = 1$ mm?

P8.6. A method proposed by Rayleigh, allows us to measure the index of a gas. It consists of placing two identical glass containers of length l filled with this gas, in front of a Young's slit (or a more convenient equivalent setup, such as a Michelson interferometer). The slits are separated by a distance d and illuminated normally and the fringes are observed on a screen placed at the focal length f behind a converging lens. **a)** Calculate the spacing between fringes i as a function of λ, d and f. How do we determine the fringe of order 0? **b)** The gas in one of the containers is pumped out slowly and completely. How do the interference fringes move? Express the index n of the gas as a function of the number N_f of fringes, which pass by the position of the fringe of order 0. What is the index of air if $\lambda = 589.3$ nm, $l = 30$ cm and $N_f = 148$. **c)** We place the containers filled with the gas in the same conditions. If the temperature of one of them is increased, the index of the gas that it contains decreases. How do the fringes move? Calculate the variation of the index which produces a displacement equal to the spacing i.

Thin films

P8.7. a) Show that the phase shift between the first two reflected waves by a thin film and that of the first two transmitted waves is given by [8.22]. **b)** We illuminate almost normally a wedge-shaped air film between two glass plates, which make a small angle α. Evaluate the phase shift of the first two reflected waves at a distance x from the wedge. Deduce that the fringes are parallel to the wedge with a spacing $i \approx \lambda/2\alpha$. **c)** A planar convex lens is placed on an optical flat and illuminated normally. The gap between the spherical surface of the lens of radius R and the optical flat constitutes a circularly symmetric, wedge-shaped air film. Show that we observe bright circular fringes of radii $r_p \approx \sqrt{(p+1/2)\lambda R}$ called Newton's rings. Describe what we observe if this setup is illuminated by sunlight?

Diffraction

P8.8. a) Light of wavelength $\lambda = 0.6$ μm is incident normally on a slit of width 0.2 mm. The diffraction pattern is observed on a screen placed at the focal length $f = 100$ cm of a converging lens. What are the distances from the first minimum and the first secondary maximum to the centre of the principal maximum? **b)** Describe the observed pattern if the slit is illuminated by two wavelengths $\lambda_1 = 0.6$ μm and $\lambda_2 = 0.5$ μm. **c)** The slit is illuminated by a single wavelength $\lambda = 0.6$ μm, but in a direction making an angle $\theta' = 30°$ with the normal to the slit. Show that the intensity in the direction θ is $\mathcal{I} = \mathcal{I}_o \mathcal{F}_d(\Phi)$, where $\Phi = 2\pi(d/\lambda)(\sin\theta - \sin\theta')$ and $\mathcal{F}_d(\Phi) = \sin^2(\Phi/2)/(\Phi/2)^2$. Deduce that the principal maximum is in the direction of geometrical optics. Determine the positions of the first minimums situated on both sides of the principal maximum.

P8.9. A sound wave of frequency 5 kHz propagates in a pipe of radius 5 cm and undergoes diffraction at its open end. What is the direction θ_{min} corresponding to the first minimum? What would the value of this angle be if the pipe were terminated by a truncated cone whose base has a radius of 20 cm?

P8.10. A microwave is diffracted by a slit of width d and observed on a screen at a long distance D. **a)** Express the angles θ_1 and θ_2 of the first two secondary maximums (measured from the direction of the axis Oz) as functions of d and the wavelength λ. **b)** Determine λ if $\theta_2 = 40°$ and $d = 1$ cm. **c)** What is the angular width of the principal maximum, defined as the angle between the directions, at which the intensity of the wave is $\mathcal{I}_o/2$, where $\mathcal{I}_o$ is the intensity of the central maximum?

P8.11. a) Show that the diffracted wave by N identical apertures, that are distributed at random, is $\underline{u}_d(M) = \underline{u}_{d1}(M)\,[1 + \Sigma_{j\neq 1} e^{i\phi_j}]$, where $\underline{u}_{d1}$ is the diffracted wave by the first aperture and ϕ_j is the phase shift of the aperture (j) with respect to the first aperture. Deduce that the diffraction pattern produced by the N apertures is the same as that of a single aperture but with a principal maximum intensity multiplied by N^2. The same result is valid if the wave is diffracted by N identical obstacles distributed at random (powder, for instance). **b)** The Sun and the Moon appear to be surrounded by a halo when the atmosphere is slightly cloudy or dusty. Explain why. What can you conclude if the angular radius of this halo is 3° and light has an average wavelength of 0.6 μm?

Diffraction gratings

P8.12. Using the phasor diagram, analyze the interference of the diffracted waves by six slits of spacing d emitting waves of equal amplitudes that are in phase.

P8.13. In the case of radio waves, it is easy to have sources with spacing of the order of the wavelength. A rectilinear antenna, carrying an oscillating current, emits an electromagnetic wave especially in radial directions of its median plane. Consider two parallel antennas with a spacing d (Figure 8.20). Let $E_1 = E_m \sin(\omega t)$ and $E_2 = E_m \sin(\omega t + \phi)$ be the waves that they emit. **a)** Write down the expressions of the waves at a long distance R from the midpoint O and in the direction making an angle θ with the normal to the plane of the antennas. What is the resultant wave? Determine the directions corresponding to the maximum intensity of the wave and to the minimum intensity. **b)** Assuming that $d = \lambda/2$ and $\phi = 0$, determine the intensity for $\theta = 0$, 30°, 60° and 90°. What can you deduce concerning the direction of the emitted wave? **c)** Now assume that $d = \lambda/2$ and $\phi = 30°$. What is the new direction of the emitted wave? **d)** Is it possible to narrow the direction of emission by increasing the number of antennas?

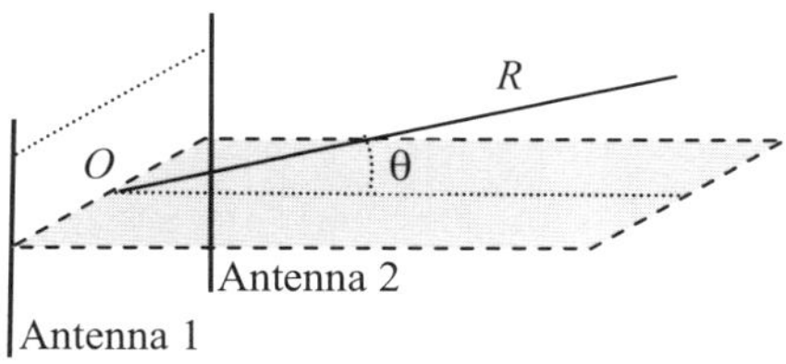

Figure 8.20. *Problem P8.13*

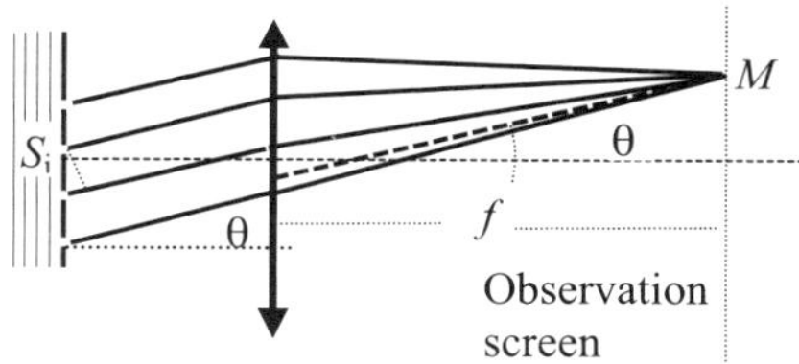

Figure 8.21. *Problem P8.14*

P8.14. Four coplanar and parallel slits have a spacing d (Figure 8.21). They are illuminated by light of wavelength λ. We observe the fringes on a screen, which coincides with the focal plane of a converging lens. **a)** Calculate the intensity of the diffracted wave at the point M in the direction θ. **b)** Plot the intensity as a function of $\phi = 2\pi(d/\lambda) \sin\theta$. Deduce the angular positions of the principal maximums and the secondary maximums. Compare these two types of maximums. **c)** Determine the positions of the principal maximums on the screen if $d = 4.5\ \lambda$ and $f = 50$ cm.

P8.15. a) In order to determine the spacing of a diffraction grating, we observe the diffraction by transmission for a beam of wavelength 589 nm. We find that one of the principal maximums is in the direction of the incident beam and the third is in the direction of 45.0°. What is the spacing of the diffraction grating? **b)** This diffraction grating is illuminated by light of an unknown wavelength; we find that the third order maximum is in the direction of 35.5°. What is the value of λ? **c)** Hydrogen and deuterium have two lines which differ by 0.18 nm close to $\lambda = 656.3$ nm. What should the minimum number of diffraction grating lines be in order to separate these two lines? **d)** This diffraction grating is illuminated normally by sunlight. The transmitted light is received on a screen located at 60 cm from the diffraction grating. In this screen we make a slit between the distances 10 cm and 11 cm from the central maximum. What is the band of the wavelength passing by this slit?

Diffraction of X-rays

P8.16*. a) Estimate the spacing d of the unit cells in ClNa, knowing that its density is 2,164 kg/m^3 and its molar mass is 58.44 g/mole. **b)** In a 0.20 nm X-ray diffraction experiment using this crystal, we find that the first maximum is deviated by 20.47°. What is the value of d? What is the spacing between the ions? Show that the most precise measurement of d is obtained if the angle θ is close to $\pi/2$. **c)** A beam of X-rays, with wavelengths varying between 0.095 nm and 0.130 nm, is incident at an angle of 45° with the crystallographic planes. What are the reflected wavelengths? **d)** A beam of X-rays of wavelength $\lambda = 0.11$ nm is incident at an angle of 45° (Figure 8.22). By what angle must the crystal be rotated in order to have a reflection on planes 1, 2 and 3?

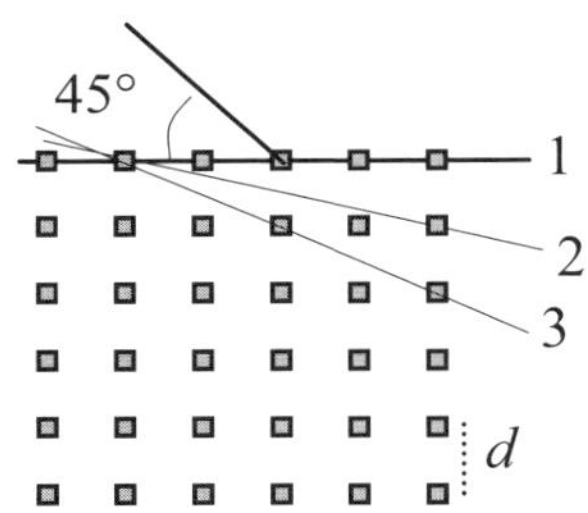

Figure 8.22. *Problem P8.16*

Chapter 9

Standing Waves and Guided Waves

In an open space, a wave propagates very often in all directions; this causes the decrease of its intensity with the travelled distance r. A *waveguide* conserves the intensity of the wave by confining it within a structure, which limits the medium of propagation in the transverse directions. Guided waves propagate in well-determined *modes*. Each mode is characterized by a cut-off frequency, a phase velocity and a group velocity, which depend on the frequency of the wave and on the geometry of the waveguide. The propagation properties in the infinite medium are recovered if the transverse dimensions of the waveguide are much larger than the wavelength. The waves on the surface of the water in a canal are an example of guided waves (section 5.10). If all the dimensions of the medium are limited, it can support only *standing* or *stationary waves* in *normal modes* of discrete frequencies (also called *normal frequencies*). In each mode, the propagation medium is a juxtaposition of wave zones with no transfer of physical quantities from one zone to the other. More generally, the guided or standing wave may be a superposition of modes.

In this chapter we study one-dimensional standing waves and we generalize to two-dimensional and three-dimensional waves. Afterwards, we analyze guided waves. We may analyze their propagation by studying the successive reflections on the guide walls. However, a more practical and general method consists of directly finding the solutions of the wave equation that satisfy the boundary conditions. First, we study the case of sound waves, which are simpler to analyze. Then, we consider the case of electromagnetic waves, which have many important applications but are complicated to analyze because of their transverse vector nature.

9.1. One-dimensional standing waves

First we consider one-dimensional waves obeying the equation

$$\ddot{u} - v^2 (\partial^2_{xx} u) = 0 , \qquad [9.1]$$

where v is the speed of propagation in the infinite medium. We have seen that it has solutions of the form $u_{pr} = u(t \pm x/v)$, which are progressive waves. The *method of separation of variables* consists of looking at solutions of the form $u_{st} = g(t)\, f(x)$. Substituting this expression in [9.1], this equation takes the form

$$f(x)\,(d^2g/dt^2) = v^2\, g(t)\,(d^2f/dx^2)\,, \qquad \text{i.e.} \qquad g''/g = v^2 f''/f. \qquad [9.2]$$

As g''/g is a function of t and v^2f''/f is a function of x, this equation is only identically verified (i.e. for any x and t) if both sides are equal to a constant. If this constant is positive, the solutions $g(t)$ and $f(x)$ are exponential in t and x respectively. It is easy to show that such expressions cannot verify the boundary conditions. Thus, we consider the case that this constant is negative. Designating it by $-\omega^2$, we obtain the equations $g'' = -\omega^2 g$ and $f'' = -(\omega^2/v^2)f$, whose solutions are $g = A\cos(\omega t + \alpha)$ and $f = B\cos(\omega x/v + \phi)$. Thus, the solution of [9.1] may be written as

$$u_{st} = A\cos(\omega x/v + \phi)\cos(\omega t + \alpha) = A\cos(2\pi\, x/\lambda + \phi)\cos(\omega t + \alpha), \qquad [9.3]$$

where $\lambda = 2\pi v/\omega$ is the wavelength. It is always possible to choose the phases α and ϕ so that the amplitude A is positive.

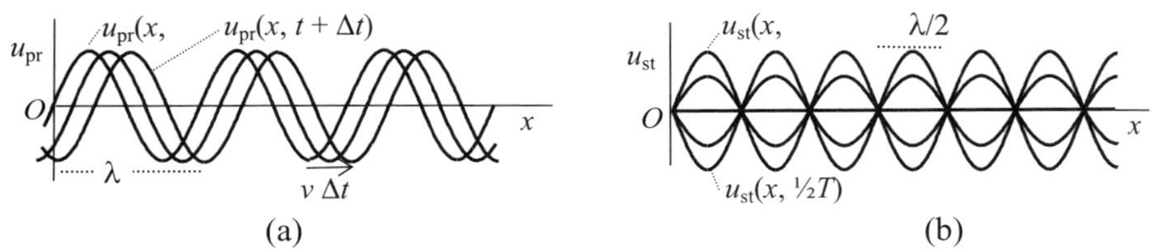

Figure 9.1. *a) Progressive wave, and b) standing wave*

The solutions $u_{pr} = A\cos[\omega(t - x/v)]$ and $u_{st} = A\cos(\omega t)\sin(\omega x/v)$ on a string are illustrated in Figures 9.1a and 9.1b at successive instants of time. In the case of u_{pr}, all the points of the string oscillate with the same amplitude A. The wave function and all the corresponding physical quantities are continuously transmitted from one point of the string to another. In the case of u_{st}, some equally spaced points of the string remain at rest; they are called *nodes*. At the halfway point between two nodes, the amplitude of vibration is at a maximum; these points are the *antinodes*. Thus, the string is divided by the nodes into *wave zones* in which the wave function and the physical quantities oscillate in the course of time, apparently with no transfer from one zone to another through the nodes. We say that u_{pr} is a *progressive wave*, while u_{st} is a *standing wave* (i.e. that remains confined to a finite region). A progressive wave must be sustained by an emitter, which acts at the entry of an infinite medium. The emitter continuously supplies energy, which is carried by the wave. The progressive wave vanishes if the emitter is turned-off, while a standing wave, once established, needs no emitter and it remains indefinitely over time if there is no energy dissipation.

Standing wave [9.3] of amplitude A may sometimes be considered to be the superposition of two waves u' and u'' of equal frequencies, the same polarization and amplitude $A/2$ which propagate in opposite directions (the incident and the totally reflected waves, for instance). Indeed, the sum of such waves is

$$u = u' + u'' = \tfrac{1}{2} A \cos[\omega(t - x/v) + \phi'] + \tfrac{1}{2} A \cos[\omega(t + x/v) + \phi'']$$
$$= A \cos(2\pi x/\lambda + \phi) \cos(\omega t + \alpha) \quad \text{where } \phi = \phi'/2 - \phi''/2 \quad \text{and } \alpha = \phi'/2 + \phi''/2. \quad [9.4]$$

At some points, u' and u'' are out of phase producing nodes, where $u = 0$ at any time. Their positions are given by the condition

$$\cos(2\pi x/\lambda + \phi) = 0 \qquad \text{hence,} \qquad x = \tfrac{1}{2}\lambda\,(\tfrac{1}{2} + p - \phi/\pi) \qquad (p \text{ integer}). \quad [9.5]$$

At some other points, u' and u'' are in phase producing antinodes, where u has the maximum amplitude $2A$. Their positions are given by the condition

$$|\cos(2\pi x/\lambda + \phi)| = 1 \qquad \text{hence,} \qquad x = \tfrac{1}{2}\lambda\,(p - \phi/\pi) \qquad (p \text{ integer}). \quad [9.6]$$

The antinodes are halfway between adjacent nodes and the distance between two adjacent nodes or two adjacent antinodes is $\lambda/2$. The points situated between two adjacent nodes oscillate in phase but with different amplitudes while two points situated on one side of a node and the other oscillate in phase opposition.

In the case of a progressive wave, the two forms of energy have equal densities at each point and at each time. If the wave is one-dimensional, the energy flux is $P(x, t) = vU_l(x, t)$, where U_l is the total energy density per unit length. If the wave is three-dimensional and $\mathbf{e}$ is the unit vector in the direction of propagation, the vector energy flux density is $\mathbf{S}(\mathbf{r}, t) = vU_v(\mathbf{r}, t)\,\mathbf{e}$, where U_v is the total energy density per unit volume. The time-average energy density and the intensity of the progressive wave cannot be zero at any point. On the contrary, in the case of a standing wave, the densities of the two forms of energy at any given point are not equal, not even their time-averages. The intensity of the wave is zero everywhere and so is the time-average transfer of all physical quantities. Because the two waves u' and u'', which propagate in opposite directions, transport on average the same physical quantities in opposite directions; thus there is no average global transfer of physical quantities. In the following, we are only concerned by standing waves, to simplify the notation we omit the index (st) for the wave function.

A standing wave may be established in a bounded medium; then, it results from the interference of the direct wave and the reflected wave on the boundaries. It may only be supported in discrete *modes* of normal frequencies $\tilde{\nu}_n$ and wavelength $\lambda_n = v/\tilde{\nu}_n$. Here, n stands for one or more discrete numbers n_1, n_2, etc. The modes are determined by the boundary conditions (or mathematical conditions in some cases). In order to impose these conditions, it is convenient to write the standing wave of the mode (n) in the form

$$u_n = [A_n \sin(2\pi\, x/\lambda_n) + B_n \cos(2\pi\, x/\lambda_n)]\cos(\omega_n t + \alpha_n). \qquad [9.7]$$

If only one mode corresponds to the normal frequency $\tilde{\nu}_n$, whose wave function is of the form $u_n = g_n(t)\, f_n(\mathbf{r})$, $\tilde{\nu}_n$ is said to be *non-degenerate*. In some cases, several modes $u_n^{(p)} = f_n^{(p)}(\mathbf{r})\, g_n(t)$ correspond to $\tilde{\nu}_n$, where $p = 1, 2\ldots d$ is a number used to distinguish the modes. We say then that $\tilde{\nu}_n$ is *degenerate* and its *degeneracy* is d. In the following, as examples, we consider standing elastic waves on a string and sound waves in a pipe.

1) Standing waves on a string

Consider a string of length L, tension F and mass by unit length m_l, fixed at its ends $x = 0$ and $x = L$. At these points, the displacement $u(0, t)$ and $u(L, t)$ must be zero at any time. This requires that $B = 0$ and $2\pi L/\lambda_n = n\pi$, thus

$$\lambda_n = 2L/n \quad \text{and} \quad \tilde{\nu}_n = \tfrac{1}{2}\, nv/L \quad \text{with} \quad v = \sqrt{F/m_l}\,. \qquad [9.8]$$

$$u_n = A_n \sin(n\pi x/L)\cos(\omega_n t + \alpha_n) \quad \text{where } n = 1, 2\ldots \qquad [9.9]$$

Figure 9.2. *The first three modes on a string with fixed ends*

Figure 9.2 illustrates the profile of the string if it is fixed at its ends and it oscillates in the first three modes. By necessity, the fixed ends must be nodes of displacement. The string is then formed by an integer number n of wave zones of length $\lambda_n/2$.

If one end of the string is fixed and the other is free (for instance, attached to a ring that slides without friction on a transverse rod), the string must be normal to the rod at this end. Indeed, if the string is oblique, a small element of the string that precedes the end is subject to the reaction of the rod (that is normal to the rod) and to the tension force of the string (which is tangent to the string). This element would be subject to a finite transverse force. As it may be as small as desired, its acceleration would be infinite and this is impossible. Thus, the wave function must verify the conditions $u(0, t) = 0$ and $\partial_x u = 0$ at $x = L$. The string is then formed by $(n - ½)$ wave zones of length $\lambda/2$ and the modes may be written as

$$u_n = A_n \sin[\pi x(n - ½)/L] \cos(\omega_n t + \alpha_n)$$
$$\lambda_n = 2L/(n - ½) \qquad \text{and} \qquad \tilde{\nu}_n = ½\,(n - ½)\,v/L \qquad \text{with } v = \sqrt{F/m_l} \qquad [9.10]$$

Let us analyze the energy on a string that is fixed at both ends, for instance. The kinetic energy density and the potential energy density on the string oscillating in the mode (n) (see equation [9.9]) are

$$U_{(C)\,l} = ½\, m_l\,(\dot{u}_n)^2 = ½\, m_l(A_n\omega_n)^2 \sin^2(n\pi x/L)\sin^2(\omega_n t + \alpha_n)$$
$$U_{(P)\,l} = ½\, F\,(\partial_x u_n)^2 = ½\, m_l\,(A_n\omega_n)^2 \cos^2(n\pi x/L)\cos^2(\omega_n t + \alpha_n). \qquad [9.11]$$

In contrast to the case of progressive waves, the two energy densities are not equal, not even in the time-average. However, the total average energy density at a given point x remains constant in the course of time and equal to

$$< U_l > = < U_{(C)\,l} > + < U_{(P)\,l} > = ¼\, m_l\,(A_n\omega_n)^2 = ¼\, m_l\,(n\pi v A_n/L)^2. \qquad [9.12]$$

It is the same at all the points of the string. At the antinodes, the string is not lengthened or shortened but the amplitude of the velocity is maximum, and the energy at these points is purely kinetic. At the nodes, the velocity of the string is always zero but the amplitude of the variation of its length is maximum, and the energy at these points is purely potential. The total energy of the string in the mode (n) is

$$U_n = L < U_l > = ¼\, L m_l\,(A_n\omega_n)^2 = ¼\, \pi^2 (F/L)\, n^2 A_n^{\,2}. \qquad [9.13]$$

We note that, for the same amplitude of vibration in the modes, the energy in the mode (n) is proportional to n^2.

The energy flux that is carried by the wave at a point x and at time t is

$$P(x, t) = -F\,\dot{u}\,(\partial_x u_n) = ¼ m_l v\, A_n^{\,2}\omega_n^{\,2} \sin(2\omega_n x/v)\sin(2\omega_n t + 2\alpha_n). \qquad [9.14]$$

At the nodes and the antinodes, this flux is zero at any time. At the other points, the flux is positive during a quarter of a period and negative during the next quarter of a period and its average is zero. Thus, there is some energy transfer in each half-zone, but no transfer from one half-zone to another.

We can also regard the string as a system of an infinite number of oscillators oscillating in the modes of angular frequencies ω_n, arbitrary amplitudes A_n and arbitrary phases α_n. The general standing wave is a superposition of these modes

$$u(x, t) = \Sigma_{n\,a} = \Sigma_n A_n \sin(\omega_n x/v) \cos(\omega_n t + \alpha_n). \qquad [9.15]$$

2) Standing sound waves in a pipe

A standing sound wave in a pipe is of the form

$$\underline{u} = [A \sin(kx) + C \cos(kx)]\, e^{i\omega t}, \qquad \underline{p}_a = -B(\nabla.\mathbf{u}) = \omega Z_s[-A \cos(kx) + C \sin(kx)]\, e^{i\omega t},$$

where $Z_s = \sqrt{Bm_v}$ is the acoustic impedance. If the pipe is closed at one end by a rigid obstacle (infinite impedance), the displacement is zero at this end at any time; thus, it is a node of displacement and an antinode of pressure. On the contrary, at an open end, the acoustic pressure is zero; thus, it is a node of pressure and an antinode of displacement. A pipe that is closed (or open) at both ends must contain an integer number (n) of wave zones. If it is closed at both ends, we must have

$$u_n = A_n \sin(n\pi x/L) \cos(\omega_n t + \alpha_n), \qquad p_{a,\,n} = -\omega Z_s A_n(n\pi x/L) \cos(\omega_n t + \alpha_n)$$

$$\lambda_n = 2L/n \quad \text{and} \quad \tilde{\nu}_n = \tfrac{1}{2}\, n v_s/L \qquad \text{where} \quad v_s = \sqrt{B/m_v}\,. \qquad [9.16]$$

If the pipe is open at both ends, we must have $p_a(0, t) = p_a(L, t) = 0$. Thus, the normal frequencies are $\tilde{\nu}_n = \tfrac{1}{2}\, n v_s/L$ and the wave may be written in the form

$$u_n = C_n \cos(n\pi x/L) \cos(\omega_n t + \phi_n), \quad p_{a,\,n} = C_n\omega_n Z_s \sin(n\pi x/L) \cos(\omega_n t + \phi_n). \quad [9.17]$$

If the pipe is closed at the origin and open at the other extremity, it must contain a half-integer number ($n - \tfrac{1}{2}$) of wave zones of length $\lambda_n/2$, and the wave must verify the conditions $u(0, t) = 0$ and $p_a(L, t) = 0$. Thus the displacement is of the form [9.10] and the pressure wave is $p_{a,\,n} = -A_n\omega_n Z_s \cos[(n-\tfrac{1}{2})\pi x/L) \cos(\omega_n t + \alpha_n)$.

For instance, if $v_s = 332$ m/s, a 30 cm long pipe has the normal frequencies 553 Hz, 1 107 Hz, 1 660 Hz, etc., if it is closed or open at both ends and the normal frequencies 277, 830, 1,383 Hz, etc., if it is open at one end and closed at the other.

As in the case of standing waves on a string, the analysis of the energy shows that the two energy densities are not equal and that there is no transfer of energy from one half-wave zone to another.

9.2. Standing waves on a membrane and in a rectangular cavity

We consider an elastic rectangular membrane of sides a and b fixed at its periphery (Figure 9.3). Let m_s be its mass per unit area and F_l its tension by unit length of the periphery. As we saw in section 5.5, the wave equation on this membrane is

$$\ddot{u} - v^2[\partial^2_{xx}u + \partial^2_{yy}u] = 0, \qquad v = \sqrt{F_l/ms} \qquad [9.18]$$

with the condition that the vibration u is zero on the periphery.

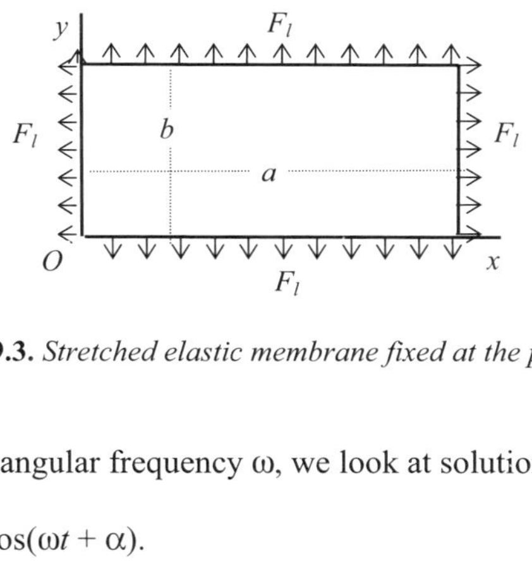

Figure 9.3. *Stretched elastic membrane fixed at the periphery*

If the wave has an angular frequency ω, we look at solutions of the form

$$u = X(x)\,Y(y)\cos(\omega t + \alpha). \qquad [9.19]$$

Substituting this expression to u in [9.18] and dividing by $X(x)\,Y(y)\cos(\omega t + \alpha)$, we obtain the equation

$$X''/X + Y''/Y + \omega^2/v^2 = 0. \qquad [9.20]$$

As the first term is a function of x and the second is a function of y, this equation is identically satisfied only if these terms are constant and their sum is equal to $-\omega^2/v^2$. If, for instance, X''/X is positive, the solution X is exponential and the condition $u = 0$ for $x = 0$ and $x = a$ cannot be satisfied. Thus, both terms X''/X and Y''/Y must be negative and we may write

$$X'' = -k_1^2 X, \quad Y'' = -k_1^2 Y \qquad \text{with} \qquad \omega^2 = v^2(k_1^2 + k_2^2). \qquad [9.21]$$

This gives for [9.18] a solution $u = A\ \sin(k_1 x + \phi_1)\ \sin(k_2 y + \phi_2)\ \cos(\omega t + \alpha)$. Imposing the condition $u = 0$ on the periphery, we obtain $\phi_1 = \phi_2 = 0$ and

$$k_1 = m\pi/a, \qquad k_2 = n\pi/b \qquad \text{and} \qquad \omega_{\mathrm{m,\,n}} = \pi v\sqrt{m^2/a^2 + n^2/b^2}\,, \tag{9.22}$$

$$u_{\mathrm{m,\,n}} = A_{\mathrm{m,\,n}} \sin(m\pi x/a)\ \sin(n\pi y/b)\ \cos(\omega_{\mathrm{m,\,n}}\, t + \alpha_{\mathrm{m,\,n}}), \tag{9.23}$$

where m and n are integers that specify the mode. Each integer may be independently equal to 1, 2, 3, ... The energy density averaged over a period in the mode (m, n) is

$$\begin{aligned} < U_{\mathrm{s}} >_{\mathrm{T}} &= < \tfrac{1}{2}\, m_{\mathrm{s}}\, (\dot{u}_{\mathrm{m,n}})^2 + \tfrac{1}{2}\, F_l\, (\partial_{\mathrm{x}} u_{\mathrm{m,\,n}})^2 + \tfrac{1}{2}\, F_l\, (\partial_{\mathrm{y}} u_{\mathrm{m,\,n}})^2 >_{\mathrm{T}} \\ &= \tfrac{1}{4}\, A_{\mathrm{m,\,n}}{}^2 \pi^2 F_l\, [(m/a)^2 \sin^2(n\pi y/b) + (n/b)^2 \sin^2(m\pi x/a))]. \end{aligned}$$

Taking the average on the surface of the membrane, we obtain the average energy density

$$< U_{\mathrm{s}} > = (1/8)\, A_{\mathrm{m,\,n}}{}^2 \pi^2 F_l\, (m^2/a^2 + n^2/b^2)]. \tag{9.24}$$

Now consider a rectangular cavity, that is a rigid box occupying the volume $0 < x < a$, $0 < y < b$ and $0 < z < c$. It is convenient to describe the sound harmonic wave by the velocity potential $\Psi(\mathbf{r}, t) = F(\mathbf{r})\cos(\omega t + \alpha)$, which verifies d'Alembert wave equation with a speed of propagation $v_{\mathrm{s}} = \sqrt{B/m_{\mathrm{v}}}$. $F(\mathbf{r})$ must be a solution of the *Helmholtz equation*

$$v_{\mathrm{s}}{}^2\, (\partial^2{}_{\mathrm{xx}} F + \partial^2{}_{\mathrm{yy}} F + \partial^2{}_{\mathrm{zz}} F) + \omega^2 F = 0. \tag{9.25}$$

The corresponding velocity, displacement and acoustic pressure are:

$$\dot{\mathbf{u}} = -\nabla F \cos(\omega t + \alpha), \qquad \mathbf{u} = -(\nabla F/\omega)\sin(\omega t + \alpha), \qquad p_{\mathrm{a}}(\mathbf{r}, t) = -m_{\mathrm{v}}\omega F \sin(\omega t + \alpha).$$

Using the method of separation of variables with $F = X(x)\ Y(y)\ Z(z)$, we may write the wave in the cavity as

$$\begin{aligned} F &= A \cos(hx + \phi) \cos(ly + \phi') \cos(kz + \phi'') \\ \dot{\mathbf{u}} &= A\ [h \sin(hx + \phi) \cos(ly + \phi') \cos(kz + \phi'')\, \mathbf{e}_{\mathrm{x}} \\ &\quad + l \cos(hx + \phi) \sin(ly + \phi') \cos(kz + \phi'')\, \mathbf{e}_{\mathrm{y}} \\ &\quad + k \cos(hx + \phi) \cos(ly + \phi') \sin(kz + \phi'')\, \mathbf{e}_{\mathrm{z}}] \cos(\omega t + \alpha). \end{aligned} \tag{9.26}$$

Helmholtz equation is verified if $h^2 + l^2 + k^2 = \omega^2/v_s^2$. Imposing the condition that the normal component of the velocity vanishes on the rigid walls, we obtain

$$\phi = \phi' = \phi'' = 0, \quad h = m\frac{\pi}{a}, \quad l = n\frac{\pi}{b}, \quad k = p\frac{\pi}{c}, \quad \omega_{m,\,n,\,p} = \pi v_s\sqrt{\frac{m^2}{a^2}+\frac{n^2}{b^2}+\frac{p^2}{c^2}}$$

$$\Psi_{m,\,n,\,p} = A\cos(m\pi x/a)\cos(n\pi y/b)\cos(p\pi z/c)\cos(\omega_{m,\,n,\,p}\,t + \alpha_{m,\,n,\,p}). \qquad [9.27]$$

$\Psi_{m,\,n,\,p}$ is a three-dimensional standing wave. The integers m, n and p, which can be taken as positive or zero, specify the normal mode of the cavity. In particular, the modes $m = n = 0$ are uniform on any section of the cavity that is parallel to the plane Oxy. The modes in a pipe that is closed at both ends are particular modes of the cavity.

For instance, in the case of a cubic cavity of sides a, the normal frequencies are $\tilde{\nu}_{n,m,p} = (v/2a)\,(m^2 + n^2 + p^2)^{½}$. If $a = 30$ cm and $v_s = 332$ m/s, the first frequencies are 553, 783, 958, 1,107 Hz, etc., while for $a = 3$ m (a room, for instance), they are 55, 78, 96, 111 Hz, etc.

EXAMPLE 9.1.– Determine the angular frequency of the first four modes of a square membrane of sides a if the speed of propagation is v. What is their degeneracy? Determine the nodal lines of the modes u_{12} and u_{21}. Determine the nodal lines of the linear superposition $u_{12} + u_{21}$ if they have the same amplitude and the same phase.

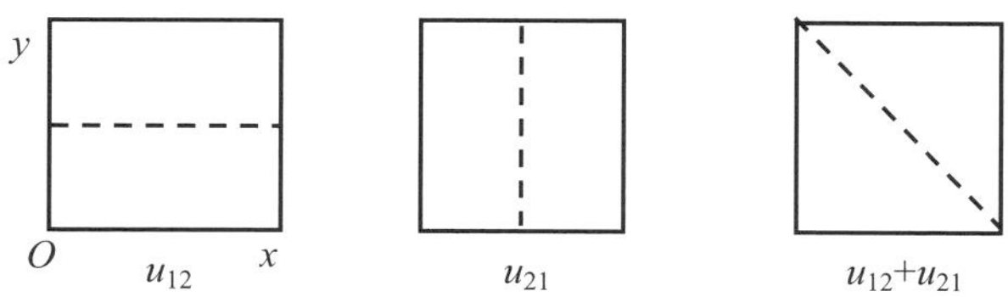

Figure 9.4. *Example 9.1*

SOLUTION– The normal frequencies are given by [9.22], i.e.

$$\omega_{11} = \sqrt{2}\,\pi v/a, \qquad \omega_{12} = \omega_{21} = \sqrt{5}\,\pi v/a, \qquad \omega_{22} = \sqrt{8}\,\pi v/a, \qquad \omega_{13} = \omega_{31} = \sqrt{10}\,\pi/a.$$

ω_{11} corresponds to one wave function u_{11}, and ω_{22} corresponds to one wave function u_{22}; thus, they are not degenerate. The angular frequency $\omega_{12} = \omega_{21}$ corresponds to two different wave functions u_{12} and u_{21}; thus, it has degeneracy 2. Any linear superposition of these modes is also a standing wave with the same angular frequency ω_{12}. Similarly, $\omega_{13} = \omega_{31}$ has degeneracy 2.

Taking the origin of coordinates at the vertex O, we may write:

$$u_{12} = A\sin(\pi x/a)\sin(2\pi y/a)\cos(\omega_{12}t+\alpha)$$
$$u_{21} = A\sin(2\pi x/a)\sin(\pi y/a)\cos(\omega_{12}t+\alpha)$$
$$u_{12}+u_{21} = 4A\sin(\pi x/a)\sin(\pi y/a)\cos[\pi(x+y)/2a]\cos[\pi(x-y)/2a]\cos(\omega_{12}t+\alpha).$$

There is one nodal line (other than the periphery) $y = a/2$ for u_{12}, the line $x = a/2$ for u_{21} and the line $x + y = a$ for $u_{12} + u_{21}$ (Figure 9.4).

9.3. Fourier analysis of standing waves*

The wave functions of the standing modes for any system have two important properties:

i) The u_n are *orthogonal* in the sense that the integral of the product $\underline{u}_{n'}\,\underline{u}_n^*$ on the whole system is zero if $n' \neq n$.

ii) The u_n form a *complete* set in the sense that any wave verifying the boundary conditions of the system is a linear superposition of the wave functions u_n.

For instance, in the case of a string fixed at both ends, the wave functions of [9.9] are real and verify the orthogonality relation

$$\int_0^L dx\ u_{n'}(x,t)\,u_n(x,t) = 0 \qquad \text{if} \quad n' \neq n. \qquad [9.28]$$

The amplitudes A_n may always be chosen so that the integrals be equal to 1 if $n' = n$. On the other hand, any superposition of the modes is a possible state of vibration of the string fixed at its ends. Conversely, any state of vibration $u(x, t)$ of the string is a superposition of the normal modes of the form

$$u(x,t) = \Sigma_n\, u_n(x,t) \equiv \Sigma_n \sin(n\pi\, x/L)\,[B_n\cos(n\pi\, vt/L) + C_n\sin(n\pi\, vt/L)]. \quad [9.29]$$

This equation constitutes the *Fourier analysis* of the wave function $u(x, t)$. If we know the function $u(x_o, t)$ at a point x_o (which is not a node), the general method of section 2.5 allows us to determine the coefficients B_n and C_n:

$$B_n\sin(n\pi x_o/L) = (v/L)\int_0^{2L/v} dt\ u(x_o,t)\cos(n\pi vt/L)$$
$$C_n\sin(n\pi x_o/L) = (v/L)\int_0^{2L/v} dt\ u(x_o,t)\sin(n\pi vt/Lt), \qquad [9.30]$$

where we used the relations

$$\int_0^{2\pi} d\varphi \cos(n'\varphi)\cos(n\varphi) = \pi \quad \text{and} \quad \int_0^{2\pi} d\varphi \cos(n'\varphi)\cos(n\varphi) = 0 \quad (\text{if } n' \neq n). \quad [9.31]$$

Sometimes, we know the initial conditions (displacement and velocity at an initial time t_o) for every point of the string. We obtain the relationships

$$B_n \cos(n\pi v t_o/L) + C_n \sin(n'\pi v t_o/L) = (2/L) \int_0^L dx\ u(x, t_o) \sin(n\,\pi x/L), \quad [9.32]$$

$$-B_n \sin(\frac{n\pi v t_o}{L}) + C_n \cos(\frac{n\pi v t_o}{L}) = \frac{2}{n\pi v} \int_0^L dx\ \dot{u}(x.t_o) \sin(\frac{n\pi x}{L}) \quad [9.33]$$

which determine the B_n and $C_{n.}$. In particular, if $t_o = 0$, we obtain

$$B_n = \frac{2}{L} \int_0^L dx\ u(x, 0) \sin(\frac{n\pi x}{L}), \qquad C_n = \frac{2}{n\pi v} \int_0^L dx\ \dot{u}(x.0) \sin(\frac{n\pi x}{L}). \quad [9.34]$$

Consider the standing wave on a string that is fixed at both ends

$$u = \Sigma_n A_n \sin(\omega_n x/v) \cos(\omega_n t + \alpha_n). \quad [9.35]$$

The energy density may be calculated using the relations of [9.11] obtaining

$$\begin{aligned} U_l &= U_{(C)l} + U_{(P)\,l} \\ &= \tfrac{1}{2}\, m_l \Sigma_{n,\,n'}\, \omega_n \omega_{n'} A_n A_{n'} [\sin(\omega_n t + \alpha_n) \sin(\omega_n x/v) \sin(\omega_{n'} t + \alpha_{n'}) \sin(\omega_{n'} x/v) \\ &\quad + \cos(\omega_n t + \alpha_n) \cos(\omega_n x/v) \cos(\omega_{n'} t + \alpha_{n'}) \cos(\omega_{n'} x/v)]. \end{aligned} \quad [9.36]$$

The energy of the whole string is obtained by integrating on x from 0 to L. The terms such that $n' \neq n$ give no contribution; the double sum on n' and n reduces to a simple sum on n:

$$U = \int_0^L dx\ U_l(x) = \tfrac{1}{4} L m_l \Sigma_n (\omega_n A_n)^2 = \Sigma_n U_n. \quad [9.37]$$

Thus, as in the case of localized systems with several degrees of freedom, *the total energy is the sum of the energies of the normal modes*. In the case of a continuous system, the number of degrees of freedom is infinite and the system is equivalent to an infinite set of independent oscillators, corresponding to the modes u_n.

In general, the amplitudes and the phases of the modes, which constitute the wave, depend on how the system is excited. For instance, if a string is fixed at its ends and struck in the middle, the wave is a superposition of odd-n modes (which have an antinode at this point). If it is struck at the point of coordinate $L/4$, the wave is a superposition of the modes $n = 4p + 2$. Conversely, if one "fingers" a vibrating string at the point $x = L/3$, for instance, all the modes that do not have a node at this point are eliminated.

EXAMPLE 9.2.– A string that is fixed at both ends is initially deformed by displacing its middle by a distance d before releasing it without an initial velocity (Figure 9.5). **a)** Write down the expression of the produced wave. **b)** Calculate the total energy of the excited modes and verify that it is equal to the work done in displacing the middle of the string by a distance d.

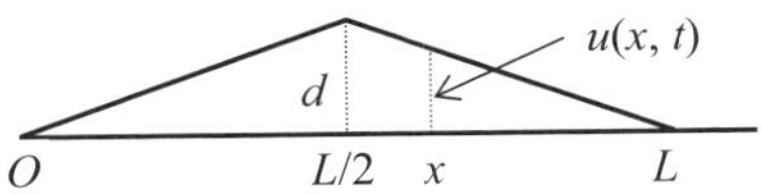

Figure 9.5. *Example 9.2*

SOLUTION –

a) The initial conditions (for $t = 0$) are in this case $\dot{u}(x,0) = 0$ and

$u(x, 0) = 2xd/L$ (for $0 < x < L/2$) and $u(x, 0) = 2d(1 - x/L)$ (for $L/2 < x < L$).

The expressions of [9.34] give the Fourier coefficients

$B_n = (-1)^p\, 8d/\pi^2(2p+1)^2$ if $n = 2p + 1$ and $B_n = 0$ if $n = 2p$ and $C_n = 0$.

Thus, the wave can be written as

$$u = \frac{8d}{\pi^2} \Sigma_{p\geq 0} \frac{(-1)^p}{(2p+1)^2} \cos[(2p+1)\pi v \frac{t}{L}] \sin[(2p+1)\pi \frac{x}{L}].$$

We note that the amplitude of the modes decreases like $1/(2p+1)^2$, this means that only the first few modes are effectively excited.

b) The total energy of the string is given by expression [9.37], i.e.

$$U = \frac{1}{4} L m_l \Sigma_n (\omega_n A_n)^2 = \frac{16d^2}{\pi^4} \Sigma_p \frac{1}{(2p+1)^4} m_l \omega_{2p+1}{}^2 = \frac{16d^2 F}{L\pi^2} \Sigma_p \frac{1}{(2p+1)^2}.$$

The sum $\Sigma\,(2p + 1)^{-2}$ is equal to $\pi^2/8$; thus, the total energy is $2d^2F/L$. The initial total energy is the potential energy. This is the work done to lengthen the string from L to $2\sqrt{L^2/4 + d^2}$, i.e. $F\,[\sqrt{L^2/4 + d^2} - L] \approx 2Fd^2/L$. Thus, the total energy of the standing wave on the string is equal to the work done to excite the string.

9.4. Resonance and standing waves

A medium that occupies a finite region of space is equivalent to a system of an infinite number of coupled oscillators whose normal frequencies are $\tilde{\nu}_n$ and normal variables are u_n. If it is subject to excitation by a harmonic force, it resonates at excitation frequencies $\tilde{\nu}$ equal to one of the normal frequencies $\tilde{\nu}_n$. Then, the standing wave of the system has a large amplitude for the mode u_n (thus, for u). The power that the generator of the force supplies is then at a maximum. The amplitude of the excited mode may be very large, but there are always some dissipative forces reducing it. If these forces are not too large, a steady state of forced oscillation is reached rapidly so that the supplied power is completely dissipated.

For instance, let us consider an electrically sustained tuning fork acting at the origin O of a string of length L fixed at the other end (Figure 9.6a). The amplitude of vibration of the tuning fork is always small, compared to that of the string; thus, O is nearly a node of vibration. If the frequency of the fork is equal to one of the normal frequencies of the string, $\tilde{\nu}_n = (n/2L)\sqrt{F/m_l}$, the string resonates and it vibrates with a large amplitude. In fact, there is always some energy loss because of the frictions. This gives each mode a certain resonance width. If the excitation frequency falls within the bandwidth of a normal frequency, the normal mode is excited. In principle, all the normal modes may be excited. However, the excitation of a mode of high frequency $\tilde{\nu}_n$ is less likely because it requires a high energy. The variation of the absorbed power as a function of the excitation frequency is illustrated in Figure 9.6b. It is formed by a series of resonance peaks with a certain width. At low excitation frequencies, the resonances are well separated. At high frequencies, the resonances are closer and their width is comparable to their frequency spacing. The response of the system to the excitation becomes a slowly varying function of the frequency.

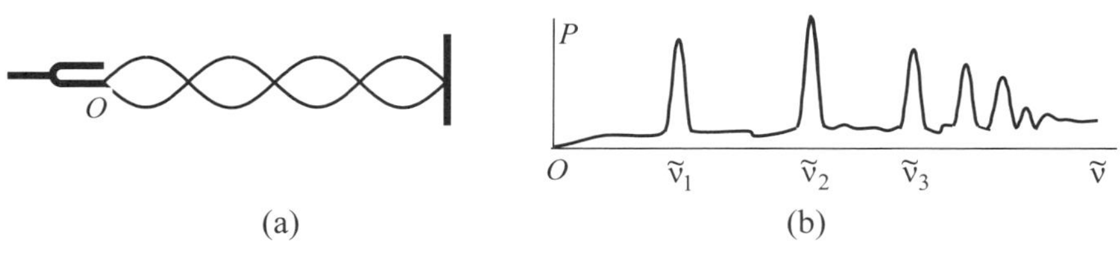

Figure 9.6. *a) A String excited by a tuning fork, and b) the absorbed power*

A quantum system, such as atoms or molecules, has an infinite number of stationary modes, exactly like a cavity. If it is exposed to an electromagnetic field, it may absorb energy. The variation of the absorbed power versus the wave frequency is similar to that of Figure 9.6b with sharp peaks at the normal frequencies $\tilde{\nu}_n$. If, for instance, a gas is exposed to a beam of light of bandwidth $\Delta\tilde{\nu}$, the frequencies of this beam, which are equal to the normal frequencies of the gas molecules, are absorbed. Analyzing the spectrum of the emerging beam using a spectrometer, the absorbed frequencies appear as dark lines (absorption spectrum).

9.5. Sound wave guided by two parallel plates

In our study of the propagation of sound in a pipe (section 5.4) and standing waves in a pipe (section 9.1), we have assumed that the wave is uniform on the pipe cross-sections; the wave vector **k** is then parallel to the pipe axis. In fact, this is one of many modes of propagation in the pipe. In this section we analyze the propagation of sound waves between two plane and rigid plates (that is, which absorb no sound energy). The component of the vector velocity $\dot{\mathbf{u}}$ of the gas, which is normal to the plates, must be zero on the plates.

We chose the axes of coordinates such that one of the plates is positioned in the plane *Oyz* and the other is positioned in the plane $x = a$. While propagating toward the positive *z*, the wave may undergo successive reflections on the plates (Figure 9.7a). We assume that the wave vector lies in the *Oxz* plane. This situation may be effectively realized in a rectangular pipe if the wave vector is parallel to the pair of lateral surfaces that are parallel to *Ozx* (Figure 9.7b). Then, this pair has no effect in guiding the wave, which undergoes successive reflections on the other pair of surfaces, exactly as if they were infinite planes.

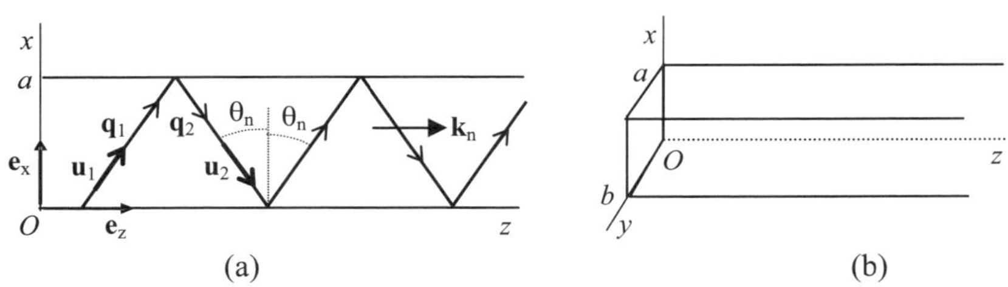

Figure 9.7. *a) Guided sound wave by reflection on two plane and parallel plates. b) Rectangular sound waveguide*

In the guide, the successive incident and reflected waves may be written as

$$\mathbf{u}_1 = \mathbf{e}_1 A_1 \sin(\omega t - \mathbf{q}_1.\mathbf{r} + \phi_1), \qquad \mathbf{u}_2 = \mathbf{e}_2 A_2 \sin(\omega t - \mathbf{q}_2.\mathbf{r} + \phi_2), \qquad [9.38]$$

where $\mathbf{e}_1 = -\cos\theta\,\mathbf{e}_x + \sin\theta\,\mathbf{e}_z$ and $\mathbf{e}_2 = \cos\theta\,\mathbf{e}_x + \sin\theta\,\mathbf{e}_z$ are the unit vectors in the direction of the vectors $\mathbf{q}_1$ and $\mathbf{q}_2$ respectively. The waves [9.38] verify the wave equation in the medium between the plates if $\mathbf{q}_1$ and $\mathbf{q}_2$ have a magnitude $q = \omega/v_s$, where v_s is the speed of sound (which appears in the wave equation for the infinite medium). The resulting displacement wave and the velocity of the gas at a point (x, z) are

$$\begin{aligned}\mathbf{u} &= [-A_1\sin(\omega t - \mathbf{q}_1.\mathbf{r} + \phi_1) + A_2\sin(\omega t - \mathbf{q}_2.\mathbf{r} + \phi_2)]\cos\theta\,\mathbf{e}_x \\ &\quad + [A_1\sin(\omega t - \mathbf{q}_1.\mathbf{r} + \phi_1) + A_2\sin(\omega t - \mathbf{q}_2.\mathbf{r} + \phi_2)]\sin\theta\,\mathbf{e}_z. \\ \dot{\mathbf{u}} &= \omega\,[-A_1\cos(\omega t - \mathbf{q}_1.\mathbf{r} + \phi_1) + A_2\cos(\omega t - \mathbf{q}_2.\mathbf{r} + \phi_2)]\cos\theta\,\mathbf{e}_x \\ &\quad + \omega\,[A_1\cos(\omega t - \mathbf{q}_1.\mathbf{r} + \phi_1) + A_2\cos(\omega t - \mathbf{q}_2.\mathbf{r} + \phi_2)]\sin\theta\,\mathbf{e}_z.\end{aligned} \qquad [9.39]$$

On the first plate ($x = 0$), we must have

$$\dot{u}_x = \omega[-A_1\cos(\omega t - qz\sin\theta + \phi_1) + A_2\cos(\omega t - qz\sin\theta + \phi_2)]\cos\theta = 0.$$

This condition may be verified for any t and z in two cases:

1) $\cos\theta = 0$, that is, $\theta = \pi/2$. In this mode, the wave vector is parallel to Oz. The condition $\dot{u}_x = 0$ is also satisfied on the second plate ($x = a$). This is the mode of propagation that we considered in Chapter 5. Then, we have $q = \omega/v_s$ without any restriction on ω. The wave propagates without dispersion with a phase velocity and a group velocity equal to v_s exactly as if the medium were infinite.

2) $\phi_1 = \phi_2 \equiv \phi$ and $A_1 = A_2 \equiv A$. Then, the resulting wave may be written as

$$\begin{aligned}\mathbf{u} &= -2A\cos\theta\sin(qx\cos\theta)\cos(\omega t + \phi - qz\sin\theta)\,\mathbf{e}_x \\ &\quad + 2A\sin\theta\cos(qx\cos\theta)\sin(\omega t + \phi - qz\sin\theta)\,\mathbf{e}_z.\end{aligned}$$

Imposing the condition $\dot{u}_x = 0$ on the second plate ($x = a$), we get the condition $\cos\theta\sin(qa\cos\theta) = 0$. It is satisfied if

$$qa\cos\theta_n = n\pi, \qquad \text{i.e.} \quad \omega\cos\theta_n = \omega_n \equiv n\pi v_s/a \qquad (n \text{ integer}). \qquad [9.40]$$

This corresponds to discrete modes of propagation specified by n. As $\cos\theta_n < 1$, the wave of angular frequency ω may propagate in the mode n only if ω is higher than ω_n. Thus ω_n is a *cut-off frequency* for this mode. The wave, which propagates in the mode n, undergoes reflections on the plates at the incidence angle θ_n such that $\cos\theta_n = \omega_n/\omega$. Thus, it may be written as

$$\begin{aligned}\mathbf{u}_n &= 2A\,[-(\omega_n/\omega)\sin(n\pi x/a)\cos(\omega t - k_n z + \phi)\,\mathbf{e}_x \\ &\quad + \gamma_n\cos(n\pi\,x/a)\sin(\omega t - k_n z + \phi)\,\mathbf{e}_z],\end{aligned} \qquad [9.41]$$

where we have set

$$\gamma_n = \sin\theta_n = \sqrt{1 - \omega_n^2/\omega^2} \qquad \text{and} \quad k_n = q \sin\theta_n = \omega\gamma_n/v_s. \qquad [9.42]$$

We designate by $\lambda = 2\pi/q = 2\pi v_s/\omega$ the wavelength in the infinite medium. The *guided wavelength*, i.e. the distance in the direction z between two points in the same state of oscillation is $\lambda_n = 2\pi/k_n$, the phase *velocity* is $v_{(p),\,n} = \omega/k_n$ and the *group velocity* is $v_{(g),\,n} = \partial\omega/\partial k_n$. These quantities are given by the expressions

$$\lambda_n = \lambda/\gamma_n > \lambda, \qquad v_{(p),\,n} = v_s/\gamma_n > v_s, \qquad v_{(g),\,n} = v_s\gamma_n < v_s. \qquad [9.43]$$

For instance, if the plates are separated by a distance of 10 cm and the speed of sound in open air is 332 m/s, the cut-off frequency of the first mode is $\tilde{\nu}_1 = v_s/2a =$ 1,66 kHz. For $\tilde{\nu} = 3$ kHz, the wavelength in open air is $\lambda = v_s/\tilde{\nu} = 11.07$ cm and $\gamma_1 = [1 - (\tilde{\nu}_1/\tilde{\nu})^2]^{½} = 0.833$. Thus, the guided wavelength is $\lambda_1 = \lambda/\gamma_1 =$ 13,29 cm, its phase velocity and group velocity are respectively $v_{(p),\,1} = v_s/\gamma_1 = 399$ m/s and $v_{(g),\,n} = \gamma_1 v_s = 277$ m/s.

9.6. Guided sound waves in a rectangular pipe

In this section we analyze the propagation of sound waves in a pipe by using the wave equation and the boundary conditions on the walls. It is convenient to introduce the *velocity potential* $\Psi(\mathbf{r}, t)$ such that $\dot{\mathbf{u}} = -\nabla\Psi(\mathbf{r}, t)$. It obeys the wave equation $\ddot{\Psi}(\mathbf{r}.t) - v_s^2\,\Delta\Psi(\mathbf{r}, t) = 0$ (see section 5.6). The acoustic pressure is related to the displacement by the equation $p_a = -B\nabla.\mathbf{u}$. In addition, the sound wave must verify the condition that the normal component of the velocity $\dot{\mathbf{u}}$ vanishes on the walls of the pipe. A solution, which represents a wave of angular frequency ω, is of the form

$$\underline{\Psi}(\mathbf{r}, t) = \underline{F}(\mathbf{r})\, e^{i\omega t}, \quad \underline{\dot{\mathbf{u}}} = -\nabla\underline{F}\, e^{i\omega t}, \quad \underline{\mathbf{u}} = (i/\omega)\nabla\underline{F}\, e^{i\omega t}, \quad \underline{p}_a(\mathbf{r}, t) = i m_v \omega \underline{F}\, e^{i\omega t}.$$

Substituting the expression $\underline{\Psi}$ in the wave equation, we find that the function $\underline{F}$ obeys *Helmholtz equation*

$$\Delta\underline{F}(\mathbf{r}) + (\omega^2/v_s^2)\,\underline{F}(\mathbf{r}) = 0. \qquad [9.44]$$

We use the Cartesian coordinates x, y and z with Oz parallel to the direction of the pipe, and the axes Ox and Oy perpendicular to the lateral walls. Let a and b be the transverse dimensions of the pipe (Figure 9.7b). Using the method of separation of variables, we look at solutions of the Helmholtz equation of the form

$\underline{F}(x, y, z) = \underline{X}(x)\ \underline{Y}(y)\ \underline{Z}(z)$. Substituting this expression in [9.44] and dividing by $\underline{X}\,\underline{Y}\,\underline{Z}$, we obtain

$$\underline{X}''/\underline{X} + \underline{Y}''/\underline{Y} + \underline{Z}''/\underline{Z} + \omega^2/v_s^{\,2} = 0. \qquad [9.45]$$

This equation is only verified identically if each of its terms are constant. Thus, we have

$$\underline{X}'' = -h^2\underline{X}, \qquad \underline{Y}'' = -l^2\underline{Y}, \qquad \underline{Z}'' = -\eta\underline{Z} \quad \text{with} \quad h^2 + l^2 + \eta = \omega^2/v_s^{\,2}, \quad [9.46]$$

where we have taken negative signs for $\underline{X}''/\underline{X}$ and $\underline{Y}''/\underline{Y}$, otherwise the wave would be exponential as a function of x and y and it would not satisfy the boundary conditions on the walls.

If $\eta = \omega^2/v_s^{\,2} - h^2 - l^2$ is positive, we designate it by k^2. The solutions of [9.46] are then of the form $\underline{X} = Ae^{ihx}$, $\underline{Y} = Be^{ily}$ and $\underline{Z} = Ce^{ikz}$ or their complex conjugate. We may linearly combine these functions and their complex conjugate to form a velocity potential of the form

$$\underline{\Psi} = \Psi_o \cos(hx + \phi)\cos(ly + \phi')\ e^{i(\omega t - kz)}.$$

The velocity of the gas is then

$$\underline{\dot{\mathbf{u}}} = -\nabla\underline{\Psi}(\mathbf{r}, t) = \Psi_o\big[h\sin(hx + \phi)\cos(ly + \phi')\,\mathbf{e}_x + l\cos(hx + \phi)\sin(ly + \phi')\,\mathbf{e}_y + ik\cos(hx + \phi)\cos(ly + \phi')\,\mathbf{e}_z\big]\ e^{i(\omega t - kz)}.$$

The normal component of the velocity vanishes on the walls if $\phi = \phi' = 0$, $h = m\pi/a$ and $l = n\pi/b$, where m and n are integers. The *mode* is specified by m and n and its wave function of this mode can then be written as

$$\underline{\Psi}_{m,n} = \Psi_o \cos(m\pi\, x/a)\cos(n\pi\, y/b)\, e^{i(\omega t - k_{m,n} z)}, \qquad [9.47]$$

where the amplitude Ψ_o is arbitrary. Changing the sign of m and n comes down to change the sign of Ψ_o; this will not change the mode. Thus, we may take m and n as positive or zero. Then, the last relation of [9.46] gives

$$k_{m,n} = (\omega/v_s)\,\gamma_{m,n} \qquad \text{where} \qquad \gamma_{m,n} = \sqrt{1 - (\omega_{m,n}/\omega)^2}$$
$$\text{with } \omega_{m,n} = \pi v_s\,\sqrt{m^2/a^2 + n^2/b^2}\,. \qquad [9.48]$$

Expression [9.47] represents a wave propagating in the direction Oz with a wave vector $k_{m,n}$ and a modulated amplitude, which depends on the transverse coordinates x and y.

If $\eta = \omega^2/v_s^2 - h^2 - l^2$ is negative, we write $\eta = -\delta_{m,n}^2$. In this case, the wave is attenuated in the direction Oz with an attenuation coefficient

$$\delta_{m,n} = (\omega/v_s)\sqrt{(\omega_{m,n}/\omega)^2 - 1}\,. \qquad [9.49]$$

Thus, $\omega_{m,n}$ is a cut-off angular frequency for the mode (m, n). The pipe is *reactive* if $\omega < \omega_{m,n}$ and *dispersive* if $\omega > \omega_{m,n}$. In this last case, the phase velocity of the wave in the direction Oz and the group velocity are respectively

$$v_{(p)m,n} = \omega/k = v_s/\gamma_{m,n} > v_s \qquad \text{and} \qquad v_{(g)m,n} = \partial\omega/\partial k = v_s\gamma_{m,n} < v_s. \qquad [9.50]$$

The phase velocity is always larger than the speed v_s in the infinite medium, while the group velocity is less than v_s. We also have the relation $v_{(p)m,n}v_{(g)m,n} = v_s^2$.

In the mode $m = n = 0$, the functions $\underline{\Psi}$, $\dot{\underline{\mathbf{u}}}$, $\underline{\mathbf{u}}$ and the acoustic pressure $\underline{p}_a$ are independent on the transverse coordinates x and y and the cut-off frequency is equal to zero. The wave propagates in this mode with the speed v_s, as if the medium were infinite. This is the wave that we considered in sections 5.4 and section 9.1.

9.7. Transmission lines

An electromagnetic wave may be guided by a conducting structure, such as two parallel conductors or coaxial cables. We consider first a chain of LC circuits formed by inductors of self-inductance L in series and coupled by capacitors of capacitance C at $z_1 = d$, $z_2 = 2d$, ..., $z_p = pd$, ... (Figure 9.8a). We say that this is a *lumped transmission line*. We assume that a generator acts at the entry and produces a distribution of potential $\underline{V}_p$ across the capacitors and current intensities $\underline{I}_p$ in the inductors. The charge of the pth capacitor is $\underline{Q}_p = C\underline{V}_p$. Using the complex representation for the potentials and the currents and Kirchhoff laws for nodes and loops in electrical networks, we get the equations

$$\dot{\underline{Q}}_p = C\dot{\underline{V}}_p = \underline{I}_p - \underline{I}_{p+1}, \quad L\dot{\underline{I}}_p + (\underline{V}_p - \underline{V}_{p-1}) = 0. \qquad [9.51]$$

Differentiating the first equation with respect to t and using the second equation to eliminate the intensities, we obtain the equation

$$\ddot{\underline{V}}_p + \omega_o^2(2\underline{V}_p - \underline{V}_{p-1} - \underline{V}_{p+1}) = 0 \qquad \text{where} \quad \omega_o = 1/\sqrt{LC}\,. \qquad [9.52]$$

Assuming that the excitation is simple harmonic at angular frequency ω, an analysis similar to that of section 5.1 shows that the solution may be written as

$$\underline{V}_p = V_m\, e^{i(\omega t - pkd)}, \qquad \underline{I}_p = \sqrt{C/L}\; V_m\, e^{i(\omega t - pkd + kd/2)}. \qquad [9.53]$$

k is related to ω by the dispersion relation $\omega = 2\omega_o\,|\sin(kd/2)|$, which determines k if ω is lower than the cut-off frequency $2\omega_o$. The chain of coupled *LC* circuits is then dispersive and the expressions of [9.53] represent a wave propagating towards the positive z. On the other hand, if $\omega > 2\omega_o$, k is imaginary of the form $-i\delta$; the wave is then attenuated and the transmission line is *reactive*.

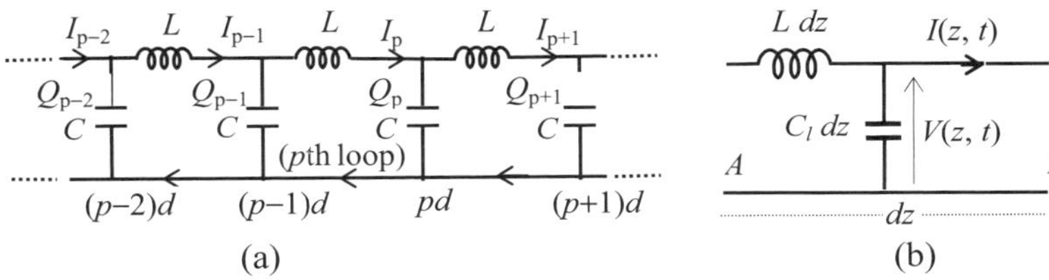

Figure 9.8. *a) Chain of LC circuits, and b) element of a transmission line*

A *distributed transmission line*, formed by two long and parallel conductors, may be considered to be the limit of a lumped line if $d \to 0$, $C \to C_l d$ and $L \to L_l d$ where L_l is the self-inductance per unit length and C_l is the capacitance per unit length. We may also analyze the transmission line directly. An element *AB* of length dz is formed by a self-inductance $L_l\,dz$ in the direction of the line and a transverse capacitance $C_l\,dz$ (Figure 9.8b). We neglect the resistance of the conductors and the escape current between them. The potential $\underline{V}$ between the conductors and the current intensity $\underline{I}$ are then functions of z and t. The potential drop in going from A to B is $-d\underline{V} = \dot{\underline{I}}\, L_l\, dz$, and the charge of the equivalent capacitor is $\underline{V} C_l\, dz$. Thus, the current intensity is $C_l \dot{\underline{V}}\, dz$ across the capacitor. The decrease of the current intensity in going from A to B is $-\,d\underline{I} = C_l \dot{\underline{V}}\, dz$. We deduce the equations

$$\dot{\underline{I}}\, L_l + (\partial_z \underline{V}) = 0, \qquad \dot{\underline{V}}\, C_l + (\partial_z \underline{I}) = 0. \qquad [9.54]$$

We differentiate the first equation with respect to t and the second with respect to z and eliminate $\underline{V}$ between them. Similarly, we derive the first with respect to z and the second with respect to t and eliminate $\underline{I}$ with them. We obtain the wave equations

$$\ddot{\underline{I}} - v^2 (\partial^2_{zz} \underline{I}) = 0, \qquad \ddot{\underline{V}} - v^2 (\partial^2_{zz} \underline{V}) = 0, \qquad [9.55]$$

where we set $v = 1/\sqrt{C_l L_l}$. These equations may be obtained as the limits of equations [9.51] if $d \to 0$. We may show that $C_l L_l = \mu\varepsilon$; the speed of propagation is thus $v = 1/\sqrt{\mu\varepsilon}$. This result is valid for any geometric form of the transmission line and even in the case of an infinite medium. In the limit $d \to 0$, the cut-off frequency ω_o becomes infinite and the dispersion relation becomes $\omega = vk$. At any frequency, the wave propagates on the line with the same phase and group velocities equal to v. Identifying pd to z, solution [9.53] becomes

$$\underline{V}(z, t) = \underline{V}_m\, e^{i(\omega t - kz)}, \qquad \underline{I}(z, t) = (i/\omega L_l)\,(\partial_z \underline{V}) = (\underline{V}_m/Z_l)\, e^{i(\omega t - kz)}, \qquad [9.56]$$

where $Z_l = \sqrt{L_l/C_l}$ is the impedance of the line. This solution represents a wave propagating towards the positive z. We may also consider solutions with the phase $(\omega t + kz)$ representing a wave propagating towards the negative z and a superposition of both types of solutions. In particular, if we consider the real solutions, the electric energy density and the magnetic energy density per unit length are respectively $U_{(E)\,l} = ½\, C_l\, V^2$ and $U_{(M)\,l} = ½\, L_l\, I^2$. The power that is carried at point z at time t is $P = VI$. The line constitutes an electromagnetic waveguide.

9.8. Electromagnetic waveguides*

In general, an electromagnetic waveguide is any structure designed to guide the wave. It may be a two-conductor transmission line (coaxial cable, two-wire or mictrostrip transmission lines), a hollow metallic pipe that is empty or filled with a dielectric or an optical fiber. We can imagine the wave in the waveguide to be travelling in a zigzag path, undergoing repeated reflections on the opposite walls of the guide.

Within an ideal conductor, there are no fields; so the electromagnetic wave is totally reflected on its surface and there is no energy loss as Joule heat. In a good conductor (such as copper), this property is approximately true. In the following, we assume that the conductors are perfect and that the guide has translation symmetry in the direction Oz (Figure 9.9a). In the insulator, which fills the waveguide, the fields obey Maxwell's equations [6.15] to [6.18], and thus the wave equation [6.19]. On the surface of the conductor the fields obey the boundary conditions:

$$\mathbf{B}_{\perp} = 0 \qquad \text{and} \qquad \mathbf{E}_{//} = 0, \qquad [9.57]$$

where $\mathbf{B}_{\perp}$ is the component of $\mathbf{B}$ that is perpendicular to this surface and $\mathbf{E}_{//}$ is the component of $\mathbf{E}$ that is parallel to this surface. It should be noted that $\mathbf{E}_{\perp}$ and $\mathbf{B}_{//}$ may not be zero because the surface of the conductor may carry variable charge and current densities.

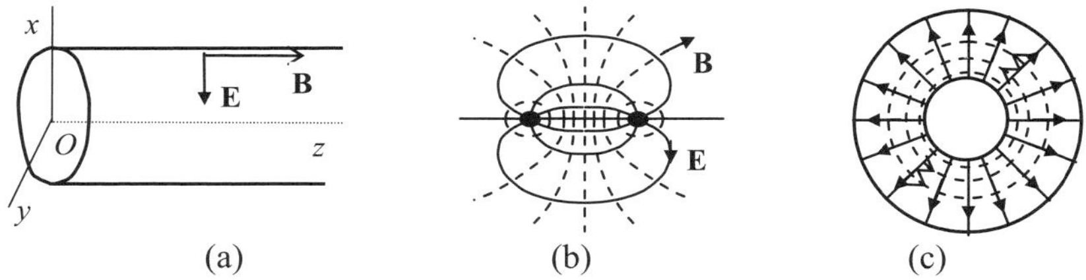

Figure 9.9. *a) Electromagnetic waveguide in the form of a hollow conductor, b) waveguide formed by two parallel conducting wires, and c) coaxial cable. The lines of* **E** *are drawn as solid lines and those of* **B** *are drawn as dashed lines*

These conditions infer that the wave can propagate in the guide only in certain modes with a lower cut-off frequency. In each mode, the wave has specific orientations for the fields **E** and **B** with respect to the walls, and it propagates with a phase velocity and a group velocity, which depend on the frequency and the geometry of the guide. In general, the conditions of [9.57] are compatible with Maxwell's equations for only three types of waves:

a) *Transverse magnetic* (TM) waves such that

$$B_z = 0 \quad \text{(everywhere)} \quad \text{and} \quad \mathbf{E}_{//} = 0 \quad \text{(on the conductors).} \qquad [9.58]$$

b) *Transverse electric* (TE) waves such that

$$E_z = 0 \quad \text{(everywhere)} \quad \text{and} \quad \mathbf{B}_{\perp} = 0 \quad \text{(on the conductors).} \qquad [9.59]$$

c) *Transverse electromagnetic* (TEM) waves such that

$$E_z = 0 \quad \text{(everywhere)} \quad \text{and} \quad B_z = 0 \quad \text{(everywhere).} \qquad [9.60]$$

The *TEM* waves propagate only in free space or along transmission lines formed by two conductors, such as two parallel wires (Figure 9.9b) or a coaxial cable (Figure 9.9c). The fields **E** and **B** lie in the normal sections and are perpendicular to each other. Near the conductor, **E** is perpendicular to the conductor and **B** is tangential.

In the case of a wave at angular frequency ω, which propagates in the direction *Oz* of a waveguide, which has translation symmetry in this direction, the fields have the form

$$\underline{\mathbf{E}} = \hat{\underline{\mathbf{E}}}(x, y)\, e^{i(\omega t - kz)}, \qquad \underline{\mathbf{B}} = \hat{\underline{\mathbf{B}}}(x, y)\, e^{i(\omega t - kz)}, \qquad [9.61]$$

where $\hat{\underline{\mathbf{E}}}(x,y)$ and $\hat{\underline{\mathbf{B}}}(x,y)$ are two vector functions which depend on the geometry of the waveguide. Substituting these expressions to $\underline{\mathbf{E}}$ and $\underline{\mathbf{B}}$ in the wave equations [6.19], we find that $\hat{\underline{\mathbf{E}}}$ and $\hat{\underline{\mathbf{B}}}$ verify Helmholtz equations

$$\partial^2_{xx}\hat{\underline{\mathbf{E}}} + \partial^2_{xx}\hat{\underline{\mathbf{E}}} + [(\omega^2/v^2) - k^2]\hat{\underline{\mathbf{E}}} = 0, \qquad \partial^2_{xx}\hat{\underline{\mathbf{B}}} + \partial^2_{xx}\hat{\underline{\mathbf{B}}} + [(\omega^2/v^2) - k^2]\hat{\underline{\mathbf{B}}} = 0. \quad [9.62]$$

We note that the functions $\hat{\underline{\mathbf{E}}}(x,y)$ and $\hat{\underline{\mathbf{B}}}(x,y)$ are not independent because $\underline{\mathbf{E}}$ and $\underline{\mathbf{B}}$ are related by Maxwell's equations.

9.9. Waveguides formed by two planes and parallel plates*

This is the easiest waveguide to study mathematically. We chose the axes of coordinates so that one of the plates lies in the plane Oyz and the other in the plane $x = a$ (Figure 9.10). The fields do not depend on y because of the translation symmetry in this direction. Thus, the fields may be written as

$$\underline{\mathbf{E}} = \hat{\underline{\mathbf{E}}}(x)\, e^{i(\omega t - kz)}, \qquad \underline{\mathbf{B}} = \hat{\underline{\mathbf{B}}}(x)\, e^{i(\omega t - kz)}, \qquad [9.63]$$

where $\hat{\underline{\mathbf{E}}}(x)$ and $\hat{\underline{\mathbf{B}}}(x)$ are two vector functions of x only. In this case, the Helmholtz equations of [9.62] become the simple differential equations

$$\frac{d^2\hat{\underline{\mathbf{E}}}}{dx^2} + q^2\hat{\underline{\mathbf{E}}}(x) = 0, \qquad \frac{d^2\hat{\underline{\mathbf{B}}}}{dx^2} + q^2\hat{\underline{\mathbf{B}}}(x) = 0 \qquad \text{with} \quad q^2 = \frac{\omega^2}{v^2} - k^2. \quad [9.64]$$

The equation for $\hat{\underline{\mathbf{B}}}(x)$ has the general solution

$$\hat{\underline{\mathbf{B}}}(x) = (\underline{B}_1\,\mathbf{e}_x + \underline{B}_2\,\mathbf{e}_y + \underline{B}_3\,\mathbf{e}_z)\, e^{iqx} + (\underline{B}'_1\,\mathbf{e}_x + \underline{B}'_2\,\mathbf{e}_y + \underline{B}'_3\,\mathbf{e}_z)\, e^{-iqx}. \qquad [9.65]$$

The corresponding magnetic field is

$$\underline{\mathbf{B}} = [(\underline{B}_1\,\mathbf{e}_x + \underline{B}_2\,\mathbf{e}_y + \underline{B}_3\,\mathbf{e}_z)\, e^{iqx} + (\underline{B}'_1\,\mathbf{e}_x + \underline{B}'_2\,\mathbf{e}_y + \underline{B}'_3\,\mathbf{e}_z)\, e^{-iqx}]e^{i(\omega t - kz)}. \quad [9.66]$$

Maxwell's equation $\nabla.\underline{\mathbf{B}} = 0$ is verified if

$$q\underline{B}_1 = k\underline{B}_3, \qquad q\underline{B}'_1 = -k\underline{B}'_3. \qquad [9.67]$$

The electric field is related to the magnetic field by the third Maxwell equation $\nabla \times \underline{\mathbf{B}} = \dot{\underline{\mathbf{E}}}/v^2$, which gives in this case

$$\underline{\mathbf{E}} = (v^2/\omega)\ \{[k\underline{B}_2\mathbf{e}_x - (k\underline{B}_1 + q\underline{B}_3)\mathbf{e}_y + q\underline{B}_2\mathbf{e}_z]\, e^{iqx} + [k\underline{B}'_2\mathbf{e}_x - (k\underline{B}'_1 - q\underline{B}'_3)\mathbf{e}_y - q\underline{B}'_2\,\mathbf{e}_z]\, e^{-iqx}\}\ e^{i(\omega t - kz)}. \quad [9.68]$$

The other Maxwell equations $\nabla.\underline{\mathbf{E}} = 0$ and $\nabla \times \underline{\mathbf{E}} = -\dot{\underline{\mathbf{B}}}$ are identically verified. Using [9.67] and $q^2 + k^2 = \omega^2/v^2$, we may write

$$\underline{\mathbf{B}} = \{[\underline{B}_1\mathbf{e}_x + \underline{B}_2\mathbf{e}_y + (q/k)\ \underline{B}_1\mathbf{e}_z]\, e^{iqx} + [\underline{B}'_1\mathbf{e}_x + \underline{B}'_2\mathbf{e}_y - (q/k)\,\underline{B}'_1\mathbf{e}_z]\, e^{-iqx}\}\ e^{i(\omega t - kz)}$$

$$\underline{\mathbf{E}} = (v^2/\omega)\{[k\underline{B}_2\,\mathbf{e}_x - (\omega^2/kv^2)\ \underline{B}_1\,\mathbf{e}_y + q\underline{B}_2\,\mathbf{e}_z]\, e^{iqx} + [k\underline{B}'_2\,\mathbf{e}_x - (\omega^2/kv^2)\ \underline{B}'_1\,\mathbf{e}_y - q\underline{B}'_2\mathbf{e}_z]\, e^{-iqx}\}\ e^{i(\omega t - kz)}. \quad [9.69]$$

The boundary conditions of [9.57] are verified on the plate $x = 0$ if

$$\underline{B}'_1 = -\,\underline{B}_1, \qquad \underline{B}'_2 = \underline{B}_2. \quad [9.70]$$

Thus, we may write the fields in the form

$$\underline{\mathbf{B}} = 2\big[i\underline{B}_1 \sin(qx)\ \mathbf{e}_x + \underline{B}_2 \cos(qx)\ \mathbf{e}_y + (q/k)\underline{B}_1\ \cos(qx)\ \mathbf{e}_z\big]\, e^{i(\omega t - kz)}$$

$$\underline{\mathbf{E}} = 2(v^2/\omega)\big[k\underline{B}_2\ \cos(qx)\ \mathbf{e}_x - i(\omega^2/kv^2)\underline{B}_1\ \sin(qx)\mathbf{e}_y + iq\underline{B}_2\ \sin(qx)\mathbf{e}_z\big]\, e^{i(\omega t - kz)}.$$

The boundary conditions of [9.57] on the plate $x = a$ are verified if

$$\underline{B}_1 \sin qa = 0 \qquad \text{and} \qquad q\underline{B}_2 \sin qa = 0. \quad [9.71]$$

These equations may be verified in the following three cases:

a) $q = 0$, then $k = \omega/v$ and the fields have the form

$$\underline{\mathbf{E}} = v\underline{B}_o\,\mathbf{e}_x\ e^{i(\omega t - kz)}, \qquad \underline{\mathbf{B}} = \underline{B}_o\,\mathbf{e}_y\ e^{i(\omega t - kz)}, \quad [9.72]$$

where we have redefined the amplitude $\underline{B}_o = 2\underline{B}_2$. This is a *TEM* wave with k related to ω by the dispersion relation

$$\omega = vk. \quad [9.73]$$

The phase velocity and the group velocity are

$$v_{(p)} = \omega/k = v \qquad \text{and} \qquad v_{(g)} = \partial\omega/\partial k = v. \quad [9.74]$$

Thus, this waveguide transmits *TEM* waves at any frequency with a phase velocity and a group velocity equal to the speed of propagation v, exactly as if the medium were infinite. Figure 9.10a shows the lines of the fields **E** and **B** for this *TEM* wave.

b) $\sin(qa) = 0$ (with $q \neq 0$) and $\underline{B}_1 = 0$. Then q may have one of the values

$$q_m = m\pi/a, \qquad m = 1, 2, 3, \ldots \qquad [9.75]$$

The integer number m specifies the mode. The corresponding value of k is given by the dispersion relation

$$k_m^{\,2} = \omega^2/v^2 - q_m^{\,2}. \qquad [9.76]$$

We may write also

$$k_m = \gamma_m\omega/v, \quad q_m = \omega_m/v, \quad \text{where } \gamma_m = \sqrt{1-\omega_m^{\,2}/\omega^2} \quad \text{and } \omega_m = m\pi/a \qquad [9.77]$$

If $\omega > \omega_m$, k_m is real; the wave is progressive and the guide is dispersive. On the other hand, if $\omega < \omega_m$, k_m is imaginary, the guide is reactive and the wave is attenuated with an attenuation coefficient in the direction Oz

$$\delta_m = (1/v)\sqrt{\omega_m^{\,2} - \omega^2}\,. \qquad [9.78]$$

Thus, ω_m is a lower *cut-off frequency* for the mode m. The fields in this mode are

$$\underline{\mathbf{E}} = v\underline{B}_o[\gamma_m \cos(m\pi\, x/a)\,\mathbf{e}_x + i(\omega_m/\omega)\sin(m\pi\, x/a)\,\mathbf{e}_z]\,e^{i(\omega t - k_m z)}$$
$$\underline{\mathbf{B}} = \underline{B}_o \cos(m\pi\, x/a)\; e^{i(\omega t - k_m z)}\,\mathbf{e}_y, \qquad [9.79]$$

where we have redefined the amplitude $\underline{B}_o = 2\underline{B}_2$. The corresponding real fields are

$$\mathbf{E} = vB_o[\gamma_m \cos(m\pi\frac{x}{a})\cos(\omega t - k_m z + \phi)\,\mathbf{e}_x - \frac{\omega_m}{\omega}\sin(m\pi\frac{x}{a})\sin(\omega t - k_m z + \phi)\mathbf{e}_z]$$
$$\mathbf{B} = B_o\,\mathbf{e}_y \cos(m\pi\,\frac{x}{a})\cos(\omega t - k_m z + \phi). \qquad [9.80]$$

As $B_z = 0$ everywhere, this is a *TM* wave. Figures 9.10b and 9.10c illustrate the fields of a *TM* wave ($m = 1$) between the plates.

c) $\sin(qa) = 0$ (with $q \neq 0$) and $B_2 = 0$. q may have one of the values of [9.75] with the same dispersion relation as [9.76] and a cut-off frequency [9.77]. The fields in the mode m are

$$\underline{\mathbf{B}} = \underline{B}_o[-\sin(m\pi x/a)\,\mathbf{e}_x + i(\omega_m/\omega\gamma_m)\cos(m\pi\, x/a)\,\mathbf{e}_z]\; e^{i(\omega t - k_m z + \phi)}$$
$$\underline{\mathbf{E}} = \underline{B}_o\,(v/\gamma_m)\sin(m\pi\, x/a)\; e^{i(\omega t - k_m z + \phi)}\;\mathbf{e}_y \qquad [9.81]$$

or the real fields

$$\mathbf{B} = B_o[-\sin(m\pi\frac{x}{a})\cos(\omega t - k_m z + \phi)\,\mathbf{e}_x - \frac{\omega_m}{\omega\gamma_m}\cos(m\pi\frac{x}{a})\sin(\omega t - k_m z + \phi)\mathbf{e}_z]$$

$$\mathbf{E} = B_o\frac{v}{\gamma_m}\sin(m\pi\, x/a)\cos(\omega t - k_m z + \phi)\,\mathbf{e}_y. \qquad [9.82]$$

As $E_z = 0$ everywhere, this is a *TE* wave. Figures 9.10d and 9.10e illustrate the fields **E** and **B** of a *TE* (m = 1) wave.

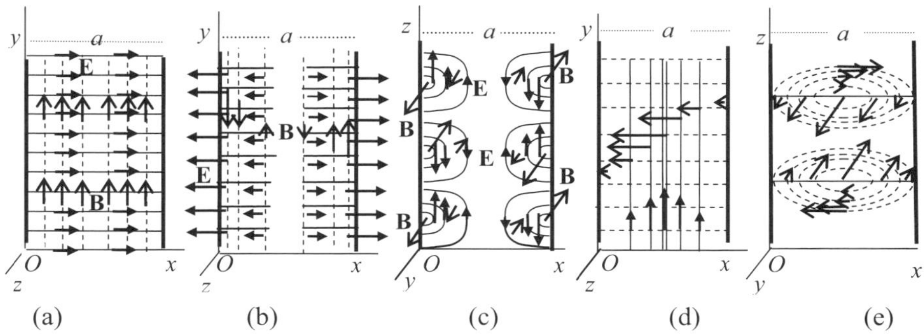

Figure 9.10. *A Waveguide formed by two plane parallel plates: a) fields* **E** *and* **B** *of a TEM wave; b) fields* **E** *and* **B** *of a TM wave (m = 1) in a transverse section, and c) its fields in a longitudinal section; d) fields* **E** *and* **B** *of a TE wave (m = 1) in a transverse section, and e) its fields in a longitudinal section. The lines of* **E** *are solid lines and those of* **B** *are dashed lines*

On the contrary to *TEM* waves, a *TM* wave or a *TE* wave can only propagate in the guide in a mode m if its angular frequency ω is higher than the angular cut-off frequency of this mode $\omega_m = m\pi v/a$. Using the dispersion relation [9.77], we obtain the phase velocity and the group velocity of the mode m

$$v_{(p)m} = \omega/k_m = v/\gamma_m > v \qquad \text{and} \qquad v_{(g)m} = \partial\omega/\partial k_m = v\,\gamma_m < v. \qquad [9.83]$$

We always have $v_{(p)m} > v$, $v_{(g)m} < v$ and $v_{(p)m}v_{(g)m} = v^2$.

9.10. Guided electromagnetic waves in a hollow conductor*

At hyper frequencies ($\tilde{\nu}$ of the order of the GHz), two-conductor waveguides are not practical. A single conductor waveguide (i.e. a hollow conductor) may be used. This type of waveguide cannot support *TEM* waves at any frequency. It may support a *TE* wave or a *TM* wave in a given mode if the frequency of the wave is higher than the cut-off frequency of the mode, which is determined by the geometry of the waveguide. The *TE* or *TM* mode, which has the lowest cut-off frequency, is called the *dominant mode*. Thus, the frequency of the dominant mode is the minimum frequency of a wave, which can propagate in the waveguide without attenuation. In this section we study, in detail, rectangular waveguides and we provide some results for circular waveguides.

The simplest waveguide to analyze has a rectangular cross-section with sides a and b (Figure 9.7b). Such a guide is practical for the transmission of linearly polarized waves. If we write the fields in the form [9.61], each component of the vector functions $\underline{\hat{\mathbf{E}}}(x,y)$ and $\hat{\mathbf{B}}(x,y)$ obeys Helmholtz equation [9.62]. Let us write, for instance

$$\underline{\hat{E}}_{x}(x,y) = \underline{X}(x)\, \underline{Y}(y). \qquad [9.84]$$

Substituting this expression into Helmholtz equation and dividing by $\underline{X}(x)\, \underline{Y}(y)$, we obtain

$$\underline{X}'' / \underline{X} + \underline{Y}'' / \underline{Y} + \omega^2/v^2 - k^2 = 0. \qquad [9.85]$$

Each term of this equation must be constant, that is,

$$\underline{X}'' = -p^2 \underline{X}, \quad \underline{Y}'' = -q^2 \underline{Y} \quad \text{with} \quad p^2 + q^2 + k^2 = \omega^2/v^2, \qquad [9.86]$$

where we chose the (–) sign in order not to have exponential solutions, which cannot respect the boundary conditions. Thus, we have

$$X = A \sin(px + \phi), \qquad Y = B \sin(qy + \psi) \qquad [9.87]$$

and similarly for the other components $\underline{\hat{E}}_{y}(x,y)$ and $\underline{\hat{E}}_{z}(x,y)$. The condition $\nabla.\mathbf{E} = 0$ may be verified only if the components of $\mathbf{E}$ are simple harmonic functions of x and y with the same parameters p and q. Thus, omitting the global factor $e^{i(\omega t - kz)}$, we may write

$$E_x = A_1 \sin(px + \phi_1) \sin(qy + \psi_1), \qquad E_y = A_2 \sin(px + \phi_2) \sin(qy + \psi_2)$$
$$E_z = A_3 \sin(px + \phi_3) \sin(qy + \psi_3). \qquad [9.88]$$

where p and the ϕ_i are not all equal to zero, this is also the case for q and the ψ_i (because this is equivalent to a field equal to zero). We may take the phases lying between 0 and π (π excluded) and the amplitudes to be positive, negative or zero. The boundary condition $\mathbf{E}_{//} = 0$ on the surfaces $x = 0$ and $y = 0$ gives the conditions

$$A_1 \sin \psi_1 = A_2 \sin \phi_2 = A_3 \sin \phi_3 = A_3 \sin \psi_3 = 0. \qquad [9.89]$$

Redefining the amplitudes, we may write the electric field in the form

$$\mathbf{E} = A_1 \sin(px+\phi_1) \sin(qy)\, \mathbf{e}_x + A_2 \sin(px) \sin(qy+\psi_2)\, \mathbf{e}_y + A_3 \sin(px) \sin(qy)\, \mathbf{e}_z.$$

The condition $\mathbf{E}_{//} = 0$ on the surfaces $z = a$ and $y = b$ is verified if

$$A_1 \sin qb = A_2 \sin pa = A_3 \sin pa = A_3 \sin qb = 0. \qquad [9.90]$$

The magnetic field is given by the equation $\nabla \times \mathbf{E} = -\dot{\mathbf{B}}$, that is,

$$\underline{B}_x = (1/\omega)\sin(px)\, [\mathrm{i}qA_3 \cos(qy) - kA_2 \sin(qy + \psi_2)]$$
$$\underline{B}_y = (1/\omega) \sin(qy)\, [\, kA_1 \sin(px + \phi_1) - \mathrm{i}pA_3 \cos(px)]$$
$$\underline{B}_z = (1/\omega)\, [\, \mathrm{i}pA_2 \cos(px) \sin(qy + \psi_2) - \mathrm{i}qA_1 \sin(px + \phi_1) \cos(qy)]. \qquad [9.91]$$

The condition $\mathbf{B}_\perp = 0$ on the surfaces $x = 0$, $x = a$, $y = 0$ and $y = b$ is verified if the conditions [9.89] and [9.90] are verified. Maxwell's equation $\nabla.\mathbf{E} = 0$ is verified if

$$pA_1 \cos \phi_1 = qA_2 \cos \psi_2 = pA_1 \sin \phi_1 + qA_2 \sin \psi_2 + \mathrm{i}kA_3 = 0. \qquad [9.92]$$

while the other Maxwell equations are verified. All the conditions [9.89], [9.90] and [9.92] can only be verified for the following types of waves:

a) *Transverses electric* (TE) waves:

$$\underline{\mathbf{B}} = A[pk \sin(px) \cos(qy)\, \mathbf{e}_x + kq \sin(qy) \cos(px)\, \mathbf{e}_y - \mathrm{i}(p^2+q^2) \cos(px) \cos(qy)\, \mathbf{e}_z].$$
$$\underline{\mathbf{E}} = \omega A\, [q \cos(px) \sin(qy)\, \mathbf{e}_x - p \sin(px) \cos(qy)\, \mathbf{e}_y] \qquad [9.93]$$

b) *Transverse magnetic* (TM) waves:

$$\underline{\mathbf{E}} = B[pk \cos(px) \sin(qy)\, \mathbf{e}_x + qk \sin(px) \cos(qy)\, \mathbf{e}_y + \mathrm{i}(p^2+q^2) \sin(px) \sin(qy)\, \mathbf{e}_z]$$
$$\underline{\mathbf{B}} = B(\omega/v^2)\, [-q \sin(px) \cos(qy)\, \mathbf{e}_x + p \sin(qy) \cos(px)\, \mathbf{e}_y]. \qquad [9.94]$$

In expressions [9.93] and [9.94], p and q take only the values:

$$p = m\pi/a, \qquad q = n\pi/b \qquad \text{where} \quad m = 0, 1, 2, \ldots \text{ and } n = 0, 1, 2, \ldots \quad [9.95]$$

The integers m and n specifying the mode, cannot both be equal to zero. Then, the last relation [9.86] gives

$$k = \gamma_{m,n}\,\omega/v, \qquad \gamma_{m,n} = \sqrt{1-\omega_{m,n}{}^2/\omega^2}\ , \tag{9.96}$$

where $\omega_{m,n}$ is the angular cut-off frequency given by:

$$\omega_{m,n} = \pi v\sqrt{m^2/a^2 + n^2/b^2}\ . \tag{9.97}$$

If the angular frequency ω of the wave is higher than $\omega_{m,n}$, $k_{m,n}$ is real; the wave is then progressive without attenuation with the dispersion relation [9.96]. The phase velocity and the group velocity are

$$v_{(P)m,n} = \omega/k = v/\gamma_{m,n} > v \qquad \text{and} \qquad v_{(g)m,n} = \partial\omega/\partial k = \gamma_{m,n} v < v. \tag{9.98}$$

If ω is lower than $\omega_{m,n}$, $k_{m,n}$ is imaginary of the form $-i\delta_{m,n}$; the wave is then attenuated in the Oz direction with an attenuation coefficient

$$\delta_{m,n} = (1/v)\sqrt{\omega_{m,n}{}^2 - \omega^2}\ . \tag{9.99}$$

Figure 9.11a illustrates the dispersion relation for the mode (m, n): ω increases from the cut-off angular frequency $\omega_{m,n}$ and asymptotically approaches the straight-line $\omega = vk$, which is the dispersion relation in an infinite medium. Figure 9.11b illustrates the attenuation coefficient $\delta_{m,n}$ and the wave number $k_{m,n}$ versus ω: $\delta_{m,n}$ and $k_{m,n}$ are equal to zero at the cut-off frequency $\omega_{m,n}$. Figure 9.11c illustrates the phase velocity and the group velocity versus ω: we always have $v_{(P)m,n} > v$, $v_{(g)m,n} < v$ and $v_{(P)m,n}\, v_{(g)m,n} = v^2$. At high frequency, $v_{(P)m,n}$ and $v_{(g)m,n}$ asymptotically approach the speed of propagation v in the infinite medium.

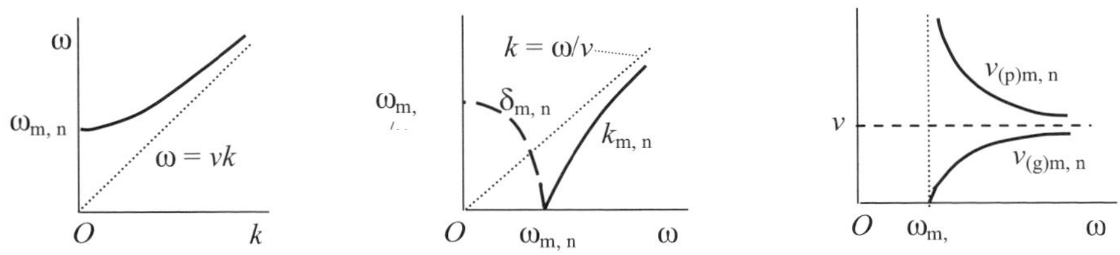

Figure 9.11. *a) Dispersion relation of a waveguide, b) attenuation coefficient and wave number versus ω, and c) phase velocity and group velocity versus ω for a mode (m, n)*

In the case of a circular waveguide of radius R, the fields can be expressed in terms of Bessel functions; the cut-off frequencies are not the same for the TE modes and the TM modes; they are given by

$$\omega_{m,j}(TM) = x_{m,j}\, v/R \qquad [9.100]$$

$$\omega_{m,j}(TE) = x'_{m,j}\, v/R, \qquad [9.101]$$

where $x_{m,j}$ is the jth zero of the Bessel function $J_m(x)$. The first zeros are

$$\begin{array}{llll} \text{For } m = 0, & x_{0,1} = 2.405, & x_{0,2} = 5.520, & x_{0,3} = 8.654.... \\ \text{For } m = 1, & x_{1,2} = 3.832, & x_{1,2} = 7.076, & x_{1,3} = 10.173... \\ \text{For } m = 2, & x_{2,1} = 5.136, & x_{2,2} = 8.417, & x_{2,3} = 12.620... \end{array} \qquad [9.102]$$

$x'_{m,j}$ is the jth zero of the function $J'_m(x) \equiv dJ_m/dx$. The first zeros are

$$\begin{array}{llll} \text{For } m = 0, & x'_{0,1} = 3.832, & x'_{0,2} = 7.016, & x'_{0,3} = 10.174... \\ \text{For } m = 1, & x'_{1,1} = 1.841, & x'_{1,2} = 5.331, & x'_{1,3} = 8.536... \\ \text{For } m = 2, & x'_{2,1} = 3.054, & x'_{2,2} = 6.706, & x'_{2,3} = 9.970... \end{array} \qquad [9.103]$$

The lowest cut-off frequency corresponds to the TE mode $m = 1$ and $j = 1$

$$\omega_{1,1}(TE) = x'_{1,1}\, v/R = 1.84\, v/R. \qquad [9.104]$$

9.11. Applications of waveguides

An electromagnetic wave of any frequency $\tilde{\nu}$ may propagate in the TEM mode only in waveguides formed by two conductors. It may propagate in a TE or TM mode if $\tilde{\nu}$ is higher than the cut-off frequency $\tilde{\nu}_i$ of this mode. Thus, a wave with a frequency lower than the frequency $\tilde{\nu}_1$ of the dominant mode cannot propagate in any TE or TM mode. If $\tilde{\nu}_1 < \tilde{\nu} < \tilde{\nu}_2$, the wave may propagate in the dominant mode and, if $\tilde{\nu}_2 < \tilde{\nu} < \tilde{\nu}_3$, it may propagate in the dominant mode and the second mode and so on. As the velocity depends on the frequency and the mode, there is a dispersion phenomenon. If a signal is a superposition of several frequencies, each spectral component propagates with its proper speed and the signal is inevitably deformed, even if a single mode is used for the transmission. The deformation is more important if more than one mode is used. Thus, to transmit a signal in a waveguide constituted by a hollow conductor with minimum deformation only the dominant mode must be used. This is achieved by using frequencies that lie between the cut-off frequency of the dominant mode and that of the second mode. Other frequency bands may be used with appropriate filters to enable a single mode.

As the cut-off frequencies depend on the geometry of the waveguide, it must be chosen so that the wave propagates in the dominant mode. The transverse dimensions of the waveguide are then comparable to the wavelength. For instance, the mode m of a waveguide formed by two plane and parallel plates has a cut-off frequency $\tilde{\nu}_m = mv/2a$. A wave of frequency $\tilde{\nu}$ may only propagate in the dominant mode if $v/2a < \tilde{\nu} < v/a$, i.e. $a < \lambda < 2a$, where $\lambda = v/\tilde{\nu}$ is the wavelength in the infinite medium. Similarly, to transmit light in an optical fiber, its radius must be of the order of the micrometer. As the transverse dimensions of waveguides are of the order of λ, an important part of the wave is diffracted at its extremity and another part is reflected back. If the waveguide ends in an antenna, the transmitted wave may be directed and the reflection at its end reduced.

One of the most important uses of waveguides is to transmit signals and information. In order to only allow the dominant mode, the fields at the entry of the guide must be established in this mode configuration by using appropriate mode filters. In order to transmit waves of frequency between several tens of kHz and a few hundred MHz, we use coaxial cables, constituted by a wire surrounded by an insulating medium and a braided copper sheathing; this also allows us to protect the signal from the interference of other unwanted signals. In such a cable, the *TEM* wave can be used if its frequency is lower than the cut-off frequency of the *TM* and *TE* modes. Coaxial cables are used, for instance, to connect the parabolic antenna to a television set and the amplifier of a radar system to its antenna. In telephony, a bundle of some 20 cables of this type allows us to transmit more than 130,000 simultaneous telephone calls. At hyper-frequencies (several GHz), we may use the *TE* or *TM* modes if the *TEM* can be eliminated. This can be achieved with the use of a rectangular or cylindrical single-conductor waveguide. For instance, a circular waveguide with a 5 cm radius allows us to use a bandwidth of several GHz near 40 GHz. This enables the transmission of some 50,000 telephone calls or 50 television channels.

A transparent fiber is an optical waveguide. This is a very thin cylindrical fiber made of a very transparent material with a high refraction index. It is surrounded by a layer with a lower index, which is specially chosen to increase the reflection coefficient for the used light and to protect the fiber. For short distance transmissions, we may use plastic fibers (polystyrene, for instance), which are economical, light and flexible but have a high attenuation. We also use fibers constituted by a core of pure silica surrounded by a silicone layer. They are stiffer and more costly than plastic fibres, but they have a lower attenuation. We may also use fluoride glass which is more difficult to manufacture but that has an extremely low attenuation.

A large number of fibers may be assembled to form an optical fiber cable, which may be bent with a radius as small as 1 cm, since light propagates separately in each fiber. This type of cable may be used to transmit images, thus allowing us to observe normally inaccessible places, such as internal organs of the body (endoscopy). It may be used to transmit an intense laser beam to be used in medical surgery and in industry. Because of the high frequency of light waves, fibers and lasers may be associated with telecommunications for the transmission of an enormous quantity of information as optical signals for distances of up to thousands of kilometers.

9.12. Summary

A *standing wave* may be established in a bounded medium of propagation. Its amplitude vanishes at the *nodes* and it is at a maximum at the *antinodes*. The region between two adjacent nodes form a *wave zone*. All the points of the same wave zone oscillate in phase and points of adjacent wave zones oscillate in opposite phase.

The one-dimensional wave equation $\ddot{u} - v^2(\partial^2_{xx}u) = 0$ has solutions of the form $u_{st} = f(x)\, g(t)$. The wave equation is then equivalent to two independent equations for g and f. The harmonic solution, which satisfies the boundary conditions, is a standing wave $u = A \cos(\omega x/v + \phi) \cos(\omega t + \alpha)$. The two energy densities are not equal. The energy flux vanishes at the nodes and antinodes; thus, there is no transfer of physical quantities from one half-zone to the other. The intensity of the wave is zero at all points.

The boundary conditions can be satisfied only in *discrete modes* of the form $u_n = f_n(\mathbf{r})\, g_n(t)$, where $n \equiv (n_1, n_2, \text{etc.})$ and the n_i are integer numbers. The functions u_n are *orthogonal* and *complete*: any function $u(\mathbf{r}, t)$, that verifies the same boundary conditions, is a linear superposition of the modes: $u(\mathbf{r}, t) = \sum_n a_n u_n(\mathbf{r}, t)$.

If a string is fixed at both ends, the condition $u(0, t) = u(L, t) = 0$ is verified if the string is formed by an integer number n of wave zones of length $\lambda/2$; thus, $L = n\lambda/2$ and $u_n = A_n \sin(\omega_n x/v) \cos(\omega_n t + \alpha_n)$, where $n = 1, 2\ldots$, $\omega_n = n\pi v/L$ and $v = \sqrt{F/m_l}$. The total energy of the string in the mode (n) is $U_n = \frac{1}{4}\pi^2(F/L)\, n^2 A_n^2$. A string that is fixed at one end and free at the other is formed by $(n - \frac{1}{2})$ wave zones of length $\lambda/2$; thus, $\tilde{\nu}_n = (v/4L)(2n - 1)$ with $n = 1, 2, 3\ldots$

A closed end of a pipe is a node of displacement (antinode of pressure) and an open end is a node of pressure (antinode of displacement). If a pipe is closed (or open) at both ends, its length must be $L = n\lambda_n/2$; hence, $\tilde{\nu}_n = nv/2L$ with $n = 1, 2, 3\ldots$. If a pipe is open at one end and closed at the other, its length must be $L = (n - \frac{1}{2})\, \lambda_n/2$; hence, $\tilde{\nu}_n = (n - \frac{1}{2})v/2L$ with $n = 1, 2, 3\ldots$

The modes of standing waves on a rectangular membrane of sides a and b are

$$u_{mn} = A_{mn} \sin(m\pi x/a) \sin(n\pi y/b) \cos(\omega_{mn} t + \alpha_{mn}) \quad \text{with } \omega_{mn} = \pi v\sqrt{m^2/a^2 + n^2/b^2},$$

where m and n are integers and $v = \sqrt{F_l/m_s}$.

The modes of standing waves in a rectangular cavity of sides a, b and c are

$$\Psi_{m,n,p} = A \cos(m\pi x/a) \cos(n\pi y/b) \cos(p\pi z/c) \cos(\omega_{m,n,p}\, t + \alpha_{m,n,p}),$$

where $\omega_{m,n,p} = \pi v\,(m^2/a^2 + n^2/b^2 + p^2/c^2)^{1/2}$. m, n and p are non-negative integers.

A bounded medium is equivalent to an infinite number of oscillators with angular frequencies ω_n. The total energy is the sum of the excitation energies of the modes.

The standing wave on a string is a superposition of normal modes

$$u = \Sigma_n\, [B_n \cos(n\pi vt/L) + C_n \sin(n\pi vt/L)] \sin(n\pi x/L).$$

The coefficients B_n and C_n may be evaluated if u is known as a function of time at a point x_o or if we know the initial conditions (displacement and velocity at $t = 0$) at each point of the string. The lowest normal mode is the *fundamental mode*, while the others are the *harmonics*.

If a bounded medium is excited with a frequency equal to one of its normal frequencies, it resonates; the amplitude of the corresponding mode is large and the received power is large. High n modes have a large energy; thus, they are more difficult to excite than the lower n modes.

The waves propagate in a waveguide in *modes*; each mode is characterized by a cut-off frequency ω_n. If $\omega > \omega_n$, the wave may propagate in the mode (n) with a wave vector k_n in the direction of the waveguide, a guided wavelength $\lambda_n = 2\pi/k_n$, a phase velocity $v_{(p)n} = \omega/k_n$ and a group velocity $v_{(g)n} = d\omega/dk_n$ given by

$$k_n = \omega\gamma_n/v, \quad \lambda_n = \lambda/\gamma_n, \quad v_{(p)n} = v/\gamma_n > v, \quad v_{(g)n} = v\gamma_n < v, \quad \text{where } \gamma_n = [1 - \omega_n^2/\omega^2]^{1/2}.$$

If $\omega < \omega_n$, the wave has an attenuation coefficient $\delta_n = (1/v)[\omega_n^2 - \omega^2]^{1/2}$.

A sound wave may propagate between two parallel plates separated by a distance a, in a direction parallel to the plates as in the infinite medium. It may also propagate by undergoing successive reflections on the plates at an angle of incidence θ_n. The condition $\dot{u}_\perp = 0$ on the plates is verified if $\omega \cos\theta_n = \omega_n$, where $\omega_n \equiv n\pi v/a$.

A guided sound wave in a pipe whose section is rectangular of sides a and b is of the form $\underline{\Psi}(\mathbf{r}, t) = \underline{F}(\mathbf{r})e^{\mathrm{i}\omega t}$. $\underline{F}$ verifies the *Helmholtz equation* $(\Delta + \omega^2/v^2)\underline{F}(\mathbf{r}) = 0$. Then, $\dot{\underline{\mathbf{u}}} = -\nabla\underline{\Psi} = -\nabla\underline{F}e^{\mathrm{i}\omega t}$, $\underline{\mathbf{u}} = (\mathrm{i}/\omega)\nabla\underline{F}e^{\mathrm{i}\omega t}$ and $\underline{p}_{\mathrm{a}} = \mathrm{i}m_{\mathrm{v}}\omega\underline{F}e^{\mathrm{i}\omega t}$. The condition $\dot{\underline{u}}_{\perp} = 0$ on the walls requires that

$$\underline{\Psi}_{\mathrm{m,\,n}} = \Psi_{\mathrm{o}} \cos(m\pi x/a)\cos(n\pi y/b)\; e^{\mathrm{i}(\omega t - kz)} \quad \text{and} \quad \omega_{\mathrm{m,\,n}} = \pi v\,(m^2/a^2 + n^2/b^2)^{1/2}.$$

An electromagnetic wave may propagate on a line of self-inductance L_l and capacitance C_l per unit length, obeying the equations

$$\dot{\underline{I}}\, L_l + (\partial\underline{V}/\partial z) = 0 \quad \text{and} \quad \dot{\underline{V}}\, C_l + (\partial\underline{I}/\partial z) = 0.$$

The speed of propagation is $v = 1/\sqrt{L_l C_l} = 1/\sqrt{\mu\varepsilon}$. The potential and the intensity are

$$V(z, t) = V_{\mathrm{m}} \cos(\omega t - kz) \quad \text{and} \quad I(z, t) = (C_l/L_l)\, V_{\mathrm{m}} \cos(\omega t - kz).$$

The fields of an electromagnetic wave in a waveguide verifies Maxwell's equations. On the walls made of a perfect conductor, the tangential component of **E** and the normal component of **B** must vanish. We may have three types of waves:

- *Transverse electric* (*TE*) waves: $E_z = 0$ everywhere and $\mathbf{B}_{\perp} = 0$ on the walls.
- *Transverse magnetic* (*TM*) waves: $B_z = 0$ everywhere and $\mathbf{E}_{//} = 0$ on the walls.
- *Transverse electromagnetic* (*TEM*) waves: $E_z = 0$ and $B_z = 0$ everywhere.

TEM waves at any frequency may be guided only between two conductors constituting a *transmission line*. Then, **E** and **B** lie in the normal section and are perpendicular to each other. Near a conductor, the field **E** is perpendicular to the conductor and **B** is tangential. *TEM* waves propagate with velocities $v_{(\mathrm{p})} = v_{(\mathrm{g})} = v$.

A waveguide formed by two parallel plates separated by a distance a, may support *TE* waves and *TM* waves in modes (m), whose cut-off frequency is $\omega_{\mathrm{m}} = m\pi v/a$.

A hollow conductor waveguide, whose section is rectangular of sides a and b, support *TE* and *TM* waves with cut-off frequencies $\omega_{\mathrm{m,\,n}} = \pi v(m^2/a^2 + n^2/b^2)^{1/2}$. The lowest cut-off frequency in a circular waveguide corresponds to *TE* waves and it is given by $\omega_{1,\,1}(TE) = x'_{1,\,1}\, v/R = 1.84\; v/R$.

Waveguides have many applications in telecommunications. To avoid important deformation of signals, a single transmission mode (the dominant mode) must be used. Light may be guided by a transparent fiber. Bundles of fibers are used in telecommunications, industry and medicine, etc.

9.13. Problem solving suggestions

1) A standing wave may always be written as a product of a function of space coordinates and time: $u(\mathbf{r}, t) = f(\mathbf{r})\, g(t)$. This is the only form that verifies the boundary conditions. It vanishes at any time at the *nodes* that are the zeros of $f(\mathbf{r})$. A one-dimensional standing wave may be written in the form $u(x, t) = f(x)\, g(t)$ and a three-dimensional wave may be written in the form $u(x, t) = X(x)\, Y(y)\, Z(z)\, g(t)$.

2) In a waveguide, three velocities must be considered: the speed of propagation v (which appears in the wave equation), the phase velocity $v_{(P)} = \omega/k_z$ where k_z is the component of the wave vector in the direction of the waveguide and the group velocity $v_{(g)} = \partial\omega/\partial k_z$. In an infinite non-dispersive medium, $v_{(P)}$ and $v_{(g)}$ do not depend on the frequency and they are equal to v. In an infinite dispersive medium, it is not possible to write a wave equation that is independent on ω; thus, v is not defined. The phase velocity $v_{(P)}$ and the group velocity $v_{(g)}$ depend on the frequency. A waveguide is always dispersive, even if it is empty or filled with a non-dispersive medium. Both the phase and the group velocities $v_{(P)}$ and $v_{(g)}$ depend on the frequency and the geometry of the waveguide. If the waveguide is filled with a medium that is itself dispersive, the dispersion is due to both the nature of the medium and the geometry of the waveguide. We did not consider this general case in this chapter.

3) In the limit of short wavelengths, compared to the transverse dimensions, the cut-off frequency approaches zero and the factor $\gamma_n = (1 - \omega_n^2/\omega^2)^{1/2}$ approaches 1. The phase velocity $v_{(P)n} = v/\gamma_n$ and the group velocity $v_{(g)n} = v\gamma_n$ approach v for all the modes. In this case, the wave propagates exactly as in an infinite medium.

4) In an infinite medium, the two energy densities of progressive waves are equal at each point of the medium and at any time. The energy flux density is $\mathbf{S} = U_v v\mathbf{e}$, where U_v is the total energy density. This is not the case for standing waves. In the case of a progressive guided wave, the energy densities are only equal in the average in space and over time and the energy flux density may only be written in the form $<\mathbf{S}> = < U_v > v_{(g)}\mathbf{e}_z$ for the average values. Note that this equation states that the energy propagates with the group velocity $v_{(g)}$. In the case of an attenuated wave, there is a time-independent energy density but there is no energy flux.

5) In a waveguide formed by a hollow conductor, it is possible to have only *TM* waves (i.e. such that $B_z = 0$ everywhere and $\mathbf{E}_{//} = 0$ on the conductors) and *TE* waves (i.e. such that $E_z = 0$ everywhere and $\mathbf{B}_\perp = 0$ on the conductors). In the case of a two-

conductor waveguide, it is possible to also have *TEM* waves (i.e. such that $E_z = 0$ and $B_z = 0$ everywhere). In an infinite isotropic medium it is possible to only have *TEM* waves.

9.14. Conceptual questions

Q1. What are the essential differences between a progressive wave and a standing wave, with regard to the mathematical form, wavelength, frequency and oscillation at given points and the energy? The superposition of two progressive waves of equal amplitudes that propagate in opposite directions is a standing wave. Is it possible to consider a progressive wave as the superposition of two standing waves?

Q2. Here is an example of standing waves in a medium without boundaries in the direction of propagation. Consider an elastic ring of radius *R*. Let *v* be the speed of propagation of elastic transverse waves on this ring. What are the frequencies $\tilde{\nu}_n$ of the normal modes? Draw the form of this ring in the mode *n* at a given time.

Q3. Explain what is meant by the modes of a waveguide and a cavity, the cut-off frequency of a mode in a waveguide, and the normal frequency of a cavity. Can a wave propagate in a waveguide exactly like in an infinite medium?

Q4. What are the causes of dispersion in an infinite medium and in a waveguide? For instance, consider light propagating in an infinite vacuum, in an empty waveguide, in "infinite" glass, and in a glass fiber.

Q5. Determine the cut-off frequencies of sound in a pipe whose section is a square of sides *a*. A wave of frequency $\tilde{\nu}$ is incident on the entry. Discuss the modes of propagation of this wave depending on the value of $\tilde{\nu}$. Assume that the incident wave has a frequency lower than the cut-off frequency of the second mode. Which frequencies propagate in the pipe? What happens to the other frequencies?

Q6. Assume that you speak at one end of a long pipe and another person listens at the other end. Which wave does he receive first, the wave propagating in the pipe or the wave propagating in the open air? Is the voice he hears deformed?

Q7. Is the field **E** always perpendicular to the field **B** in a waveguide? Are **E** and **B** always perpendicular to the direction of propagation? Draw the field lines in the case of the *TE* and *TM* waves in a rectangular waveguide.

Q8. Consider an electromagnetic wave in the mode (*m, n*) in a rectangular waveguide. Write down the expressions for the electric energy density, the magnetic energy density and the Poynting vector. Calculate the electric energy density and the magnetic energy density per unit length in the waveguide. Calculate their mean values over a period. Calculate the average energy flux on the section of the waveguide and its average over a period. What can you conclude?

9.15. Problems

Standing waves and normal modes

P9.1. Two waves $y_1 = 0.1 \sin(0.5\,x - 50t)$ and $y_2 = 0.1 \sin(0.5\,x + 50t)$ propagate on a string where x, y_1 and y_2 are in cm. **a)** Determine the frequency, wavelength and propagation speed of these waves. **b)** Determine the positions of the nodes and the antinodes. **c)** What would be the positions of the nodes and the antinodes if the second wave was $y_2 = 0.1 \sin(0.5\,x + 50t + \phi)$?

P9.2. a) A string of length 0.5 m is fixed at both ends. Its fundamental frequency is 200 Hz. What is the speed of propagation on this string? **b)** Verify that the period of the fundamental mode is equal to the time that it takes for a signal to go up and back along the entire string. Is it possible to make a similar interpretation for the period of higher modes? **c)** The mass density of this string is 7.9 g/cm^3 and its diameter is 0.8 mm. What is its tension?

P9.3*. A string that is fixed at both ends vibrates in the mode n. **a)** Calculate the kinetic energy density $U_{(C)l}$ and the potential energy density $U_{(P)l}$ per unit length. Plot them as functions of the position at a given time t in the case $n = 3$. Verify that $U_{(C)l}$ is always zero at the nodes and $U_{(P)l}$ is zero at the antinodes. **b)** Plot $U_{(C)l}$ and $U_{(P)l}$ as functions of time at a given point. Show that these densities oscillate between a maximum and a minimum with a period $T/2$. Show that, when one of these densities is at a maximum, the other is at a minimum and that the average total energy density over a half-period is constant. **c)** Calculate the energy flux. Verify that the total energy stored between a node and an antinode remains constant in the course of time. Thus, there is no transfer of energy from a half-zone to the adjacent zone. **d)** Calculate the total energy of the string in the mode n. Verify that, for a given vibration amplitude A, the total energy is proportional to A^2 and $\omega_n^{\ 2}$ (that is, proportional to n^2).

P9.4.** The results for this problem of the polarization of waves on a string are valid for any transverse waves, such as electromagnetic waves. A circularly polarized wave may be considered to be the superposition of two linearly polarized waves in perpendicular directions and in quadrature:

$$\mathbf{u}^{\pm} = u_o[\cos(\omega t)\mathbf{e}_x + \cos(\omega t \pm \pi/2)\,\mathbf{e}_y]\sin(kz).$$

a) Show that the points of the string describe circles, and that the points at a distance $\lambda/2$ apart describe circles of equal radius. Verify that the nodes correspond to a radius equal to zero while antinodes correspond to a maximum radius u_o. What is the direction of rotation? **b)** Show that, if a progressive circularly polarized wave is totally reflected on a fixed obstacle, the reflected wave is circularly polarized in the opposite direction to the incident wave. Show that the wave resulting from the superposition of the incident and reflected waves is a circularly polarized standing wave. **c)** Show that the circular polarization of the reflected wave in the opposite direction to the incident wave in the case of circular polarization, results from the conservation of angular momentum at the reflection point.

P9.5. A pipe of length 30 cm is open at both ends. **a)** Write down the displacement expressions and the acoustic pressure for the *n*th harmonic. What are the frequencies of the first three harmonics? **b)** What is the relative variation of the frequency of the *n*th harmonic if the temperature increases from 0° to 1°? **c)** What are the harmonics, which may be stationary if a hole is pierced in the middle of the pipe, and if two holes are pierced: the first at one-third and the other at the middle?

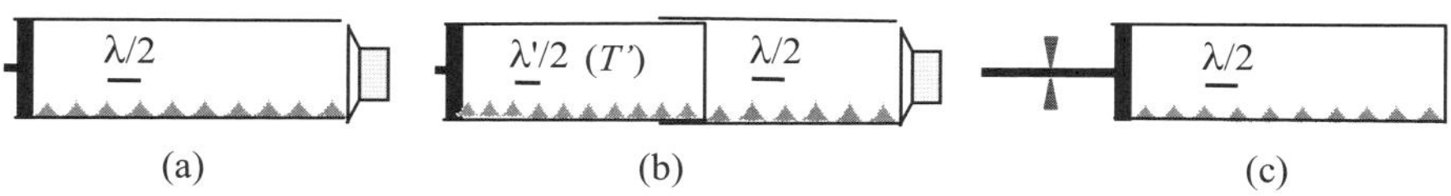

Figure 9.12. *Problem P9.6*

P9.6. *Kundt tube* is an acoustic pipe that is closed at one end by a piston whose position may be adjusted to have standing waves. It serves to measure the speed of sound in a gas. It is excited at the other open end by a loudspeaker that is sustained electrically at a known frequency (Figure 9.12a). In the old version, the tube is in glass and filled by air or the gas under study, and a light visible powder is spread along the pipe. The powder accumulates in small heaps at the nodes of displacement of the standing wave. In the modern version, a small microphone is displaced along the pipe and the induced electric potential is observed on the screen of an oscilloscope to determine the positions of nodes and antinodes. **a)** Is it possible to measure the speed of sound v_s in air using this tube? To measure the speed of sound v'_s in another gas, we use two joint Kundt tubes separated by a very light membrane (Figure 9.12b): the tube T is filled with air and the tube T' is filled with the gas. The air tube serves to set the gas tube oscillating. Show that $v_s/v'_s = \lambda/\lambda'$.**b)** To measure the speed of elastic waves in a rod, we fix its middle and let one of its ends be free, while the other is coupled to the piston of a Kundt tube (Figure 9.12c). If the rod is rubbed longitudinally, it resonates mostly in its fundamental mode and its vibration is transmitted to the gas. What is the relation of the speed of elastic waves in the rod and the observed wavelength in the Kundt tube?

P9.7* A pipe of length L has one end closed by a rigid obstacle and the other end closed by a piston. The piston is displaced according to $u_P = f(t)$ of period T. Determine the wave which is established in the pipe.

Standing waves on a membrane and in a cavity

P9.8. A square elastic membrane of sides L and mass m_s per unit area is taut with a tension F_l per unit length along its periphery. **a)** Calculate the frequency of the first four modes. What is their degeneracy? **b)** Calculate the total energy of the membrane oscillating in the mode (m, n) with an amplitude $A_{m,n}$. Verify that this energy is proportional to $A_{m,n}^2$ and to $\omega_{m,n}^2$.

P9.9. What are the frequencies of the first six modes of a cubic cavity of sides L? What is the degeneracy of these frequencies?

Fourier analysis of standing waves

P9.10*. Any wave on a string that is fixed at its ends may be written in the form [9.29]. **a)** If we know $u(x_o, t)$ at a given point x_o (which is not a node), we may calculate the Fourier coefficient B_n and C_n. For this, multiply both sides of equation [9.29] by $\cos(n'\pi vt/L)$ and integrate from $t = 0$ to $t = 2L/v$; do the same, multiplying it by $\sin(m\pi vt/L)$. Deduce the relationships of [9.30]. **b)** If we know the initial conditions (displacement and velocity at a given time t_o) at each point of the string, multiply both sides of [9.29] by $\sin(n'\pi x/L)$ and integrate on x from 0 to L. Deduce equation [9.32]. Differentiate both sides of this equation with respect to time and obtain equation [9.33]. Deduce the relationships of [9.34].

P9.11*. A string of length L is fixed at both ends. Let v be the speed of propagation of the transverse elastic waves on this string. One strikes it in such a way as to give the segment $L/3 < x < L/2$ an initial velocity v_o. Determine the amplitudes of the first three excited harmonics.

P9.12*. A string of length 1 m and mass 100 g is fixed at both ends with a tension of 1,000 N. **a)** Determine the frequency of the first four modes and the corresponding wavelength. Write down the expression of the vibration at the point x and at time t. Draw the profile of the string in each mode. **b)** Consider the superposition of the first two modes in phase, equal amplitudes and the same linear polarization. Write down the expression of the resulting wave and draw the profile of the string at a given time. Assuming that the string is in an arbitrary state of vibration, we "finger" the point situated at 25 cm from its end. Which modes of vibration are not eliminated? **c)** The string vibrates in the nth mode with an amplitude A. Calculate its total vibration energy. **d)** Show that, if the string vibrates in a superposition of the modes m and n with the amplitudes A_m and A_n respectively, and a phase difference ϕ, the total vibration energy is the sum of the energies of each mode. This is a consequence of the orthogonality of the normal modes.

P9.13*. A string of length L is fixed at both ends. It is initially deformed as in Figure 9.13 and released without initial velocity. Let v be the speed of propagation of the transverse elastic waves along this string. **a)** Show that the standing wave established on this string is of the form $u(t, x) = \Sigma_{n\geq 0} A_n \cos(n\pi vt/L) \sin(n\pi x/L)$, where $A_n = 4d/\pi n$ if n is odd and $A_n = 0$ if n is even. **b)** Draw the profile of the string at $t = 0$ if we only keep the first three terms of this series.

Figure 9.13. *Problem P9.13*

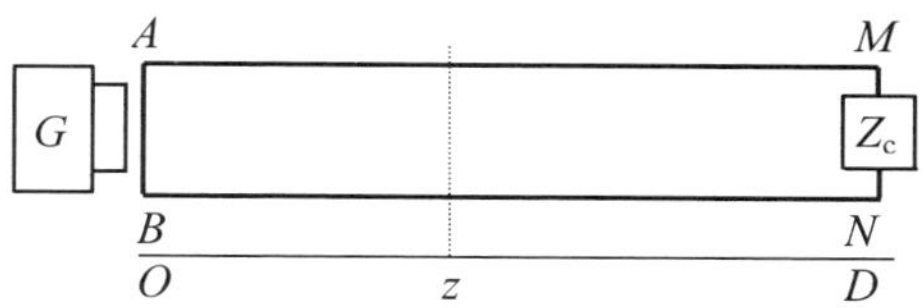

Figure 9.14. *Problem P9.15*

Guided acoustic waves

P9.14. The dispersion relation of sound or electromagnetic waves in a rectangular waveguide, for instance, is $k = \gamma_{m,n}\, \omega/v$; where we have set $\gamma_{m,n} = [1 - \omega_{m,n}^2/\omega^2]^{½}$ with $\omega_{m,n}^2 = (\pi v)^2(m^2/a^2 + n^2/b^2)$. **a)** We may also write $\omega^2 = v^2 k^2 + \omega_{m,n}^2$. This relation is similar to the dispersion relation of the Klein-Gordon equation (in the ionosphere, for instance). Show that, if the wave has an angular frequency $\omega > \omega_{m,n}$, it propagates without attenuation (the waveguide is then dispersive). On the contrary, if $\omega < \omega_{m,n}$, the wave is attenuated. Calculate the attenuation coefficient as a function of ω and $\omega_{m,n}$. **b)** What is the cut-off frequency of the fundamental mode of an acoustic pipe, whose section is a square of sides 10 cm? What is the wavelength at the frequency of 2 kHz? Compare it to the wavelength in open-air. **c)** What is the attenuation coefficient of the wave at 1 kHz? How much is the intensity reduced after travelling 30 cm? What happens to the lost energy?

Standing electromagnetic waves

P9.15. Assume that an electric line is short-circuited between A and B where it is excited at high frequency while its end at M and N ($z = D$) is connected to an impedance $\underline{Z}_c$ (Figure 9.14). **a)** Write the general solutions for $\underline{V}$ and $\underline{I}$, impose the boundary conditions and show that the modes may be written as:

$$V = A\cos(\omega t + \phi)\cos(\omega z/v) \quad \text{and} \quad I = (A/Z_l)\sin(\omega t + \phi)\sin(\omega z/v)$$

where the normal frequencies are given by the condition $\tan(2\pi D/\lambda) = \mathrm{i}(\underline{Z}_c/Z_l)$. **b)** Determine the normal modes if the line is short-circuited and if it is open-ended.

P9.16*. A transmission line is formed by two long, metallic, plane and parallel plates. It is short-circuited at the point z_o by a transverse metallic plate of negligible resistance. **a)** Show that standing waves are established for $z < z_o$, such that the nodes of the intensity I are the antinodes of the potential V. **b)** Show that this short-circuited transmission line is equivalent to an open transmission line at the point $z_o - \lambda/4$.

P9.17.** Assume that an electric field $\mathbf{E} = A\cos(kz)\cos(\omega t)\,\mathbf{e}_x$ is established in a region of a dielectric. **a)** Calculate the magnetic field. **b)** Calculate the electric energy density, the magnetic energy density and the Poynting vector. **c)** Describe the fields **E** and **B** as functions of z in a wave zone situated between two nodal planes of **E** at the instants $t = 0$, $T/8$, $T/4$, $3T/8$ and $T/2$. Analyze the distribution of the electric energy, magnetic energy and total energy, the magnitude and the direction of the Poynting vector at these instants of time. **d)** Show that the total energy that is stored in a half-zone (of length $\lambda/4$ between a nodal plane and an antinodal plane) remains constant in the course of time. Thus, there is no energy transfer from one half-zone to the adjacent half-zone (standing waves). Calculate the total electromagnetic energy that is stored in a half-zone.

Electromagnetic waveguides formed by two conductors

P9.18. A waveguide is formed by two plane and parallel plates separated by a distance a. **a)** Verify that the expressions of the electromagnetic waves in this waveguide are the limits of the expressions of the waves in a hollow conductor whose section is rectangular of sides a and b if we take the limit $b \to \infty$. **b)** In the case of a TE wave, for instance, verify that the wave may be written as

$$\underline{\mathbf{E}} = -\mathrm{i}v\underline{B}_o\,\mathbf{e}_y\,[\,e^{\mathrm{i}(\omega t+px-kz)} - e^{\mathrm{i}(\omega t-px-kz)}$$
$$\underline{\mathbf{B}} = \mathrm{i}\underline{B}_o\{[\mathbf{e}_x + (\omega_m/\gamma_m\omega)\,\mathbf{e}_z]\,e^{\mathrm{i}(\omega t+px-kz)} + [-\mathbf{e}_x + (\omega_m/\gamma_m\omega)\,\mathbf{e}_z]\,e^{\mathrm{i}(\omega t-px-kz)}\,\}$$

Deduce that this wave may be considered to be the superposition of an oblique wave and the corresponding reflected wave on the plates (zigzag wave).

P9.19. A cable is formed by a cylindrical conductor of radius a for the current in one direction, and a cylindrical metallic shell of internal radius b for the current in the opposite direction. We assume that the current has the form $\underline{I} = \underline{I}_m\,e^{\mathrm{i}(\omega t-kz)}$ and we consider only TEM waves. Calculate the magnetic field. Deduce the electric field, the Poynting vector and the transported power by this line.

Single-conductor electromagnetic waveguides

P9.20. An electromagnetic waveguide has a rectangular section of sides $a = 2$ cm and $b = 1$ cm. **a)** What is the lowest cut-off frequency for *TE* waves? Write down the expressions of the electric and magnetic fields in this mode. Determine the frequency of the first five modes. **b)** Calculate the phase velocity and the group velocity for a wave with a frequency of 10 GHz that propagates in the dominant mode.

P9.21. What is the cut-off frequency of the dominant mode in a waveguide whose section is a square of side a. Consider a fiber of this type with index 1.5. If light has a frequency that is less than the cut-off frequency, it cannot propagate. What should the minimum value a_m of a be to allow the propagation of light, whose wavelength is $\lambda = 5{,}000$ nm, in vacuum? Determine the phase velocity and the group velocity for the dominant mode if $a = 2a_m$.

P9.22. Assume that a cylindrical hollow conductor of internal radius 1.5 cm is used as a waveguide to transmit electromagnetic waves. Determine the cut-off frequency of the first three modes. To reduce the deformation of signals, only the frequencies lying between the cut-off frequency of the dominant mode and the second mode must be used. What are the limits of this frequency band? Estimate the number of simultaneous telephone calls that may be transmitted by this waveguide. Estimate the number of TV channels that may be transmitted by this waveguide.

Answers to the Problems

Chapter 1

P1.1. a) $x_1 = 2\cos(40\pi t) + 0.7958\sin(40\pi t)$, $x_2 = -2\cos(40\pi t) + 2.387\sin(40\pi t)$ and $x_3 = 3\cos(40\pi t) + 3.979\sin(40\pi t) = 4.98\cos(40\pi t - 0.925)$; hence, $x_{max} = 4.98$ cm and $v_{max} = 6.26$ m/s.

P1.2. a) $A_1 = \mathcal{Re}\,\underline{C}$, $A_2 = -\mathcal{Im}\,\underline{C}$, $\underline{C} = A_1 - \mathrm{i}A_2 = A\,e^{\mathrm{i}\phi}$, where $A = \sqrt{A_1^{\,2} + A_2^{\,2}}$, $\cos\phi = A_1/A$ and $\sin\phi = A_1/A$ (i.e. $\tan\phi = -A_2/A_1$).

P1.3. $\tilde{\nu} = 1.5$ rad/s, $\phi = -2.356$ rad, $\underline{x} = \underline{A}\,e^{3\mathrm{i}\pi t}$, $\dot{x} = \underline{A}'\,e^{3\mathrm{i}\pi t}$, $\ddot{x} = \underline{A}''\,e^{3\mathrm{i}\pi t}$, where $\underline{A} = 5\,e^{-3\mathrm{i}\pi/4}$, $\underline{A}' = 15\pi\,e^{-\mathrm{i}\pi/4}$ and $\underline{A}'' = 45\pi^2\,e^{\mathrm{i}\pi/4}$.

P1.4. a) $F = -Kx$, where $K = 9\pi^2 m = 44.4$ N/m. **b)** $U_{(P)} = 0.888 \times 10^{-2}\cos^2(3\pi t)$ J, $U_{(C)} = 0.00888\sin^2(3\pi t)$ J, $U_{(T)} = 0.00888$ J. **c)** $U_{(P)} = 22.2\,x^2$, $U_{(C)} = 0.00888 - 22.2\,x^2$.

P1.5. $F = -Kx$; hence $\omega = \sqrt{K/m}$.

P1.6. a) $T = 2\pi\sqrt{mK/K_1K_2}$, where $K = K_1 + K_2$. **b)** The lengths of the springs are: $a = (K_1/K)L_1 + (K_2/K)(L - L_2)$, $b = (K_2/K)L_2 + (K_1/K)(L - L_1)$ and $T = 2\pi\sqrt{m/K}$.

P1.7. a) $x = (v_o/\omega)\sin(\omega t)$, where $\omega = \sqrt{K/m}$. **b)** $\dot{x} = v_o\cos(\omega t)$, $\ddot{x} = -v_o\omega\sin(\omega t)$, $U_{(C)} = \frac{1}{2}\,mv_o^{\,2}\cos^2(\omega t)$, $U_{(P)} = \frac{1}{2}\,mv_o^{\,2}\sin^2(\omega t)$. **c)** $< x > = 0$, $< \dot{x} > = 0$, $< x^2 > = \frac{1}{2}\,(v_o/\omega)^2$, $< \dot{x}^{\,2} > = \frac{1}{2}\,v_o^{\,2}$ and $< U_{(C)} > = < U_{(P)} > = \frac{1}{4}\,mv_o^{\,2}$.

P1.8. a) $W = \frac{1}{2} Kx^2$ is independent of the path and the intermediary velocity. **b)** $dm = m d\xi/(x+L)$, $v(\xi) = \xi \dot{x}/(x+L)$, $dU_{(C)} = \frac{1}{2}\, dm\, v^2(\xi)$ and $U_{(C)} = \frac{1}{2}\mu \dot{x}^2$, where $\mu = (M + m/3)$. The equation of motion is $\mu \ddot{x} + K x = 0$, $T = 2\pi \sqrt{\mu/K}$.

P1.9. $T = 2\pi \sqrt{J/K}$, where $K = K_1 + K_2$.

P1.10. a) $\omega = \sqrt{g/L} = 2.215$, $\theta = \theta_o \cos(\omega t)$, $\dot{\theta} = -\omega\theta_o \sin(\omega t)$, $\ddot{\theta} = -\omega^2 \theta_o \cos(\omega t)$, where $\theta_o = 15° = 0.2618$ rad, $v = 1.160\,|\sin(\omega t)|$ (in m/s), $a_c = 0.6724 \sin^2(\omega t)$ (in m/s^2) and $F \approx 10.48 - 1.009 \cos^2\theta$ (in N). **b)** The law of conservation of energy gives $v^2 = 2gL(\cos\theta - \cos\theta_o)$; hence, $v = 1.090$ m/s, while the result of question (a) gives $v = 1.093$ m/s.

P1.11. a) Treated as a simple pendulum, $T = 2.056$ s and as a physical pendulum, $T = 2.007$ s. **b)** The velocity of the elevator has no effect on T. If its acceleration is 1 m/s^2, $T' = T\sqrt{g/(g+a)} = 1.912$ s. **c)** $T' = 2.164$ s. **d)** $T' = T\sqrt{g/g'} = 2.010$ s.

P1.12. a) $x = \frac{1}{2} A e^{i\phi} e^{\eta_1 t} + \frac{1}{2} A e^{-i\phi} e^{\eta_2 t}$, where $\eta_1 = -\delta + i\hat{\omega}$ and $\eta_2 = -\delta - i\hat{\omega}$. **b)** $\delta = 0.069$ s^{-1}, $\tau = 14.4$ s and $f_q = 22.7$.

P1.13. a) $\dot{\theta} = 2.47\, e^{-0.15t} \cos(9.42t + 1.59)$. **b)** $\theta_m = 0.262$ rad, $\hat{T} = 0.667$ s, $\theta(0) = 0.262$ rad, $\dot{\theta}(0) = -0.04$ rad/s. **c)** $f_q = 31.4$. **d)** $t = 15.4$ s.

P1.14. a) $x = (v_o/\tilde{\omega})e^{-\beta t}\sin(\tilde{\omega} t)$, $\dot{x} = (v_o/\tilde{\omega})e^{-\beta t}[\tilde{\omega}\cos(\tilde{\omega} t) - \beta\sin(\tilde{\omega} t)]$, $\beta = b/2m$, $\tilde{\omega}^2 = K/m - (b/2m)^2$. **b)** $U_{(T)} = \frac{1}{2} m v_o^2 e^{-2\beta t}[1 + 2(\beta/\tilde{\omega})^2 \sin^2(\tilde{\omega} t) - (\beta/\tilde{\omega})\sin(2\tilde{\omega} t)]$. **c)** $U_{\text{lost}} = \frac{1}{2} m v_o^2 - \frac{1}{2} m v_o^2 e^{-2\beta t}[1 + 2(\beta/\tilde{\omega})^2 \sin^2(\tilde{\omega} t) - (\beta/\tilde{\omega})\sin(2\tilde{\omega} t)]$. **d)** $\beta = 3$ s^{-1}, $\tilde{\omega} = 4$ rad/s, $U_o = 1.0 \times 10^{-5}$ J and $U(5 \times 10^{-2}\text{ s}) = 0.5573 \times 10^{-5}$ J.

P1.15. a) $L\ddot{Q} + R\dot{Q} + Q/C = 0$, $Q = (\omega_o/\tilde{\omega})Q_o e^{-\beta t}\cos(\tilde{\omega} t + \phi)$, where $\omega_o = 1/\sqrt{LC}$, $\tilde{\omega} = \sqrt{1/LC - R^2/4L^2}$, $\beta = R/2L$ and $\tan\phi = -(\beta/\tilde{\omega})$ such that $-\pi/2 < \phi < 0$.
b) $I = -\omega_o Q_o\, e^{-\beta t}[\sin(\tilde{\omega} t + \phi) + (\beta/\tilde{\omega})\cos(\tilde{\omega} t + \phi)] = -(\omega_o^2/\tilde{\omega})Q_o e^{-\beta t}\sin(\tilde{\omega} t)$,
$U_{(E)} = \frac{1}{2}(Q_o^2/C)(\omega_o/\tilde{\omega})^2 e^{-2\beta t}\cos^2(\tilde{\omega} t + \phi)$,
$U_{(M)} = (Q_o^2/2C)e^{-2\beta t}[\sin^2(\tilde{\omega} t + \phi) + (\beta/\tilde{\omega})^2\cos^2(\tilde{\omega} t + \phi) + (\beta/\tilde{\omega})\sin(2\tilde{\omega} t + 2\phi)]$,
$U_{(T)} = U_{(E)} + U_{(M)} = (Q_o^2/2C)e^{-2\beta t}[1 + 2(\beta/\tilde{\omega})^2\cos^2(\tilde{\omega} t + \phi) + (\beta/\tilde{\omega})\sin(2\tilde{\omega} t + 2\phi)]$.
c) $U_{(J)} = (Q_o^2/2C)\{1 - e^{-2\beta t}[1 + 2(\beta/\tilde{\omega})^2\cos^2(\tilde{\omega} t + \phi) + (\beta/\tilde{\omega})\sin(2\tilde{\omega} t + 2\phi)]\} = U(0) - U(t)$. **d)** $\tau = 1/\beta = 2 \times 10^{-3}$ s, $f_q = \omega_o\tau/2 = 0.707$, $t = 9.90 \times 10^{-3}$ s.

P1.16. a) $Q = Q_o \cos(\omega_o t)$ and $I = Q_o\omega_o\cos(\omega_o t + \pi/2)$. $U_{(EM)} = Q_o^2/2C$. The energy conservation gives the equation $\frac{1}{2} Q^2/C + \frac{1}{2} LI^2$; hence, $\ddot{Q} + (R/L)\dot{Q} + Q/LC = 0$.
b) $Q = A\,e^{-\beta t}\cos(\tilde{\omega} t + \phi)$, $I = \dot{Q} = -A\,e^{-\beta t}\,[\beta\cos(\tilde{\omega} t + \phi) + \tilde{\omega}\sin(\tilde{\omega} t + \phi)]$.
$U_{(EM)} = \frac{1}{2} LA^2 e^{-2\beta t}\,[\tilde{\omega}^2 + 2\beta^2\cos^2(\tilde{\omega} t+\phi) + \beta\tilde{\omega}\sin 2(\tilde{\omega} t+\phi)]$,
where $\tilde{\omega} = \sqrt{1/LC - R^2/4L^2}$. **c)** $f_q = \sqrt{L/CR^2}$

P1.17. $h' = h\rho/\rho_o$, $T = 2\pi\sqrt{h\rho/\rho_o g}$.

P1.18. Let $x(t)$ be the displacement of the liquid. We find $U_{(C)} = \frac{1}{2} L S m_v \dot{x}^2$ and $U_{(P)} = g S m_v x^2$; hence, the equation of motion $\ddot{x} + 2(g/L)\,x = 0$ and $T = \pi\sqrt{2L/g}$.

P1.19. $U_{(C)} = \frac{1}{2} m\dot{x}^2(1+4a^2x^2)$, $U_{(P)} = mgax^2$; hence, $\ddot{x} + 2gax = 0$ and $T = 2\pi/\sqrt{2ga}$.

P1.20. $\dot{x} = \pm\sqrt{g_{eq}(x_o - |x|)}$, where $g_{eq} = g\sin(2\alpha)$. $T = 8\sqrt{x_o/g_{eq}}$.

P1.21. a) $x = r_o$, stable. **b)** Set $X = (m_1x_1 + m_2x_2)/(m_1 + m_2)$ and $x = x_1 - x_2$, $U = \frac{1}{2} M\dot{X}^2 + \frac{1}{2}\mu\dot{x}^2 + Da^2(x - r_o)^2 - D$; hence the equation $\mu\ddot{x} + 2Da^2(x - r_o) = 0$. The motion is harmonic with a frequency $\tilde{\nu} = (a/2\pi)\sqrt{2D/\mu}$.

P1.22. a) $F = -Kx + ax^2$; hence, $m\ddot{x} + Kx - ax^2 = 0$. **c)** $U_{(T)} = \frac{1}{2} KA^2 + O(a^2)$.

P1.23. $U_{(C)} = \frac{1}{2} m\dot{y}^2$ and $U_{(P)} = K(\sqrt{a^2+y^2} - a)^2$.

P1.24. a) $U_{(T)} = \frac{1}{2} J\dot{\theta}^2 - mgL\cos\theta$; hence $J\ddot{\theta} + mgL\sin\theta = 0$.

P1.25. The general solution is: $I_1 = A_{(1)}\cos(\omega_{(1)}t + \phi_{(1)}) + A_{(2)}\cos(\omega_{(2)}t + \phi_{(2)})$ and $I_2 = -A_{(1)}\cos(\omega_{(1)}t + \phi_{(1)}) + A_{(2)}\cos(\omega_{(2)}t + \phi_{(2)})$.

P1.26. a) The general solution is: $u_1 = A_{(1)}\cos(\omega_{(1)}t + \phi_{(1)}) + A_{(2)}\cos(\omega_{(2)}t + \phi_{(2)})$,
$u_2 = (A_{(1)}/\mu_{12})(K_1/\omega_{(1)}^2 - \mu_{11})\cos(\omega_{(1)}t+\phi_{(1)}) + (A_{(2)}/\mu_{12})(K_1/\omega_{(2)}^2 - \mu_{11})\cos(\omega_{(1)}t+\phi_{(1)})$.
b) $I_1 = A_{(1)}\cos(\omega_{(1)}t + \phi_{(1)}) + A_{(2)}\cos(\omega_{(2)}t + \phi_{(2)})$,
$I_2 = -A_{(1)}\cos(\omega_{(1)}t + \phi_{(1)}) + A_{(2)}\cos(\omega_{(2)}t + \phi_{(2)})$, where $\omega_{(1)} = \omega_o$ and $\omega_{(2)} = \omega_o/\sqrt{3}$.

c) The normal variables are: $y_1 = \frac{1}{2}\sqrt{2L+2M}\,(Q_1 + Q_2)$ and $y_2 = \frac{1}{2}\sqrt{2L-2M}\,(Q_1 - Q_2)$,
$U_{(EM)} = \frac{1}{2}(\dot{y}_1^2 + \dot{y}_2^2) + \frac{1}{2}\omega_{(1)}^2 y_1^2 + \frac{1}{2}\omega_{(2)}^2 y_2^2$.

P1.27. a) $\ddot{x}_1 + 2\beta\,\dot{x}_1 + (\omega_o^2 + \omega_3^2)x_1 - \omega_3^2 x_2 = 0,\ \ddot{x}_2 + 2\beta\,\dot{x}_2 + (\omega_o^2 + \omega_3^2)x_2 - \omega_3^2 x_1 = 0.$

c) – If $\beta^2 > \omega_o^2 + 3\omega_3^2$, setting $\sigma_{(1)} = \sqrt{\beta^2 - \omega_o^2 - 2\omega_3^2}$ and $\sigma_{(2)} = \sqrt{\beta^2 - \omega_o^2}$, we find

$x_1 = e^{-\beta t}\,[A_1 \exp(\sigma_{(1)}t) + A_2 \exp(-\sigma_{(1)}t)] + e^{-\beta t}\,[B_1 \exp(\sigma_{(2)}t) + B_2 \exp(-\sigma_{(2)}t)],$

$x_2 = -e^{-\beta t}\,[A_1 \exp(\sigma_{(1)}t) + A_2 \exp(-\sigma_{(1)}t)] + e^{-\beta t}\,[B_1 \exp(\sigma_{(2)}t) + B_2 \exp(-\sigma_{(2)}t)].$

– If $\omega_o^2 < \beta^2 < \omega_o^2 + 2\omega_3^2$, setting $\omega_{(1)} = \sqrt{\omega_o^2 + 2\omega_3^2 - \beta^2}$, we find

$x_1 = A\ e^{-\beta t} \cos(\omega_{(1)}t + \phi_{(1)}) + e^{-\beta t}\,[B_1 \exp(\sigma_{(2)}t) + B_2 \exp(-\sigma_{(2)}t)],$

$x_2 = -A\ e^{-\beta t} \cos(\omega_{(1)}t + \phi_{(1)}) + e^{-\beta t}\,[B_1 \exp(\sigma_{(2)}t) + B_2 \exp(-\sigma_{(2)}t)].$

– If $\beta^2 < \omega_o^2$, setting $\omega_{(1)} = \sqrt{\omega_o^2 + 2\omega_3^2 - \beta^2}$ and $\omega_{(2)} = \sqrt{\omega_o^2 - \beta^2}$, we find

$x_1 = A\,e^{-\beta t} \cos(\omega_{(1)}t) + \phi_{(1)}) + B\ e^{-\beta t}\ \cos(\omega_{(2)}t) + \phi_{(2)}),$

$x_2 = -A\,e^{-\beta t} \cos(\omega_{(1)}t) + \phi_{(1)}) + B\ e^{-\beta t}\ \cos(\omega_{(2)}t) + \phi_{(2)}).$

d) The equations of y_1 and y_2 are $\ddot{y}_1 + 2\beta\dot{y}_1 + (\omega_o^2 + 2\omega_3^2)\,y_1 = 0,\ \ddot{y}_2 + 2\beta\dot{y}_2 + \omega_o^2\,y_2 = 0.$

– If $\beta < \sqrt{\omega_o^2 + 2\omega_3^2}$, $y_1 = 2A\,e^{-\beta t} \cos(\omega_{(1)}t + \phi_{(1)})$.

– If $\beta > \sqrt{\omega_o^2 + 2\omega_3^2}$, $y_1 = 2\,e^{-\beta t}\,(A_1\,e^{\sigma_{(1)}t} + A_2\,e^{-\sigma_{(1)}t})$.

– If $\beta < \omega_o$, $y_2 = 2B\,e^{-\beta t} \cos(\omega_{(2)}t + \phi_{(2)})$ and if $\beta > \omega_o$, $y_2 = 2\,e^{-\beta t}\,(B_1\,e^{\sigma_{(2)}t} + B_2\,e^{-\sigma_{(2)}t})$.

P1.28. a) $\ddot{I}_1 + \omega_o^2(I_1 - I_2) = 0,\ \ddot{I}_2 + \omega_o^2(-I_1 + 2I_2 - I_3),\ \ddot{I}_3 + \omega_o^2(-I_2 + I_3) = 0,$
where $\omega_o = 1/\sqrt{LC}$. Normal angular frequencies: $\omega_{(1)} = 0$, $\omega_{(2)} = \omega_o$ and $\omega_{(3)} = \sqrt{3}\,\omega_o$.

b) Mode 1: $I_1 = I_2 = I_3 = a + \alpha t$. Mode 2: $I_1 = -I_3 = b \cos(\omega_o t + \phi_2)$, $I_2 = 0$,
Mode 3: $I_1 = I_3 = c \cos(\sqrt{3}\,\omega_o t + \phi_3)$, $I_2 = -2c \cos(\sqrt{3}\,\omega_o t + \phi_3)$. General solution:

$I_1 = a + \alpha t + b \cos(\omega_o t + \phi_2) + c \cos(\sqrt{3}\,\omega_o t + \phi_3)$, $I_2 = a + \alpha t - 2c \cos(\sqrt{3}\,\omega_o t + \phi_3)$,

$I_3 = a + \alpha t - b \cos(\omega_o t + \phi_2) + c \cos(\sqrt{3}\,\omega_o t + \phi_3)$.

P1.29. a) $(\mu_1\eta^2 + b_{11}\eta + a_{11})\,(\mu_2\eta^2 + b_{22}\eta + a_{22}) - b_{12}\,b_{21}\eta^2 = 0;$

$U_{(T)}(t) = \tfrac{1}{2}\,\mu_1\,\dot{u}_1^2 + \tfrac{1}{2}\,a_{11}\,u_1^2 + \tfrac{1}{2}\,\mu_2\,\dot{u}_2^2 + \tfrac{1}{2}a_{22}\,u_2^2$

$dU_{(T)}/dt = -\,b_{11}\,\dot{u}_1^2 - b_{22}\,\dot{u}_2^2 - (b_{12} + b_{21})\,\dot{u}_1\,\dot{u}_2$

P1.30. a) $U_{(C)} = \tfrac{1}{2}\,m\,\dot{x}_1^2 + \tfrac{1}{2}\,M\dot{x}_2^2 + \tfrac{1}{2}\,m\,\dot{x}_3^2$ and $U_{(P)} = \tfrac{1}{2}\,K(x_1 - x_2)^2 + \tfrac{1}{2}\,K(x_3 - x_2)^2$.
The equations of motion are $m\,\ddot{x}_1 + K(x_1 - x_2) = 0$, $M\,\ddot{x}_2 + K(2x_2 - x_1 - x_3) = 0$
and $m\,\ddot{x}_3 + K(x_3 - x_2) = 0$. The normal modes are

$x_{(1)1} = x_{(1)2} = x_{(1)3} = v_o t + a.$

$x_{(2)1} = -x_{(2)3} = A_{(2)} \cos(\omega_{(2)}\,t + \phi_{(2)})$, $x_{(2)3} = 0$, where $\omega_{(2)} = \sqrt{K/m}$

$x_{(3)1} = x_{(3)3} = A_{(3)} \cos(\omega_{(3)}\,t + \phi_{(3)})$, $x_{(3)2} = -\,(2m/M)\,A_{(3)} \cos(\omega_{(3)}t + \phi_{(3)})$,
where $\omega_{(3)} = \sqrt{K/m + 2K/M}$.

b) $y_1 = \alpha_1(mx_1 + Mx_2 + mx_3)$, $y_2 = m\,\alpha_2(x_1 - x_3)$, $y_3 = \alpha_3 m\,(x_1 - 2x_2 + x_3)$, where $\alpha_1 = (2m+M)^{-1/2}$, $\alpha_2 = (2m)^{-1/2}$ and $\alpha_3 = (2m+4m^2/M)^{-1/2}$. The energy may be written as $U_{(T)} = \frac{1}{2}(\dot{y}_1^{\,2} + \dot{y}_2^{\,2} + \dot{y}_3^{\,2}) + \frac{1}{2}(\omega_1^2 y_1^2 + \omega_2^2 y_2^2 + \omega_3^2 y_3^2)$.

P1.31. a) $\ddot{x}_1 + 2\beta\,\dot{x}_1 + (\omega_o^2 + \omega_3^2)x_1 - \omega_3^2 x_2 = 0$, $\ddot{x}_2 + 2\beta\,\dot{x}_2 + (\omega_o^2 + \omega_3^2)x_2 - \omega_3^2 x_1 = 0$.
c) – If $\beta^2 > \omega_o^2 + 3\omega_3^2$, setting $\sigma_{(1)} = \sqrt{\beta^2 - \omega_o^2 - 2\omega_3^2}$ and $\sigma_{(2)} = \sqrt{\beta^2 - \omega_o^2}$, we find
$x_1 = e^{-\beta t}\,[A_1\exp(\sigma_{(1)}t) + A_2\exp(-\sigma_{(1)}t)] + e^{-\beta t}\,[B_1\exp(\sigma_{(2)}t) + B_2\exp(-\sigma_{(2)}t)]$,
$x_2 = -\,e^{-\beta t}\,[A_1\exp(\sigma_{(1)}t) + A_2\exp(-\sigma_{(1)}t)] + e^{-\beta t}\,[B_1\exp(\sigma_{(2)}t) + B_2\exp(-\sigma_{(2)}t)]$.
– If $\omega_o^2 < \beta^2 < \omega_o^2 + 2\omega_3^2$, setting $\omega_{(1)} = \sqrt{\omega_o^2 + 2\omega_3^2 - \beta^2}$, we find
$x_1 = A\,e^{-\beta t}\cos(\omega_{(1)}t + \phi_{(1)}) + e^{-\beta t}\,[B_1\exp(\sigma_{(2)}t) + B_2\exp(-\sigma_{(2)}t)]$,
$x_2 = -\,A\,e^{-\beta t}\cos(\omega_{(1)}t + \phi_{(1)}) + e^{-\beta t}\,[B_1\exp(\sigma_{(2)}t) + B_2\exp(-\sigma_{(2)}t)]$.
– If $\beta^2 < \omega_o^2$, setting $\omega_{(1)} = \sqrt{\omega_o^2 + 2\omega_3^2 - \beta^2}$ and $\omega_{(2)} = \sqrt{\omega_o^2 - \beta^2}$, we find
$x_1 = A\,e^{-\beta t}\cos(\omega_{(1)}t) + \phi_{(1)}) + B\,e^{-\beta t}\cos(\omega_{(2)}t) + \phi_{(2)})$,
$x_2 = -\,A\,e^{-\beta t}\cos(\omega_{(1)}t) + \phi_{(1)}) + B\,e^{-\beta t}\cos(\omega_{(2)}t) + \phi_{(2)})$.
d) The equations of y_1 and y_2 are $\ddot{y}_1 + 2\beta\,\dot{y}_1 + (\omega_o^2 + 2\omega_3^2)y_1 = 0$, $\ddot{y}_2 + 2\beta\,\dot{y}_2 + \omega_o^2 y_2 = 0$.
– If $\beta < \sqrt{\omega_o^2 + 2\omega_3^2}$, $y_1 = 2A\,e^{-\beta t}\cos(\omega_{(1)}t + \phi_{(1)})$.
– If $\beta > \sqrt{\omega_o^2 + 2\omega_3^2}$, $y_1 = 2\,e^{-\beta t}\,(A_1 e^{\sigma_{(1)}t} + A_2 e^{-\sigma_{(1)}t})$.
– If $\beta < \omega_o$, $y_2 = 2B\,e^{-\beta t}\cos(\omega_{(2)}t + \phi_{(2)})$ and if $\beta > \omega_o$, $y_2 = 2\,e^{-\beta t}\,(B_1 e^{\sigma_{(2)}t} + B_2 e^{-\sigma_{(2)}t})$.

P1.32. a) $\ddot{Q}_1 + (M/L)\,\ddot{Q}_2 + \omega_o^2 Q_1 = 0$ and $(M/L)\,\ddot{Q}_1 + \ddot{Q}_2 + \omega_o^2 Q_2 = 0$. The general solution is $Q_1 = A\cos(\omega_{(1)} + \phi) + B\cos(\omega_{(1)} + \phi')$ and $Q_2 = A\cos(\omega_{(1)} + \phi) - B\cos(\omega_{(1)} + \phi')$.
b) Normal variables: $y_1 = \frac{1}{2}\sqrt{2L+2M}\,(Q_1 + Q_2)$ and $y_2 = \frac{1}{2}\sqrt{2L - 2M}\,(Q_1 - Q_2)$.

Chapter 2

P2.1. a) $x_1 = 2.01\cos(\omega t + 2.72)$. **b)** $x_2 = 2.91\cos(\omega t + 1.40)$. **c)** $x_3 = 5\cos(\omega t - 0.643)$.

P2.2. $U = \frac{1}{2}m\omega^2[a_1^2 + a_2^2 + 2\,a_1 a_2\cos(\phi_1 - \phi_2)] = U_1 + U_2 + m\omega^2\,a_1 a_2\cos(\phi_1 - \phi_2)$, $U = U_1 + U_2$ if $\phi_1 - \phi_2 = \pi/2 + n\pi$ (motions in quadrature).

P2.3. a) $y = x$; the trajectory is the first diagonal of the square of sides $2A_1$. **b)** $y = -x$; the trajectory is the second diagonal of the square of sides $2A_1$. **c)** $x^2 + y^2 = A_1^2$; the trajectory is the clockwise circle of radius A_1 (right-circularly polarized motion). **d)** $x^2 + y^2 = A_1^2$; the trajectory is the anticlockwise circle of radius A_1 (left-circularly polarized motion). **e)** $y = x/2$; the trajectory is the first diagonal of the rectangle of sides $2A_1$ and A_1. **f)** $x^2/4A_2^2 + y^2/A_2^2 = 1$; the trajectory is a clockwise right ellipse inscribed within a rectangle of sides $4A_2$ in the direction Ox and $2A_2$ in the direction Oy (right elliptically polarized motion). **g)** $x^2/A_1^2 + y^2/A_2^2 - 2(xy/A_1A_2)\cos\phi = \sin^2\phi$; the trajectory is a clockwise right ellipse inscribed within a rectangle of sides $2A_1$ in the direction Ox and $2A_2 = A_2$ in the direction Oy (right elliptically polarized motion).

P2.4. a) $\mathbf{F} \approx -Kx\,\mathbf{e}_x - Ky\,\mathbf{e}_y \rightarrow m\ddot{x} + 2Kx = 0$, $m\ddot{y} + 2Ky = 0$. **b)** $x = A_1\cos(\omega t + \phi_1)$, $y = A_2\cos(\omega t + \phi_2)$, where $\omega = \sqrt{2K/m}$. In general, the motion is right or left elliptically polarized depending on the values of $\phi = \phi_2 - \phi_1$. It is polarized linearly in the direction of Oy if $A_1 = 0$, polarized linearly in the direction of Ox if $A_2 = 0$, polarized linearly in the direction of the first bisector if $\phi = 2n\pi$ and polarized linearly in the direction of the second bisector if $\phi = (2n + 1)\pi$. It is right-circularly polarized if $A_1 = A_2$ and $\phi = \pi/2$ and left-circularly polarized if $A_1 = A_2$ and $\phi = -\pi/2$.

P2.5. a) $\mathbf{u} = \mathbf{u}^+ + \mathbf{u}^- = 2A\cos\omega t\,\mathbf{e}_x$. **b)** $\mathbf{u} = 2A\cos(\phi/2)\cos(\omega t + \phi/2)\,\mathbf{e}_x + 2A\sin(\phi/2)\cos(\omega t + \phi/2)\,\mathbf{e}_y$. The resulting vibration has an amplitude $2A$, and it is linearly polarized in the direction which makes an angle $\phi/2$ with Ox.

P2.6. b) $\overrightarrow{OM} = a'\cos(\omega t + \phi'_1)\,\mathbf{e}'_x + b'\cos(\omega t + \phi'_1 + \phi')\,\mathbf{e}'_y$, where
$a' = [a^2\cos^2\alpha + b^2\sin^2\alpha + ab\sin 2\alpha\cos\phi]^{1/2}$,
$b' = [a^2\sin^2\alpha + b^2\cos^2\alpha - ab\sin 2\alpha\cos\phi]^{1/2}$,
$\cos\phi'_1 = [a\cos\alpha + b\sin\alpha\cos\phi]/a'$, $\sin\phi'_1 = (b/a')\sin\alpha\sin\phi$,
$\cos\phi' = [(b^2 - a^2)\sin 2\alpha + 2ab\cos 2\alpha\cos\phi]/2a'b'$, $\sin\phi' = (ab/a'b')\sin\phi$.
c) The area of the ellipse is the same in both systems of axes.
d) $\tan 2\alpha = [2ab/(a^2 - b^2)]\cos\phi$.

P2.7. The period of motion is 2 s.

P2.8. $x = 2A\cos(t\Delta\omega - \phi/2)\cos(\omega t + \phi/2)$.

P2.9. $u(t) = \Sigma_{n\geq 1} b_n \sin(n\omega t)$, where $\omega = \pi/2$ and $b_n = (8/n^2\pi^2)\sin n\pi/2$.

P2.10. $\omega_o = 2\pi/T$, $a_o = 2V_o/\pi$, $a_1 = \frac{1}{2}V_o$, $a_{2n+1} = 0$ and $a_{2n} = (-1)^{n+1}2V_o/\pi(4n^2 - 1)$; hence: $u_a(t) = (V_o/\pi)\,[1 + \frac{1}{2}\pi\cos(\omega_o t) + (2/3)\cos(2\omega_o t) - (2/15)\cos(4\omega_o t) + \ldots]$, $a'_{(P)} = -4V_o/\pi(4p^2 - 1)$; hence $u_b(t) = 2(V_o/\pi)[1 - (2/3)\cos(2\omega_o t) - (2/15)\cos(4\omega_o t) + \ldots]$

P2.12. $\underline{u}(t) = \Delta t\,\mathrm{sinc}(t\,\Delta\tilde{\nu})[a_o\cos(2\pi t\tilde{\nu}_o) + b_o\sin(2\pi t\tilde{\nu}_o)]$

P2.13. $u(t) = \int_0^\infty d\tilde{\nu}\; A(\tilde{\nu})\cos(\omega t)$, where $A(\tilde{\nu}) = (a/\tau\pi^2\tilde{\nu}^2)[1 - \cos(2\pi\tilde{\nu}\tau)]$, $\Delta\tilde{\nu} = 1/\tau$.

P2.19. a) $u(t) = \frac{2A}{\pi} \sum_{n=0}^{n=+\infty} \frac{1}{2p+1} \sin[\pi(2p+1)\frac{t}{\tau}]$ or $\frac{-iA}{\pi} \sum_{n=-\infty}^{n=+\infty} \frac{1}{2p+1} e^{i(2p+1)\pi t/\tau}$

$\underline{\mathcal{U}}(\tilde{\nu}) = (-iA/\pi)\, \Sigma_{-\infty<p<\infty} [1/(2p+1)]\, \delta[\tilde{\nu} - (p+½)/\tau]$.

Chapter 3

P3.1. $u_{(P)} = F_o/m\omega_o^2$ and $u_{tr} = (F_o/m\omega_o^2)[1-(\omega_o/\tilde{\omega})\, e^{-\beta t} \cos(\tilde{\omega} t - \phi)]$, where $\tilde{\omega} = \sqrt{\omega_o^2 - \beta^2}$ and $\phi = \tan^{-1}(\beta/\tilde{\omega})$. If $t >> 1/\beta$, $u \rightarrow u_{(P)}$. It takes a time of 23 s to reach 0.99 x_{lim}. The necessary work is W=10 kJ.

P3.2. $A_{max} = F_m/2\mu\beta\sqrt{\omega_o^2 - \beta^2}$ for $\Omega_{max} = \sqrt{\omega_o^2 - 2\beta^2}$,

$<P> = \frac{\beta\Omega^2 F_m^2/\mu}{(\Omega^2 - \omega_o^2)^2 + 4\beta^2\Omega^2}$, $<P>_{max} = \frac{F_m^2}{4\beta\mu}$ for $\Omega^2 = \omega_o^2$.

$<P> = ½ <P>_{max}$ for $\Omega_{\pm} = \sqrt{\omega_o^2 + \beta^2} \pm \beta$, hence $\Gamma \equiv \Omega_+ - \Omega_- = 2\beta$.

P3.3. a) $A_{el} = A \cos\phi = (\mu F_m/Z^2)(\omega_o^2/\Omega^2 - 1)$ and $A_{ab} = A \sin\phi = 2\beta\mu F_m/\Omega Z^2$.
c) $A_{el} = 0$ for $\Omega = \omega_o$, while A_{ab} has a maximum $\approx F_m/4\mu\beta\omega_o$ for $\Omega = \omega_o\sqrt{1 - 2\beta/\omega_o} \approx \omega_o - \beta$.

P3.4. a) $\omega_o = 20$ rad.s^{-1}, $\beta = 4.86$ s^{-1} and $b = 1.94$ Kg/s, $\tilde{T} = 0.324$ s. For $t = 10\tilde{T}$, $A = 1.45 \times 10^{-7} A_m$. **b)** $A = 1.73 \times 10^{-2}$ m, $\phi = 2.61$ rad, $<P> = 0.26$ W, $Z = 3.86$ Kg/s and $\cos\phi' = \sin\phi = 0.51$.

P3.5. a) $\ddot{x} + 2\beta\dot{x} = 0$, where $\beta = 10$ s^{-1}, $x = x_o + (x'_o/2\beta)(1 - e^{-2\beta t})$, $\dot{x} = x'_o e^{-2\beta t}$.
b) $\ddot{x} + 2\beta\dot{x} + \omega_o^2 x = 0$, where $\omega_o = \sqrt{K/m} = 20$ rad/s, $x = A\, e^{-\beta t} \cos(\tilde{\omega} t + \phi)$, where $\tilde{\omega} = (\omega_o^2 - \beta^2)^{½} = 17.3$ rad/s, $\tilde{T} = 0.363$ s. **c)** $\ddot{x} + 2\beta\dot{x} = (F_m/m)\cos(\Omega t)$, $x = x_o + (x'_o/2\beta)(1 - e^{-2\beta t}) + [F_m/m(\Omega^2 + 4\beta^2)]\, e^{-2\beta t} + [F_m/m\Omega\Omega']\cos(\Omega t - \phi)$, where $\Omega' = (\Omega^2 + 4\beta^2)^{½}$ and $\tan\phi = -2\beta/\Omega$ $(\pi/2 < \phi < \pi)$. In the steady motion, x approaches $x_o + x'_o/2\beta + [F_m/m\Omega\Omega']\cos(\Omega t - \phi)$. This is an oscillatory motion back and forth near the equilibrium position $x_o + x'_o/2\beta$. **d)** $x = A\cos(\Omega t - \phi)$, where $A = 6.46$ mm and $\phi = 2.64$ rad, $U_{(C)} = ½\, mA^2\Omega^2 \sin^2(\Omega t - \phi)$, $U_{(P)} = ½\, mA^2\omega_o^2 \cos^2(\Omega t - \phi)$, $U_{(T)} = ½\, mA^2\Omega^2 \sin^2(\Omega t - \phi) + ½\, mA^2\omega_o^2 \cos^2(\Omega t - \phi)$, $< U_{(C)} > = ¼\, mA^2\Omega^2 = 0.528 \times 10^{-2}$ J, $< U_{(P)} > = ¼\, mA^2\omega_o^2 = 0.104 \times 10^{-2}$ J, $< U_{(T)} > = 0.633 \times 10^{-2}$ J. $< P > = ½\, A\Omega F_m \sin\phi = \beta m\Omega^2 A^2 = 0.2113$ W, $Z = 10.3$ kg.s^{-1} and $\sin\phi = 0.485$.

P3.6. a) $\ddot{Q} + 2\beta\dot{Q} + \omega_o^2 Q = \mathcal{E}_o/L$ where $\beta = R/2L$ and $\omega_o^2 = 1/LC$.
– If $R^2 < 4L/C$, we find $Q(t) = C\mathcal{E}_o[1 + (\xi - 1)e^{-\beta t}[\cos(\tilde{\omega}t) + (\beta/\tilde{\omega})\sin(\tilde{\omega}t)]$,
$I(t) = (\omega_o^2/\tilde{\omega})C\mathcal{E}_o(1 - \xi)\ e^{-\beta t}\sin(\tilde{\omega}t)$, where $\xi = Q_o/C\mathcal{E}_o$ and $\tilde{\omega} = \sqrt{\omega_o^2 - \beta^2}$
– If $R^2 > 4L/C$, the solution is $Q(t) = C\mathcal{E}_o[1 + (\xi - 1)e^{-\beta t}[\mathrm{ch}(\sigma t) + (\beta/\sigma)\,\mathrm{sh}(\sigma t)]$,
$I(t) = (\omega_o^2/\sigma)C\mathcal{E}_o(1-\xi)e^{-\beta t}\,\mathrm{sh}(\sigma t)$.
b) – If $R^2 < 4L/C$, $U_{(E)} = U_o\{1 - (1 - \xi)e^{-2\beta t}[\cos(\tilde{\omega}t) + (\beta/\tilde{\omega})\sin(\tilde{\omega}t)]\}^{2,}$ where
$U_o = ½\, C\mathcal{E}_o^2$. $U_{(M)} = U_o(\omega_o/\tilde{\omega})^2(1 - \xi)^2\ e^{-2\beta t}\sin^2(\tilde{\omega}t)$,
$U_{(T)} = U_o\{1 + (1 - \xi)^2\ e^{-2\beta t}\ [(\omega_o/\tilde{\omega})^2 - (\beta/\tilde{\omega})^2\cos(2\tilde{\omega}t) + (\beta/\tilde{\omega})\sin(2\tilde{\omega}t)]$
$\quad -2(1 - \xi)e^{-\beta t}[\cos(\tilde{\omega}t) + (\beta/\tilde{\omega})\sin(\tilde{\omega}t)]\}$.
– If $R^2 > 4L/C$, $U_{(E)} = U_o\{1 - (1 - \xi)e^{-2\beta t}[\mathrm{ch}(\sigma t) + (\beta/\sigma)\,\mathrm{sh}(\sigma t)]\}^2$,
$U_{(M)} = U_o(\omega_o/\sigma)^2(1 - \xi)^2 e^{-2\beta t}\,\mathrm{sh}^2(\sigma t)$,
$U_{(EM)} = U_o\{1 + (1 - \xi)^2 e^{-2\beta t}[-(\omega_o/\sigma)^2 + (\beta/\sigma)^2\mathrm{ch}(2\sigma t) + (\beta/\sigma)\,\mathrm{sh}(2\sigma t)]$
$\quad - 2(1 - \xi)e^{-\beta t}[\mathrm{ch}(\sigma t) + (\beta/\sigma)\,\mathrm{sh}(\omega t)]\}$.
c) $\ddot{Q} + 2\beta\dot{Q} + \omega_o^2 Q = (\mathcal{E}_m/L)\cos(\Omega t)$. The steady state solution is
$Q = A\cos(\Omega t - \phi)$ and $I = -\Omega A\sin(\Omega t - \phi)$, where $A = \mathcal{E}_m/\Omega Z$, with
$Z = \sqrt{R^2 + (1/C\Omega - L\Omega)^2}$ and $\tan\phi = 2\beta\Omega/(\omega_o^2 - \Omega^2)$ such that $0 < \phi < \pi$.
$P_J = RI^2 = R\Omega^2 A^2\sin^2(\Omega t - \phi)$, $< P_J > = ½R\mathcal{E}_m^2/Z^2$. If L, C and Ω are fixed, $<P_J>$ has a maximum $\mathcal{E}_m^2/4R$ for $R = |L\Omega - 1/\Omega C|$. If R, L and C are fixed, $<P_J>$ has a maximum $\mathcal{E}_m^2/2R$ for $\Omega = \omega_o$ (resonance).

P3.7. a) $I(t) = I\cos(\omega t - \phi')$, $Q(t) = (I/\omega)\cos(\omega t - \phi' - \pi/2)$, where $I = \mathcal{E}_m/Z$,
$Z = \sqrt{R^2 + (1/C\omega - L\omega)^2}$ and $\tan\phi' = (L\omega - 1/C\omega)/R$ $\quad (-½\pi < \phi' < ½\pi)$.

b) $V_{bm}/\mathcal{E}_m = \sqrt{1 + L/CR^2} = 4.58$

P3.8. a) $I(t) = I_m\cos(\Omega t - \phi')$, $Q(t) = (I_m/\Omega)\cos(\Omega t - \phi' - \pi/2)$, where $I_m = \mathcal{E}_m/Z$, with
$Z = \sqrt{R^2 + (1/C\Omega - L\Omega)^2}$ and $\tan\phi' = (L\Omega - 1/C\Omega)/R$ $(-\pi/2 < \phi' < \pi/2)$.
b) $U_{(EM)} = ½\, L(\mathcal{E}_m/\Omega Z)^2\,[\Omega^2\cos^2(\Omega t - \phi') + \omega_o^2\sin^2(\Omega t - \phi')]$.
c) $dU_G = \mathcal{E}I dt$, $dU_J = I^2 R dt$, $dU_{(EM)} = I[Q/C + L\dot{I}\,]dt$, $dU_G = dU_{(EM)} + dU_J$.
d) $<P_G> = <P_J> = ½R\mathcal{E}_m^2/Z^2$.
e) $P_{max} = ½\,\mathcal{E}_o^2/R$ for $\Omega = \omega_o$. $< P > = ½\,P_{max}$ for $\Omega = \Omega_{(\pm)} = \sqrt{\omega_o^2 - \beta^2} \pm \beta$.

P3.9. $I_m = \mathcal{E}_m[(1 - LC\Omega^2)^2 + (RC\Omega)^2]^{½}/[R^2 + L^2\Omega^2]^{½}$, $\tan\phi_z = (\Omega/R)(L - L^2C\Omega^2 - R^2C)$ $(-\pi/2 < \phi_z < \pi/2)$ and $<P_{(ex)}> = ½\,\mathcal{E}_m I_m\cos\phi_z = ½\,R\mathcal{E}_m^2/[R^2 + L^2\Omega^2]$

P3.10. a) $\underline{x}_1 = (F_m/m\Delta)(\omega_r^2 + \omega_o^2 - \Omega^2)\,e^{i\Omega t}$, $\underline{x}_2 = (F_m/m\Delta)\omega_r^2\,e^{i\Omega t}$, where $\omega_o^2 = g/L$, $\omega_r^2 = K/m$ and $\Delta = (\omega_o^2 + 2\omega_r^2 - \Omega^2)(\omega_o^2 - \Omega^2)$. The normal angular frequencies are $\omega_{(1)} = \omega_o$ and $\omega_{(2)} = \sqrt{\omega_o^2 + 2\omega_r^2}$. The pass-band is $(\omega_{(1)}, \omega_{(2)})$ so that $(|A_2/A_1| > 1)$.
b) If L is large, the system is a low-pass filter with $A_2/A_1 = \omega_r^2/(\omega_r^2 - \Omega^2)$.
c) If $K >> mg/2L$, the system is a high-pass filter with $A_2 = A_1$.

P3.11. b) $x = At + B + \Sigma_{(P)\geq 0}\, a_{2(P)+1} \sin(\Omega_{2(P)+1}t)$, where A and B are arbitrary constants, $\Omega_{2(P)+1} = (2p + 1)\Omega$ and $a_{2(P)+1} = -4F_m/m\pi\Omega^2(2p + 1)^3$.
c) $x_{2(P)+1} = A_{2(P)+1}\cos(\Omega_{2(P)+1}t - \pi/2 - \phi_{2(P)+1})$, where

$$A_{2(P)+1} = \frac{4F_m}{m\pi(2p+1)}\{[\omega_o^2 - \Omega_{2p+1}^2]^2 + 4\beta^2\Omega_{2p+1}^2\}^{-1/2},\ \tan\phi_{2(P)+1} = \frac{2\beta\Omega_{2p+1}}{\omega_o^2 - \Omega_{2p+1}^2}.$$

P3.12. a) Mode 1: $I_{(1)1} = -I_{(1)2} = A_{(1)}\cos(\omega_{(1)}t + \phi_{(1)})$, where $\omega_{(1)} = \omega_o = \sqrt{1/LC}$.
Mode 2: $I_{(2)1} = I_{(2)2} = A_{(2)}\cos(\omega_{(2)}t + \phi_{(2)})$, where $\omega_{(2)} = \sqrt{3}\,\omega_o$. The general solution is: $I_1 = A_{(1)}\cos(\omega_{(1)}t+\phi_{(1)}) + A_{(2)}\cos(\omega_{(2)}t+\phi_{(2)})$, $I_2 = -A_{(1)}\cos(\omega_{(1)}t+\phi_{(1)}) + A_{(2)}\cos(\omega_{(2)}t+\phi_{(2)})$.
b) $\underline{I}_1 = i\underline{\mathcal{E}}\Omega(2\omega_o^2 - \Omega^2)/L\Delta$, $\underline{I}_2 = -i\underline{\mathcal{E}}\Omega\omega_o^2/L\Delta$ and $\underline{I}_3 = i\underline{\mathcal{E}}\Omega/L(\Omega^2 - \omega_{(2)}^2)$, where $\Delta = (\Omega^2 - \omega_{(1)}^2)(\Omega^2 - \omega_{(2)}^2)$.

P3.13. a) $\ddot{x}_1 + 2\beta\,\dot{x}_1 + \omega_o^2 x_1 + \omega_r^2(x_1 - x_2) = (F_m/m)\cos(\Omega t)$,
$\ddot{x}_2 + \omega_o^2 x_2 + \omega_r^2(x_2 - x_1) + \omega_r^2(x_2 - x_3) = 0$, $\ddot{x}_3 + 2\beta\,\dot{x}_3 + \omega_o^2 x_3 + \omega_r^2(x_3 - x_2) = 0$.
c) $A_1 = (F_m/m|\Delta|)\{[(2\omega_r^2 + \omega_o^2 - \Omega^2)(\omega_r^2 + \omega_o^2 - \Omega^2) - \omega_r^4]^2 + 4\beta^2\Omega^2(2\omega_r^2 + \omega_o^2 - \Omega^2)^2\}^{1/2}$,
$A_2 = (F_m\omega_r^2/m|\Delta|)[(\omega_r^2 + \omega_o^2 - \Omega^2)^2 + 4\beta^2\Omega^2]^{1/2}$, $A_3 = F_m\omega_r^4/m|\Delta|$
where $|\Delta|^2 = [(\omega_{(1)}^2 - \Omega^2)^2 + 4\delta_{(1)}^2\Omega^2][(\omega_{(2)}^2 - \Omega^2)^2 + 4\delta_{(2)}^2\Omega^2][(\omega_{(3)}^2 - \Omega^2)^2 + 4\delta_{(3)}^2\Omega^2]$.

P3.14. V increases from 0 to V_i, decreases to V_e, increases to V_i, decreases to V_e, increases to V_i and so on. The period of these oscillations is
$T = \tau_2 \ln[(V_i - \mathcal{E}')/(V_e - \mathcal{E}')] + \tau_1 \ln[(V_e - \mathcal{E})/(V_i - \mathcal{E})]$.

P3.15. $M_1\ddot{\underline{u}}_1 + b\,\dot{\underline{u}}_1 + K_1\underline{u}_1 + K_2(\underline{u}_1 - \underline{u}_2) = F_m\,e^{i\Omega t}$, $M_2\ddot{\underline{u}}_2 + b\,\dot{\underline{u}}_2 + K_2(\underline{u}_2 - \underline{u}_1) = 0$.
$\underline{u}_1 = -i\Omega\underline{z}_2(F_m/\Delta)\,e^{i\Omega t}$, $\underline{u}_2 = -K_2(F_m/\Delta)\,e^{i\Omega t}$, where $\underline{z}_1 = b + i(M_1\Omega - K_1/\Omega)$, $\underline{z}_2 = b + i(M_2\Omega - K_2/\Omega)$ and $\Delta = \Omega^2\underline{z}_1\underline{z}_2 + K_2^2 - i\Omega\,\underline{z}_2 K_2$. The impedance at the entry is $\underline{Z}_1 = \underline{z}_1 + K_2^2/\underline{z}_2\Omega^2 - iK_2/\Omega$.

P3.16. a) $m_1\ddot{\underline{u}}_1 + K(u_1 - u_2) = F_m\,e^{i\Omega t}$, $m_2\ddot{\underline{u}}_2 + K(u_2 - u_1) = 0$,
$\underline{u}_1 = (F_m/\Delta)(K - m_2\Omega^2)\,e^{i\Omega t}$, $\underline{u}_2 = (KF_m/\Delta)e^{i\Omega t}$, where $\Delta = (K - m_1\Omega^2)(K - m_2\Omega^2) - K^2$.
b) The vibration of m_1 is small ($|A_1/A_2| << 1$) if $K \approx m_2\Omega^2$.
c) The vibration of m_2 is small ($|A_2/A_1| << 1$) if $K << m_2\Omega^2$.
d) $|A_2/A_1| = |K/(K - m_2\Omega^2)| < 10^{-2}$ if $K/m_2 < 352\ \mathrm{s}^{-2}$, $L = 2.79$ cm.

P3.17 $A = m\Omega A_T/Z$, where $Z = \sqrt{4m^2\beta^2 + (K/\Omega - m\Omega)^2}$. A is large for a given Ω if K is close to $m\Omega^2$.

Chapter 4

P4.1. $u(x, t) = 10^{-2} \sin(10\pi t - \pi x)$.

P4.2. $u(x, t) = f(t - x/v) \equiv A \sin[2\pi(t - x/v)/T]\, \theta(t - x/v)\, \theta(T/2 - t + x/v)$. At time t, the disturbance has the profile of a sinusoidal function for $v(t - T/2) < x < vt$.

P4.5. a) $u = 4 \sin[\omega(t - \mathbf{e}.\mathbf{r}/v)] = 4 \sin[\omega(t - \sqrt{2}\,(x + y))]$.
b) $u = A \cos(800\,\pi t - 2.962\,x - 5.130\,y - 5.924\,z)$.

P4.7. b) $\mathbf{e} = \mathbf{k}/k = (m\,\mathbf{e}_x - n\,\mathbf{e}_x + p\,\mathbf{e}_z)/k$, $\lambda = 2\pi/k$, $v_{(p)} = \omega/k$ and $k = (m^2 + n^2 + p^2)^{½}$.

P4.8. $\mathbf{u}_{\text{before O}} = (\mathbf{r}/r)\, u_o\, [(\omega/r)\cos[\omega(t + r/v))] + (v/r^2)\sin[\omega(t + r/v)]$
$\mathbf{U}_{\text{after O}} = (\mathbf{r}/r)\, u_o\, [(\omega/r)\cos[\omega(t - r/v))] + (v/r^2)\sin[\omega(t - r/v)]$
r changes sign as the wave passes by the focal point O.

P4.9. d) $v_{(p)} = c/(A + B/\lambda^2) = 1.85\times10^8$ m/s and $v_{(g)} = c(A - B/\lambda^2)/(A + B/\lambda^2)^2 = 1.77\times10^8$ m/s.

P4.11. a) $u(t) = \int_0^\infty d\omega\, [A(\omega)\cos(\omega t) + B(\omega)\sin(\omega t)]$, where $A(\omega) = (a/\pi\omega)\sin(\omega\tau)$, $B(\omega) = (a/\pi\omega)[1 - \cos(\omega\tau)]$. **b)** $u(x, t) = u(0, t - x/v)$. **c)** $v_{(g)} = d\omega/dk = v_o$.
d) $u(x, t) = ½a \cos(\alpha x/v_o)[\varepsilon(t - x/v_o) - \varepsilon(t - x/v_o - \tau)]$, where $\varepsilon(x) = x/|x|$.

P4.12. a) At the entry, $u(0,t) = \mathcal{R}e\, \underline{u}(0,t) = 2\sqrt{\pi}\,\kappa A_m\, e^{-(\kappa v_g t)^2} \cos(\omega_o t)$; hence $\Delta t.\Delta\omega = \Delta t.(d\omega/dk)\,\Delta k = v_{(g)}\,\Delta t.\Delta k \approx 2$. After traveling a distance x, the wave becomes $\underline{u}(x, t) = 2\kappa A_m\, [\pi/(1 - 2it\beta\kappa^2]^{½}\, e^{-i\phi(t)}\, e^{-(x - v_g t)^2/4\xi^2}\, e^{i(\omega_o t - k_o x)}$, where $\phi = \beta(\kappa^2/2\xi^2)\, t(x - v_g t)^2$ and $\xi^2 = ¼\kappa^{-2} + \beta^2\kappa^2 t^2$. Thus, we have the uncertainty relation $\Delta x.\Delta k \approx 2(1 + 4\beta^2\kappa^4 t^2)^{½}$. This relation shows that the wave packet spreads.
b) $u(\mathbf{r}, t) = \kappa_x\kappa_y\kappa_z\, A(\pi)^{3/2}\, e^{-\kappa_x{}^2(x - v_{gx}t)^2/4 - \kappa_y{}^2(x - v_{gy}t)^2/4 - \kappa_z{}^2(x - v_{gz}t)^2/4}\, e^{i(\omega_o t - \mathbf{k}_o.\mathbf{r})}$.

P4.13. $x_{center} = v_{(g)}t$ and $\Delta x = 2\pi/k_1$.

P4.14. a) $y = 2a\cos(2x - 15t)\sin(52x - 215t)$. **b)** $v_m = 7.5$ m/s. **c)** $x_{n+1} - x_n = 1.571$ m.
d) $y = 2a\cos(2x - 15t + \pi/4)\sin(52x - 215t + \pi/4)$. The points, where the amplitude is zero are displaced by a distance of 0.392 m.

P4.15. a) $(a + c/2)$, $b/2$, $b/2$, $c/4$ and $c/4$ respectively.

P4.16. a) $V = I_m R_o \cos(\omega t)\,[1 + a_m \cos(\omega_m t)]$
$= I_m R_o \cos(\omega t) + \tfrac{1}{2} I_m R_o a_m \cos[(\omega - \omega_m)t] + \tfrac{1}{2} I_m R_o a_m \cos[(\omega + \omega_m)t)]$.
b) $V_s = \tfrac{1}{2}\beta(A^2 + a^2) + \alpha a \cos(\omega_m t) + \tfrac{1}{2}\beta A^2 \cos(2\omega t) + \tfrac{1}{2}\beta a^2 \cos(2\omega_m t)$
$+ [\alpha A + 2\beta a A \cos(\omega_m t)] \cos(\omega t)$.
c) $V_e = A \cos(\omega t) + \tfrac{1}{2} a \cos[\omega - \omega_m)t] + \tfrac{1}{2} a \cos[\omega + \omega_m)t]$,
$V_s = (\tfrac{1}{2}\beta A^2 + \tfrac{1}{4}\beta a^2) + \beta A a \cos(\omega_m t) + \tfrac{1}{4}\beta a^2 \cos(2\omega_m t) + \tfrac{1}{2}\alpha a \cos[(\omega - \omega_m)t]$
$+ \alpha A \cos(\omega t) + \tfrac{1}{2} \alpha a \cos[(\omega + \omega_m)t] + (\beta a^2/8) \cos[(2\omega - 2\omega_m)t]$
$+ \tfrac{1}{2}\beta A a \cos[(2\omega - \omega_m)t] + \tfrac{1}{4}\beta(2A^2 + a^2)\cos(2\omega t)$
$+ \tfrac{1}{2}\beta A a \cos[(2\omega + \omega_m)t] + (\beta a^2/8) \cos[(2\omega + 2\omega_m)t]$.

P4.17. a) $V = V_m \cos(\omega t + \alpha)$, where $\omega = \omega_o[1 + a \cos(\omega_m t)]$.
b) $V = V_m \cos(\omega_o t + \alpha) + \tfrac{1}{2} V_m \cos[(\omega_m + \omega_o + a\omega_o)t + \alpha] + \tfrac{1}{2} V_m \cos[(\omega_o + a\omega_o - \omega_m)\, t + \alpha]$
$- \tfrac{1}{2} V_m \cos[(\omega_o + \omega_m)t + \alpha] - \tfrac{1}{2} V_m \cos[(\omega_o - \omega_m)\, t + \alpha\,]$.

P4.18. c) The system becomes a high-pass filter ($\omega > \omega_o$) with the dispersion relationship $k^2 = (\omega^2 - \omega_o^2)/v^2$. The phase velocity is $v_{(P)} = v(1 - \omega_o^2/\omega^2)^{-\frac{1}{2}}$ and the group velocity is $v_g = v(1 - \omega_o^2/\omega^2)^{\frac{1}{2}}$. If $\omega < \omega_o$, the wave is attenuated with $\delta = (\omega_o^2 - \omega^2)^{\frac{1}{2}}/v$. **d)** $\ddot{u} + 2\beta\,\dot{u} - v^2\partial^2_{xx}u + \omega_o^2 u = 0$.

P4.19. a) $\tilde{\nu} = 568.5$ Hz, $\lambda' = \lambda = 0.584$ m. **b)** $\tilde{\nu} = 568.5$ Hz, $\lambda' = \lambda = 0.664$ m.

P4.20. a) $\tilde{\nu}_p = 3\,274$ Hz. **b)** $\tilde{\nu}_2 = 2\,768$ Hz. **c)** As the ambulance moves towards the observer, he perceives two sounds of frequencies $\tilde{\nu}_1 = \tilde{\nu}_4 = 3{,}274$ Hz and, as the ambulance moves away, he perceives two sounds of frequencies $\tilde{\nu}_2 = 2{,}768$ Hz and $\tilde{\nu}_4 = 3{,}274$ Hz, and ,thus, beats of frequency 506 Hz.

P4.21. a) $\Delta\tilde{\nu} / \tilde{\nu}_e = (\tilde{\nu}_r / \tilde{\nu}_e - 1) = -2\,(v^2 t/c)(h^2 + v^2 t^2)^{-\frac{1}{2}}$. **b)** $h = 75$ km.

P4.22. $\Delta\tilde{\nu} = \pm\, 3.72 \times 10^{9\ \mathrm{Hz}}$, i.e. $\Delta\tilde{\nu} / \tilde{\nu} = 6.83 \times 10^{-6}$.

P4.23. a) The wave is conic with a half-angle $\alpha = 30°$. **b)** $v = v_{airplane} \sin\alpha = v_s$.
c) $t = 2.61$ s. The airplane is then at 1.732 km.

Chapter 5

P5.1. $t = 2\sqrt{L/g} = 0.903$ s.

P5.2. a) $v_{(p)} = \omega d/2 \sin^{-1}(\omega/2\omega_o)$, $v_{(g)} = \omega_o d\sqrt{1-\omega^2/4\omega_o^2}$.
b) If $\omega << \omega_o$, $v_{(p)} \to \omega_o d\,[1 - (1/24)(kd)^2 + ...]$, $v_{(g)} \to \omega_o d\,[1 - (1/8)(kd)^2 + ...]$.
d) $\underline{Z} = m\omega_o \exp(ikd/2)$, $Z = m\omega_o$ and $<P> = (F_m^2/2m\omega_o)\sqrt{1-\omega^2/4\omega_o^2}$.

P5.3. a) $v_{(p)} = v_{(g)} = L\sqrt{K/M}$ =15.6 m/s.**b)** $\Delta F = K_l\,\Delta x\,\partial^2_{xx}u$; hence, $m_l\,\ddot{u} = K_l(\partial^2_{xx}u)$.

P5.5. c) $v_s(30°) = 1.05\,v_s(0°)$, $\lambda(30°) = 1.05\,\lambda(0°)$ independently of $\tilde{\nu}$.

P5.6. $u(x,t) = (1/B\omega^2 m_v)\,p_m \sin(\omega t - kx)$, $T(x,t) - \overline{T} = \overline{T}\,(1 - 1/\gamma)(p_m/\,\overline{p}\,)\cos(\omega t - kx)$, $m_v - \overline{m}_v = (1/\gamma)\,\overline{m}_v\,(p_m/\,\overline{p}\,)\cos(\omega t - kx)$, T_m (audible) = 7.7×10^{-8} K, $m_{v,m}$(audible) = 0.91×10^{-9} kg/m³, T_m (ultrasound) = 38.5 K, $m_{v,m}$ (ultrasound) = 0.45 kg/m³.

P5.8. $<U_l> = ½m_l\,\omega^2A^2$. The velocity of the energy = v in a non-dispersive medium.

P5.9. a) $v = 40$ m/s, $u(x, t) = 3\times10^{-3}\sin[\pi(800t - 20x + 1)]$, $\lambda = 0.1$ m and $<P> = 56.8$ W.
b) $Z = \sqrt{Fm_l} = 2$ Kg/s, $U_l = 2.84\cos^2(20\pi x - 800\,\pi t)$ (in joules/m) and $<P> = 56.8$ W.

P5.10. a) $U_{(C)\,v} = ½\,m_v\,\omega^2\,u_o^2\cos^2(\omega t - kx)$, $U_{(P)\,v} = ½m_v\,\omega^2\,u_o^2\cos^2(\omega t - kx)$.
c) $<P> = ½\mathcal{S}\sqrt{Ym_v}\,\omega^2\,u_o^2$. **d)** $F(x, t) = \mathcal{S}\omega\sqrt{Ym_v}\;u_o\sin(\omega t - kx + \pi/2)$.
d) $F(x, t) = \mathcal{S}\omega\,\sqrt{Ym_v}\;u_o\sin(\omega t - kx + \pi/2)$.

P5.11. $p_a(x, t) = -\,\omega m_v v_s u_m \sin(\omega t - kx)$, $\Psi(x, t) = v_s u_m \cos(\omega t - kx)$, $p_m = 0.270$ Pa, $u'_m = 6.28 \times 10^{-4}$ m.s⁻¹, $<U_v = ½m_v\,\omega^2\,u_m^2 = 2.552 \times 10^{-7}$ J.m⁻³, $\mathcal{I} = ½m_v\,v_s\,\omega^2\,u_m^2 = 8.47 \times 10^{-5}$ W/m², $\mathcal{I}_p = 79$ dB.

P5.12. $u(x, t) = u_m\cos(\omega t - kx)$, $\Pi(x, t) = p_m\cos(\omega t - kx + \pi/2)$, where $u_m = 10^{-5}$ m $\omega = 10^3\,\pi$ rad/s, $k = 9.46$ m⁻¹ and $p_m = 13.5$ Pa. **b)** $U_v = m_v\,\omega^2\,u_m^2\sin^2(\omega t - kx)$, $S = v_s\,m_v\,\omega^2\,u_m^2\sin^2(\omega t - kx)$, $\mathcal{I}_s = 0.212$ W.m⁻², $\mathcal{I}_p = 113$ dB, $P = 1.06 \times 10^{-4}$ W.

P5.13. $\mathcal{I}_s = 10^{-9}$ W.m⁻², $P_r = \mathcal{I}_s/v_s = 3.01\times10^{-12}$ Pa, $p_m = \sqrt{2m_v v_s\,\mathcal{I}_s} = 9.26 \times 10^{-4}$ Pa $F = P_r\mathcal{S} =\ 6.02 \times 10^{-15}$ N and 1.20×10^{-14}N respectively.

P5.14. a) $\xi(t) = A\cos(\Omega t - \phi)$, $u(x, t) = A\cos[\Omega(t - x/v_s) - \phi]$.
$P = m_v v_s\mathcal{S}\,\Omega^2A^2\sin^2[\Omega(t - x/v_s) - \phi]$, $f = -\,m_v v_s\mathcal{S}\dot{\xi} \equiv -\,b'\,\dot{\xi}$, $b' = m_v v_s\mathcal{S}$,
$M\ddot{\xi} + (b + b')\,\dot{\xi} + k\xi = F_m\cos(\Omega t)$,

$A = F_m/\Omega\sqrt{(b+b')^2 + (k/\Omega - M\Omega)^2}$, $\tan\phi = \Omega(b+b')/(k - M\Omega^2)$,

$p_a(\mathbf{r}, t) = m_v\, v_s\, \Omega\, A \cos[\Omega(t - x/v_s) - \phi + \pi/2]$, $Z = \sqrt{(b+b')^2 + (k/\Omega - M\Omega)^2}$.

b) $<P> = ½b'\Omega^2 A^2 = ½(b'F_m/Z)^2$, $<P_T> = (b+b')F_m^2/2Z^2$, $\rho = <P>/<P_T> = b'/(b+b')$.

c) $b''_{water} \approx 3\,500\, b'_{air}$. Thus $\rho_{water} >> \rho_{air}$.

P5.15. a) $\underline{u} = -i\Psi_m(R/\omega r^2)[1 + ikr]\, e^{i(\omega t - kr)}\, \mathbf{e}_r$, $\underline{p}_a(r, t) = im_v\Psi_m\omega(R/r)\, e^{i(\omega t - kr)}$, $\mathbf{S} = m_v\Psi_m^2\, \omega^2(R^2/r^3 v_s)\sin^2(\omega t - kr)\,\mathbf{r}$ and $<P> = 2\pi Z_a(R^2/v_s^2)\Psi_m^2\omega^2 = 2\pi Z_a(ru_m\omega)^2 = 2\pi\,(rp_m)^2/Z_a$. b) $P = \pi R^2\, \dot{u}(0,t)\;\; p_a(0,t) = \pi R^2\omega^2\, Z_a\, u_m^2 \sin^2(\omega t - kx)$ and $Z = b + im(\omega - \omega_o^2/\omega) = \pi R^2\, Z_a\,[1 - i\pi R^2\, Z_a/4m\omega]$.

P5.16. c) $\mathbf{u} = A\mathbf{e}_r[(1/\omega r^2)\sin\xi + (1/rv_s)\cos\xi]$ and $p_a = -(A\, m_v\omega/r)\sin\xi$, where $\xi = \omega(t - r/v_s)$.

d) $\mathbf{S} = \mathbf{e}\,(m_v A^2/r^3)[\cos\xi - (r\omega/v_s)\sin\xi]\,[(2v_s^2/\omega r^2 - \omega)\sin\xi + 2(v_s/r)\cos\xi]$.

At a long distance, $P \to 4\pi\,(\omega^2 m_v A^2/v_s)\sin^2\xi$.

P5.17. a) $\mathbf{u}(\mathbf{r}, t) = \mathbf{u}_m\cos(\omega t - \mathbf{k.r})$, $\dot{\mathbf{u}}(\mathbf{r}, t) = \mathbf{u}'_m\cos(\omega t - \mathbf{k.r} + \pi/2)$, $\mathbf{u}'_m = \omega\mathbf{u}_m$; $p_a(\mathbf{r}, t) = p_m\cos(\omega t - \mathbf{k.r} + \pi/2)$, $k = \omega/v_s = \omega\sqrt{m_v/B}$ and $p_m = Bku_m = \sqrt{Bm_v}\;\omega u_m$.

b) $dU_{(C)} = ½\, m_v\, \omega^2\, u_m^2\, d\mathcal{V}\cos^2(\omega t - \mathbf{k.r})$, $dW \approx B\omega\, k^2 u_m^2\sin(2\omega t - 2\mathbf{k.r})\, dl\, d\mathcal{S}\, dt$, $\mathbf{S}(\mathbf{r}, t) = B\omega\,(\mathbf{k.u}_m)\,\mathbf{u}_m\sin^2(\omega t - \mathbf{k.r})$. **c)** In the case of air: p_m (500 Hz) = 27.0 Pa, p_m(50 kHz) = 2.70×10^3 Pa, $\mathcal{I}$(500 Hz) = 0.85 W/m^2, $\mathcal{I}$(50 kHz) = 8.5 kW/m^2. In the case of water: p_m (500 Hz) = 9.4×10^4 Pa, p_m (50 kHz) = 9.4×10^6 Pa. $\mathcal{I}$(500 Hz) = 3.0×10^3 W/m^2, $\mathcal{I}$(50 kHz) = 3.0×10^7 W/m^2.

P5.18. a) $\mathcal{I}_p(r) = \mathcal{I}_p(r_o) - 20\log(r/r_o)$, $r = 1580$ m.

b) $\mathcal{I}_p(r) = 130 - 20\log(r/r_o) - 7\,r$ (r in km), $\mathcal{I}_p$(5 km) = 55 db.

P5.19. At the audibility threshold, $u_m = 1.48 \times 10^{-11}$ m, $\mathcal{I} = 4.64 \times 10^{-13}$ W/m^2 and $\mathcal{I}_p = -3.3$ db. At the pain threshold, $u_m = 2.22 \times 10^{-5}$ m, $\mathcal{I} = 1.04$ W/m^2 and $\mathcal{I}_p = 120$ db.

P5.20. $u_m = 1.09 \times 10^{-8}$ m, $p_m = 2.93 \times 10^{-2}$ Pa, $P = 1.26 \times 10^{-5}$ W, $P' = 147$ W.

P5.22. $v_{(g)} = ½\sqrt{(g/k)\,\mathrm{th}\,(kh)}\,[1 + 2kh/\mathrm{sh}(2kh)]$.

P5.23. a) If the liquid is very deep, $A = (g/\omega)\, s_m\, e^{kz} \equiv A_o\, e^{kz}$. The amplitude is reduced to 20 cm at $z = -4.77$ m. **b)** Setting $u'_m = kgs_m/\omega\,\mathrm{ch}(kh)$, we find $u'_x = u'_m\,\mathrm{ch}[k(z+h)]\cos(\omega t - kx)$, $u'_y = 0$, $u'_z = -u'_m\,\mathrm{sh}[k(z+h)]\sin(\omega t - kx)$.

P5.24. For $\lambda = 1$ cm, for instance, we find that $v_{(p)} = \sqrt{gh} = 7$ m/s, $v_{(p)} = \sqrt{(\lambda g/2\pi)\,\mathrm{th}\,(2\pi h/\lambda)} = 0.125$ m/s, $v_{(p)} = \sqrt{(\lambda g/2\pi + 2\pi F/\lambda m_v)\,\mathrm{th}\,(2\pi h/\lambda)} =$ 0.247 m/s and $v_{(p)} = \sqrt{2\pi F/\lambda m_v} = 0.213$ m/s.

P5.25. a) $\lambda_m = 2\pi\sqrt{F/m_v} = 1.71$ cm, $\tilde{\nu} = 13.5$ Hz. **b)** The relative importance of the surface tension term compared to the gravity term in the expression of the phase velocity is given by the ratio $\rho = (2\pi F/\lambda m_v)/\,(\lambda g/2\pi) = 2.92 \times 10^{-4}/\lambda^2$.

Chapter 6

P6.1. a) $\mathbf{E} = A\cos(\omega t - kz)\,\mathbf{e}_x + B\cos(\omega t - kz)\,\mathbf{e}_y$ (linear polarization).
b) $\mathbf{E} = A\cos(\omega t - kz)\,\mathbf{e}_x + A\cos(\omega t - kz - \pi/2)\,\mathbf{e}_y$ (left-circular polarization).
c) $\mathbf{E} = A\cos(\omega t - kx)\,\mathbf{e}_y + A\cos(\omega t - kx + \pi/2)\,\mathbf{e}_z$ (right-circular polarization).
d) $\mathbf{E} = vA\,[\cos(\omega t - kz - \pi/2)\,\mathbf{e}_x + \cos(\omega t - kz - \pi/2 - \pi/6)\,\mathbf{e}_y]$ (left-elliptic polarization).
e) $\mathbf{E} = vA[\cos(\omega t - ky + \pi)\,\mathbf{e}_z + \cos(\omega t - ky + \pi - \pi/2)\mathbf{e}_x]$ (left-circular polarization).
f) $\mathbf{E} = vA\,[\cos(\omega t + kx + \pi/2)\,\mathbf{e}_z + \cos(\omega t + kx + \pi/2 + \pi/2)\,\mathbf{e}_y\,]$ (right-circular polarization).

P6.2. a) $\mathbf{E} = E\mathbf{e}_1\cos\xi$, $\xi = \omega t - kx\sin\theta - kz\cos\theta + \phi$ and $\mathbf{e}_1 = \cos\theta\,\mathbf{e}_x - \sin\theta\,\mathbf{e}_z$.
b) $\mathbf{E} = E\mathbf{e}_y\cos\xi$. **c)** $\mathbf{E} = E\cos\xi\,\mathbf{e}_1 - E\sin\xi\,\mathbf{e}_y$. **d)** $\mathbf{E} = E\cos\xi\,\mathbf{e}_1 + E\sin\xi\,\mathbf{e}_y$.

P6.3. b) $U_{(E)\,v} = U_{(M)\,v} = \frac{1}{2}\,\varepsilon E_m^2\cos^2(\omega t - \mathbf{k.r})$.

P6.4. a) $E = V/d$, $U_{(E)\,v} = \frac{1}{2}\varepsilon V^2/d^2$ and $U_{(E)} = \frac{1}{2}\varepsilon\mathcal{S}V^2/d = \frac{1}{2}CV^2$. **b)** $B(r') = \frac{1}{2}\mu r' I/\mathcal{S}$ and $\mathbf{S} = (IE/2\pi r)\,\mathrm{n}_l$; hence $P = dE_{(E)}/dt$.

P6.5. $\mathbf{S} = -\,(j^2 r/2\sigma)\,\mathbf{e}_r$.

P6.6. a) $\mathbf{E} = E_m\cos\xi\,[-\cos\theta\,\mathbf{e}_y + \sin\theta\,\mathbf{e}_z]$, $\xi = \omega t - ky\sin\theta + kz\cos\theta + \phi$ and $k = n\omega/c$, $\mathbf{B} = (nE_m/c)\cos\xi\,\mathbf{e}_x$. **b)** $U_{(EM)v} = n^2\varepsilon_o E_m^2\cos^2\xi$, $\mathbf{S} = (n/\mu_o c)E_m^2\cos^2\xi\,\mathbf{e}$, $\mathcal{I} = (n/2\mu_o c)E_m^2$ and $\langle dP/d\mathcal{S}\rangle = \mathcal{I}\cos\theta$, $\langle P_s\rangle = \mathcal{I}\cos\theta$.

P6.7. $E_m = (2\mu_o c\mathcal{I})^{1/2} = 8.7$ kV.m^{-1}, $P = 4\pi r^2\mathcal{I} = 1.26$ MW, $r = 2.6 \times 10^{-8}$ m.

P6.8. a) $U_{(EM)\,v} = \varepsilon E_m^2\cos^2\xi$ and $\mathbf{S} = (E_m^2/\mu v)\cos^2\xi\,\mathbf{e}_z = U_{(EM)\,v}\,v\,\mathbf{e}_z$, $\xi = \omega t - kz$.
b) $E_\gamma = h\omega/2\pi$, $p_\gamma = h\omega/2\pi c$, $\langle N_\gamma\rangle = \pi\varepsilon E_m^2/h\omega$. $N_S = c\,\langle N_\gamma\rangle$, $P_s = p_r = \frac{1}{2}\,\varepsilon E_m^2$.

P6.10. a) $N = 5.86 \times 10^{28}$ electrons/m^3, $\tilde{\nu}_p = 2.17 \times 10^{15}$ Hz.
b) $\delta = (\omega_p - \omega^2)^{1/2}/v = 4.4 \times 10^7$ m^{-1}. $l = 0.023$ μ.
c) $x_{0.5} = 7.9 \times 10^{-3}$ μ, $x_{0.01} = 0.052$ μ, $\lambda < 2\pi c/\omega_p = 0.138$ μ (X-rays).

P6.11. b) $\omega_{(C)} = \sigma/\varepsilon = 6.65 \times 10^{18}$ rad/s. For $\lambda = 0.6$ μm and $l = 1/\delta = 2.94$ nm.

P6.12. a) $\tau = \varepsilon/\sigma$. **c)** In copper, $\tau = 1{,}5 \times 10^{-17}$ s.

P6.13. a) $m\dot{v} + bv + eE = 0$, $v = -eE_o/(b + \mathrm{i}m\omega)\, e^{\mathrm{i}\omega t}$. **b)** $\langle P_v \rangle = \frac{1}{2}\, bNe^2E_o^2/(b^2+m^2\omega^2)$.
c) $b = 3.7 \times 10^{-17}$ N.s/m. **d)** $\omega >> 4 \times 10^{13}$ (i.e. $\lambda << 50$ μm).

P6.14. a) $\mathbf{E} = E_o\, e^{-\delta z - ikz}\, \mathbf{e}_x$, $\mathbf{B} = -(\mathrm{i}/\omega)(\delta + \mathrm{i}k)\, E_o\, e^{-\delta z - ikz}\, \mathbf{e}_y$ and $\mathbf{j} = \sigma E_o\, e^{-\delta z - ikz}\, \mathbf{e}_x$, where $\delta^2 = \frac{1}{2}(\omega/c)^2[(1 + \omega_o^2/\omega^2)^{1/2} + 1]$, $\omega_o = \sigma/\varepsilon_o$ and $k = \sigma\omega\mu_o/2\delta$. **b)** The charge density is very small if $\nabla.\mathbf{j}/\omega\varepsilon_o << \nabla.\mathbf{E}$, i.e. $\omega >> \omega_o$, then $\delta \approx (\omega/c)$ and $k = \frac{1}{2}\,\sigma\mu_o c$.

P6.15. a) $\mathbf{E} = E_m \cos\xi\, \mathbf{e}_x$, $\mathbf{B} = (E_m/c)\cos\xi\, \mathbf{e}_y$ and $\mathbf{S} = (E_m^2/\mu c)\cos^2\xi\, \mathbf{e}_z$, where $\xi = \omega t - kz$ and $k = \omega/c$. **b)** $V = -2(E_m/k \tan\alpha)\sin(\frac{1}{2}kL\sin\alpha)\cos(\omega t - kz_I)$, where z_I is the coordinate of the middle of the antenna. V_m is maximum if $\alpha = 0$, then $V_m = LE_m$.

P6.16. a) $\mathcal{I}(r) = P/\Omega r^2$. **b)** $P_i = 0.428\, \mathcal{S}PR^4/\lambda^2D^4$.

P6.17. $\lambda = 121.5023$ nm, $U_{(C)e} = 1.6349 \times 10^{-18}$ J and $U'_{(C)e} = 2.1799 \times 10^{-18}$ J.

Chapter 7

P7.1. a) $v_1 = 66.9$ m/s, $v_2 = 71.4$ m/s, $u = u_m \cos[2\pi\tilde{\nu}\,(t - x/v_1) + \phi]$, $u_m = 1$ cm, $u' = 0.033\, u'_m \cos[2\pi\tilde{\nu}\,(t + x/v_1) + \phi]$ and $u'' = 1.03\cos[2\pi\tilde{\nu}\,(t - x/v_2) + \phi]$ (in cm).
b) $P = 11.81\sin^2[40\pi(t - x/66.9) + \phi]$, $P' = -1.28 \times 10^{-2}\sin^2[40\pi(t + x/66.9) + \phi]$, $P'' = 11.80\sin^2[40\pi(t - x/71.4) + \phi]$ (in watts). $1 - \mathcal{R}^2 = (v_1/v_2)\,\mathcal{T}^2$.

P7.2 a) The position of the ring is $\underline{y}(t) = \underline{u} + \underline{u}'|_{x=0}$. Its equation of motion gives $\omega(-\omega M + \mathrm{i}b)(\underline{u}_m + \underline{u}'_m) - \mathrm{i}kF(\underline{u}_m - \underline{u}'_m) = 0$; hence $\mathcal{R} = -(M\omega^2 - \mathrm{i}b\omega + \mathrm{i}kF)/(M\omega^2 - \mathrm{i}b\omega - \mathrm{i}kF)$. **c)** There is no reflected wave if $M = 0$ and $\omega b = kF$. Then all the power carried by the wave $P = \omega kF\, u_m^2 \sin^2(\omega t)$ is dissipated by the friction force exerted on the ring.

P7.3. $u'_m/u_m = -0.9994$, $u''_m/u_m = 5.72 \times 10^{-4}$, $\mathcal{I}'/\mathcal{I} = 0.9989$ and $\mathcal{I}''/\mathcal{I} = 1.144 \times 10^{-3}$.

P7.4. a) $\underline{Z}_m = b + iX$, where $X = m\omega - K/\omega$ and $Z_m = \sqrt{b^2+X^2}$. **b)** $\underline{\mathcal{R}} \equiv \underline{u}'_m/\underline{u}_m = \mathcal{R}\, e^{i\phi}$, where $\mathcal{R} = [(Y-b)^2 + X^2]^{½}/[(Y+b)^2 + X^2]^{½}$ and $\phi = \mathcal{A}rg(Y^2 - b^2 - X^2 - 2iYX)$.
c) $P_i = Y\omega^2 u_m{}^2 \sin^2[\omega(t - x/v_s)]$, $P_r = -Y\mathcal{R}^2\omega^2 u_m{}^2 \sin^2[\omega(t + x/v_s) + \phi]$,
$P_m = P_i - |P_r||_{x=L} = ½Y\omega^2 u_m{}^2\{1 - \mathcal{R}^2 - \cos(2\omega t) + \mathcal{R}^2\cos(2\omega t + 2\phi)\}$.
d) $X = 15.39$ N.s.m^{-1}, $Z_a = 429.3$ N.s.m^{-3}, $Y = 2.1463$ N.s.m^{-1}, $Z_m = 18.3$ N.s.m^{-1}, $\mathcal{R} = 0.8813$, $<P_i> = 1.059$ mW, $<P_r> = -0.82$ mW and $<P_m> = 0.236$ mW.

P7.6. a) $\sin\theta'' = 4.5180 \sin\theta$, $\theta_L = 12.78°$.
b) $\mathcal{R}_{a\text{-}w} = u'_{m,a}/u_{m,a} = (3496\cos\theta - \Delta)/(3496\cos\theta + \Delta)$, where $\Delta = (1 - 20.412\sin^2\theta)^{½}$, $\mathcal{R}_{w\text{-}a} = u'_{m,w}/u_{m,w} = (\cos\theta - 3496\,\Delta')/(\cos\theta + 3496\,\Delta')$, where $\Delta' = (1 - 0.04897\sin^2\theta)^{½}$.

P7.7. a) $\underline{u}$(air) = $\underline{u}$(glass) and p_a(air) = p_a(glass).
b) $\mathcal{R} = (\mu^2 - 1)\sin(k_2 e)/\Delta$ and $\mathcal{T} = 2i\mu e^{ike}/\Delta$, where $\Delta = (\mu^2+1)\sin(k_2 e) - 2i\mu\cos(k_2 e)$.
$e = (2p+1)\lambda_2/4$. The thinner plate has a thickness $e = \lambda_2/4$.

P7.8. b) $\theta'' = 40°.6$. $\mathcal{R}_{//} = 0.066$; $\mathcal{T}_{//} = 0.702$; $\mathcal{R}_\perp = -0.337$; $\mathcal{T}_\perp = 0.6$; $\alpha' = 79°$ and $\alpha'' = 40°$.

P7.9. a) $\mathcal{I}' = \mathcal{R}^2\mathcal{I}$, $\mathcal{I}'' = \mathcal{T}^2(Z_1/Z_2)\mathcal{I}$. Energy conservation on the surface requires that $(1 - \mathcal{R}^2)\cos\theta' = \mathcal{T}^2(Z_1/Z_2)\cos\theta''$, which is verified by $\mathcal{R}_{//}$ and $\mathcal{T}_{//}$ and by $\mathcal{R}_\perp$ and $\mathcal{T}_\perp$, and, thus, in the case of any wave. **b)** $I_3 = 16n^2 I_1/(1+n)^4 I_1 = 0.9216\, I_1$. After four plates, the intensity becomes $I_f = (0.9216)^4\, I_1 = 0.7214\, I_1$.

P7.10. $f_{R//} = 8.464 \times 10^{-3}$, $f_{T//} = 0.9915$, $f_{R\perp} = 9.199 \times 10^{-2}$ $f_{T\perp} = 0.9081$. The degrees of polarization are $\mathcal{P}' = -0.8315$ and $\mathcal{P}'' = 0.0439$.

P7.11. $\mathcal{R}_{//} = 0$, $\mathcal{T}_{//} = 1/n$, $\mathcal{R}_\perp = -(n^2-1)/(n^2+1)$ and $\mathcal{T}_\perp = 2/(n^2+1)$; hence $f_R = ½(n^2 - 1)^2/(n^2+1)^2$ and $f_T = ½(n^4 + 6n^2 + 1)^2/(n^2+1)^2$, thus $f_R + f_T = 1$.

P7.12. We set $\Delta_1 = [(n^2 - \sin^2\theta]^{½}$ and $\Delta_2 = [(1 - n^2\sin^2\theta]^{½}$.
a) $\mathcal{I}'_\perp(\theta)/\mathcal{I} = [\Delta_1 - \cos\theta]^4/(n^2-1)^2$, $\mathcal{I}'_\perp(0) = 0.04\,\mathcal{I}$, $\mathcal{I}'_\perp(\pi/2) = \mathcal{I}$, $\mathcal{I}'_\perp(\pi/4) = 0.092\,\mathcal{I}$.
b) $\mathcal{I}'_{//}(\theta)/\mathcal{I} = [\Delta_1 - n^2\cos\theta]^2/[\Delta_1 + n^2\cos\theta]^2$, $\mathcal{I}'_{//}(0)/\mathcal{I} = 0.04$, $\mathcal{I}'_{//}(\pi/2)/\mathcal{I} = 1$, $\mathcal{I}'_{//}(\theta_B)/\mathcal{I} = 0$, $\mathcal{I}'_{//}(\pi/4)/\mathcal{I} = 8.47 \times 10^{-3}$. **c)** $\mathcal{I}'_\perp/\mathcal{I} = (n\cos\theta - \Delta_2)^2/(n\cos\theta + \Delta_2)^2$, $\mathcal{I}'_{//}/\mathcal{I} = (n\Delta_2 - \cos\theta)^2/(n\Delta_2 + \cos\theta)^2$. $\mathcal{I}'_\perp/\mathcal{I}$ increases from $(n-1)^2/(n+1)^2 = 0.04$ for $\theta = 0$ to 1 for $\theta = i_L = 41.8°$. $\mathcal{I}'_{//}/\mathcal{I} = \mathcal{R}_{//}{}^2$ decreases from $(n-1)^2/(n+1)^2 = 0.04$ for $\theta = 0$, vanishes for $\theta = \theta'_B = \arctan(1/n)$ and then increases to 1 for $\theta = i_L$.

P7.13. a) $\mathbf{k} = k(\sin\theta\, \mathbf{e}_x + \cos\theta\, \mathbf{e}_z)$ and $\mathbf{k}' = k'(\sin\theta'\, \mathbf{e}_x - \cos\theta'\, \mathbf{e}_z)$
$\underline{\mathbf{B}} = \underline{B}_m e^{i(\omega t - \mathbf{k.r})}\, \mathbf{e}_y$, $\underline{\mathbf{E}} = v_1 \underline{B}_m (\cos\theta\, \mathbf{e}_x - \sin\theta\, \mathbf{e}_z)\, e^{i(\omega t - \mathbf{k.r})}$
$\underline{\mathbf{B}}' = -\underline{B}'_m\, e^{i(\omega t - \mathbf{k}'.\mathbf{r})}\, \mathbf{e}_y$, $\underline{\mathbf{E}}' = v_1 \underline{B}'_m (\cos\theta'\, \mathbf{e}_x + \sin\theta'\, \mathbf{e}_z)\, e^{i(\omega t - \mathbf{k}'.\mathbf{r})}$.
b) $\underline{\mathbf{B}}'' = \underline{B}''_m\, e^{-\delta z}\, e^{i(\omega t - px)}\, \mathbf{e}_y$, $\mathbf{E}'' = -(v_2^2/\omega)\underline{B}''_m e^{-\delta z}\, (i\delta\, \mathbf{e}_x + p\, \mathbf{e}_z)\, e^{i(\omega t - px)}$, where $p = (\omega/v_1)\sin\theta$ and $\delta = (2\pi/\lambda_2)\sqrt{\sin^2\theta/\sin^2 i_L - 1}$. For $\theta = 60°$, $\delta = 3.6\, \lambda_2^{-1}$.

c) $\underline{B}'_m = -\underline{B}_m e^{2i\phi}$ and $B''_m = 2\underline{B}_m \cos\phi\; e^{i\phi}$, $\tan\phi = (n_1/n_2\cos\theta)\sqrt{\sin^2\theta/\sin^2 i_L - 1}$.
d) $\langle\mathbf{S}\rangle = (c/2\mu_o n_1)\, B_m^2 (\sin\theta\, \mathbf{e}_x + \cos\theta\, \mathbf{e}_z)$, $\langle\mathbf{S}'\rangle = (c/2\mu_o n_1)\, B_m^2(\sin\theta\, \mathbf{e}_x - \cos\theta\, \mathbf{e}_z)$ and $\langle\mathbf{S}''\rangle = (2cn_1/\mu_o n_2^2)\, B_m^2\, e^{-2\delta z}\, \cos^2\phi\, \sin\theta\, \mathbf{e}_x$.

P7.14. a) $i_L \approx 41.8°$, $\delta = 5.927\ \mu^{-1}$, $l \approx 1\,\mu$m. **b)** Setting $\eta = \omega\cos\theta/\delta c n$, we find $\mathcal{R} = -(1/\Delta)(1+\eta^2)\,\mathrm{sh}(L\delta)$ and $\mathcal{T} = 2i\eta/\Delta$, where $\Delta = (1-\eta^2)\,\mathrm{sh}(L\delta) + 2i\,\mathrm{ch}(L\delta)$. Numerical values: $|\mathcal{R}| = 1 - 1.31\times 10^{-5}$ and $|\mathcal{T}| = 5.13\times 10^{-3}$.

P7.16. a) $\mathcal{R} = (k-p)/(k+p)$ and $\mathcal{T} = 2k/(k+p)$, where $p^2 = (\omega^2 - \omega_o^2 - 2i\beta\omega)/v_2^2$ and $k = \omega/v_1$. **b)** If the medium is reactive ($\beta = 0$) and $\omega > \omega_o$, $\mathcal{R} < 1$ and if $\omega < \omega_o$, $|\mathcal{R}| = 1$ (total reflection). **c)** If the medium is purely resistive ($\beta \neq 0$) but $\omega_o = 0$, $|\mathcal{R}| < 1$ (a certain fraction of the wave penetrates in the medium).

P7.17. a) $\mathbf{k} = k(\sin\theta\, \mathbf{e}_x + \cos\theta\, \mathbf{e}_z)$ and $\mathbf{k}' = k'(\sin\theta'\, \mathbf{e}_x - \cos\theta'\, \mathbf{e}_z)$
$\underline{\mathbf{E}} = \underline{E}_m\, \mathbf{e}_y\, e^{i(\omega t - \mathbf{k.r})}$, $\underline{\mathbf{B}} = (\underline{E}_m/v_1)(-\cos\theta\, \mathbf{e}_x + \sin\theta\, \mathbf{e}_z)\, e^{i(\omega t - \mathbf{k.r})}$
$\underline{\mathbf{E}}' = \underline{E}'_m\, \mathbf{e}_y\, e^{i(\omega t - \mathbf{k}'.\mathbf{r})}$, $\underline{\mathbf{B}}' = (\underline{E}'_m/v_1)(\cos\theta'\, \mathbf{e}_x + \sin\theta'\, \mathbf{e}_z)\, e^{i(\omega t - \mathbf{k}'.\mathbf{r})}$
b) $\underline{\mathbf{E}}'' = \underline{E}''_m\, \mathbf{e}_y\, e^{-\delta z}\, e^{i(\omega t - px - hz)}$, $\underline{\mathbf{B}}'' = (\underline{E}''_m/\omega)(-\underline{s}\, \mathbf{e}_x + p\, \mathbf{e}_z)\, e^{-\delta z}\, e^{i(\omega t - px - hz)}$.
c) $$\frac{\underline{E}'_m}{\underline{E}_m} = \frac{\mu_2\omega\cos\theta - \mu_1 v_1(h - i\delta)}{\mu_2\omega\cos\theta + \mu_1 v_1(h - i\delta)},\quad \frac{\underline{E}''_m}{\underline{E}_m} = \frac{2\mu_2\omega\cos\theta}{\mu_2\omega\cos\theta + \mu_1 v_1(h - i\delta)}.$$

P7.18. a) $\mathbf{E} = \mathbf{E}_m\, e^{i(\omega t - \mathbf{q.r})}$, $\mathbf{B} = \mathbf{B}_m\, e^{i(\omega t - \mathbf{q.r})}$, $q^2 = \omega^2\mu_o\underline{\varepsilon} = \omega^2\mu_o(\varepsilon_2 - i\sigma_c/\omega)$, $\mathbf{q.E}_m = 0$, $\mathbf{B}_m = (1/\omega)\, \mathbf{q}\times\mathbf{E}_m$ and $\mathbf{E}_m = -(1/\mu_o\underline{\varepsilon}\omega\, (\mathbf{q}\times\mathbf{B}_m)$. **b)** No, since the ratio $\sin\theta''/\sin\theta$ is not constant. **c)** $f_R > 95\%$ if $\omega < 4.07\times 10^{17}$ rad/s. For $\lambda = 500$ nm, $|\mathcal{R}| = 0.97$ and $\phi = 3.11$ rad. The intensity is reduced to less than 10% if $x > 3.0$ nm.

Chapter 8

P8.1. b) $\mathcal{I}(P)$ is minimum if $PA - PB = (p + \frac{1}{2})\lambda$. For a given p, the locus of P is a branch of a hyperbola of foci A and B. They intersect Ox at $x = (2p+1)\,\lambda/4$, i.e. ± 0.415 m, ± 1.245 m and ± 2.075 m. **c)** $\tilde{\nu} = 166\, p$, where p is a half-integer, i.e. $\tilde{\nu} = 83 - 249 - 415$ Hz, etc.

P8.2. The bright points lie in planes that are parallel to the mirror at $x_{(P)} = 0.1(2p + 1)$ (in μm), i.e. $x = 0.1$ μm, 0.3 μm and 0.5 μm, etc.

P8.3. a) $\lambda = 600$ nm (order = 2) and $\lambda = 429$ nm (order 3). **b)** $\lambda = 476 \pm 12$ nm. **c)** $\delta p = ad/b\lambda = 0.16$.

P8.4. a) $\mathcal{I} = 4\mathcal{I}_o[1 + \cos(2\pi x/i)\cos(\pi x\,\delta\lambda/i\lambda)]$, where $i = \lambda D/d$. **b)** $\mathcal{C} = |\cos(\pi x d\,\delta\lambda/D\lambda^2)|$. One has only to measure i and the positions x_o, where $\mathcal{C} = 0$; then $\lambda = id/D$ and $\delta\lambda = i\lambda/2x_o$.

P8.5. a) 2×10^3 fringes. In fact, the number of sharp fringes is much less. **b)** $\phi_B = 2\pi Rd/\lambda(a^2 + d^2/4)^{1/2}$, $\Delta x = iRd/\lambda(a^2 + d^2/4)^{1/2}$, $\Delta x < i/4$ if $a > 4$ m. **c)** $\Delta i/i = \Delta\lambda/\lambda$. If $\Delta i/i < 1/4$, we can observe $\Delta\lambda = 0.1$ μ. **d)** $x \approx 10^3\, i$.

P8.6. a) $i = \lambda f/d$. If we use sunlight, the fringe of order 0 is white. **b)** If the container is paced in front of the lower slit, the fringes are displaced upward; $\Delta\phi = 2\pi(n - 1)l/\lambda$, $n = 1 + N\lambda/l = 1.000291$. **c)** $\delta x = i$ if $dn = \lambda/l$.

P8.7. b) $\phi_R = 4\pi\alpha x/\lambda - \pi$. **c)** Each wavelength (color) gives its own bright rings, whose radii are $r_{(P)} \approx \sqrt{(p+1/2)\lambda R}$. The centre is dark and surrounded by the various colors.

P8.8. a) x (first minimum) = 3 mm, x (first secondary maximum) = 4.5 mm. **b)** The principal maximums coincide at $x = 0$, the first minimum for $\lambda = 0.5$ μ is at $x' = 2.5$ mm and its first secondary maximum is at $x' = 3.75$ mm. **c)** The first minimum is at $x = 57.74 \pm 0.46$ cm.

P8.9. The direction of the first minimum is given by $\sin\theta_{min} = 0.61\,\lambda/R$. Thus, $\theta_{min} = 54.10°$ without the cone and $\theta_{min} = 11.68°$ using the cone.

P8.10. a) $\sin\theta_1 = 3\lambda/2d$ and $\sin\theta_2 = 5\lambda/2d$. **b)** $\lambda = 0.26$ cm. **c)** $2\theta = 0.2282$ rad.

P8.11. b) The radius of the diffracting centers is $R \approx 7$ μm.

P8.12. We find maximums of amplitude and intensity for $\phi = 2\pi(d/\lambda)\sin\theta = \pi/2$, $5\pi/6$, $7\pi/6$ and $4\pi/3$. The corresponding amplitudes are $A = \sqrt{2}\,a$, $A \approx 1.03\,a$, and $A = \sqrt{2}\,a$.

P8.13. a) $E_1 = E_m \sin[\omega(t - r_1/c)]$, $E_2 = E_m \sin[\omega(t - r_2/c) + \phi]$, $E = 2E_m \cos(\pi\Delta r/\lambda - \phi/2)\sin[\omega(t - r_1/2c - r_2/2c) + \phi/2]$, where $\Delta r = r_2 - r_1 \approx d\sin\theta$. $\mathcal{I}$ is maximum if $\sin\theta_{max} = (p + \phi/2\pi)\lambda/d$ and minimum if $\sin\theta_{min} = (p + \phi/2\pi + 1/2)\lambda/d$.

To modify the direction of the maximums, we have only to modify the phase ϕ. **b)** If $d = \lambda/2$ and $\phi = 0$, $\mathcal{I}(\theta) = 4\mathcal{I}_o\cos^2(\frac{1}{2}\pi \sin\theta)$; hence $\mathcal{I}(0) = 4\mathcal{I}_o$, $\mathcal{I}(30°) = 2\mathcal{I}_o$, $\mathcal{I}(60°) = 0.175\,\mathcal{I}_o$ and $\mathcal{I}(90°) = 0$. **c)** If $d = \lambda/2$ and $\phi = 30°$, $\mathcal{I}$ is maximum for $\theta_{max} = 9.6°$. **d)** With N antennas, $\mathcal{I} = N^2\mathcal{I}_o \sin^2(N\phi_T/2)/N^2 \sin^2(\phi_T/2)$ where $\phi_T = \phi - 2\pi(d/\lambda)\sin\theta$. The width of the principal maximum is $\delta\theta \approx 2\lambda/Nd$.

P8.14. a) $\mathcal{I} = \mathcal{I}_o \sin^2(2\phi)/\sin^2(\phi/2)$, where $\mathcal{I}_o = A^2/2\mu_o c$, $\phi = 2\pi(d/\lambda)\sin\theta$.
b) $\mathcal{I}$ has principal maximums $\mathcal{I}_{max} = 16\,\mathcal{I}_o$ for $\phi = 2p\pi$, i.e. $\sin\theta = p\lambda/d$, where $p = 0, 1, \ldots$ ($p < d/\lambda$). Between the principal maximums of orders p and $p+1$, $\mathcal{I}$ vanishes for $\phi_{min} = 2p\pi + \pi/2$, $2p\pi + \pi$, $2p\pi + 3\pi/2$ and it has two secondary maximums for $\phi \approx 2p\pi + 3\pi/4$ and $2p\pi + 5\pi/4$ with an intensity $\approx 1.17\,\mathcal{I}_o$.
c) $x_{(P)} = f\tan\theta_{(P)} = pf\lambda/(d^2 - p^2\lambda^2)^{1/2}$, i.e. 0, 11.4 cm, 24.8 cm, 44.7 cm and 97.0.

P8.15. a) $d = 2.50$ µm. **b)** $\lambda' = 484$ nm. **c)** $N > 3650/p$, where the order is p.
d) 411.0 nm $< \lambda <$ 450.8 nm.

P8.16. a) $d = 0.564$ nm. **b)** $d = \lambda/2\sin\theta_1 = 0.554$ *nm*. **c)** $\Delta d/d = \Delta\theta_1/\tan\theta_1$. **d)** To have a reflection on plane 1, the crystal must be turned by −39.39°; −33.73°; −27.95; −21.99°; −15.74°; −9.09°; −1.83°; 6.43°; 16.6° or 32.80°. To have a reflection on plane 2, it must be turned by −7.20°or 22,74. To have a reflection on plane 3, it must be turned by −5.81°; 7.49°; 22.54° or 42.54°.

Chapter 9

P9.1. a) $\tilde{\nu} = 7.96$ Hz, $\lambda = 12.57$ cm, $v = 100$ cm/s. **b)** $x_p(\text{nodes}) = 6.28\,p$ (in cm), $x'_p(\text{antinodes}) = 6.28(p+\frac{1}{2})$. **c)** The nodes would be displaced by $\delta x = -(\phi/4\pi)\lambda$.

P9.2. a) $v = 200$ m/s. **b)** T_n is the time it takes for the signal to go there and back n times along the string. **c)** $F = 158.84$ N.

P9.3. a) $u_n(x, t) = A_n \sin(n\pi x/L)\sin(n\pi vt/L + \phi_n)$, $v = \sqrt{F/m_l}$,
$U_{(C)l}(x, t) = \frac{1}{2}F(n\pi/L)^2 A_n^2 \sin^2(n\pi x/L)\cos^2(n\pi vt/L + \phi_n)$,
$U_{(P)l}(x, t) = \frac{1}{2}F(n\pi/L)^2 A_n^2 \cos^2(n\pi x/L)\sin^2(n\pi vt/L + \phi_n)]$.
c) $P(x, t) = -FA_n^2(n\pi/L)^2 v\cos(n\pi x/L)\sin(n\pi vt/L + \phi_n)\sin(n\pi x/L)\cos(n\pi vt/L + \phi_n)$.
d) $U_n = \pi^2 n^2 A_n^2 F/4L$.

P9.4. a) If $\mathbf{u}^{\pm} = u_o[\cos(\omega t)\,\mathbf{e}_x + \cos(\omega t \pm \pi/2)\,\mathbf{e}_y]\sin(kz)$, $(\mathbf{u}^{\pm})^2 = u_o^2\sin^2(kz)$. The motion of the point z of the string is thus circular, of radius $u_o|\sin(kz)|$.
b) $\mathbf{u}^{\pm}_i = u_o[\cos(\omega t - kz)\mathbf{e}_x + \cos(\omega t - kz \pm \pi/2)\,\mathbf{e}_y]$ and

$\mathbf{u}^{\pm}{}_{r} = u_o\,[\cos(\omega t + kz + \pi)\,\mathbf{e}_x + \cos(\omega t + kz + \pi \pm \pi/2)\mathbf{e}_y]$ imply that $\mathbf{u}^{\pm} = \mathbf{u}^{\pm}{}_{i} + \mathbf{u}^{\pm}{}_{r} = 2u_o\,[\cos(\omega t - \pi/2)\mathbf{e}_x + \cos(\omega t - \pi/2 \pm \pi/2)\mathbf{e}_y]\sin(kz)$. **c)** The flux of the angular momentum of the incident wave and that of the reflected wave are $\mathbf{L}^{\pm}{}_{i} = v\mathbf{l}_{i}{}^{\pm}{}_{,l} = v m_l \mathbf{u}^{\pm}{}_{i} \times \dot{\mathbf{u}}^{\pm}{}_{i} = \mp\,\omega u_o{}^2 m_l \mathbf{e}_z$ and $\mathbf{L}^{\pm}{}_{r} = -v\mathbf{l}_{r}{}^{\pm}{}_{,l} = -v m_l \mathbf{u}^{\pm}{}_{i} \times \dot{\mathbf{u}}^{\pm}{}_{i} = \pm\omega u_o{}^2 m_l \mathbf{e}_z$ Thus, the total flux of angular momentum is zero at each point of the string.

P9.5. a) $u_n(x, t) = A_n \cos(n\pi x/L)\, g_n(t)$ and $p_{a,n}(x, t) = (Bn\pi/L)\, A_n \sin(n\pi x/L)\, g_n(t)$, where $g_n(t) = \sin(n\pi v_s t/L + \phi_n)$. $\lambda_n = 2L/n$, $\tilde{\nu}_n = n v_s/2L = 553\, n$, i.e. 553 Hz, 1107 Hz … **b)** $v_s = \sqrt{\gamma RT/M} = v_o\sqrt{T/T_o} \Rightarrow \delta\tilde{\nu}_n/\tilde{\nu}_n = \delta v_s/v_s = ½\delta T/T = 1.8 \times 10^{-3}$. **c)** The standing waves in the pipe with one hole are the even-n modes. In the pipe with two holes the modes are $n = 2^p 3^q$, where p and q are positive integers.

P9.6. a) We must have $\tilde{\nu}_n = n v_s/2L$. **b)** $v_{rod}/v_s = 2L/\lambda_{air}$.

P9.7. Write $f(t) = \Sigma_n\, a_n \exp(in\omega t)$, where $\omega = 2\pi/T$. The boundary conditions $u(L, t) = 0$ and $u(0,t) = f(t)$ are satisfied if $\underline{u} = \Sigma_n A_n \sin[n\omega(L - x)/v_s]\exp(in\omega t)$, where the amplitudes are $A_n = a_n/\sin(n\omega L/v_s)$. They become infinite if $n\omega L/v_s = p\pi$, i.e. $n\omega$ equal to one of the normal angular frequencies $p\pi v_s/L$ of the pipe. In fact, it remains finite if we take into account the energy dissipation.

P9.8. a) The modes are $u_{m,n} = A_{m,n} \sin(m\pi x/L)\sin(n\pi y/L)\sin(2\pi\tilde{\nu}_{m,n} t + \phi_{m,n})$, where m and n are positive integers and $\tilde{\nu}_{m,n} = (v_{el}/2L)(m^2 + n^2)^{½}$ with $v_{el} = \sqrt{F_l/m_s}$. The first frequencies are $\tilde{\nu}_{0,1} = \tilde{\nu}_{1,0} = v_{el}/2L$ (degeneracy 2), $\tilde{\nu}_{1,1} = v_{el}/\sqrt{2}L$ (non-degenerate), $\tilde{\nu}_{0,2} = \tilde{\nu}_{2,0} = v_{el}/L$ (degeneracy 2), $\tilde{\nu}_{1,2} = \tilde{\nu}_{2,1} = \sqrt{5}\, v_{el}/2L$ (degeneracy 2), $\tilde{\nu}_{2,2} = \sqrt{2}\, v_{el}/L$ (non-degenerate). **b)** $U_{m,n} = ½\pi^2 m_s L^2 \tilde{\nu}_{m,n}{}^2 A_{m,n}{}^2$.

P9.9. $u_{m,n,p}(x, y, z, t) = A_{m,n,p} \sin(m\pi x/L)\sin(n\pi y/L)\sin(p\pi z/L)\cos(\omega_{m,n,p} t + \phi_{m,n,p})$ $\tilde{\nu}_{m,n,p} = (v/2L)(m^2 + n^2 + p^2)^{½}$: $\tilde{\nu}_{1,0,0} = \tilde{\nu}_{0,1,0} = \tilde{\nu}_{0,0,1} = v/2L$ (degeneracy 3), $\tilde{\nu}_{1,1,0} = \tilde{\nu}_{1,0,1} = \tilde{\nu}_{0,1,1} = v/\sqrt{2}\,L$ (degeneracy 3) and $\tilde{\nu}_{1,1,1} = \sqrt{3}\, v/2L$ (degeneracy 1).

P9.11. The general vibration of the string is a superposition of the modes: $u(x, t) = \Sigma_{n\geq 1} \sin(n\pi x/L)\,[B_n \cos(n\pi vt/L) + C_n \sin(n\pi vt/L)]$. The initial conditions, $u(x, 0) = 0$ and $\dot{u}(x,0) = v(x)$, where $v(x) = v_o$ for $L/3 < x < L/2$ give

$$u(x, t) = (Lv_o/\pi^2 v)[\sin(\pi x/L)\sin(\pi vt/L) + (1/4)\sin(2\pi x/L)\sin(2\pi vt/L) - (2/9)\sin(3\pi x/L)\sin(3\pi vt/L) + \ldots].$$

P9.12. a) $\tilde{\nu}_n = 50\,n$, $\lambda_n = 2/n$ (in meters), $u_n(x, t) = A_n \sin(n\pi\, x/L)\sin(n\pi\, vt/L + \phi_n)$. **b)** $u(x, t) = A\,[\sin(\pi x/L)\sin(\pi vt/L + \phi) + \sin(2\pi\, x/L)\sin(2\pi\, vt/L + \phi)]$. By fingering the string at the point $L' = 25$ cm, the modes that are not eliminated are $u_{4p}(x, t)$, where $p = 1, 2, 3\ldots$ **c)** $U_n = \pi^2 F\, n^2 A_n^2/4L$.

P9.14. a) If $\omega > \omega_{m,n}$, k^2 is positive and k is real. Then, the wave propagates without attenuation, but if $\omega < \omega_{m,n}$, $k = \pm\, i\delta$, where $\delta = (\omega/v_s)\sqrt{\omega_{m,n}{}^2/\omega^2 - 1}$.

b) $\tilde{\nu}_{m,n} = (v_s/2a)\sqrt{m^2 + n^2}$, thus, $\tilde{\nu}_{0,1} = \tilde{\nu}_{1,0} = 1.66$ kHz. If $\tilde{\nu} = 2$ kHz, $k = 21.11\ \text{m}^{-1}$ and $\lambda = 0.298$ m; while, in the open space $\lambda_\infty = 0.166$ m. **c)** If $\tilde{\nu} = 1$ kHz, $\delta = 25.08\ \text{m}^{-1}$. After traveling a distance of 30 cm, the intensity is divided by 3.4×10^6.

P9.15. b) If the line is short-circuited, $\underline{Z}_c = 0$; then $\omega_n = \pi n v/D$ and
$V_n = A\cos(\omega_n t + \phi)\cos(n\pi z/D)$ and $I_n = (A/Z_l)\sin(\omega_n t + \phi)\sin(n\pi z/D)$
If the line is open ($\underline{Z}_c = \infty$); then $\omega_n = \pi(n + ½)v/D$ and
$V_n = A\cos[\omega_n t + \phi]\cos[(n + ½)\pi z/D]$ and $I_n = (A/Z_l)\sin[\omega_n t + \phi]\sin[(n + ½)\pi z/D]$.

P9.16. a) $Z = (L_l/C_l)^{½}$ and $v = \omega/k = (L_l C_l)^{-½}$. $V_T = V_m\cos(\omega t - kz) + V'_m\cos(\omega t + kz + \alpha)$
$I_T = (V_m/Z)\cos(\omega t - kz) - (V'_m/Z)\cos(\omega t + kz + \alpha)$. The short-circuit at $z = z_o$ implies that $V_T(z_o, t) = 0$; hence $V'_m = V_m$ and $\alpha = \pi - 2kz_o$, then
$V_T(z, t) = 2V_m\sin(\omega t - kz_o)\sin(kz - kz_o)$, $I_T(z, t) = 2(V_m/Z)\cos(\omega t - kz_o)\cos(kz - kz_o)$.
b) V_T and I_T are the same as if the cable is open at $z_o - \lambda/4$; then $I_T(z_o - \lambda/4, t) = 0$.

P9.17. a) $\mathbf{B} = (A/v)\sin(\omega t)\sin(kz)\mathbf{e}_y$, $v = \omega/k$. **b)** $U_{(E),v} = ½\varepsilon A^2\cos^2(\omega t)\cos^2(kz)$,
$U_{(M),v} = ½\varepsilon\, A^2\sin^2(\omega t)\sin^2(kz)$, **c)** $\mathbf{S} = (A^2/4\mu v)\sin(2\omega t)\sin(2kz)\,\mathbf{e}_z$.
d) $U_{(EM)}$ (half wave zone) $= (1/16)\,\varepsilon A^2\lambda S$.

P9.19. $\mathbf{B} = (\mu_o/2\pi\rho)\, I_m\, e^{i(\omega t - kz)}\,\mathbf{e}_\varphi$, $\mathbf{E} = (c\mu_o/2\pi\rho)\, I_m\, e^{i(\omega t - kz)}\,\mathbf{e}_\rho$,
$\mathbf{S} = (\mu_o c/4\pi^2\rho^2)\, I_m^2\cos^2(\omega t - kz)\,\mathbf{e}_z$, $P = (\mu_o c/4\pi)\, I_m^2\ln(b/a)$.

P9.20. a) $\tilde{\nu}_{m,n} = ½c(m^2/a^2 + n^2/b^2)^{½}$. $\tilde{\nu}_{1,0} = 7.5$ GHz, $k = (\omega/c)\,\gamma_{1,0}$,
where $\gamma_{1,0} = (1 - \omega_{10}{}^2/\omega^2)^{½}$. The fields are $\mathbf{E}(\text{TE}) = E_o\sin(\pi x/a)\, e^{i(\omega t - kz)}\,\mathbf{e}_y$ and
$\mathbf{B}(\text{TE}) = (E_o/c)\gamma_{1,0}[-\sin(\pi x/a)\,\mathbf{e}_x + i(\pi/ak)\cos(\pi x/a)\,\mathbf{e}_z]e^{i(\omega t - kz)}$.
$\tilde{\nu}_{1,0} = 7.5$, $\tilde{\nu}_{2,0} = \tilde{\nu}_{0,1} = 15.0$, $\tilde{\nu}_{1,1} = 16.8$, $\tilde{\nu}_{2,1} = 21.2$, $\tilde{\nu}_{3,0} = 22.5$ (in GHz).
b) $\omega^2 = \omega_{1,0}{}^2 + c^2k^2$; thus $v_{(p)} = \omega/k = c(1 - \tilde{\nu}_{1,0}{}^2/\tilde{\nu}^2)^{-½} = 4.54 \times 10^8$ m/s,
$v_{(g)} = d\omega/dk = c(1 - \tilde{\nu}_{1,0}{}^2/\tilde{\nu}^2)^{½} = 1.98 \times 10^8$ m/s.

P9.21. $\omega_{p,q} = (\pi v/a)(p^2 + q^2)^{1/2}$, $a_{min} = \lambda/2n = 0.1667$ μm. $v_{(p)} = \omega/k = v/\gamma_{p,q}$, $v_{(g)} = v\gamma_{p,q} \Rightarrow \gamma_{0,1} = \gamma_{1,0} = 0.866$, $v_{(p)} = 2.31 \times 10^8$ m/s, $v_{(g)} = 1.73 \times 10^8$ m/s.

P9.22. $\tilde{\nu}_{m,j}(TM) = (c/2\pi R)\, x_{m,j}$ where $x_{0,1} = 2.40$, $x_{1,1} = 3.83$, $x_{2,1} = 5.14$, $x_{0,2} = 5.52$ $\tilde{\nu}'_{m,j}(TE) = (c/2\pi R)x'_{m,j}$, where $x'_{1,1} = 1.841$, $x'_{2,1} = 3.054$, $x'_{0,1} = 3.832$, $x'_{1,2} = 5.331$. For $R = 1.5$ cm, $\tilde{\nu}'_{1,1}(TE) = 5.856$, $\tilde{\nu}_{0,1}(TM) = 7.650$ and $\tilde{\nu}'_{2,1}(TE) = 9.714$ (in GHz). The bandwidth is $\tilde{\nu}_{0,1}(TM) - \tilde{\nu}'_{1,1}(TE) = 1.794$ GHz. This corresponds to 1.8×10^5 telephone calls or 287 TV channels.

APPENDICES

Appendix A

Mathematical Review

In this appendix, we designate the natural or Napierian logarithm by $\ln(x)$, the hyperbolic functions by $\sinh(x)$, $\cosh(x)$ and $\tanh(x)$. The inverse functions are designated by $\sinh^{-1}(x)$, $\cosh^{-1}(x)$, $\tanh^{-1}(x)$, $\sin^{-1}(x)$, $\cos^{-1}(x)$ and $\tan^{-1}(x)$, instead of Arcsin x, etc. The unit of angles is the radian. To simplify the notations, the partial derivatives (or derivatives) are designated by $\partial_x f$ for $\partial f/\partial x$, $\partial^2_{xy} f$ for $\partial^2 f/\partial x\,\partial y$ etc.

A.1. Expansion formulas

Taylor series near $x = 0$ and $x = a$ are respectively

$$f(x) = f(0) + \partial_x f\,|_{x=0}\, x/1! + \partial^2_x f\,|_{x=0}\, x^2/2! + \partial^3_x f\,|_{x=0}\, x^3/3! + \ldots$$

$$f(x) = f(a) + \partial_x f\,|_{x=a}\, (x-a)/1! + \partial^2_x f\,|_{x=a}\, (x-a)^2/2! + \partial^3_x f\,|_{x=a}\, (x-a)^3/3! + \ldots$$

Examples:

$$(1+x)^n = 1 + n\,x + n(n-1)\,x^2/2! + n(n-1)(n-2)\,x^3/3! + \ldots \qquad (|x| < 1)$$

$$(1+x)^{-1} = 1 - x + x^2 - x^3 + x^4 \ldots \qquad (|x| < 1)$$

$$(1+x)^{1/2} = 1 + (1/2\times 1!)\,x - (1/2^2\times 2!)x^2 + (1\times 3/2^3\times 3!)x^3 \ldots \qquad (|x| < 1)$$

$$(x+y)^n = x^n + n\,x^{n-1}y + n(n-1)\,x^{n-2}y^2/2! + n(n-1)(n-2)\,x^{n-3}y^3/3! + \ldots \qquad (|y| < |x|)$$

A.2. Logarithmic, exponential and hyperbolic functions

$y = e^x = 1 + x/1! + x^2/2! + x^3/3! + ...,$ $\qquad \ln(1 + x) = x - x^2/2! + x^3/3! - ... \quad (x^2 < 1)$

$\sinh(x) = ½(e^x - e^{-x}) = x/1! + x^3/3! + x^5/5!\ ...,$ $\qquad \text{coch}(x) = ½(e^x + e^{-x}) = 1 + x^2/2! + x^4/4!\ ...$

$\tanh(x) = \sinh(x)/\text{coch}(x) = x - x^3/3 + 2x^5/15...,$ $\qquad \cosh^2(x) - \sinh^2(x) = 1$

$\sinh(x \pm y) = \sinh x \cosh y \pm \cosh x \sinh y,$ $\qquad \cosh(x \pm y) = \cosh x \cosh y \pm \sinh x \sinh y$

$\cosh(2x) = 2\cosh^2 x - 1 = 2\sinh^2 x + 1,$ $\qquad \sinh(2x) = 2 \sinh x \cosh x$

$$\tanh(x \pm y) = \frac{\tanh x \pm \tanh y}{1 \pm \tanh x . \tanh y}, \qquad \tanh(2x) = \frac{2\tanh x}{1 + \tanh^2 x}$$

$$\cosh x + \cosh y = 2\cosh\frac{x+y}{2}\cosh\frac{x-y}{2}, \quad \cosh x - \cosh y = 2\sinh\frac{x+y}{2}\sinh\frac{x-y}{2}$$

$$\sinh x \pm \sinh y = 2\sinh\frac{x \pm y}{2}\cosh\frac{x \mp y}{2}, \quad \tanh x \pm \tanh y = \frac{\text{sh}(x \pm y)}{\text{ch } x \text{ ch } y}$$

$$\sinh^{-1} x = \ln(x + \sqrt{x^2+1}) = \cosh^{-1}\sqrt{x^2+1} = x - \frac{1}{2\times 3}x^3 + \ldots = \Sigma_{k\geq 0}\frac{(-1)^k (2k)!\, x^{2k+1}}{(2k+1)\times 2^{2k}(k!)^2} \quad (x^2 < 1)$$

$$= \ln 2x - \Sigma_{k\geq 1}(-1)^k \frac{(2k)!}{2k \times 2^{2k}(k!)^2 x^{2k}} \quad (x^2 > 1)$$

$$\cosh^{-1} x = \ln(x + \sqrt{x^2-1}) = \sinh^{-1}\sqrt{x^2-1} = \ln 2x - \Sigma_{k\geq 1}\frac{(2k)!}{2k \times 2^{2k}(k!)^2 x^{2k}} \quad (x^2 > 1)$$

$$\tanh^{-1} x = ½ \ln[(1+x)/(1-x)] = x + x^3/3 + x^5/5 + \ldots = \Sigma_{k\geq 0}\, x^{2k+1}/(2k+1) \quad (x^2 < 1)$$

A.3. Trigonometric functions

$\sin x = x/1! - x3/3! + x^5/5!\ ...,$ $\qquad \cos x = 1 - x^2/2! + x^4/4!\ ...$

$\cos x = \sin(\pi/2 - x) = -\cos(\pi - x),$ $\qquad \sin x = \cos(x - \pi/2) = \sin(\pi - x)$

$\tan x = \sin x/\cos x = x + x^3/3 + 2x^5/15 + \ldots|_{|x|<\pi/2},$ $\qquad \cos^2(x) + \sin^2(x) = 1$

$\sin(x \pm y) = \sin x \cos y \pm \cos x \sin y,$ $\quad \cos(x \pm y) = \cos x \cos y \mp \sin x \sin y$

$\sin(2x) = 2 \sin x \cos x,$ $\quad \cos(2x) = 2 \cos^2 x - 1 = 1 - 2 \sin^2 x$

$\cos x + \cos y = 2 \cos \frac{x+y}{2} \cos \frac{x-y}{2},$ $\quad \cos x - \cos y = 2 \sin \frac{x+y}{2} \sin \frac{y-x}{2}$

$\sin x + \sin y = 2 \sin ½(x+y) \cos ½(x-y),$ $\quad \sin x - \sin y = 2 \sin ½(x-y) \cos ½(x+y)$

$\tan(x \pm y) = (\tan x \pm \tan y)/(1 \mp \tan x.\tan y),$ $\quad \tan x \pm \tan y = \sin(x \pm y)/\cos x \,.\, \cos y$

$$\sin^{-1}(x) = \frac{\pi}{2} - \cos^{-1}(x) = \frac{1}{1}x + \frac{1}{2.3}x^3 + \frac{1.3}{2.4.5}x^5 + \ldots = \Sigma_{k\geq 0} \frac{(2k)!\, x^{2k+1}}{(2k+1)\times 2^{2k}(k!)^2} \qquad (x^2 < 1)$$

$\tan^{-1}(x) = x - x^3/3 + x^5/5 - x^7/7 + \ldots|_{\text{if } |x|<1}$ or $\pi/2 - 1/x + 1/3x^3 - 1/5x^5 + \ldots|_{\text{if } |x| \geq 1}$

A.4. Integrals

$\partial_x \int dx\, f(x) = f(x),$ $\quad \int dx\, \partial_x f \cdot g(x) = f(x)\, g(x) - \int dx\, \partial_x g \,.\, f(x)$

$\int dx\, x^{-1} = \ln(x)$ $\quad \int dx\, x^n = x^{n+1}/(n+1) \qquad (n \neq -1),$

$\int dx\, \partial_x f / f = \ln f(x),$ $\quad \partial_x \int_{a(x)}^{b(x)} dt\, f(t) = f[b(x)]\, \partial_x b - f[a(x)] \partial_x a$

$\int dx\, e^{ax} = e^{ax}/a,$ $\quad \int dx\, x\, e^{ax} = (ax-1)\, e^{ax}/a^2$

$\int dx \sinh x = \cosh x,$ $\quad \int dx \cosh x = \sinh x$

$\int dx \tanh x = \ln(\cosh x),$ $\quad \int dx/\tanh x = \ln(\sinh x)$

$\int dx \sin x = -\cos x,$ $\quad \int dx \cos x = \sin x$

$\int dx \tan x = -\ln(\cos x),$ $\quad \int dx/\tan x = \ln(\sin x)$

$\int dx \sin^2 x = ½x - ¼ \sin 2x,$ $\quad \int dx \cos^2 x = ½x + ¼ \sin 2x$

$\int dx \sin^3 x = ⅓ \cos^3 x - \cos x,$ $\quad \int dx \cos^3 x = \sin x - ⅓ \sin^3 x$

$\int dx/\sin x = \ln(\tan x/2)$, $\int dx/\cos x = \ln[\tan(x/2 + \pi/4)]$

$\int dx/\sin^2 x = -\text{cotan}\, x$, $\int dx/\cos^2 x = \int dx(1 + \tan^2 x) = \tan x$

$\int dx\, x \sin x = \sin x - x \cos x$, $\int dx\, x \cos x = \cos x + x \sin x$

$\int dx\, x \sin^2 x = \frac{1}{4} x^2 - \frac{1}{4} x \sin 2x - \frac{1}{8} \cos 2x$, $\int dx\, x \cos^2 x = \frac{1}{4}x^2 + \frac{1}{4}x \sin 2x + \frac{1}{8} \cos 2x$

$\int dx\, x^2 \sin^2 x = \frac{1}{6} x^3 - (\frac{1}{4} x^2 - \frac{1}{8}) \sin 2x - \frac{1}{4} x \cos 2x$

$\int dx\, x^2 \cos^2 x = \frac{1}{6} x^3 + (\frac{1}{4} x^2 - \frac{1}{8}) \sin 2x + \frac{1}{4} x \cos 2x$

$\int dx\, (a + b \cos x)^{-1} = 2(a^2 - b^2)^{-1/2} \tan^{-1}[f/(a+b)]$ (where $f = |a^2 - b^2|^{1/2} \tan(x/2)$ if $a^2 > b^2$)

or $(b^2 - a^2)^{-1/2} \ln[(f+a+b)/(f-a-b)]$ (where $f = |b^2 - a^2|^{1/2} \tan(x/2)$ if $a^2 < b^2$)

$$\int dx \sin mx \sin nx = \frac{\sin x(m-n)}{2(m-n)} - \frac{\sin x(m+n)}{2(m+n)} \qquad (\text{if } m^2 \neq n^2)$$

$$\int dx \cos mx \cos nx = \frac{\sin x(m-n)}{2(m-n)} + \frac{\sin x(m+n)}{2(m+n)} \qquad (\text{if } m^2 \neq n^2)$$

$$\int dx \sin mx \cos nx = -\frac{\cos x(m-n)}{2(m-n)} - \frac{\cos x(m+n)}{2(m+n)} \qquad (\text{if } m^2 \neq n^2)$$

$$\int dx\, e^{ax} \cos mx = \frac{e^{ax}}{m^2 + a^2}(a \cos mx + m \sin mx)$$

$$\int dx\, e^{ax} \sin mx = \frac{e^{ax}}{m^2 + a^2}(a \sin mx - m \cos mx)$$

$\int dx\, (a^2 + x^2)^{-1} = (1/a) \tan^{-1}(x/a)$ $\int dx (a^2 - x^2)^{-1} = (1/2a) \ln[(a+x)/(a-x)]$

$\int dx (x^2 \pm a^2)^{-1/2} = \ln[x + (x^2 \pm a^2)^{1/2}]$ $\int dx (a^2 - x^2)^{-1/2} = \sin^{-1}(x/a) = \pi/2 - \cos^{-1}(x/a)$

Integrals involving $v = a + bx + cx^2$ (with $w = b + 2cx$ and $\Delta = b^2 - 4ac$)

$\int dx\, w^{-1/2} = (1/c)\, w^{1/2}$ $\int dx\, w^{-3/2} = -(1/c)\, w^{-1/2}$

$\int dx/v = \Delta^{-1/2} \ln[(w - \Delta^{1/2})/(w + \Delta^{1/2})]|_{\text{if } \Delta > 0}$ or $-2/w|_{\text{if } \Delta = 0}$ or $= 2|\Delta|^{-1/2} \tan^{-1}(w|\Delta|^{-1/2})|_{\text{if } \Delta < 0}$

$$\int \frac{dx}{x\sqrt{v}} = -\frac{1}{\sqrt{a}} \ln \frac{bx+2a+2\sqrt{av}}{x} \Big|_{\text{if } a>0} \quad \text{or} = \frac{1}{\sqrt{-a}} \sin^{-1}\left(\frac{bx+2a}{x\sqrt{\Delta}}\right)\Big|_{\text{if } a<0 \text{ and } \Delta>0}$$

$$= \frac{1}{\sqrt{-a}} \tan^{-1} \frac{bx+2a}{2\sqrt{-av}} \Big|_{\text{if } a<0} \quad \text{or} \quad = -\frac{1}{\sqrt{a}} \sinh^{-1}\left(\frac{bx+2a}{x\sqrt{-\Delta}}\right)\Big|_{\text{if } a>0 \text{ and } \Delta<0}$$

$$= -\frac{1}{\sqrt{a}} \tanh^{-1}\left(\frac{bx+2a}{2\sqrt{av}}\right)\Big|_{\text{if } a>0} \quad \text{or} \quad = -\frac{1}{\sqrt{a}} \ln\left(b+2\frac{a}{x}\right)\Big|_{\text{if } a>0 \text{ and } \Delta=0}$$

$$= -\frac{2\sqrt{bx+cx^2}}{bx} \Big|_{\text{if } a=0 \text{ and } b\neq 0}$$

$$\int \frac{dx}{\sqrt{v}} = c^{-1/2} \ln(w+2\sqrt{cv})\Big|_{\text{if } c>0} \quad \text{or} \quad = c^{-1/2} \sinh^{-1}[w|\Delta|^{-1/2}]\Big|_{\text{if } c>0 \text{ and } \Delta<0}$$

$$= -|c|^{-1/2} \sin^{-1}(w\Delta^{-1/2})\Big|_{\text{if } c<0 \text{ and } \Delta>0} \quad \text{or} \quad = c^{-1/2} \ln w \Big|_{\text{if } c>0 \text{ and } \Delta=0}$$

$$\int dx\, x\sqrt{v} = \frac{v^{3/2}}{3c} - \frac{bw\sqrt{v}}{8c^2} + \frac{b\Delta}{16c^2} \int \frac{dx}{\sqrt{v}}, \qquad \int \frac{dx}{v^{n+1}} = -\frac{w}{n\Delta v^n} + (2-4n)\frac{c}{n\Delta} \int \frac{dx}{v^n} \quad (\text{if } n\neq 0)$$

$$\int dx \frac{x}{v} = \frac{1}{2c} \ln v - \frac{b}{2c} \int \frac{dx}{v}, \qquad \int \frac{dx}{xv} = \frac{1}{2a} \ln \frac{x^2}{v} - \frac{b}{2a} \int \frac{dx}{v}$$

$$\int dx \frac{x^2}{v} = \frac{x}{c} - \frac{b}{2c^2} \ln v + \frac{b^2-2ac}{2c^2} \int \frac{dx}{v}, \qquad \int dx\sqrt{v} = \frac{w\sqrt{v}}{4c} - \frac{\Delta}{8c} \int \frac{dx}{\sqrt{v}}$$

$$\int dx \frac{\sqrt{v}}{x} = \sqrt{v} + a\int \frac{dx}{x\sqrt{v}} + \frac{b}{2} \int \frac{dx}{\sqrt{v}}, \qquad \int dx \frac{\sqrt{v}}{x^2} = -\frac{\sqrt{v}}{x} + \frac{b}{2} \int \frac{dx}{x\sqrt{v}} + c \int \frac{dx}{\sqrt{v}}$$

$$\int dx\, v^{-3/2} = -2w/\Delta v^{1/2}, \qquad \int dx\, xv^{-3/2} = 2(2a+bx)/\Delta v^{1/2}$$

$$\int dx \frac{x^2}{v^{3/2}} = 2\frac{x(2ac-b^2)-ab}{c\Delta\sqrt{v}} + \frac{1}{c} \int \frac{dx}{\sqrt{v}}, \qquad \int dx \frac{x}{\sqrt{v}} = \frac{\sqrt{v}}{c} - \frac{b}{2c} \int \frac{dx}{\sqrt{v}}$$

$$\int dx \frac{x^2}{\sqrt{v}} = \frac{2cx-3b}{4c^2} \sqrt{v} + \left(\frac{3b^2}{8c^2} - \frac{a}{2c}\right) \int \frac{dx}{\sqrt{v}}$$

Some finite integrals

$\int_0^\infty \frac{dx}{x} \tan(x) = \frac{1}{2}\pi,$ $\qquad \int_0^\infty \frac{dx}{x} e^{-px} \cos(mx) = \infty$

$\int_0^u \frac{dx}{x} \sin(x) = \mathrm{Si}(u),$ $\qquad \int_0^\infty \frac{dx}{x} e^{-px} \sin(mx) = \tan^{-1} \frac{m}{p} \qquad (m > 0)$

$\int_0^\infty dx\, e^{-a^2x^2} = \frac{\sqrt{\pi}}{2a},$ $\qquad \int_0^\infty dx\, x^{2n} e^{-a^2x^2} = \frac{(2n-1)!!}{2^{n+1}} \frac{\sqrt{\pi}}{a^{2n+1}}$

$\int_0^\infty dx\, x^{2n+1} e^{-a^2x^2} = \frac{n!}{2a^{2n+2}},$ $\qquad \int_0^\infty dx\, x^n e^{2qx-a^2x^2} = \frac{\sqrt{\pi}}{2^{n-1}a^3} \frac{d^{n-1}}{dq^{n-1}} \left(q e^{q^2/a^2}\right)$

A.5. Complex numbers

A *complex number* (represented by an underlined symbol) is the association of two real numbers:

$$\underline{z} = x + \mathrm{i}y \quad \text{where} \quad \mathrm{i}^2 = -1$$

x and y are respectively the *real part* and the *imaginary part* of $\underline{z}$:

$$x = \mathcal{Re}\, \underline{z} \quad \text{and} \quad y = \mathcal{Im}\, \underline{z}$$

The *complex conjugate* of $\underline{z} \equiv x + \mathrm{i}y$ is $\underline{z}^* \equiv x - \mathrm{i}y$. A function of $\underline{z}$ is defined by the same Taylor series as for a real variable. For instance,

$$e^{\mathrm{i}\phi} = 1 + \frac{1}{1!}(\mathrm{i}\phi) + \frac{1}{2!}(\mathrm{i}\phi)^2 + \frac{1}{3!}(\mathrm{i}\phi)^3 + \frac{1}{4!}(\mathrm{i}\phi)^4 \ldots$$
$$= 1 - \frac{1}{2!}\phi^2 + \frac{1}{4!}\phi^4 \ldots + \mathrm{i}\left(\frac{1}{1!}\phi - \frac{1}{3!}\phi^3 + \ldots\right) = \cos\phi + \mathrm{i}\sin\phi \text{ (Euler equation).}$$

We deduce that

$$\sinh(x) = -\mathrm{i}\sin(\mathrm{i}x), \quad \cosh(x) = \cos(\mathrm{i}x), \quad \tanh(x) = -\mathrm{i}\tan(\mathrm{i}x)$$

$$\sinh(\mathrm{i}x) = \mathrm{i}\sin(x), \quad \cosh(\mathrm{i}x) = \cos(x), \quad \tanh(\mathrm{i}x) = \mathrm{i}\tan(x).$$

A complex number $\underline{z} = x + \mathrm{i}y$ may be represented by a point of coordinates x and y in the (x, y) (Argand diagram, Figure A1). We may use the polar coordinates ρ and ϕ for this point; then,

$$\underline{z} = x + \mathrm{i}y = \rho \cos \phi + \mathrm{i}\, \rho \sin \phi = \rho\, e^{\mathrm{i}\phi}$$

with the relations

$$x = \rho \cos\phi, \qquad y = \rho \sin\phi, \qquad \tan \phi = y/x, \qquad \rho = |\underline{z}| = (x^2 + y^2)^{1/2}$$

ρ is the *modulus* of $\underline{z}$ and ϕ is its *argument* or its *phase* (determined up to 2π).

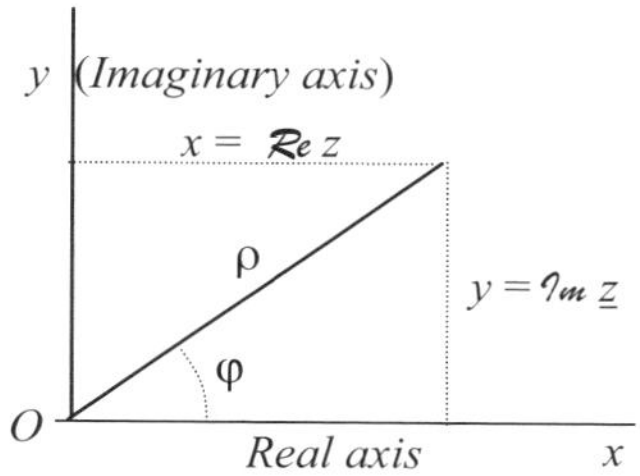

Figure A.1. *Argand diagram for complex numbers*

The sum of two complex numbers is:

$$\underline{z}_1 + \underline{z}_2 = (x_1 + \mathrm{i}y_1) + (x_2 + \mathrm{i}y_2) = (x_1 + x_2) + \mathrm{i}\,(y_1 + y_2)$$

The product and the ratio of two complex numbers may easily be evaluated using the exponential form:

$$\underline{z}_1\,\underline{z}_2 = (\rho_1\, e^{\mathrm{i}\varphi_1})(\rho_2\, e^{\mathrm{i}\varphi_2}) = \rho_1\rho_1\; e^{\mathrm{i}(\varphi_1+\varphi_2)}, \quad \frac{\underline{z}_1}{\underline{z}_2} = \frac{\rho_1 e^{\mathrm{i}\varphi_1}}{\rho_2 e^{\mathrm{i}\varphi_2}} = \frac{\rho_1}{\rho_2}\, e^{\mathrm{i}(\varphi_1-\varphi_2)}.$$

If the algebraic form is used, we find

$$\underline{z}_1\,\underline{z}_2 = (x_1 + \mathrm{i}y_1)(x_2 + \mathrm{i}y_2) = x_1 x_2 - y_1 y_2 + \mathrm{i}(x_1 y_2 + x_2 y_1),$$

$$\frac{\underline{z}_1}{\underline{z}_2} = \frac{x_1 + \mathrm{i}y_1}{x_2 + \mathrm{i}y_2} = \frac{x_1 + \mathrm{i}y_1}{x_2 + \mathrm{i}y_2}\,\frac{x_2 - \mathrm{i}y_2}{x_2 - \mathrm{i}y_2} = \frac{x_1x_2 + y_1y_2 + \mathrm{i}(x_2y_1 - x_1y_2)}{{x_2}^2 + {y_2}^2}.$$

A.6. Vector analysis in Cartesian coordinates

To write coordinate summations, we designate the Cartesian coordinates by $x_1 \equiv x$, $x_2 \equiv y$ and $x_3 \equiv z$ and, for instance, the partial derivative with respect to x_1 by $\partial_1 f \equiv \partial_x f \equiv \partial f/\partial x$. The unit vectors of the axes are $\mathbf{e}_x$, $\mathbf{e}_y$ and $\mathbf{e}_z$; then, for instance

$$\Sigma_i \partial_i f\, \mathbf{e}_i \equiv (\partial f/\partial x_1)\mathbf{e}_1 + (\partial f/\partial x_2)\mathbf{e}_2 + (\partial f/\partial x_3)\mathbf{e}_3 \equiv (\partial f/\partial x)\mathbf{e}_x + (\partial f/\partial y)\mathbf{e}_y + (\partial f/\partial z)\mathbf{e}_z.$$

A vector $\mathbf{V}$ may be specified by its Cartesian components such that

$$\mathbf{V} = V_x\, \mathbf{e}_x + V_y\, \mathbf{e}_y + V_z\, \mathbf{e}_z = \Sigma_i\, V_i\, \mathbf{e}_i.$$

The sum of two vectors $\mathbf{V}$ and $\mathbf{W}$ is:

$$\mathbf{V} + \mathbf{W} = (V_x + W_x)\, \mathbf{e}_x + (V_y + W_y)\, \mathbf{e}_y + (V_z + W_z)\, \mathbf{e}_z = \Sigma_i\, (V_i + W_i)\mathbf{e}_i.$$

If $\mathbf{V}$ and $\mathbf{W}$ have magnitudes V and W, and make an angle θ, their *scalar product* is:

$$\mathbf{V}.\mathbf{W} = VW \cos\alpha = V_x W_x + V_y W_y + V_z W_z = \Sigma_i\, V_i\, W_i.$$

The *cross products* of the base vectors are $\mathbf{e}_y \times \mathbf{e}_z = \mathbf{e}_x$, $\mathbf{e}_z \times \mathbf{e}_x = \mathbf{e}_y$ and $\mathbf{e}_x \times \mathbf{e}_y = \mathbf{e}_z$). Therefore, the cross product (also called the *vector product*) of $\mathbf{V}$ and $\mathbf{W}$ is:

$$\mathbf{V} \times \mathbf{W} = (V_x\, \mathbf{e}_x + V_y\, \mathbf{e}_y + V_z\, \mathbf{e}_z) \times (W_x\, \mathbf{e}_x + W_y\, \mathbf{e}_y + W_z\, \mathbf{e}_z) = \begin{bmatrix} \mathbf{e}_x & \mathbf{e}_y & \mathbf{e}_z \\ V_x & V_y & V_z \\ W_x & W_y & W_z \end{bmatrix}$$

$$= (V_y W_z - V_z W_y)\, \mathbf{e}_x + (V_z W_x - V_x W_z)\, \mathbf{e}_y + (V_x W_y - V_y W_x)\, \mathbf{e}_z.$$

The *double cross product* of three vectors is

$$\mathbf{U} \times (\mathbf{V} \times \mathbf{W}) = (\mathbf{U}.\mathbf{W})\, \mathbf{V} - (\mathbf{U}.\mathbf{V})\, \mathbf{W}.$$

The vector operator of derivation ∇ (called *del* or *nabla*) is

$$\nabla = \Sigma_i\, \mathbf{e}_i\, \partial_i \equiv \mathbf{e}_x\, \partial_x + \mathbf{e}_y\, \partial_y + \mathbf{e}_z\, \partial_z.$$

The *gradient* of a scalar field f is a vector field:

$$\nabla f \equiv \Sigma_i\, \partial_i f\, \mathbf{e}_i \equiv \partial_x f\, \mathbf{e}_x + \partial_y f\, \mathbf{e}_y + \partial_z f\, \mathbf{e}_z, \quad .$$

It verifies the properties

$$df = d\mathbf{r}.\nabla f, \qquad \nabla(fg) = f\nabla g + g\nabla f.$$

The *divergence* of a vector field **V** is the scalar field:

$$\nabla.\mathbf{V} = \Sigma_i\, \partial_i V_i \equiv \partial_x V_x + \partial_y V_y + \partial_z V_z.$$

The *curl* of a vector field **V** is the vector field:

$$\nabla \times \mathbf{V} = (\partial_y V_z - \partial_z V_y)\, \mathbf{e}_x + (\partial_z V_x - \partial_x V_z)\, \mathbf{e}_y + (\partial_x V_y - \partial_y V_x)\, \mathbf{e}_z.$$

The *Laplacian* is the operator:

$$\Delta = \nabla^2 = \partial^2_{xx} + \partial^2_{yy} + \partial^2_{zz}.$$

Here are some useful vector analysis relationships

$$\nabla \times (\nabla f) = 0 \qquad \nabla.(\nabla f) = \Delta f$$

$$\nabla.(f\mathbf{V}) = f\nabla.\mathbf{V} + \nabla f.\mathbf{V} \qquad \nabla.(\mathbf{V} \times \mathbf{U}) = \mathbf{U}.(\nabla \times \mathbf{V}) - \mathbf{V}.(\nabla \times \mathbf{U})$$

$$\Delta(fg) = f\Delta g + 2\,\nabla f.\nabla g + g\,\Delta f \qquad \nabla.(\nabla \times \mathbf{V}) = 0$$

$$\nabla \times (\nabla \times \mathbf{V}) = \nabla\,(\nabla.\mathbf{V}) - \Delta\mathbf{V} \qquad \nabla \times (f\mathbf{V}) = \nabla f \times \mathbf{V} + f\nabla \times \mathbf{V})$$

The *circulation* of a vector field **V** on a path $\mathcal{C}$ is:

$$\int_{\mathcal{C}} d\mathbf{r}.\mathbf{V} = \int_{\mathcal{C}} \Sigma_i\, dx_i\, V_i \equiv \int_{\mathcal{C}} (dx\, V_x + dy\, V_y + dz\, V_z).$$

The *flux* of a vector field **V** through a surface $\mathcal{S}$ (of orthogonal unit vector **n**) is:

$$\iint_{\mathcal{S}} d\mathcal{S}\, \mathbf{n}.\mathbf{V}.$$

Stokes theorem allows us to write the circulation of a vector field **V** on a closed and oriented path $\mathcal{C}$ as the flux of $\nabla \times \mathbf{V}$ through any surface $\mathcal{S}$ bounded by $\mathcal{C}$. The normal unit vector **n** is oriented according to the right-hand rule:

$$\int_{\mathcal{C}} d\mathbf{r}.\mathbf{V} = \iint_{\mathcal{S}} d\mathcal{S}\, \mathbf{n}.(\nabla \times \mathbf{V}).$$

Gauss-Ostrogradski theorem states that the flux of a vector field **V** outgoing from a closed surface $\mathcal{S}$ is equal to the integral of $\nabla.\mathbf{V}$ over the volume $\mathcal{V}$ enclosed by $\mathcal{S}$:

$$\iint_{\mathcal{S}} d\mathcal{S}\, \mathbf{n}.\mathbf{V} = \iiint_{\mathcal{V}} d^3r\, \nabla.\mathbf{V}.$$

A.7. Vector analysis in curvilinear coordinates

Cylindrical coordinates

If a physical system has rotation symmetry around the z-axis, it is more convenient to use cylindrical coordinates (ρ, φ, z) (Figure A.2a). The cylindrical coordinates of a point M are related to its Cartesian coordinates by the equations:

$$\rho = \sqrt{x^2 + y^2}, \qquad \tan\varphi = y/x, \qquad x = \rho\cos\varphi, \qquad y = \rho\sin\varphi$$

Note that the whole space corresponds to $0 < \rho < \infty$, $0 < \varphi < 2\pi$ and $-\infty < z < \infty$. The position of a point M is written as:

$$\mathbf{r} = \mathbf{OM} = \rho\cos\varphi\,\mathbf{e}_x + \rho\sin\varphi\,\mathbf{e}_\varphi + z\,\mathbf{e}_z$$

The unit vectors tangent to the coordinate curves are related to the Cartesian vectors $\mathbf{e}_i$ by the equations:

$$\mathbf{e}_\rho = \cos\varphi\,\mathbf{e}_x + \sin\varphi\,\mathbf{e}_y, \qquad \mathbf{e}_\varphi = -\sin_x\varphi\,\mathbf{e}_x + \cos\varphi\,\mathbf{e}_y$$

$$\text{or} \quad \mathbf{e}_x = \cos\varphi\,\mathbf{e}_\rho - \sin\varphi\,\mathbf{e}_\varphi, \qquad \mathbf{e}_y = \sin\varphi\,\mathbf{e}_\rho + \cos\varphi\,\mathbf{e}_\varphi$$

The displacement vector $d\mathbf{r}$ generated by the variations $d\rho$, $d\varphi$ and dz is given by:

$$d\mathbf{r} = d\rho\,\mathbf{e}_\rho + \rho\,d\varphi\,\mathbf{e}_\varphi + dz\,\mathbf{e}_z$$

A vector field $\mathbf{A}$ is specified by its cylindrical components A_ρ, A_φ and A_z or by its Cartesian components A_x, A_y and A_z according to:

$$\mathbf{A} = A_x\,\mathbf{e}_x + A_y\,\mathbf{e}_y + A_z\,\mathbf{e}_z = A_\rho\,\mathbf{e}_\rho + A_\varphi\,\mathbf{e}_\varphi + A_z\mathbf{e}_z$$

Thus, the components are related by the equations:

$$A_\rho = A_x\cos\varphi + A_y\sin\varphi, \qquad A_\varphi = -A_x\sin\varphi + A_y\cos\varphi$$

Conversely:

$$A_x = A_\rho\cos\varphi - A_\varphi\sin\varphi, \qquad A_y = A_\rho\sin\varphi + A_\varphi\cos\varphi$$

A symmetry in rotation around Oz requires the components A_ρ, A_φ and A_z to be independent of φ.

The element of area ds_{ij} generated by two displacements $d\mathbf{r}_i$ and $d\mathbf{r}_j$ along the lines of coordinates i and j and the element of volume $d\mathcal{V}$ are:

$$ds_{23} = \rho\, d\varphi\, dz, \qquad ds_{31} = d\rho\, dz, \qquad ds_{12} = \rho\, d\rho\, d\varphi, \qquad d\mathcal{V} = \rho\, d\rho\, d\varphi\, dz$$

The gradient of a scalar field f, the divergence of a vector field $\mathbf{A}$ and the curl of $\mathbf{A}$, as well as the Laplacian of f and $\mathbf{A}$ are given by:

$$\nabla f = \partial_\rho f\ \mathbf{e}_\rho + \rho^{-1}\, \partial_\varphi f\ \mathbf{e}_\varphi + \partial_z f\ \mathbf{e}_z$$

$$\nabla.\mathbf{A} = \rho^{-1}\, \partial_\rho(\rho A_\rho) + \rho^{-1}\, \partial_\varphi A_\varphi + \partial_z A_z$$

$$\nabla\times \mathbf{A} = [\, \rho^{-1}\, \partial_\varphi A_z - \partial_z A_\varphi]\ \mathbf{e}_\rho + [\partial_z A_\rho - \partial_\rho A_z]\ \mathbf{e}_\varphi + \rho^{-1}[\partial_\rho(\rho A_\varphi) - \partial_\varphi A_\rho)]\ \mathbf{e}_z$$

$$\Delta f = \rho^{-1}\, \partial_\rho(\rho\, \partial_\rho f) + \rho^{-2}\, \partial^2_{\varphi\varphi} f + \partial^2_{zz} f$$

$$\begin{aligned}\Delta\mathbf{A} = {} & [\partial^2_{\rho\rho}\, A_\rho - \rho^{-2} A_\rho + \rho^{-1}\, \partial_\rho A_\rho + \rho^{-2}\, \partial^2_{\varphi\varphi} A_\rho - 2\, \rho^{-2}\, \partial_\varphi A_\varphi + \partial^2_{zz} A_\rho\,]\ \mathbf{e}_\rho \\ & + [\, \rho^{-1}\, \partial^2_{\rho\rho}(\rho A_\varphi) - \rho^{-2}\, \partial_\rho(\rho A_\varphi) + 2\, \rho^{-2}\, \partial_\varphi A_\rho + \rho^{-2}\, \partial^2_{\varphi\varphi} A_\varphi + \partial^2_{zz} A_\varphi\,]\ \mathbf{e}_\varphi \\ & + [\partial^2_{\rho\rho} A_z + \rho^{-1}\, \partial_\rho A_z + \rho^{-2}\, \partial^2_{\varphi\varphi} A_z + \partial^2_{zz} A_z]\ \mathbf{e}_z\end{aligned}$$

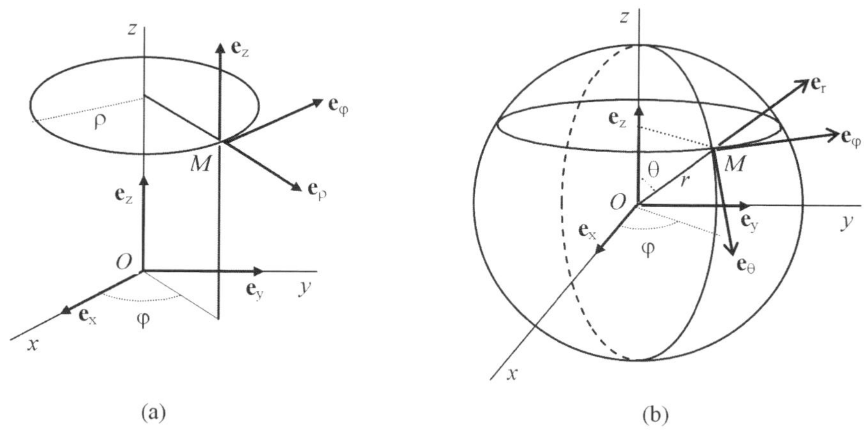

Figure A.2. *a) Cylindrical coordinates, and b) spherical coordinates*

Spherical coordinates:

If a physical system has a rotation symmetry around a point O, it is convenient to use spherical coordinates r, θ and φ (Figure A.2b). The whole space corresponds to $0 < r < \infty$, $0 < \theta < \pi$ and $0 < \varphi < 2\pi$. The spherical coordinates of point M are related to its Cartesian coordinates by the equations:

$$r=\sqrt{x^2+y^2+z^2}, \qquad \tan\theta=\sqrt{x^2+y^2}/z,$$

$$\cos\varphi = x/\sqrt{x^2+y^2}, \qquad \sin\varphi = y/\sqrt{x^2+y^2}, \qquad \tan\varphi = y/x,$$

$$x = r\sin\theta\cos\varphi, \qquad y = r\sin\theta\sin\varphi, \qquad z = r\cos\theta.$$

The position of a point M is written as

$$\mathbf{r} = \mathbf{OM} = r\sin\theta\cos\varphi\,\mathbf{e}_x + r\sin\theta\sin\varphi\,\mathbf{e}_y + r\cos\theta\,\mathbf{e}_z.$$

The unit vectors tangent to the coordinate curves are related to the Cartesian vectors $\mathbf{e}_i$ by the equations:

$$\mathbf{e}_r = \sin\theta\cos\varphi\,\mathbf{e}_x + \sin\theta\sin\varphi\,\mathbf{e}_y + \cos\theta\,\mathbf{e}_z,$$

$$\mathbf{e}_\theta = \cos\theta\cos\varphi\,\mathbf{e}_x + \cos\theta\sin\varphi\,\mathbf{e}_y - \sin\theta\,\mathbf{e}_z,$$

$$\mathbf{e}_\varphi = -\sin\varphi\,\mathbf{e}_x + \cos\varphi\,\mathbf{e}_y.$$

Conversely:

$$\mathbf{e}_x = \sin\theta\cos\varphi\,\mathbf{e}_r + \cos\theta\cos\varphi\,\mathbf{e}_\theta - \sin\theta\,\mathbf{e}_\varphi,$$

$$\mathbf{e}_y = \sin\theta\sin\varphi\,\mathbf{e}_r + \cos\theta\sin\varphi\,\mathbf{e}_\theta + \cos\theta\,\mathbf{e}_\varphi,$$

$$\mathbf{e}_z = \cos\theta\,\mathbf{e}_r - \sin\theta\,\mathbf{e}_{\theta c}.$$

The displacement vector $d\mathbf{r}$ generated by the variations dr, $d\theta$ and $d\varphi$ is

$$d\mathbf{r} = dr\,\mathbf{e}_r + r\,d\theta\,\mathbf{e}_\theta + r\sin\theta\,d\varphi\,\mathbf{e}_\varphi.$$

A vector field $\mathbf{A}$ is specified by its spherical components A_r, A_θ and A_φ or its Cartesian components A_x, A_y and A_z according to:

$$\mathbf{A} = A_x\,\mathbf{e}_x + A_y\,\mathbf{e}_y + A_z\,\mathbf{e}_z = A_r\,\mathbf{e}_r + A_\theta\,\mathbf{e}_\theta + A_\varphi\,\mathbf{e}_\varphi.$$

Thus, the components are related by the equations

$$A_r = -A_x\sin\theta\cos\varphi + A_y\sin\theta\sin\varphi + A_z\cos\theta,$$

$$A_\theta = A_x\sin\theta\sin\varphi + A_y\sin\theta\cos\varphi - A_z\sin\theta, \qquad A_\varphi = -A_x\sin\theta + A_y\cos\theta.$$

Conversely:

$$A_x = A_r \sin\theta \cos\varphi + A_\theta \cos\theta \cos\varphi - A_\varphi \sin\varphi,$$

$$A_y = A_r \sin\theta \sin\varphi + A_\theta \cos\theta \sin\varphi + A_\varphi \cos\varphi, \qquad A_z = A_r \cos\theta - A_\theta \sin\theta.$$

A rotation symmetry about O requires the components A_r, A_θ and A_φ to be independent of θ and φ.

The element of area ds_{ij} generated by two displacements $d\mathbf{r}_i$ and $d\mathbf{r}_j$ along the lines of coordinates i and j and the element of volume $d\mathcal{V}$ are:

$$ds_{23} = r^2 \sin\theta \, d\theta \, d\varphi, \;\; ds_{31} = r \sin\theta \, dr \, d\varphi, \;\; ds_{12} = r \, dr \, d\theta, \;\; d\mathcal{V} = r^2 \sin\theta \, d\theta \, d\varphi.$$

The gradient of a scalar field f, the divergence of a vector field $\mathbf{A}$ and the curl of $\mathbf{A}$, as well as the Laplacian of f and $\mathbf{A}$ are given by:

$$\nabla f = \partial_r f \; \mathbf{e}_r + r^{-1} \partial_\theta f \, \mathbf{e}_\theta + (r \sin\theta)^{-1} \partial_\varphi f \; \mathbf{e}_\varphi$$

$$\nabla.\mathbf{A} = r^{-2} \partial_r(r^2 A_r) + (r \sin\theta)^{-1} \partial_\theta(\sin\theta \, A_\theta) + (r \sin\theta)^{-1} \partial_\varphi A_\varphi$$

$$\nabla\times\mathbf{A} = (r \sin\theta)^{-1}[\partial_\theta(\sin\theta \, A_\varphi) - \partial_\varphi A_\theta] \, \mathbf{e}_r + (r \sin\theta)^{-1}[\partial_\varphi A_r - \sin\theta \, \partial_r(rA_\varphi)] \, \mathbf{e}_\theta$$
$$+ r^{-1} [\partial_r(rA_\theta) - \partial_\theta A_r] \, \mathbf{e}_\varphi$$

$$\Delta f = r^{-2} \partial_r (r^2 \partial_r f) + (r^2 \sin\theta)^{-1} \partial_\theta (\sin\theta \, \partial_\theta f) + (r \sin\theta)^{-2} \partial^2_{\varphi\varphi} f$$

$$\begin{aligned}\Delta\mathbf{A} = r^{-2}\{&[r^2 \partial^2_{rr} A_r + 2r \, \partial_r A_r - 2A_r + \partial^2_{\theta\theta} A_r - 2 \, \partial_\theta A_\theta + (\tan\theta)^{-1}(\partial_\theta A_r - 2A_\theta) \\ &\quad -2 (\sin\theta)^{-1} \partial_\varphi A_\varphi + (\sin\theta)^{-2} \partial^2_{\varphi\varphi} A_r] \, \mathbf{e}_r \\ &+ [r^2 \partial^2_{rr} A_\theta + 2r \, \partial_r A_\theta + \partial^2_{\theta\theta} A_\theta + 2 \, \partial_\theta A_r \\ &+ (\tan\theta)^{-1} \partial_\theta A_\theta - 2 \cos\theta \, (\sin\theta)^{-2} \partial_\varphi A_\varphi + (\sin\theta)^{-2} (\partial^2_{\varphi\varphi} A_\theta - A_\theta) \,] \, \mathbf{e}_\theta \\ &+ [r^2 \partial^2_{rr} A_\varphi + 2r \, \partial_r A_\varphi - 2A_\varphi + \partial^2_{\theta\theta} A_\varphi + (\sin\theta)^{-1}(\partial^2_{\varphi\theta} A_r + 2\partial_\varphi A_r - r\partial^2_{r\varphi} A_r) \\ &\quad + 3 (\tan\theta)^{-1} \partial_\theta A_\varphi + \partial^2_{\varphi\varphi} A_\varphi + A_\varphi \,] \, \mathbf{e}_\varphi\}.\end{aligned}$$

Appendix B

Units in Physics

B.1. Multiples and submultiples of units

The multiples of units and their submultiples by powers of 10 are designated follows in Table B.1.

10^n	**Prefix**	**Example**	10^n	**Prefix**	**Example**
10^{18}	exa- (E)	exajoule, EJ	10^{-1}	deci- (d)	decibel, dB
10^{15}	peta- (P)	petasecond, Ps	10^{-2}	centi- (c)	centimeter, cm
10^{12}	tera- (T)	terahertz, THz	10^{-3}	milli- (m)	millimeter, mm
10^{9}	giga- (G)	gigavolt, GV	10^{-6}	micro- (μ)	microgram, μg
10^{6}	miga- (M)	megawatt, MW	10^{-9}	nano- (n)	nanometer, nm
10^{3}	kilo- (K)	kilogram, kg	10^{-12}	pico- (p)	picofarad, pF
10^{2}	hecto- (H)	(rarely used)	10^{-15}	femto- (f)	femtometer, fm
10^{1}	deca- (Da)	(rarely used)	10^{-18}	atto- (atto)	attocoulomb, aC

Table B.1. *Multiples and submultiples*

B.2. Fundamental and derived SI units

The units of all physical quantities are defined in terms of six *fundamental* (or *basic*) *units*, which are defined by convention: *length*, *time*, *mass*, *current intensity*, *temperature* and *luminous intensity*. The units of the other physical quantities (the so-called *derived units*) are defined with reference to the fundamental units based on the dimensional homogeneity of laws of physics. The International System of Units (SI) uses *meters* (m), seconds (s), *kilograms* (kg), *ampere* (A), *kelvin* (K) and *candela* (cd) as fundamental units. Some branches of science and engineering continue to use the CGS system (based on the *centimeter*, *gram* and *second*) for mechanical quantities. As the whole circumference is divided into 2π *radians*, the unit of angles is the radian. We also use the *degree* (°), *minute* (′) and *second* (″), and sometimes *revolution* (2π radians).

B.3. Mechanical units

Quantity (dimension)	Unit (SI)	Unit (CGS)	Remarks
Length (L)	meter (m)	centimeter (cm)	1 m = 100 cm, 1 μm = 10^{-6} m, 1 F (fermi) = 10^{-15} m
Time (T)	second (s)	second (s)	1 year = $3{,}155812 \times 10^7$ s
Mass (M)	kilogram (kg)	gram (g)	1 kg = 1 000 g, 1 u = $1{,}660\ 056\ 55 \times 10^{-27}$ kg, 1 MeV/c^2 = $1{,}782\ 68 \times 10^{-30}$ kg
Area (L^2)	m^2	cm^2	1 m^2 = 10^4 cm^2
Volume (L^3)	m^3	cm^3	1 m^3 = 10^6 cm^3, 1 liter = 10^{-3} m^3
Frequency (T^{-1})	hertz (Hz) = 1 s^{-1}		
Velocity (LT^{-1})	m/s	cm/s	1 km/h = 0,2778 m/s
Acceleration (LT^{-2})	m/s^2	cm/s^2	1 m/s^2 = 100 cm/s^2
Angular velocity (T^{-1})	rad/s	rad/s	1 tour/s = 6,283 18 rad/s
Angular acceleration (T^{-2})	rad/s^2	rad/s^2	1 tour/s^2 = 6,283 18 rad/s^2
Force (MLT^{-2})	newton (N)	dyne	1 N = 10^5 dyne
Moment of force (ML^2T^{-2})	N.m	dyne.cm	1 N.m = 10^7 dyne.cm
Momentum (MLT^{-1})	kg.m/s	g.cm/s	1 kg.m/s = 10^5 g.cm/s
Angular momentum (ML^2T^{-1})	kg.m^2/s	g.cm^2/s	1 kg.m^2/s = 10^7 g.cm^2/s
Moment of inertia (ML^2)	kg.m^2	g.cm^2	1 kg.m^2 = 10^7 g.cm^2
Mass density (ML^{-3})	kg/m^3	g/cm^3	1 kg/m^3 = 10^{-3} g/cm^3
Pressure ($ML^{-1}T^{-2}$)*	pascal (N/m^2)	barye = dyne/cm^2	1 N/m^2 = 10 dyne/cm^2 1 atm = $1{,}013\ 25 \times 10^5$ N/m^2 = 76 cm Hg 1 cm Hg = $1{,}316 \times 10^{-2}$ 1 atm = 1 333,22 N/m^2 1 bar = 10^5 N/m^2
Work, energy (ML^2T^{-2})	joule (J)	erg	1 J = 10^7 erg, 1 calorie = 4,186 J, 1 eV = $1{,}602\ 189\ 2 \times 10^{-19}$ J
Power (ML^2T^{-3})	watt(W)	erg/s	1 W = 10^7 ergs, 1 horse-power = 745,7 W
Surface tension (MT^{-2})	kg/s^2	g/s^2	1 kg/s^2 = 1 000 g/s^2
Molar mass (M mole^{-1})	kg/k mole	g/mole	1 g/mole = 1 kg/k mole
Specific volume (L^3M^{-1})	m^3/kg	cm^3/g	1 m^3/kg = 10^3 cm^3/g

Table B.2. *Mechanical units*

B.4. Electromagnetic units

Quantity	Dimensions	Unit (abbreviation)	Remarks
Current intensity	I	ampere (A)	
Electric charge	$Q = TI$	coulomb (C) = 1A.s	1 has – h = 3,600 C
Potential, e.m.f.	$\Phi = L^2MT^{-3}I^{-1}$	volt (V)	
Electric field	$E = LMT^{-3}I^{-1}$	V/m	
Capacitance	$C = L^{-2}M^{-1}T^4I^2$	farad (F)	
Displacement field	$D = L^{-2}TI$	C/m^2	
Electric displacement flux	$\Phi_D = TI$	C	
Electric dipole moment	$p_e = LTI$	C.m	
Polarization density	$P = L^{-2}TI$	C/m^2	
Electric permittivity	$\varepsilon = L^{-3}M^{-1}T^4I^2$	F/m	
Resistance	$R = L^2MT^{-3}I^{-2}$	ohm (Ω)	
Resistivity	$\rho = L^3MT^{-3}I^{-2}$	Ω.m	
Electric conductivity	$\sigma = L^{-3}M^{-1}T^3I^2$	$\Omega^{-1}m^{-1}$	
Magnetic flux	$\Phi_B = L^2MT^{-2}I^{-1}$	weber (Wb)	1 Maxwell = 10^{-8} Wb
Magnetic induction field	$B = MT^{-2}I^{-1}$	tesla (T) = Wb/m^2	1 Gauss = 10^{-4} T
Magnetic field strength	$H = L^{-1}I$	A/m	1 Oersted = $10^3/4\pi$ A/m
Magnetic moment	$M = L^2I$	A/m^2	
Intensity of magnetization	$M = L^{-1}I$	A.m	A.m
Inductance	$\mathsf{L} = L^2MT^{-2}I^{-2}$	henry (H)	
Magnetic permeability	$\mu = LMT^{-2}I^{-2}$	H/m	

Table B3. *Electromagnetic units*

Appendix C

Some Physical Constants

C.1. Mechanical and thermodynamic constants

Speed of light in a vacuum (exact value)	$c = 299{,}792{,}458 \times 10^8$ m/s
Gravitational constant	$G = 6.67259\ (85) \times 10^{-11}$ N.m^2/kg^2
Standard acceleration of free fall	$g = 9.80665$ m/s^2
Sideral year	365.2564 days
Normal pressure	$p_o = 1$ At $= 1.01325 \times 10^5$ N/m^2
Normal temperature	$T_o = 273.16$ K
Mechanical equivalent of the calorie	J = 4.1855 (4) J/Cal
Mass density of dry air (STP)	$\rho_o = 1.2929$ kg/m^3
Speed of sound in air (STP)	$v_s = 331.36$ m/s
Molar volume of ideal gases (STP)	$V_m = 2.241383\ (70) \times 10^{-2}$ m^3/mole
Avogadro number	$N_o = 6.0221367\ (36) \times 10^{23}$ mole^{-1}
Boltzmann constant	$K = 1.380658\ (12) \times 10^{-23}$ J/K
Gas constant	$R = 8.314510\ (70)$ J/K.mole
Acoustic impedance of air (STP)	$Z_a = 428$ N.s/m^3

Threshold of hearing sound intensity	$\mathcal{I}_o = 10^{-12}$ W/m^2
Stefan-Boltzmann constant	$\sigma = 5.67051\ (7) \times 10^{-8}$ J/m^2.s.K^4
Mass density of mercury	13,595.5 kg/m^3

Table C.1. *Mechanical and thermodynamic quantities. Standard g means at sea level and at latitude of 45°. STP means standard temperature (0 °C) and pressure (760 mm Hg)*

C.2. Electromagnetic and atomic constants

Elementary charge	$e = 1.60217733\ (49) \times 10^{-19}$ C
Electron-volt	eV $= 1.6021892\ (46) \times 10^{-19}$ J
Planck constant	$h = 6.626075\ (40) \times 10^{-34}$ J.s $= 4.135669\ (11)\ 10^{-15}$ eV.s
Reduced Plank constant	$\hbar = h/2\pi = 1.054572\ 66\ (63) \times 10^{-34}$ J.s $= 6.582122\ (17) \times 10^{-16}$ eV.s
Electron mass	$m_e = 9.1093897\ (54) \times 10^{-31}$ kg $= 0.51099906\ (15)$ MeV/c^2
Proton mass	$m_p = 1.672623\ (10) \times 10^{-27}$ kg $= 938.2723\ (27)$ MeV/c^2
Neutron mass	$m_n = 1.6749286\ (10) \times 10^{-27}$ kg $= 939.56563\ (28)$ MeV/c^2
Atomic mass unit	$u = \mathrm{m}(\mathrm{C}^{12})/12 = 931.49432\ (28)$ MeV/c^2 $= 1.6605402\ (10) \times 10^{-27}$ kg
Bohr radius	$a_o = 4\pi\varepsilon_o\hbar^2/m_e e^2$ $= 0.529177249\ (24) \times 10^{-10}$ m
Bohr magneton	$\mu_B = e\hbar/2m_e = 9.2741 \times 10^{-24}$ A.m^2
Electric permittivity of vacuum	$\varepsilon_o = 8.854187817 \times 10^{-12}$ A^{2}s^4/kg m^3 $1/4\pi\varepsilon_o = 8.987551787 \times 10^{9}$ N.m^2.C^{-2}
Magnetic permeability of vacuum	$\mu_o = 4\pi \times 10^{-7}$ kg m/A^{2}s^2

Table C.2. *Constants*

Further Reading

General physics texts and mathematical methods

FEYNMAN R.P., LEIGHTON R.B., SANDS M., *Lectures on Physics*, Addison-Wesley, Reading, 1964.

HECHT E., *Physics*, Brooks/Cole Publishing Company, 1996.

JEFFREY A., *Handbook of Mathematical Formulas and Integrals* (2nd edition), Academic Press, 2000.

MATHEWS J., WALKER R.L., *Mathematical Methods of Physics* (2nd edition), W. A. Benjamin, 1970

SOMMERFELD A., *Partial Differential Equations in Physics*, volume 1, Academic Press, New York, 1949.

Vibrations and waves in general

BÉCHERRAWY T., *Vibrations, ondes et optique: vibrations mécaniques et électromagnétiques*, Volume 1, Hermès-Lavoisier, Paris, 2007.

BENSON H.T., *Principles of Vibration*, Oxford University Press, New York, 2002.

CRAWFORD F.S., *Berkeley Physics Course: Waves Volume 3*, McGraw-Hill, New York, 1968.

FRENCH A.P., *Vibrations and Waves*, Norton and Co., New York, 1971.

KING G., *Vibrations and Waves*, Wiley-Blackwell, UK, 2009

KNEUBÜHL F.K., *Oscillations and Waves*, Springer, 1997

MAIN I.G., *Vibrations and Waves in Physics* (3rd edition), Cambridge University Press, 1993

PAIN H.J., *The Physics of Vibrations and Waves*, Wiley, New York, 1968.

Mechanical waves

BÉCHERRAWY T., *Vibrations, ondes et optique: ondes mécaniques*, Volume 2, Hermès-Lavoisier, Paris, 2008.

KELLY S.G., *Fundamentals of Mechanical Vibrations*, McGraw-Hill, New York, 1993.

LIGHTHILL J., *Waves in Fluids*, Cambridge University Press, Cambridge, 1978.

Electromagnetic waves

BÉCHERRAWY T., *Vibrations, ondes et optique: ondes électromagnétiques*, Volume 3, Hermès-Lavoisier, Paris, 2008.

CORSON D., LORRAIN P., *Electromagnetic Fields and Waves*, Freeman & Cie, New York, 1988.

CHENG D.K, *Field and Wave Electromagnetics* (International edition) (2nd edition), Pearson Education, 1989.

JACKSON J.D., *Electrodynamique classique*, Dunod, Paris, 2001.

Optical waves

BÉCHERRAWY T., *Vibrations, ondes et optique: ondes optiques*, volume 4, Hermès-Lavoisier, Paris, 2008.

BORN M., WOLF E., *Principles of Optics*, Pergamon Press, Oxford, 1975.

BRUHAT G., KASTLER A., BOUCHAREINE P., *Optique*, Dunod, Paris, 2005.

HECHT E., *Optics*, Addison-Wesley, London, 1987.

JENKINS F.A., WHITE, H.E., *Fundamentals of Optics*, McGraw-Hill, New York, 1976.

SILVAST W.T., *Laser Fundamentals*, Cambridge University Press, 1996.

TAILLET R., *Optique physique*, De Boeck University, 2006.

Index

A, B

C

D

E

F

G, H

I, J, K

L

M, N

O

W, X

Y, Z